Lecture Notes in Artificial Intelligence 1919

Subseries of Lecture Notes in Computer Science
Edited by J. G. Carbonell and J. Siekmann

Lecture Notes in Computer Science
Edited by G. Goos, J. Hartmanis and J. van Leeuwen

Springer
Berlin
Heidelberg
New York
Barcelona
Hong Kong
London
Milan
Paris
Singapore
Tokyo

Manuel Ojeda-Aciego   Inma P. de Guzmán
Gerhard Brewka   Luís Moniz Pereira (Eds.)

Logics in
Artificial Intelligence

European Workshop, JELIA 2000
Málaga, Spain, September 29 – October 2, 2000
Proceedings

 Springer

Series Editors

Jaime G. Carbonell,Carnegie Mellon University, Pittsburgh, PA, USA
Jörg Siekmann, University of Saarland, Saarbrücken, Germany

Volume Editors

Manuel Ojeda-Aciego
Inma P. de Guzmán
University of Málaga
Department of Applied Mathematics, E.T.S.I. Informática
29071 Málaga, Spain
E-mail: aciego@uma.es; guzman@ctima.uma.es

Gerhard Brewka
University of Leipzig
Intelligent Systems Department, Computer Science Institute
Augustusplatz 10-11, 04109 Leipzig, Germany
E-mail: brewka@informatik.uni-leipzig.de

Luís Moniz Pereira
University of Lisbon
Center for Artificial Intelligence, Department of Computer Science
2825-114 Caparica, Portugal
E-mail: lmp@fct.unl.pt

Cataloging-in-Publication Data applied for

Die Deutsche Bibliothek - CIP-Einheitsaufnahme

Logics in artificial intelligence : European workshop ; proceedings /
JELIA 2000, Málaga, Spain, September 29 - October 2, 2000. Manuel
Ojeda-Aciego ... (ed.). - Berlin ; Heidelberg ; New York ; Barcelona ;
Hong Kong ; London ; Milan ; Paris ; Singapore ; Tokyo : Springer, 2000
  (Lecture notes in computer science ; Vol. 1919 : Lecture notes in
  artificial intelligence)
  ISBN 3-540-41131-3

CR Subject Classification (1998): I.2, F.4.1, D.1.6

ISBN 3-540-41131-3 Springer-Verlag Berlin Heidelberg New York

Springer-Verlag Berlin Heidelberg New York
a member of BertelsmannSpringer Science+Business Media GmbH
© Springer-Verlag Berlin Heidelberg 2000
Printed in Germany

Typesetting: Camera-ready by author, data conversion by Boller Mediendesign
Printed on acid-free paper   SPIN 10722808   06/3142   5 4 3 2 1 0

Preface

Logics have, for many years, laid claim to providing a formal basis for the study of artificial intelligence. With the depth and maturity of methodologies, formalisms, procedures, implementations, and their applications available today, this claim is stronger than ever, as witnessed by increasing amount and range of publications in the area, to which the present proceedings accrue.

The European series of Workshops on Logics in Artificial Intelligence (or *Journées Européennes sur la Logique en Intelligence Artificielle – JELIA*) began in response to the need for a European forum for the discussion of emerging work in this burgeoning field. JELIA 2000 is the seventh such workshop in the series, following the ones held in Roscoff, France (1988); Amsterdam, Netherlands (1990); Berlin, Germany (1992); York, U.K. (1994); Évora, Portugal (1996); and Dagstuhl, Germany (1998).

JELIA 2000 will take place in Málaga, Spain, from 29 September to 2 October 2000. The workshop is organized and hosted by the Research Group of Mathematics Applied to Computing of the Department of Applied Mathematics of the University of Málaga.

As in previous workshops, the aim is to bring together researchers involved in all aspects of logic in artificial intelligence. Additional sponsorship was provided by the ESPRIT NOE Compulog-Net.

This volume contains the papers selected for presentation at the workshop along with abstracts and papers from the invited speakers. The programme committee selected these 23 papers, from 12 countries (Australia, Austria, Belgium, Canada, Finland, Germany, Hong Kong, Italy, The Netherlands, Portugal, Spain, and the United Kingdom), out of 60 submissions, from 22 countries (submissions were also received from Argentina, Brazil, Czech Republic, France, Japan, Mexico, Poland, Slovakia, Sweden, and Switzerland). We would like to thank all authors for their contributions as well as the invited speakers Johan van Benthem from the University of Amsterdam (The Netherlands), Thomas Eiter from the Vienna University of Technology (Austria), Reiner Hähnle from the Chalmers University of Technology (Sweden), and Frank Wolter from the University of Leipzig (Germany).

Papers were reviewed by the programme committee members with the help of the additional referees listed overleaf. We would like thank them all for their valuable assistance. It is planned that a selection of extended versions of the best papers will be published in the journal Studia Logica, after being subjected again to peer review.

September 2000

Gerd Brewka
Inma P. de Guzmán
Manuel Ojeda-Aciego
Luís Moniz Pereira

Logics in Artificial Intelligence

Conference Chairs

Inma P. de Guzmán, Málaga

Manuel Ojeda-Aciego, Málaga

Programme Chairs

Gerhard Brewka, Leipzig

Luis Moniz Pereira, Lisbon

Programme Committee Members

José Alferes	Universidade Nova de Lisboa, Portugal
Gerhard Brewka	University of Leipzig, Germany
Jürgen Dix	University Koblenz-Landau, Germany
Patrice Enjalbert	University of Caen, France
Luis Fariñas del Cerro	Institut de Recherche en Informatique de Toulouse (IRIT), France
Klaus Fischer	German Research Center for Artificial Intelligence (DFKI), Germany
Ulrich Furbach	University of Koblenz-Landau, Germany
Michael Gelfond	University of Texas at El Paso, USA
Inma P. de Guzmán	University of Málaga, Spain
Petr Hájek	Academy of Sciences, Czechoslovakia
Maurizio Lenzerini	University of Roma, Italy
John-Jules Meyer	University of Utrecht, The Netherlands
Bernhard Nebel	University of Freiburg, Germany
Ilkka Niemelä	Helsinki University of Technology, Finland
Manuel Ojeda-Aciego	University of Málaga, Spain
David Pearce	German Research Center for Artificial Intelligence (DFKI), Germany
Luís Moniz Pereira	Universidade Nova de Lisboa, Portugal
Henry Prakken	University of Utrecht, The Netherlands
Teodor C. Przymusinski	University of California at Riverside, USA
Venkatramanan S. Subrahmanian	University of Maryland, USA
Michael Thielscher	Dresden University of Technology, Germany
Mary-Anne Williams	University of Newcastle, Australia
Michael Zakharyaschev	University of Leeds, UK

Additional Referees

Grigoris Antoniou	Vincent Dugat	Dale Miller
Philippe Balbiani	Francisco Durán	Pavlos Peppas
Brandon Bennett	Manuel Enciso	Omar Rifi
Christoph Benzmueller	Mamoun Filali	Jussi Rintanen
Jean-Paul Bodeveix	Tomi Janhunen	Blas C. Ruíz Jiménez
Patrice Boizumault	Jean-Jacques Hébrard	Marco Schaerf
Hans-Jürgen Bürckert	Andreas Herzig	Dietmar Seipel
Alastair Burt	Steffen Hölldobler	Tommi Syrjänen
Françoise Clérin	Gerhard Lakemeyer	Hans Tompits
Ingo Dahn	Daniel Le Berre	Hudson Turner
Carlos V. Damásio	Thomas Lukasiewicz	Agustín Valverde
Giuseppe De Giacomo	Philippe Luquet	Luca Viganò
Alexander Dekhtyar	Viviana Mascardi	Michel Wermelinger

Sponsors

Ayuntamiento de Málaga	Ciencia Digital
Diputación Provincial de Málaga	Compulog Net
Junta de Andalucía	Estudio Informática
Patronato de Turismo de la Costa del Sol	Romero Candau, S.L.
Universidad de Málaga	

Table of Contents

Invited Talks

Regular Contributions

Knowledge Representation

Reasoning about Actions

Belief Revision

Theorem Proving

Argumentation

Agents

Decidability and Complexity

Updates

Preferences

'On Being Informed': Update Logics for Knowledge States

Johan van Benthem

ILLC Amsterdam
http://www.turing.wins.uva.nl/~johan/

Statements convey information, by modifying knowledge states of hearers and speakers. This dynamic aspect of communication goes beyond the usual role of logic as a provider of static 'truth conditions'. But it can be modelled rather nicely in so-called 'update logics', which have been developed since the 1980s. These systems provide a fresh look at standard logic, letting the usual models undergo suitable changes as agents absorb the content of successive utterances or messages. This lecture is a brief Whig history of update logics, with an emphasis on many-agent epistemic languages. We discuss straight update, questions and answers, and the delightful complexities of communication under various constraints. We hope to convey the attraction of giving a dynamic twist to well-known things, such as simple modal models, or basic epistemic formulas.

M. Ojeda-Aciego et al. (Eds.): JELIA 2000, LNAI 1919, pp. 1–1, 2000.

Considerations on Updates of Logic Programs

Thomas Eiter, Michael Fink, Giuliana Sabbatini, and Hans Tompits

Institut und Ludwig Wittgenstein Labor für Informationssysteme, TU Wien
Favoritenstraße 9–11, A-1040 Wien, Austria
{eiter,michael,giuliana,tompits}@kr.tuwien.ac.at

Abstract. Among others, Alferes et al. (1998) presented an approach
for updating logic programs with sets of rules based on dynamic logic
programs. We syntactically redefine dynamic logic programs and investigate their semantical properties, looking at them from perspectives such
as a belief revision and abstract consequence relation view. Since the approach does not respect minimality of change, we refine its stable model
semantics and present minimal stable models and strict stable models.
We also compare the update approach to related work, and find that is
equivalent to a class of inheritance programs independently defined by
Buccafurri et al. (1999).

1   Introduction

In recent years, agent-based computing has gained increasing interest. The need
for software agents that behave "intelligently" in their environment led to question for possibilities of equipping them with advanced reasoning capabilities.

The research on logic-based AI, and in particular the work on logic programming, has produced a number of approaches and methods from which we can
take advantage for accomplishing this goal (see e.g. [11]). It has been realized,
however, that further work is needed for extending them to fully support that
agents must adapt over time and adjust their decision making.

In a simple (but as for currently deployed agent systems, realistic) setting, an
agent's knowledge base KB may be modeled as a logic program. The agent may
now be prompted to adjust its KB after receiving new information in terms of an
update U, which is a clause or a set of clauses that need to be incorporated into
KB. Simply adding the rules of U to KB does not give a satisfactory solution
in practice, and will result in inconsistency even in simple cases. For example, if
KB contains the rule $a \leftarrow$ and U consists of the rule $not\ a \leftarrow$ stating that a
is not provable, then the union $KB \cup U$ is not consistent under stable semantics
(naturally generalized to programs with default negation in rule heads [21]),
which is the predominating two-valued semantics for declarative logic programs.

Most recently, several approaches for updating logic programs with (sets of)
rules have been presented [2,5,17,13]. In particular, the concept of dynamic logic
programs by Alferes et al., introduced in [2] and further developed in [3,5,4,20],
has attracted a lot of interest. Their approach has its roots, and generalizes,

M. Ojeda-Aciego et al. (Eds.): JELIA 2000, LNAI 1919, pp. 2–20, 2000.

the idea of revision programming [22], and provides the basis for LUPS, a logic-programming based update specification language [5]. The basic idea behind the approach is that in case of conflicting rules, a rule r in U (which is assumed to be correct as of the time of the update request) is more reliable than any rule r' in KB. Thus, application of r rejects application of r'. In the previous example, the rule $not\,a \leftarrow$ from U rejects the rule $a \leftarrow$ from KB, thus resolving the conflict by adopting that a is not provable. The idea is naturally extended to sequences of updates $U_1, \ldots, U_n$ by considering the rules in more recent updates as more reliable.

While uses and extensions of dynamic logic programming have been discussed, cf. [5,4,20], its properties and relationships to other approaches and related formalisms have been less explored (but see [4]). The aim of this paper is to shed light on these issues, and help us to get a better understanding of dynamic logic programming and related approaches in logic programming.

The main contributions of our work can be summarized as follows.

- We syntactically redefine dynamic logic programs to equivalent *update programs*, for which stable models are defined. Update programs are slightly less involved and, as we believe, better reflect the working of the approach than the original definition of dynamic logic programs. For this, information about rule rejection is explicitly represented at the object level through rejection atoms. The syntactic redefinition, which reduces the type of rules in update programs, is helpful for establishing formal results about properties.
- We investigate properties of update programs. We consider them from the perspective of belief revision, and review different sets of postulates that have been proposed in this area. We view update programs as nonmonotonic consequence operators, and consider further properties of general interest. As it turns out, update programs (and thus dynamic logic programs) do not satisfy many of the properties defined in the literature. This is partly explained by the nonmonotonicity of logic programs and the causal rejection principle embodied in the semantics, which strongly depends on the syntax of rules.
- Dynamic logic programs make no attempt to respect minimality of change. We thus refine the semantics of update programs and introduce minimal stable models and strict stable models. Informally, minimal stable models minimize the set of rules that need to be rejected, and strict stable models further refine on this by assigning rules from a later update higher priority.
- We compare update programs to alternative approaches for updating logic programs [13,17] and related work on inheritance programs [9]. We find that update programs are equivalent to a class of inheritance programs. Thus, update programs (and dynamic logic programs) may be semantically regarded as fragment of the framework in [9], which has been developed independently of [2,5]. Our results on the semantical properties of update programs apply to this fragment as well.

Due to space reasons, the presentation is necessarily succinct and proofs are omitted. More details will be given in the full version of this paper.

2   Preliminaries

Generalized logic programs [21] consist of rules built over a set $\mathcal{A}$ of propositional atoms where default negation *not* is available. A *literal*, L, is either an atom A (a *positive literal*) or the negation *not* A of an atom A (a *negative literal*, also called *default literal*). For a literal L, the *complementary literal*, *not* L, is *not* A if $L = A$, and A if $L = $ *not* A, for some atom A. For a set S of literals, *not* S is given by *not* $S = \{$*not* $L \mid L \in S\}$. We also denote by $Lit_{\mathcal{A}}$ the set $\mathcal{A} \cup$ *not* $\mathcal{A}$ of all literals over $\mathcal{A}$.

A *rule*, r, is a clause of the form $L_0 \leftarrow L_1, \ldots, L_n$, where $n \geq 0$ and L_0 may be missing, and each L_i $(0 \leq i \leq n)$ is a default literal, i.e., either an atom A or a negated atom *not* A. We call L_0 the *head* of r and the set $\{L_1, \ldots, L_n\}$ the *body* of r. The head of r will also be denoted by $H(r)$, and the body of r will be denoted by $B(r)$. If the rule r has an empty head, then r is a *constraint*; if the body of r is empty and the head is non-empty, then r is a *fact*. We say that r has a *negative head* if $H(r) = $ *not* A, for some atom A. The set $B^+(r)$ comprises the positive literals of $B(r)$, whilst $B^-(r)$ contains all default literals of $B(r)$.

By $\mathcal{L}_{\mathcal{A}}$ we denote the set of all rules over the set $\mathcal{A}$ of atoms. We will usually write $\mathcal{L}$ instead of $\mathcal{L}_{\mathcal{A}}$ if the underlying set $\mathcal{A}$ is fixed. A *generalized logic program* (GLP) P *over* $\mathcal{A}$ is a finite subset of $\mathcal{L}_{\mathcal{A}}$. If no rule in P contains a negative head, then P is a *normal logic program* (NLP); if no default negation whatsoever occurs in P, then P is a *positive program*.

By an *(Herbrand) interpretation* we understand any subset $I \subseteq \mathcal{A}$. The relation $I \models L$ for a literal L is defined as follows:

- if $L = A$ is an atom, then $I \models A$ iff $A \in I$;
- if $L = $ *not* A is a default literal, then $I \models$ *not* A iff $I \not\models A$.

If $I \models L$, then I is a *model* of L, and L is said to be *true* in I (if $I \not\models L$, then L is *false* in I). For a set S of literals, $I \models S$ iff $I \models L$ for all $L \in S$. Accordingly, we say that I is a model of S. Furthermore, for a rule r, we define $I \models r$ iff $I \models H(r)$ whenever $I \models B(r)$. In particular, if r is a contraint, then $I \models r$ iff $I \not\models B(r)$. In both cases, if $I \models r$, then I is a model of r. Finally, $I \models P$ for a program P iff $I \models r$ for all $r \in P$.

If a positive logic program P has some model, it has always a smallest Herbrand model, which we will denote by $lm(P)$. If P has no model, for technical reasons it is convenient to set $lm(P) = Lit_{\mathcal{A}}$.

We define the *reduct*, P^I, of a generalized program P w.r.t. to an Herbrand interpretation I as follows. P^I results from P by

1. deleting any rule r in P such that either $I \models B^-(r)$, or $I \models H(r)$ if $H(r) = $ *not* A for some atom A; and
2. replacing any remaining rule r by r^I, where $r^I = H(r) \leftarrow B^+(r)$ if $H(r)$ is positive, and $r^I = \leftarrow B^+(r)$ otherwise (r^I is called the *reduct* of r).

Observe that P^I is a positive program, hence $lm(P^I)$ is well-defined. We say that I is a *stable model* of P iff $lm(P^I) = I$. By $\mathcal{S}(P)$ we denote the set of all stable models of P. A program is *satisfiable* if $\mathcal{S}(P) \neq \emptyset$.

We regard a logic program P as the *epistemic state* of an agent. The given semantics is used for assigning a *belief state* to any epistemic state P in the following way.

Let $I \subseteq \mathcal{A}$ be an Herbrand interpretation. Define

$$Bel_{\mathcal{A}}(I) = \{r \in \mathcal{L}_{\mathcal{A}} \mid I \models r\}.$$

Furthermore, for a class $\mathcal{I}$ of interpretations, define $Bel_{\mathcal{A}}(\mathcal{I}) = \bigcap_{i \in \mathcal{I}} Bel_{\mathcal{A}}(I)$.

Definition 2.1. *For a logic program P, the belief state, $Bel_{\mathcal{A}}(P)$, of P is given by $Bel_{\mathcal{A}}(P) = Bel_{\mathcal{A}}(\mathcal{S}(P))$, where $\mathcal{S}(P)$ is the collection of all stable models of P.*

We write $P \models_{\mathcal{A}} r$ if $r \in Bel_{\mathcal{A}}(P)$. As well, for any program Q, we write $P \models_{\mathcal{A}} Q$ if $P \models_{\mathcal{A}} q$ for all $q \in Q$. Two programs, P_1 and P_2, are *equivalent* (modulo the set $\mathcal{A}$), symbolically $P_1 \equiv_{\mathcal{A}} P_2$, iff $Bel_{\mathcal{A}}(P_1) = Bel_{\mathcal{A}}(P_2)$. Usually we will drop the subscript "$\mathcal{A}$" in $Bel_{\mathcal{A}}(\cdot)$, $\models_{\mathcal{A}}$, and $\equiv_{\mathcal{A}}$ if no ambiguity can arise.

An alternative for defining the belief state would consist in considering brave rather than cautious inference, which we omit here.

Belief states enjoy the following natural properties:

Theorem 2.1. *For every logic program P, we have that:*

1. *$P \subseteq Bel(P)$;*
2. *$Bel(Bel(P)) = Bel(P)$;*
3. *$\{r \mid I \models r, \text{ for every interpretation } I\} \subseteq Bel(P)$.*

Clearly, the belief operator $Bel(\cdot)$ is nonmonotonic, i.e., in general $P_1 \subseteq P_2$ does not imply $Bel(P_1) \subseteq Bel(P_2)$.

3   Update Programs

We introduce a framework for update programs which simplifies the approach introduced in [2]. By an *update sequence*, P, we understand a series $P_1, \ldots, P_n$ of general logic programs where each P_i is assumed to update the information expressed by the initial section $P_1, \ldots, P_{i-1}$. This update sequence is translated into a single program P' representing the update information given by P. The "intended" stable models of P are identified with the stable models of P' (modulo the original language).

Let $P = P_1, \ldots, P_n$ be an update sequence over a set of atoms $\mathcal{A}$. We assume a set of atoms $\mathcal{A}^*$ extending $\mathcal{A}$ by new, pairwise distinct atoms $rej(\cdot)$, A_i, and A_i^-, where $A \in \mathcal{A}$ and $1 \leq i \leq n$. Furthermore, we assume an injective *naming function* $N(\cdot, \cdot)$, which assigns to each rule r in a program P_i a distinguished name, $N(r, P_i)$, obeying the condition $N(r, P_i) \neq N(r', P_j)$ whenever $i \neq j$. With a slight abuse of notation we shall identify r with $N(r, P_i)$ as usual.

Definition 3.1. *Given an update sequence $P = P_1, \ldots, P_n$ over a set of atoms $\mathcal{A}$ we define the update program $P_\triangleleft = P_1 \triangleleft \ldots \triangleleft P_n$ over $\mathcal{A}^*$ consisting of the following items:*

1. *all constraints in P_i, $1 \leq i \leq n$;*
2. *for each $r \in P_i$, $1 \leq i \leq n$:*

$$A_i \leftarrow B(r), not\ rej(r) \qquad\qquad if\ H(r) = A;$$
$$A_i^- \leftarrow B(r), not\ rej(r) \qquad\qquad if\ H(r) = not\ A;$$

3. *for each $r \in P_i$, $1 \leq i < n$:*

$$rej(r) \leftarrow B(r), A_{i+1}^- \qquad\qquad if\ H(r) = A;$$
$$rej(r) \leftarrow B(r), A_{i+1} \qquad\qquad if\ H(r) = not\ A;$$

4. *for each atom A occurring in P $(1 \leq i < n)$:*

$$A_i^- \leftarrow A_{i+1}^-; \qquad A_i \leftarrow A_{i+1}; \qquad A \leftarrow A_1; \qquad \leftarrow A_1, A_1^-.$$

Informally, this program expresses layered derivability of an atom A or a literal $not\ A$, beginning at the top layer P_n downwards to the bottom layer P_1. The rule r at layer P_i is only applicable if it is not refuted by a literal L that is incompatible with $H(r)$ derived at a higher level. Inertia rules propagate a locally derived value for A downwards to the first level, where the local value is made global; the constraint $\leftarrow A_1, A_1^-$ is used here in place of the rule $not\ A \leftarrow A_1^-$.

Similar to the transformation given in [2], $P_\lhd$ is modular in the sense that the transformation for $P' = P_1, \ldots, P_n, P_{n+1}$ augments $P_\lhd = P_1 \lhd \ldots \lhd P_n$ only with rules depending on $n + 1$.

We remark that $P_\lhd$ can obviously be slightly simplified, which is relevant for implementing our approach. All literals $not\ rej(r)$ in rules with heads A_n or A_n^- can be removed: since $rej(r)$ cannot be derived, they evaluate to true in each stable model of $P_\lhd$. Thus, no rule from P_n is rejected in a stable model of $P_\lhd$, i.e., all most recent rules are obeyed.

The intended models of an update sequence $P = P_1, \ldots, P_n$ are defined in terms of the stable models of $P_\lhd$.

Definition 3.2. *Let $P = P_1, \ldots, P_n$ be an update sequence over a set of atoms $\mathcal{A}$. Then, $S \subseteq \mathcal{A}$ is an (update) stable model of P iff $S = S' \cap \mathcal{A}$ for some stable model S' of $P_\lhd$. The collection of all update stable models of P is denoted by $\mathcal{U}(P)$.*

Following the case of single programs, an update sequence $P = P_1, \ldots, P_n$ is regarded as the epistemic state of an agent, and the belief state $Bel(P)$ is given by $Bel(\mathcal{U}(P))$. As well, the update sequence P is satisfiable iff $\mathcal{U}(P) \neq \emptyset$.

To illustrate Definition 3.2, consider the following example, taken from [2].

Example 3.1. Consider the update of P_1 by P_2, where

$$P_1 = \{\ r_1 : sleep \leftarrow not\ tv\_on, \quad r_2 : tv\_on \leftarrow, \quad r_3 : watch\_tv \leftarrow tv\_on\ \};$$
$$P_2 = \{\ r_4 : not\ tv\_on \leftarrow power\_failure, \quad r_5 : power\_failure \leftarrow\ \}.$$

The single stable model of $P = P_1, P_2$ is, as desired, $S = \{power\_failure, sleep\}$, since S' is the only stable model of $P_{\lhd}$:

$$S' = \{\ power\_failure_2, power\_failure_1, power\_failure,$$
$$tv\_on_2^-, tv\_on_1^-, rej(r_2), sleep_1, sleep\ \}.$$

If new information arrives in form of the program P_3:

$$P_3 = \{\ r_6 : \ not\ power\_failure \leftarrow\ \},$$

then the update sequence P_1, P_2, P_3 has the stable model $T = \{tv\_on, watch\_tv\}$, generated by the model T' of $P_1 \lhd P_2 \lhd P_3$:

$$T' = \{\ power\_failure_3^-, power\_failure_2^-, power\_failure_1^-,$$
$$rej(r_5), tv\_on_1, tv\_on, watch\_tv_1, watch\_tv\ \}.$$

Next, we discuss some properties of our approach. The first result guarantees that stable models of P are uniquely determined by the stable models of $P_{\lhd}$.

Theorem 3.1. *Let* $P = P_1, \ldots, P_n$ *be an update sequence over a set of atoms* $\mathcal{A}$, *and let* S, T *be stable models of* $P_{\lhd}$. *Then,* $S \cap \mathcal{A} = T \cap \mathcal{A}$ *only if* $S = T$.

If an update sequence P consists of a single program, the notion of update stable models of P and regular stable models of P coincide.

Theorem 3.2. *Let* P *be an update sequence consisting of a single program* P_1, *i.e.,* $P = P_1$. *Then,* $\mathcal{U}(P) = \mathcal{S}(P_1)$.

Stable models of update sequences can also be characterized in a purely *declarative* way. To this end, we introduce the following concept.

For an update sequence $P = P_1, \ldots, P_n$ over a set of atoms $\mathcal{A}$ and $S \subseteq \mathcal{A}$, we define the *rejection set* of S by $Rej(S, P) = \bigcup_{i=1}^{n} Rej_i(S, P)$, where $Rej_n(S, P) = \emptyset$, and, for $n > i \geq 1$,

$$Rej_i(S, P) = \{r \in P_i \mid \exists r' \in P_j \setminus Rej_j(S, P), \text{ for some } j \in \{i+1, \ldots, n\},$$
$$\text{such that } H(r') = not\ H(r) \text{ and } S \models B(r) \cup B(r')\}.$$

That is, $Rej(S, P)$ contains those rules from P which are rejected on the basis of rules which are not rejected themselves.

We obtain the following characterization of stable models, mirroring a similar result given in [2].

Theorem 3.3. *Let* $P = P_1, \ldots, P_n$ *be an update sequence over a set of atoms* $\mathcal{A}$, *and let* $S \subseteq \mathcal{A}$. *Then,* S *is a stable model of* P *iff* $S = lm((P \setminus Rej(S, P))^S)$.

4   Principles of Update Sequences

In this section, we discuss several kinds of postulates which have been advocated in the literature on belief change and examine to what extent update sequences satisfy these principles. This issue has not been addressed extensively in previous work [2,3]. We first consider update programs from the perspective of *belief revision*, and assess the relevant postulates from this area. Afterwards, we briefly analyze further properties, like viewing update programs as *nonmonotonic consequence operators* and other general principles.

4.1   Belief Revision

Following [14], two different approaches to belief revision can be distinguished: (i) *immediate revision*, where the new information is simply added to the current stock of beliefs and the belief change is accomplished through the semantics of the underlying (often, nonmonotonic) logic; and (ii) *logic-constrained revision*, where the new stock of beliefs is determined by a nontrivial operation which adds and retracts beliefs, respecting logical inference and some constraints.

In the latter approach, it is assumed that beliefs are sentences from some given logical language $\mathcal{L}_B$ which is closed under the standard boolean connectives. A *belief set*, K, is a subset of $\mathcal{L}_B$ which is closed under a consequence operator $Cn(\cdot)$ of the underlying logic. A *belief base* for K is a subset $B \subseteq K$ such that $K = Cn(B)$. A belief base is a special case of an *epistemic state* [10], which is a set of sentences E representing an associated belief set K in terms of a mapping $Bel(\cdot)$ such that $K = Bel(E)$, where E need not necessarily have the same language as K.

In what follows, we first introduce different classes of postulates, and then we examine them with respect to update sequences.

AGM Postulates   One of the main aims of logic-constrained revision is to characterize suitable revision operators through postulates. Alchourrón, Gärdenfors, and Makinson (AGM) [1] considered three basic operations on a belief set K:

- *expansion* $K + \phi$, which is simply adding the new information $\phi \in \mathcal{L}_B$ to K;
- *revision* $K \star \phi$, which is sensibly revising K in the light of ϕ (in particular, when K contradicts ϕ); and
- *contraction* $K - \phi$, which is removing ϕ from K.

AGM presented a set of postulates, K$\star$1–K$\star$8, that any revision operator $\star$ mapping a belief set $K \subseteq \mathcal{L}_B$ and a sentence $\phi \in \mathcal{L}_B$ into the revised belief set $K \star \phi$ should satisfy. If, following [10,8], we assume that K is represented by an epistemic state E, then the postulates K$\star$1–K$\star$8 can be reformulated as follows:

(K1) $E \star \phi$ represents a belief set.

(K2) $\phi \in Bel(E \star \phi)$.

(K3) $Bel(E \star \phi) \subseteq Bel(E + \phi)$.

(K4) $\neg\phi \notin Bel(E)$ implies $Bel(E + \phi) \subseteq Bel(E \star \phi)$.
(K5) $\bot \in Bel(E \star \phi)$ iff ϕ is unsatisfiable.
(K6) $\phi_1 \equiv \phi_2$ implies $Bel(E \star \phi_1) = Bel(E \star \phi_2)$.
(K7) $Bel(E \star (\phi \wedge \psi)) \subseteq Bel((E \star \phi) + \psi)$.
(K8) $\neg\psi \notin Bel(E \star \phi)$ implies $Bel((E \star \phi) + \psi) \subseteq Bel(E \star (\phi \wedge \psi))$.

Here, $E \star \phi$ and $E + \phi$ is the revision and expansion operation, respectively, applied to E. Informally, these postulates express that the new information should be reflected after the revision, and that the belief set should change as little as possible. As has been pointed, this set of postulates is appropriate for new information about an *unchanged world*, but not for incorporation of a change to the actual world. Such a mechanism is addressed by the next set of postulates, expressing *update* operations.

Update Postulates For update operators $B \diamond \phi$ realizing a change ϕ to a belief base B, Katsuno and Mendelzon [18] proposed a set of postulates, U∗1–U∗8, where both ϕ and B are propositional sentences over a finitary language. For epistemic states E, these postulates can be reformulated as follows.

(U1) $\phi \in Bel(E \diamond \phi)$.
(U2) $\phi \in Bel(E)$ implies $Bel(E \diamond \phi) = Bel(E)$.
(U3) If $Bel(E)$ is consistent and ϕ is satisfiable, then $Bel(E \diamond \phi)$ is consistent.
(U4) If $Bel(E) = Bel(E')$ and $\phi \equiv \psi$, then $Bel(E \diamond \phi) = Bel(E \diamond \psi)$.
(U5) $Bel(E \diamond (\phi \wedge \psi)) \subseteq Bel((E \diamond \phi) + \psi)$.
(U6) If $\phi \in Bel(E \diamond \psi)$ and $\psi \in Bel(E \diamond \phi)$, then $Bel(E \diamond \phi) = Bel(E \diamond \psi)$.
(U7) If $Bel(E)$ is complete, then $Bel(E \diamond (\psi \vee \psi')) \subseteq Bel(E \diamond \psi) \wedge Bel(E \diamond \psi'))$.[1]
(U8) $Bel((E \vee E') \diamond \psi) = Bel((E \diamond \psi) \vee (E' \diamond \psi)$.

Here, conjunction and disjunction of epistemic states are presumed to be definable in the given language (like, e.g., in terms of intersection and union of associated sets of models, respectively).

The most important differences between (K1)–(K8) and (U1)–(U8) are that revision, if ϕ is compatible with E, should yield the same result as expansion $E + \phi$, which is not desirable for update in general, cf. [24]. On the other hand, (U8) says that if E can be decomposed into a disjunction of states (e.g., models), then each case can be updated separately and the overall results are formed by taking the disjunction of the emerging states.

Iterated Revision Darwiche and Pearl [10] have proposed postulates for iterated revision, which can be rephrased in our setting as follows (we omit parentheses in sequences $(E \star \phi_1) \star \phi_2$ of revisions):

(C1) If $\psi_2 \in Bel(\psi_1)$, then $Bel(E \star \psi_2 \star \psi_1) = Bel(E \star \psi_1)$.
(C2) If $\neg\psi_2 \in Bel(\psi_1)$, then $Bel(E \star \psi_1 \star \psi_2) = Bel(E \star \psi_2)$.

---

[1] A belief set K is *complete* iff, for each atom A, either $A \in K$ or $\neg A \in K$.

(C3) If $\psi_2 \in Bel(E \star \psi_1)$, then $\psi_2 \in Bel(E \star \psi_2 \star \psi_1)$.
(C4) If $\neg\psi_2 \notin Bel(E \star \psi_1)$, then $\neg\psi_2 \notin Bel(E \star \psi_2 \star \psi_1)$.
(C5) If $\neg\psi_2 \in Bel(E \star \psi_1)$ and $\psi_1 \notin Bel(E \star \psi_2)$, then $\psi_1 \notin Bel(E \star \psi_1 \star \psi_2)$.
(C6) If $\neg\psi_2 \in Bel(E \star \psi_1)$ and $\neg\psi_1 \in Bel(E \star \psi_2)$, then $\neg\psi_1 \in Bel(E \star \psi_1 \star \psi_2)$.

Another set of postulates for iterated revision, corresponding to a sequence E of observations, has been formulated by Lehmann [19]. Here each observation is a sentence which is assumed to be consistent (i.e., falsity is not observed), and the epistemic state E has an associated belief set $Bel(E)$. Lehmann's postulates read as follows, where E, E' denote sequences of observations and "," stands for concatenation:

(I1) $Bel(E)$ is a consistent belief set.
(I2) $\phi \in Bel(E, \phi)$.
(I3) If $\psi \in Bel(E, \phi)$, then $\phi \Rightarrow \psi \in Bel(E)$.
(I4) If $\phi \in Bel(E)$, then $Bel(E, \phi, E') = Bel(E, E')$.
(I5) If $\psi \vdash \phi$ then $Bel(E, \phi, \psi, E') = Bel(E, \psi, E')$.
(I6) If $\neg\psi \notin Bel(E, \phi)$, then $Bel(E, \phi, \psi, E') = Bel(E, \phi, \psi, E')$.
(I7) $Bel(E, \neg\phi, \phi) \subseteq Cn(E + \phi)$.

Analysis of the Postulates In order to evaluate the different postulates, we need to adapt them for the setting of update programs. Naturally, the epistemic state $P = P_1, \ldots, P_n$ of an agent is subject to revision. However, the associated belief set $Bel(P)$ ($\subseteq \mathcal{L}_\mathcal{A}$) does not belong to a logical language closed under boolean connectives. Closing $\mathcal{L}_\mathcal{A}$ under conjunction does not cause much troubles, as the identification of finite GLPs with finite conjunctions of clauses permits that updates of a GLP P by a program P_1 can be viewed as the update of P with a single sentence from the underlying belief language. Ambiguities arise, however, with the interpretation of expansion, as well as the meaning of negation and disjunction of rules and programs, respectively.

Depending on whether the particular structure of the epistemic state E should be respected, different definitions of expansion are imaginable in our framework. At the "extensional" level of sentences, represented by a program or sequence of programs P, $Bel(P + P')$ is defined as $Bel(Bel(P) \cup P')$. At the "intensional" level of sequences $P = P_1, \ldots, P_n$, $Bel(P + P')$ could be defined as $Bel(P_1, \ldots, P_n \cup P')$. An intermediate approach would be defining $Bel(P + P') = Bel_\mathcal{A}(P_\lhd \cup P')$. We adopt the extensional view here. Note that, in general, adding P' to $Bel(P)$ does not amount to the semantical intersection of P' and $Bel(P)$ (nor of P and P', respectively).

As for negation, we might interpret the condition $\neg\phi \notin Bel(E)$ (or $\neg\psi \notin Bel(E \star \phi)$ in (K4) and (K8)) as satisfiability requirement for $E + \phi$ (or $(E \star \phi) + \psi$).

Disjunction $\vee$ of rules or programs (as epistemic states) appears to be meaningful only at the semantical level. The union $\mathcal{S}(P_1) \cup \mathcal{S}(P_2)$ of the sets of stable models of programs P_1 and P_2 may be represented syntactically through a program P_3, which in general requests an extended set of atoms. We thus do not consider the postulates involving $\vee$.

Postulate	Interpretation	Postulate holds
(K1)	(P_1, P_2) represents a belief set	yes
(K2), (U1)	$P_2 \subseteq Bel(P_1, P_2)$	yes
(U2)	$Bel(P_2) \subseteq Bel(P_1)$ implies $Bel(P_1, P_2) = Bel(P_1)$	no
(K3)	$Bel(P_1, P_2) \subseteq Bel(Bel(P_1) \cup P_2)$	yes
(U3)	If P_1 and P_2 are satisfiable, then (P_1, P_2) is satisfiable	no
(K4)	If $Bel(P_1) \cup P_2$ has a stable model, then $Bel(Bel(P_1) \cup P_2) \subseteq Bel(P_1, P_2)$	no
(K5)	(P_1, P_2) is unsatisfiable iff P_2 is unsatisfiable	no
(K6), (U4)	$P_1 \equiv P_1'$ and $P_2 \equiv P_2'$ implies $(P_1, P_2) \equiv (P_1', P_2')$	no
(K7), (U5)	$Bel(P_1, P_2 \cup P_3) \subseteq Bel(Bel(P_1, P_2) \cup P_3)$	yes
(U6)	If $Bel(P_3) \subseteq Bel(P_1, P_2)$ and $Bel(P_2) \subseteq Bel(P_1, P_3)$, then $Bel(P_1, P_2) = Bel(P_1, P_3)$	no
(K8)	If $Bel(P_1, P_2) \cup P_3$ is satisfiable then $Bel(Bel(P_1, P_2) \cup P_3) \subseteq Bel(P_1, P_2 \cup P_3)$	no

Table 1. Interpretation of Postulates (K1)–(K8) and (U1)–(U6).

Given these considerations, Table 1 summarizes our interpretation of postulates (K1)–(K8) and (U1)–(U6), together with indicating whether the respective property holds or fails. We assume that P_1 is a nonempty sequence of GLPs.

Thus, apart from very simple postulates, the majority of the adapted AGM and update postulates are violated by update programs. This holds even for the case where P_1 is a single program. In particular, $Bel(P_1, P_2)$ violates discriminating postulates such as (U2) for update and (K4) for revision. In the light of this, update programs neither have update nor revision flavor.

We remark that the picture does not change if we abandon extensional expansion and consider the postulates under intensional expansion. Thus, also under this view, update programs do not satisfy minimality of change.

The postulates (C1)–(C6) and (I1)–(I7) for iterated revision are treated in Table 2. Concerning Lehmann's [19] postulates, (I3) is considered as the pendant to AGM postulate K⋆3. In a literal interpretation of (I3), we may, since the belief language associated with GLPs does not have implication, consider the case where ψ is a default literal L_0 and $\phi = L_1 \wedge \cdots \wedge L_k$ is a conjunction of literals L_i, such that $\phi \Rightarrow \psi$ corresponds to the rule $L_0 \leftarrow L_1, \ldots, L_k$. Since the negation of GLPs is not defined, we do not interpret (I7).

Note that, although postulate (C3) fails in general, it holds if P_3 contains a single rule. Thus, all of the above postulates except C4 fail, already if P_1 is a single logic program, and, with the exception of C3, each change is given by a single rule.

A question at this point is whether, after all, the various belief change postulates from above are meaningful for update programs.

We can view the epistemic state $P = P_1, \ldots, P_n$ of an agent as a prioritized belief base in the spirit of [7,23,6]. Revision with a new piece of information Q is accomplished by simply changing the epistemic state to $P = P_1, \ldots, P_n, Q$. The

Postulate	Interpretation	Postulate holds
(C1)	If $P_3 \subseteq Bel(P_2)$, then $Bel(P_1, P_3, P_2) = Bel(P_1, P_2)$	no
(C2)	If $S \not\models P_3$, for all $S \in \mathcal{S}(P_2)$, then $Bel(P_1, P_3, P_2) = Bel(P_1, P_2)$	no
(C3)	If $P_3 \subseteq Bel(P_1, P_2)$, then $P_3 \subseteq Bel(P_1, P_3, P_2)$	no
(C4)	If $S \models P_3$ for some $S \in \mathcal{S}(P_1, P_2)$, then $S \models P_3$ for some $S \in \mathcal{S}(P_1, P_3, P_2)$	yes
(C5)	If $S \not\models P_3$ for all $S \in \mathcal{S}(P_1, P_2)$ and $P_2 \not\subseteq Bel(P_1, P_3)$, then $P_2 \not\subseteq Bel(P_1, P_2, P_3)$	no
(C6)	If $S \not\models P_3$ for all $S \in \mathcal{S}(P_1, P_2)$ and $S \not\models P_2$ for all $S \in \mathcal{S}(P_1, P_3)$, then $S \not\models P_2$ for all $S \in \mathcal{S}(P_1, P_2, P_3)$	no
(I1)	$Bel(P_1)$ is a consistent belief set	no
(I2)	$P_2 \subseteq Bel(P_1, P_2)$	yes
(I3)	If $L_0 \leftarrow \in Bel(P_1, \{L_1, \ldots, L_k\})$, then $L_0 \leftarrow L_1, \ldots, L_k \in Bel(P_1)$	yes
(I4)	If $P_2 \subseteq Bel(P_1)$, then $Bel(P_1, P_2, P_3, \ldots, P_n) = Bel(P_1, P_3, \ldots, P_n)$	no
(I5)	If $Bel(P_3) \subseteq Bel(P_2)$, then $Bel(P_1, P_2, P_3, P_4, \ldots, P_n) = Bel(P_1, P_3, P_4, \ldots, P_n)$	no
(I6)	If $S \models P_3$ for some $S \in \mathcal{S}(P_1, P_2)$, then $Bel(P_1, P_2, P_3, P_4, \ldots, P_n) = Bel(P_1, P_2, P_2 \cup P_3, P_4, \ldots, P_n)$	no

Table 2. Interpretation of Postulates (C1)–(C6) and (I1)–(I6).

change of the belief base is then automatically accomplished by the nonmonotonic semantics of a sequence of logic programs. Under this view, updating logic programs amounts to an instance of the immediate revision approach.

On the other hand, referring to the update program, we may view the belief set of the agent represented through a pair $\langle P, \mathcal{A} \rangle$ of a logic program P and a (fixed) set of atoms $\mathcal{A}$, such that its belief set is given by $Bel_{\mathcal{A}}(P)$. Under this view, a new piece of information Q is incorporated into the belief set by producing a representation, $\langle P', \mathcal{A} \rangle$, of the new belief set, where $P' = P \triangleleft Q$. Here, (a set of) sentences from an extended belief language is used to characterize the new belief state, which is constructed by a nontrivial operation employing the semantics of logic programs. Thus, update programs enjoy to some extent also a logic-constrained revision flavor. Nonetheless, as also the failure of postulates shows, they are more an instance of *immediate* than *logic-constrained* revision. What we naturally expect, though, is that the two views described above amount to the same *at a technical level*. However, as we shall demonstrate below, this is not true in general.

4.2 Further Properties

Belief revision has been related in [14] to nonmonotonic logics by interpreting it as an abstract consequence relation on sentences, where the epistemic state is fixed. In the same way, we can interpret update programs as abstract consequence relation $\vdash\!\!\sim$ on programs as follows. For a fixed epistemic state P and GLPs P_1 and P_2, we define

$$P_1 \vdash\!\!\sim_P P_2 \text{ if and only if } P_2 \subseteq Bel(P, P_1),$$

i.e., if the rules P_2 are in the belief state of the agent after update of the epistemic state with P_1.

Various properties for nonmonotonic inference operations have been identified in the literature (see, e.g., [14]). Among them are *Cautious Monotonicity*, *Cut*, *(Left) Conjunction*, *Rational Cautious Monotonicity*, and *Equivalence*. Except for Cut, none of these properties hold. We recall that Cut denotes the following schema:

$$\frac{A \wedge B_1 \wedge \ldots \wedge B_m \vdash\!\!\sim_P C \quad A \vdash\!\!\sim_P B_1 \wedge \ldots \wedge B_m}{A \vdash\!\!\sim_P C}$$

Additionally, we can also identify some very elemental properties which, as we believe, updates and sequences of updates should satisfy. The following list of properties is not developed in a systematic manner, though, and is by no means exhaustive. Update programs do enjoy, unless stated otherwise, these properties.

Addition of Tautologies: If the program P_2 contains only tautological clauses, then $(P_1, P_2) \equiv P_1$.

Initialization: $(\emptyset, P) \equiv P$.

Idempotence: $(P, P) \equiv P$.

Idempotence for Sequences: $(P_1, P_2, P_2) \equiv (P_1, P_2)$.

Update of Disjoint Programs: If $P = P_1 \cup P_2$ is a union of programs P_1, P_2 on disjoint alphabets, then $(P, P_3) \equiv (P_1, P_3) \cup (P_2, P_3)$.

Parallel updates: If P_2 and P_3 are programs defined over disjoint alphabets, then $(P_1, P_2) \cup (P_1, P_3) \equiv (P_1, P_2 \cup P_3)$.   (*Fails.*)

Noninterference: If P_2 and P_3 are programs defined over disjoint alphabets, then $(P_1, P_2, P_3) \equiv (P_1, P_3, P_2)$.

Augmented update: If $P_2 \subseteq P_3$ then $(P_1, P_2, P_3) \equiv (P_1, P_3)$.

As mentioned before, a sequence of updates $P = P_1, \ldots, P_n$ can be viewed from the point of view of "immediate" revision or of "logic-constrained" revision. The following property, which deserves particular attention, expresses equivalence of these views (the property is formulated for the case $n = 3$):

Iterativity: For any epistemic state P_1 and GLPs P_2 and P_3, it holds that $P_1 \triangleleft P_2 \triangleleft P_3 \equiv_A (P_1 \triangleleft P_2) \triangleleft P_3$.

However, this property fails. Informally, soundness of this property would mean that a sequence of three updates is a shorthand for iterated update of a single program, i.e., the result of $P_1 \triangleleft P_2$ is viewed as a singleton sequence. Stated another way, this property would mean that the definition for $P_1 \triangleleft P_2 \triangleleft P_3$ can be viewed as a shorthand for the nested case. Vice versa, this property reads as possibility to forget an update once and for all, by incorporating it immediately into the current belief set.

For a concrete counterexample, consider $P_1 = \emptyset$, $P_2 = \{a \leftarrow \ , not\, a \leftarrow \ \}$, $P_3 = \{a \leftarrow \ \}$. The program $P_\triangleleft = P_1 \triangleleft P_2 \triangleleft P_3$ has a unique stable model, in which a is true. On the other hand, $(P_1 \triangleleft P_2) \triangleleft P_3$ has no stable model. Informally, while the "local" inconsistency of P_2 is removed in $P_1 \triangleleft P_2 \triangleleft P_3$ by rejection of the rule $not\, a \leftarrow \ $ via P_3, a similar rejection in $(P_1 \triangleleft P_2) \triangleleft P_3$ is blocked because of a renaming of the predicates in $P_1 \triangleleft P_2$. The local inconsistency of P_2 is thus not eliminated.

However, under certain conditions, which exclude such possibilities for local inconsistencies, the iterativity property holds, given by the following result:

Theorem 4.1. *Let $P = P_1, \ldots, P_n$, $n \geq 2$, be an update sequence on a set of atoms A. Suppose that, for any rules $r_1, r_2 \in P_i$, $i \leq n$, such that $H(r_1) = not\, H(r_2)$, the union $B(r_1) \cup B(r_2)$ of their bodies is unsatisfiable. Then:*

$$(\cdots (P_1 \triangleleft P_2) \triangleleft P_3) \cdots \triangleleft P_{n-1}) \triangleleft P_n \equiv_A P_1 \triangleleft P_2 \triangleleft P_3 \triangleleft \cdots \triangleleft P_n.$$

5   Refined Semantics and Extensions

Minimal and Strict Stable Models Even if we abandon the AGM view, update programs do intuitively not respect minimality of change, as a new set of rules P_2 should be incorporated into an existing program P_1 with as little change as possible.

It appears natural to measure change in terms of the set of rules in P_1 which are abandoned. This leads us to prefer a stable model S_1 of $P = P_1, P_2$ over another stable model S_2 if S_1 satisfies a larger set of rules from P_1 than S_2.

Definition 5.1. *Let $P = P_1, \ldots, P_n$ be a sequence of GLPs. A stable model $S \in \mathcal{U}(P)$ is minimal iff there is no $T \in \mathcal{U}(P)$ such that $Rej(T, P) \subset Rej(S, P)$.*

Example 5.1. Consider $P_1 = \{r_1 : \ not\, a \leftarrow \ \}$, $P_2 = \{r_2 : \ a \leftarrow not\, c\}$, and $P_3 = \{r_3 : \ c \leftarrow not\, d, \ r_4 : \ d \leftarrow not\, c \ \}$. Then (P_1, P_2) has the single stable model $\{a\}$, which rejects the rule in P_1. The sequence (P_1, P_2, P_3) has two stable models: $S_1 = \{c\}$ and $S_2 = \{a, d\}$. S_1 rejects no rule, while S_2 rejects the rule r_1. Thus, S_1 is preferred to S_2 and S_1 is minimal.

Minimal stable models put no further emphasis on the temporal order of updates. Rules in more recent updates may be violated in order to satisfy rules from previous updates. Eliminating this leads us to the following notion.

Definition 5.2. *Let $S, S' \in \mathcal{U}(P)$ for an update sequence $P = P_1, \ldots, P_n$. Then, S is preferred to S' iff some $i \in \{1, \ldots, n\}$ exists such that (1) $Rej_i(S, P) \subset Rej_i(S', P)$, and (2) $Rej_j(S', P) = Rej_j(S, P)$, for all $j = i + 1, \ldots, n$. A stable model S of P is strict, if no $S' \in \mathcal{U}(P)$ exists which is preferred to S.*

Example 5.2. Consider $P = P_1, P_2, P_3, P_4$, where $P_1 = \{r_1 : \; not\, a \leftarrow \; \}$, $P_2 = \{r_2 : \; a \leftarrow not\, c\}$, $P_3 = \{r_3 : not\, c \leftarrow \; \}$, and $P_4 = \{r_4 : \; c \leftarrow not\, d$, $r_5 : \; d \leftarrow not\, c \; \}$. Then, P has two stable models, namely $S_1 = \{c\}$ and $S_2 = \{a, d\}$. We have $Rej(S_1, P) = \{r_3\}$ and $Rej(S_2, P) = \{r_1\}$. Thus, $Rej(S_1, P)$ and $Rej(S_2, P)$ are incomparable, and hence both S_1 and S_2 are minimal stable models. However, compared to S_2 in S_1 the more recent rule of P_3 is violated. Thus, S_2 is the unique strict stable model.

Clearly every strict stable model is minimal, but not vice versa. Unsurprisingly, minimal and strict stable models do not satisfy AGM minimality of change.

The trade-off for epistemic appeal is higher computational complexity than for arbitrary stable models. Let $Bel_{min}(P)$ (resp., $Bel_{str}(P)$) be the set of beliefs induced by the collection of minimal (resp., strict) stable models of $P = P_1, \ldots, P_n$.

Theorem 5.1. *Given a sequence of programs $P = P_1, P_2, \ldots, P_n$ over a set of atoms $\mathcal{A}$, deciding whether*

1. *P has a stable model is NP-complete;*
2. *$L \in Bel(P)$ for a given literal L is coNP-complete;*
3. *$L \in Bel_{min}(P)$ (resp. $L \in Bel_{str}(P)$) for a given literal L is Π_2^P-complete.*

Similar results have been derived by Inoue and Sakama [17]. The complexity results imply that minimal and strict stable models can be polynomially translated into disjunctive logic programming, which is currently under investigation.

Strong Negation Update programs can be easily extended to the setting of generalized extended logic programs (GELPs), which have besides *not* also strong negation $\neg$ as in [21]. Viewing, for $A \in \mathcal{A}$, the formula $\neg A$ as a fresh atom, the rules $not\, A \leftarrow \neg A$ and $not\, \neg A \leftarrow A$ emulate the interpretation of $\neg$ in answer set semantics (cf., e.g., [2]). More precisely, the consistent answer sets of a GELP P correspond one-to-one to the stable models of $P^\neg$, which is P augmented with the emulation rules for $\neg A$. Answer sets of a sequence of GELPs $P = P_1, \ldots, P_n$ can then be defined through this correspondence in terms of the stable models of $P^\neg = P_1^\neg, \ldots, P_n^\neg$, such that $Bel(P) = Bel(P^\neg)$.

Like for dynamic logic programs [3], $P^\neg$ can be simplified by removing some of the emulation rules. Let $CR(P)$ be the set of all emulation rules for atoms A such that $\neg A$ occurs in some rule head of P.

Theorem 5.2. *For any sequence of GELPs $P = P_1, \ldots, P_n$ over $\mathcal{A}$, $S \subseteq \mathcal{A} \cup \{\neg A \mid A \in \mathcal{A}\}$ is an answer set of P iff $S \in \mathcal{U}(P_1, \ldots, P_{n-1}, P_n \cup CR(P))$.*

First-Order Programs The semantics of a sequence $P = P_1, \ldots, P_n$ of first-order GLPs, i.e., where $\mathcal{A}$ consists of nonground atoms in a first order-language, is reduced to the ground case by defining it in terms of the sequence of instantiated programs $P^* = P_1^*, \ldots, P_n^*$ over the Herbrand universe of P as usual. That is, $\mathcal{U}(P) = \mathcal{U}(P^*)$. The definition of update program $P_\lhd$ can be easily generalized to non-ground programs, such that $P_\lhd \equiv P^*_\lhd$, i.e., $P_\lhd$ faithfully represents the update program for P^*.

6   Related Work

Dynamic Logic Programming Recall that our update programs syntactically redefine dynamic logic programs for update in [2,5], which generalize the idea of updating interpretations through revision programs [22]. As we feel, they more transparently reflect the working behind this approach.

The major difference between our update programs and dynamic logic programs is that the latter determine the values of atoms from the bottom level P_1 *upwards* towards P_n, using interia rules, while update programs determine the values in a downward fashion.

Denote by $\oplus P = P_1 \oplus \cdots \oplus P_n$ the dynamic logic program of [2] for updating P_1 with $P_2, \ldots, P_n$ over atoms $\mathcal{A}$, which is a GLP over atoms $\mathcal{A}_{dyn} \supseteq \mathcal{A}$. For any model M of P_n in $\mathcal{A}$, let

$$Rejected(M, P) = \bigcup_{i=1}^{n} \{r \in P_i \mid \exists r' \in P_j, \text{ for some } j \in \{i + 1, \ldots, n\}, \text{ such}$$
$$\text{that } H(r') = not\, H(r) \wedge S \models B(r) \cup B(r')\},$$
$$Defaults(M, P) = \{not\, A \mid \forall r \in P : H(r) = A \Rightarrow M \not\models B(r)\}.$$

Stable models of $\oplus P$, projected to $\mathcal{A}$, are semantically characterized as follows.

Definition 6.1. *For a sequence* $P = P_1, \ldots, P_n$ *of GLPs over atoms* $\mathcal{A}$, *an interpretation* $N \subseteq \mathcal{A}_{dyn}$ *is a stable model of* $\oplus P$ *iff* $M = N \cap \mathcal{A}$ *is a model of* U *such that*

$$M = lm(P \setminus Rejected(M, P) \cup Defaults(M, P)).$$

Here, literals *not A* are considered as new atoms, where implicitly the constraint $\leftarrow A, not\, A$ is added. Let us call any such M a *dynamic stable model* of P.

As one can see, we may replace $Rej(S, P)$ in Theorem 3.3 by $Rejected(S, P)$ and add all rules in $Defaults(S, P)$, as they vanish in the reduction by S. However, this implies that update and dynamic stable models coincide.

Theorem 6.1. *For any sequence* $P = P_1, \ldots, P_n$ *of GLPs over atoms* $\mathcal{A}$, $S \subseteq \mathcal{A}$ *is a dynamic stable model of* P *iff* $S \in \mathcal{U}(P)$.

Inheritance Programs A framework for logic programs with inheritance is introduced in [9]. In a hierarchy of objects $o_1, \ldots, o_n$, represented by a disjunctive extended logic program $P_1, \ldots, P_n$ [15], possible conflicts in determining the properties of o_i are resolved by favoring rules which are more specific according to the hierarchy, which is given by a (strict) partial order $<$ over the objects.

If we identify o_i with the indexed program P_i, an *inheritance program* consists of a set $P = \{P_1, \ldots, P_n\}$ of programs over atoms $\mathcal{A}$ and a partial order $<$ on P. The program $\mathcal{P}(P_i)$ for P_i (as an object) is given by $\mathcal{P}(P_i) = \{P_i\} \cup \{P_j \mid P_i < P_j\}$, i.e., the collection of programs at and above P_i.

The semantics of $\mathcal{P}(P_i)$ is defined in terms of answer sets. In the rest of this section, we assume that any program $P_i \in P$ is disjunction-free and we simplify definitions in [9] accordingly. Let, for each literal L of form A or $\neg A$, denote $\neg L$ its opposite, and let $Lit_{\mathcal{A}} = \mathcal{A} \cup \{\neg A \mid A \in \mathcal{A}\}$.

Definition 6.2. *Let $I \subseteq Lit_{\mathcal{A}}$ be an interpretation and $r \in P_j$. Then, r is overridden in I, if (1) $I \models B(r)$, (2) $\neg H(r) \in I$, and (3) there exists a rule $r_1 \in P_i$ for some $P_i < P_j$ such that $H(r_1) = \neg H(r)$.*

An interpretation $I \subseteq Lit_{\mathcal{A}}$ is a model of $\mathcal{P}$, if I satisfies all non-overridden rules in $\mathcal{P}$ and the constraint $\leftarrow A, not\ A$ for each atom $A \in \mathcal{A}$; moreover, I is minimal if it is the least model of all these rules. Answer sets are now as follows.

Definition 6.3. *A model M of $\mathcal{P} = \mathcal{P}(P_i)$, is a $DLP^<$-answer set of $\mathcal{P}$ iff M is a minimal model of $\mathcal{P}^M$, where $\mathcal{P}^M = \{r \in \mathcal{P} \mid r$ is not overridden in $M\}^M$ is the reduct of $\mathcal{P}$ by M.*

It is natural to view an update sequence $P = P_1, \ldots, P_n$ as an inheritance program where later updates are considered more specific. That is, we might view P as an inheritance program $P_n < P_{n-1} < \ldots < P_1$. It appears that the latter is in fact equivalent to the update program $P_1 \triangleleft \ldots \triangleleft P_n$.

For a sequence of GLPs $P = P_1, \ldots, P_n$ over $\mathcal{A}$, define the inheritance program $\mathcal{Q} = Q_n < Q_{n-1} < \cdots < Q_1$ as follows. Let P_i^- be the program resulting from P_i by replacing in rule heads the default negation not through $\neg$. Define $Q_1 = P_1^- \cup \{\neg A \leftarrow not\ A \mid A \in \mathcal{A}\}$ and $Q_j = P_j^-$, for $j = 2, \ldots, n$. Then we have the following.

Theorem 6.2. *Let $P = P_1, \ldots, P_n$ be a sequence of GLPs over atoms $\mathcal{A}$. Then, $S \in \mathcal{U}(P)$ iff $S \cup \{\neg A \mid A \in \mathcal{A} \setminus S\}$ is a $DLP^<$-answer set of $\mathcal{Q}(P_1, \ldots, P_n)$.*

Conversely, linear inheritance programs yield the same result as update programs in the extension with classical negation.

Theorem 6.3. *Let $P = P_1 < \cdots < P_n$ be an inheritance program over atoms $\mathcal{A}$. Then, S is a $DLP^<$-answer set of P iff S is an answer set of the sequence of GELPs $P_n, P_{n-1}, \ldots, P_1$.*

Thus, dynamic logic programs and inheritance programs are equivalent.

Program Updates through Abduction On the basis of their notion of *extended abduction*, Inoue and Sakama [17] define a framework for various update problems. The most general is *theory update*, which is update of an extended logic program (ELP) P_1 by another such program P_2. Informally, an abductive update of P_1 by P_2 is a largest consistent program P' such that $P_1 \subseteq P' \subseteq P_1 \cup P_2$ holds. This is formally captured in [17] by reducing the update problem to computing a minimal set of abducible rules $Q \subseteq P_1 \setminus P_2$ such that $(P_1 \cup P_2) \setminus Q$ is consistent. In terms of [16], $P_1 \cup P_2$ is considered for abduction where the rules in $P_1 \setminus P_2$ are abducible, and the intended update is realized via a minimal *anti-explanation* for falsity, which removes abducible rules to restore consistency.

While this looks similar to our minimal updates, there is a salient difference: abductive update does not respect *causal rejection*. A rule r from $P_1 \setminus P_2$ may be rejected even if no rule r' P_2 fires whose head contradicts applying r. For example, consider $P_1 = \{q \leftarrow \ , \ \neg q \leftarrow a\}$ and $P_2 = \{a \leftarrow \ \}$. Both P_1 and P_2 have consistent answer sets, while (P_1, P_2) has no stable model. In Inoue and Sakama's approach, one of the two rules in P_1 will be removed. Note that contradiction removal in a program P occurs as a special case $(P_1 = P, P_2 = \emptyset)$.

Abductive updates are, due to inherent minimality of change, harder than update programs; some abductive reasoning problems are Σ_2^P-complete [17].

Updates through Priorities Zhang and Foo [13] define update of an ELP P_1 by an ELP P_2 based on their work on preferences [12] as a two-step approach: In Step 1, each answer set S of P_1 is updated to a closest answer set S' of P_2, where distance is in terms of the set of atoms on which S, S' disagree and closeness is set inclusion. Then, a maximal set $Q \subseteq P_1$ is chosen such that $P_3 = P_2 \cup Q$ has an answer set containing S'. In Step 2, the answer sets of P_3 are computed using priorities, where rules of P_2 have higher priority than rules of Q.

This approach is different from ours. It is in the spirit of the *possible models approach* [24], which updates models of a propositional theory separately, thus satisfying the update postulate U8. However, like in Inoue and Sakama's approach, rules are not removed on the basis of causal rejection. In particular, the same result is obtained on the example there. Step 2 indicates a strong update flavor of the approach, since rules are unnecessarily abandoned. For example, update of $P_1 = \{p \leftarrow not\,q\}$ with $P_2 = \{q \leftarrow not\,p\}$ results in P_2, even though $P_1 \cup P_2$ is consistent. Since the result of an update leads to a set of programs, in general, naive handling of updates requires exponential space.

7   Conclusion

We have considered the approach to updating logic programs based on dynamic logic programs [2,3] and investigated various properties of this approach. Comparing it to other approaches and related work, we found that it is equivalent to a fragment of inheritance programs in [9].

Several issues remain for further work. A natural issue is the inverse of addition, i.e. retraction of rules from a logic program. Dynamic logic programming evolved into LUPS [3], which is a language for specifying update behavior in terms of addition and retraction of sets of rules to a logic program. LUPS is generic, however, as in principle, different approaches to updating logic programs could provide the semantical basis for an update step. Exploring properties of the general framework, as well as of particular such instantiations, would be worthwhile. Furthermore, reasoning about update programs describing the behavior of agents programmed in LUPS is an interesting issue.

Another issue are postulates for update operators on logic programs and, more generally, on nonmonotonic theories. As we have seen, several postulates from the area of logical theory change fail for dynamic logic programs (see [8] for related observations). This may partly be explained by nonmonotonicity of stable semantics and the dominant role of syntax for update embodied by causal rejection. However, similar features are not exceptional in the context of logic programming. It would be interesting to know further postulates and desiderata for update of logic programs besides the ones considered here, and an AGM style characterization of update operators compliant with them.

Acknowledgments. This work was partially supported by the Austrian Science Fund (FWF) under grants P13871-INF and N Z29-INF.

References

1. C. Alchourrón, P. Gärdenfors, and D. Makinson. On the Logic of Theory Change: Partial Meet Functions for Contraction and Revision. *Journal of Symbolic Logic*, 50:510–530, 1985.
2. J. Alferes, J. Leite, L. Pereira, H. Przymusinska, and T. Przymusinski. Dynamic Logic Programming. In A. Cohn and L. Schubert, editors, *Proc. KR'98*, pp. 98–109. Morgan Kaufmann, 1998.
3. J. Alferes, J. Leite, L. Pereira, H. Przymusinska, and T. C. Przymusinski. Dynamic Updates of Non-Monotonic Knowledge Bases. *Journal of Logic Programming*, 2000. To appear.
4. J. Alferes and L. Pereira. Updates plus Preferences. In *Proc. JELIA 2000*, LNCS, this volume. Springer, 2000.
5. J. Alferes, L. Pereira, H. Przymusinska, and T. Przymusinski. Lups: A Language for Updating Logic Programs. In *Proc. LPNMR'99*, LNAI 1730, pp. 162–176. Springer, 1999.
6. S. Benferhat, C. Cayrol, D. Dubois, J. Lang, and H. Prade. Inconsistency Management and Prioritized Syntax-Based Entailment. In *Proc. IJCAI-93*, pp. 640–645. Morgan Kaufman, 1993.
7. G. Brewka. Belief Revision as a Framework for Default Reasoning. In *The Logic of Theory Change*, LNAI 465, pp. 206–222. Springer, 1991.
8. G. Brewka. Declarative Representation of Revision Strategies. In *Proc. NMR '2000*, 2000.
9. F. Buccafurri, W. Faber, and N. Leone. Disjunctive Logic Programs with Inheritance. In *Proc. ICLP '99*, pp. 79–93, 1999. The MIT Press. Full version available as Technical Report DBAI-TR-99-30, Institut für Informationssysteme, Technische Universität Wien, Austria, May 1999.
10. A. Darwiche and J. Pearl. On the Logic of Iterated Belief Revision. *Artificial Intelligence*, 89(1-2):1-29, 1997.
11. F. Sadri and F. Toni. Computational Logic and Multiagent Systems: A Roadmap. Available from the ESPRIT Network of Excellence in Computational Logic (Compulog Net), http://www.compulog.org/, 1999.
12. N. Foo and Y. Zhang. Towards Generalized Rule-based Updates. In *Proc. IJCAI'97*, pp. 82–88. Morgan Kaufmann, 1997.
13. N. Foo and Y. Zhang. Updating Logic Programs. In *Proc. ECAI'98*, pp. 403-407. Wiley, 1998.
14. P. Gärdenfors and H. Rott. Belief Revision. In D. Gabbay, C. Hogger, and J. Robinson, editors, *Handbook of logic in Artificial Intelligence and Logic Programming*. Oxford Science Publications, 1995.
15. M. Gelfond and V. Lifschitz. Classical Negation in Logic Programs and Disjunctive Databases. *New Generation Computing*, 9:365–385, 1991.

16. K. Inoue and C. Sakama. Abductive Framework for Nonmonotonic Theory Change. In *Proc. IJCAI'95*, pp. 204–210. Morgan Kaufmann, 1995.
17. K. Inoue and C. Sakama. Updating Extended Logic Programs Through Abduction. In *Proc. LPNMR'99*, LNAI 1730, pp. 147–161. Springer, 1999.
18. H. Katsuno and A. Mendelzon. On the Difference Between Updating a Knowledge Database and Revising it. In *Proc. KR'91*, pp. 387–394, 1991.
19. D. Lehmann. Belief Revision, Revised. In *Proc. IJCAI'95*, pp. 1534–1540. Morgan Kaufmann, 1995.
20. J. Leite, J. Alferes, and L. Pereira. Multi-Dimensional Dynamic Logic Programming. In *Proc. Workshop on Computational Logic in Multi-Agent Systems (CLIMA)*, London, UK, July 2000.
21. V. Lifschitz and T. Y. C. Woo. Answer Sets in General Nonmonotonic Reasoning (Preliminary Report). In *Proc. KR '92*, pp. 603–614. Morgan Kaufmann, 1992.
22. V. W. Marek and M. Truszczyński. Revision Specifications by Means of Programs. In *Proc. JELIA'94*, LNAI 838, pp. 122–136. Springer, 1994.
23. B. Nebel. Belief Revision and Default Reasoning: Syntax-Based Approaches. In *Proc. KR'91*, pp. 417–428, 1991.
24. M. Winslett. Reasoning about Action Using a Possible Models Approach. In *Proc. AAAI'88*, pp. 89–93, 1988.

The K͜e͜Y Approach: Integrating Object Oriented Design and Formal Verification

Wolfgang Ahrendt[1], Thomas Baar[1], Bernhard Beckert[1], Martin Giese[1], Elmar Habermalz[1], Reiner Hähnle[2], Wolfram Menzel[1], and Peter H. Schmitt[1]

[1] University of Karlsruhe, Institute for Logic, Complexity and Deduction Systems,
D-76128 Karlsruhe, Germany,
http://i12www.ira.uka.de/~key
[2] Department of Computing Science, Chalmers University of Technology
S-41296 Gothenburg,
reiner@cs.chalmers.se

Abstract This paper reports on the ongoing KeY project aimed at bridging the gap between (a) object-oriented software engineering methods and tools and (b) deductive verification. A distinctive feature of our approach is the use of a commercial CASE tool enhanced with functionality for formal specification and deductive verification.

1 Introduction

1.1 Analysis of the Current Situation

While formal methods are by now well established in hardware and system design (the majority of producers of integrated circuits are routinely using BDD-based model checking packages for design and validation), usage of formal methods in software development is currently confined essentially to academic research projects. There are industrial applications of formal software development [8], but they are still exceptional [9].

The limits of applicability of formal methods in software design are not defined by the potential range and power of existing approaches. Several case studies clearly demonstrate that computer-aided specification and verification of realistic software is feasible [18]. The real problem lies in the excessive demand imposed by current tools on the skills of prospective users:

1. Tools for formal software specification and verification are not integrated into industrial software engineering processes.
2. User interfaces of verification tools are not ergonomic: they are complex, idiosyncratic, and are often without graphical support.
3. Users of verification tools are expected to know syntax and semantics of one or more complex formal languages. Typically, at least a tactical programming language and a logical language are involved. And even worse, to make serious use of many tools, intimate knowledge of employed logic calculi and proof search strategies is necessary.

M. Ojeda-Aciego et al. (Eds.): JELIA 2000, LNAI 1919, pp. 21–36, 2000.

Successful specification and verification of larger projects, therefore, is done separately from software development by academic specialists with several years of training in formal methods, in many cases by the tool developers themselves.

While this is viable for projects with high safety and low secrecy demands, it is unlikely that formal software specification and verification will become a routine task in industry under these circumstances.

The future challenge for formal software specification and verification is to make the considerable potential of existing methods and tools feasible to use in an industrial environment. This leads to the requirements:

1. Tools for formal software specification and verification must be integrated into industrial software engineering procedures.
2. User interfaces of these tools must comply with state-of-the-art software engineering tools.
3. The necessary amount of training in formal methods must be minimized. Moreover, techniques involving formal software specification and verification must be teachable in a structured manner. They should be integrated in courses on software engineering topics.

To be sure, the thought that full formal software verification might be possible without any background in formal methods is utopian. An industrial verification tool should, however, allow for *gradual* verification so that software engineers at any (including low) experience level with formal methods may benefit. In addition, an integrated tool with well-defined interfaces facilitates "outsourcing" those parts of the modeling process that require special skills.

Another important motivation to integrate design, development, and verification of software is provided by modern software development methodologies which are *iterative* and *incremental*. *Post mortem* verification would enforce the antiquated waterfall model. Even worse, in a linear model the extra effort needed for verification cannot be parallelized and thus compensated by greater work force. Therefore, delivery time increases considerably and would make formally verified software decisively less competitive.

But not only must the extra time for formal software development be within reasonable bounds, the cost of formal specification and verification in an industrial context requires accountability:

4. It must be possible to give realistic estimations of the cost of each step in formal software specification and verification depending on the type of software and the degree of formalization.

This implies immediately that the mere existence of tools for formal software specification and verification is not sufficient, rather, formal specification and verification have to be fully integrated into the software development process.

1.2   The KₑY Project

Since November 1998 the authors work on a project addressing the goals outlined in the previous section; we call it the KₑY project (read "key").

In the principal use case of the KeY system there are actors who want to implement a software system that complies with given requirements and formally verify its correctness. The system is responsible for adding formal details to the analysis model, for creating conditions that ensure the correctness of refinement steps (called proof obligations), for finding proofs showing that these conditions are satisfied by the model, and for generating counter examples if they are not. Special features of KeY are:

- We concentrate on object-oriented analysis and design methods (OOAD)— because of their key role in today's software development practice—, and on JAVA as the target language. In particular, we use the Unified Modeling Language (UML) [24] for visual modeling of designs and specifications and the Object Constraint Language (OCL) for adding further restrictions. This choice is supported by the fact, that the UML (which contains OCL since version 1.3) is not only an OMG standard, but has been adopted by all major OOAD software vendors and is featured in recent OOAD textbooks [22].
- We use a commercial CASE tool as starting point and enhance it by additional functionality for formal specification and verification. The current tool of our choice is TogetherSoft's TOGETHER 4.0.
- Formal verification is based on an axiomatic semantics of the *real* programming language JAVA CARD [29] (soon to be replaced by Java 2 Micro Edition, J2ME).
- As a case study to evaluate the usability of our approach we develop a scenario using smart cards with JAVA CARD as programming language [15,17]. JAVA smart cards make an extremely suitable target for a case study:
  - As an object-oriented language, JAVA CARD is well suited for OOAD;
  - JAVA CARD lacks some crucial complications of the full JAVA language (no threads, fewer data types, no graphical user interfaces);
  - JAVA CARD applications are small (JAVA smart cards currently offer 16K memory for code);
  - at the same time, JAVA CARD applications are embedded into larger program systems or business processes which should be modeled (though not necessarily formally verified) as well;
  - JAVA CARD applications are often security-critical, thus giving incentive to apply formal methods;
  - the high number (usually millions) of deployed smart cards constitutes a new motivation for formal verification, because, in contrast to software run on standard computers, arbitrary updates are not feasible;[1]
- Through direct contacts with software companies we check the soundness of our approach for real world applications (some of the experiences from these contacts are reported in [3]).

The KeY system consists of three main components (see the Figure below on the right):

---

[1] While JAVA CARD applets on smart cards can be updated in principle, for security reasons this does not extend to those applets that verify and load updates.

- The *modeling component*: this component is based on the CASE tool and is responsible for all user interactions (except interactive deduction). It is used to generate and refine models, and to store and process them. The extensions for precise modeling contains, e.g., editor and parser for the OCL. Additional functionality for the verification process is provided, e.g., for writing proof obligations.

- The *verification manager*: the link between the modeling component and the deduction component. It generates proof obligations expressed in formal logic from the refinement relations in the model. It stores and processes partial and completed proofs; and it is responsible for correctness management (to make sure, e.g., that there are no cyclic dependencies in proofs).
- The *deduction component*. It is used to actually construct proofs—or counter examples—for proof obligations generated by the verification manager. It is based on an interactive verification system combined with powerful automated deduction techniques that increase the degree of automation; it also contains a part for automatically generating counter examples from failed proof attempts. The interactive and automated techniques and those for finding counter examples are fully integrated and operate on the same data structures.

Although consisting of different components, the KeY system is going to be fully integrated with a uniform user interface.

A first KeY system prototype has been implemented, integrating the CASE tool TOGETHER and the system IBIJa [16] as (interactive) deduction component (it has limited capabilities and lacks the verification manager). Work on the full KeY system is in progress.

2   Designing a System with KeY

2.1   The Modeling Process

Software development is generally divided into four activities: analysis, design, implementation, and test. The KeY approach embraces verification as a fifth category. The way in which the development activities are arranged in a sequential order over time is called *modeling process*. It consists of different phases. The end of each phase is defined by certain criteria the actual model should meet (milestones).

In some older process models like the waterfall model or Boehm's spiral model no difference is made between the main activities—analysis, design, implementation, test—and the process phases. More recent process models distinguish between phases and activities very carefully; for example, the Rational Unified Process [19] uses the phases inception, elaboration, construction, and transition along with the above activities.

The KeY system does neither support nor require the usage of a *particular* modeling process. However, it is taken into account that most modern processes have two principles in common. They are *iterative* and *incremental*. The design of an iteration is often regarded as the refinement of the design developed in the previous iteration. This has an influence on the way in which the KeY system treats UML models and additional verification tasks (see Section 2.3). The verification activities are spread across all phases in software development. They are often carried out after test activities.

We do not assume any dependencies between the increments in the development process and the verification of proof obligations. On the right, progress in modeling is depicted along the horizontal axis and progress in verifying proof obligations on the vertical axis. The overall goal is to proceed from the upper left corner (empty model, nothing proven) to the bottom right one (complete model, all proof obligations verified). There are two extreme ways of doing that:

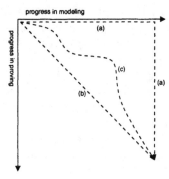

- First complete the whole modeling and coding process, only then start to verify (line (a)).
- Start verifying proof obligations as soon as they are generated (line (b)).

In practice an intermediate approach is chosen (line (c)). How this approach does exactly look is an important design decision of the verification process with strong impact on the possibilities for reuse and is the topic of future research.

2.2   Specification with the UML and the OCL

The diagrams of the Unified Modeling Language provide, in principle, an easy and concise way to formulate various aspects of a specification, however, as Steve Cook remarked [31, foreword]: "[...] there are many subtleties and nuances of meaning diagrams cannot convey by themselves."

This was a main source of motivation for the development of the Object Constraint Language (OCL), part of the UML since version 1.3 [24]. Constraints written in this language are understood in the context of a UML model, they never stand by themselves. The OCL allows to attach preconditions, postconditions, invariants, and guards to specific elements of a UML model.

When designing a system with KeY, one develops a UML model that is enriched by OCL constraints to make it more precise. This is done using the CASE tool integrated into the KeY system. To assist the user, the KeY system provides menu and dialog driven input possibility. Certain standard tasks, for example, generation of formal specifications of inductive data structures (including the common ones such as lists, stacks, trees) in the UML and the OCL can be done in a fully automated way, while the user simply supplies names of constructors and selectors. Even if formal specifications cannot fully be composed in such a schematic way, considerable parts usually can.

In addition, we have developed a method supporting the extension of a UML model by OCL constraints that is based on enriched design patterns. In the KeY system we provide common patterns that come complete with predefined OCL constraint schemata. They are flexible and allow the user to generate well-adapted constraints for the different instances of a pattern as easily as one uses patterns alone. The user needs not write formal specifications from scratch, but only to adapt and complete them. A detailed description of this technique and of experiences with its application in practice is given in [4].

As an example, consider the *composite* pattern, depicted on the right [11, p. 163ff]. This is a ubiquitous pattern in many contexts such as user interfaces, recursive data structures, and, in particular, in the model for the address book of an email client that is part of one of our case studies.

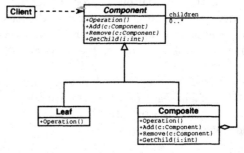

The concrete Add and Remove operations in Composite are intuitively clear but leave some questions unanswered. Can we add the same element twice? Some implementations of the *composite* pattern allow this [14]. If it is not intended, then one has to impose a constraint, such as:

context Composite::Add(c:Component)
post: self.children→select(p | p = c)→size = 1

This is a postcondition on the call of the operation Add in OCL syntax. After completion of the operation call, the stated postcondition is guaranteed to be true. Without going into details of the OCL, we give some hints on how to read this expression. The arrow "→" indicates that the expression to its left represents a collection of objects (a set, a bag, or a sequence), and the operation to its right is to be applied to this collection. The dot "." is used to navigate within diagrams and (here) yields those objects associated to the item on its left via the role name on its right. If C is the multiset of all children of the object self to which Add is applied, then the select operator yields the set $A = \{p \in C \mid p = c\}$ and the subsequent integer-valued operation size gives the number of elements in A. Thus, the postcondition expresses that after adding c as a child to self, the object c occurs exactly once among the children of self.

There are a lot of other useful (and more complex) constraints, e.g., the constraint that the child relationship between objects of class *Component* is acyclic.

2.3   The KₑY Module Concept

The KeY system supports modularization of the model in a particular way. Those parts of a model that correspond to a certain component of the modeled system are grouped together and form a *module*. Modules are a different structuring concept than iterations and serve a different purpose. A module contains all the model components (diagrams, code etc.) that refer to a certain system component. A module is not restricted to a single level of refinement.

There are three main reasons behind the module concept of the KeY system:

Structuring: Models of large systems can be structured, which makes them easier to handle.

Information hiding: Parts of a module that are not relevant for other modules are hidden. This makes it easier to change modules and correct them when errors are found, and to re-use them for different purposes.

Verification of single modules: Different modules can be verified separately, which allows to structure large verification problems. If the size of modules is limited, the complexity of verifying a system grows linearly in the number of its modules and thus in the size of the system. This is indispensable for the scalability of the KeY approach.

In the KeY approach, a hierarchical module concept with sub-modules supports the structuring of large models. The modules in a system model form a tree with respect to the sub-module relation.

Besides sub-modules and model components, a module contains the refinement relations between components that describe the same part of the modeled system in two consecutive levels of refinement. The verification problem associated with a module is to show that these refinements are correct (see Section 3.1). The refinement relations must be provided by the user; typically, they include a signature mapping.

To facilitate information hiding, a module is divided into a public part, its *contract*, and a private (hidden) part; the user can declare parts of *each* refinement level as public or private. Only the public information of a module A is visible in another module B provided that module B implicitly or explicitly *imports* module A. Moreover, a component of module B belonging to some refinement level can only *see* the visible information from module A that belongs to the same level. Thus, the private part of a module can be changed as long as its contract is not affected. For the description of a refinement relation (like a signature mapping) all elements of a module belonging to the initial model or the refined model are visible, whether declared public or not.

As the modeling process proceeds through iterations, the system model becomes ever more precise. The final step is a special case, though: the involved models—the implementation model and its realization in JAVA—do not necessarily differ in precision, but use different paradigms (specification vs. implementation) and different languages (UML with OCL vs. JAVA).[2]

---

[2] In conventional verification systems that do not use an iterative modeling process [25,27], only these final two models exist (see also the following subsection). In such

Below is a schematic example for the levels of refinement and the modules of a system model (the visibility aspect of modules is not represented here). Stronger refinement may require additional structure via (sub-)modules, hence the number of modules may increase with the degree of refinement.

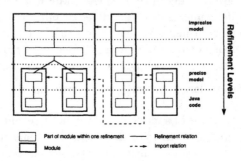

Although the import and refinement relations are similar in some respects, there is a fundamental difference: by way of example, consider a system component being (imprecisely) modeled as a class DataStorage in an early iteration. It may later be *refined* to a class DataSet, which replaces DataStorage. On the other hand, the module containing DataSet could *import* a module DataList and use lists to implement sets, in which case lists are not a refinement of sets and do not replace them.

Relation of KeY Modules to other Approaches The ideas of refinement and modularization in the KeY module concept can be compared with (and are partly influenced by) the KIV approach [27] and the B Method [1].

In KIV, each module (in the above sense) corresponds to exactly two refinement levels, that is to say, a single refinement step. The first level is an algebraic data type, the second an imperative program, whose procedures intentionally implement the operations of the data type. The import relation allows the algebraic data type operations (not the program procedures!) of the imported module to appear textually in the program of the importing module. In contrast to this, the JAVA code of a KeY module directly calls methods of the imported module's JAVA code. Thus, the object programs of our method are pure JAVA programs. Moreover, KeY modules in general have more than two refinement levels.

The B Method offers (among other things) multi-level refinement of abstract machines. There is an elaborate theory behind the precise semantics of a refinement and the resulting proof obligations. This is possible, because both, a machine and its refinement, are *completely formal*, even if the refinement happens to be *less abstract*. That differs from the situation in KeY, where all but the last refinement levels are UML-based, and a refined part is typically *more formal* than its origin. KeY advocates the integrated usage of notational paradigms as opposed to, for example, prepending OOM to abstract machine specification in the B Method [21].

---

systems, modules consist of a specification and an implementation that is a refinement of the specification.

2.4   The Internal State of Objects

The formal specification of objects and their behavior requires special techniques. One important aspect is that the behavior of objects depends on their state that is stored in their attributes, however, the methods of a JAVA class can in general not be described as functions on their input as they may have side effects and change the state. To fully specify the behavior of an object or class, it must be possible to refer to its state (including its initial state). Difficulties may arise if methods for observing the state are not defined or are declared private and, therefore, cannot be used in the public contract of a class. To model such classes, *observer methods* have to be added. These allow to observe the state of a class without changing it.

Example 1. Let class Registry contain a method seen(o:Object):Boolean that maintains a list of all the objects it has "seen". It returns false, if it "sees" an object for the first time, and true, otherwise. In this example, we add the function state():Set(Object) allowing to observe the state of an object of class Registry by returning the set of all seen objects. The behavior of seen can now be specified in the OCL as follows:

context Registry::seen(o:Object)
post: result = state@pre()→includes(o) **and**
         state() = state@pre()→including(o)

The OCL key word result refers to the return value of seen, while @pre gives the result of state() before invocation of seen, which we denote by *oldstate*. The OCL expression state@pre()→includes(o) then stands for o ∈ *oldstate* and state@pre()→including(o) stands for *oldstate* ∪ {o}.

3   Formal Verification with KₔY

Once a program is formally specified to a sufficient degree one can start to formally verify it. Neither a program nor its specification need to be complete in order to start verifying it. In this case one suitably weakens the postconditions (leaving out properties of unimplemented or unspecified parts) or strengthens preconditions (adding assumptions about unimplemented parts). Data encapsulation and structuredness of OO designs are going to be of great help here.

3.1   Proof Obligations

We use constraints in two different ways: first, they can be part of a model (the default); these constraints do not generate proof obligations by themselves. Second, constraints can be given the status of a proof obligation; these are not part of the model, but must be *shown* to hold in it. Proof obligations may arise indirectly from constraints of the first kind: by checking consistency of invariants, pre- and postconditions of a superclass and its subclasses, by checking consistency of the postcondition of an operation and the invariant of its result type, etc. Even more important are proof obligations arising from iterative refinement steps. To prove that a diagram D' is a sound refinement of a diagram D requires to check that the assertions stated in D' entail the assertions in D. A particular refinement step is the passage from a fully refined specification to its realization in concrete code.

3.2 Dynamic Logic

We use Dynamic Logic (DL) [20]—an extension of Hoare logic [2]—as the logical basis of the KeY system's software verification component. We believe that this is a good choice, as deduction in DL is based on symbolic program execution and simple program transformations, being close to a programmer's understanding of JAVA CARD. For a more detailed description of our JAVA CARD DL than given here, see [5].

DL is successfully used in the KIV software verification system [27] for an imperative programming language; and Poetzsch-Heffter and Müller's definition of a Hoare logic for a JAVA subset [26] shows that there are no principal obstacles to adapting the DL/Hoare approach to OO languages.

DL can be seen as a modal predicate logic with a modality $\langle p \rangle$ for every program p (p can be any legal JAVA CARD program); $\langle p \rangle$ refers to the successor worlds (called states in the DL framework) reachable by running the program p. In classical DL there can be several such states (worlds) because the programs can be non-deterministic; here, since JAVA CARD programs are deterministic, there is exactly one such world (if p terminates) or there is none (if p does not terminate). The formula $\langle p \rangle \phi$ expresses that the program p terminates in a state in which ϕ holds. A formula $\phi \rightarrow \langle p \rangle \psi$ is valid, if for every state s satisfying precondition ϕ a run of the program p starting in s terminates, and in the terminating state the postcondition ψ holds.

The formula $\phi \rightarrow \langle p \rangle \psi$ is similar to the Hoare triple $\{\phi\}p\{\psi\}$. In contrast to Hoare logic, the set of formulas of DL is closed under the usual logical operators: In Hoare logic, the formulas ϕ and ψ are pure first-order formulas, whereas in DL they can contain programs. DL allows programs to occur in the descriptions ϕ resp. ψ of states. With is feature it is easy, for example, to specify that a data structure is not cyclic (it is impossible in first-order logic). Also, all JAVA constructs (e.g., `instanceof`) are available in DL for the description of states. So it is not necessary to define an abstract data type *state* and to represent states as terms of that type (like in [26]); instead, DL formulas can be used to give a (partial) description of states, which is a more flexible technique and allows to concentrate on the relevant properties of a state.

In comparison to classical DL (that uses a toy programming language), a DL for a "real" OO programming language like JAVA CARD has to cope with some complications: (1) A program state does not only depend on the value of (local) program variables but also on the values of the attributes of all existing objects. (2) Evaluation of a JAVA expression may have side effects, so there is a difference between expressions and logical terms. (3) Such language features as built-in data types, exception handling, and object initialisation must be handled.

3.3 Syntax and Semantics of Java Card DL

We do not allow class definitions in the programs that are part of DL formulas, but define syntax and semantics of DL formulas wrt a given JAVA CARD program (the context), i.e., a sequence of class definitions. The programs in DL formulas are executable code and comprise all legal JAVA CARD statements, including: (a) expression statements (assignments, method calls, new-statements, etc.); (b) blocks and compound statements built with if-else, switch, for, while, and do-while; (c) statements with exception handling using try-catch-finally; (d) statements that redirect the control flow (continue, return, break, throw).

We allow programs in DL formulas (not in the context) to contain logical terms. Wherever a JAVA CARD expression can be used, a term of the same type as the expression can be used as well. Accordingly, expressions can contain terms (but not vice versa). Formulas are built as usual from the (logical) terms, the predicate symbols (including the equality predicate $\doteq$), the logical connectives $\neg$, $\wedge$, $\vee$, $\rightarrow$, the quantifiers $\forall$ and $\exists$ (that can be applied to logical variables but not to program variables), and the modal operator $\langle p \rangle$, i.e., if p is a program and ϕ is a formula, then $\langle p \rangle \phi$ is a formula as well.

The models of DL consist of program states. These states share the same universe containing a sufficient number of elements of each type. In each state a (possibly different) value (an element of the universe) of the appropriate type is assigned to: (a) the program variables, (b) the attributes (fields) of all objects, (c) the class attributes (static fields) of all classes in the context, and (d) the special object variable this. Variables and attributes of object types can be assigned the special value *null*. States do not contain any information on control flow such as a program counter or the fact that an exception has been thrown.

The semantics of a program p is a state transition, i.e., it assigns to each state s the set of all states that can be reached by running p starting in s. Since JAVA CARD is deterministic, that set either contains exactly one state or is empty. The set of states of a model must be closed under the reachability relation for all programs p, i.e., all states that are reachable must exist in a model (other models are not considered).

We consider programs that terminate abnormally to be non-terminating: nothing can be said about their final state. Examples are a program that throws an uncaught exception and a return statement outside of a method invocation. Thus, for example, $\langle$throw x;$\rangle \phi$ is unsatisfiable for all ϕ.[3]

3.4 A Sequent Calculus for Java Card DL

We outline the ideas behind our sequent calculus for JAVA CARD DL and give some of its basic rules (actually, simplified versions of the rules, e.g., initialisation of objects and classes is not considered). The DL rules of our calculus operate on

---

[3] It is still possible to express and (if true) prove the fact that a program p terminates abnormally. For example, $\langle$try{p}catch{Exception e}$\rangle(\neg$ e $\doteq$ null) expresses that p throws an exception.

$$\frac{\Gamma \vdash \mathit{cnd} \doteq \mathtt{true} \quad \Gamma \vdash \langle \pi \ \mathtt{prg} \ \mathtt{while} \ (\mathit{cnd}) \ \mathtt{prg} \ \omega \rangle \phi}{\Gamma \vdash \langle \pi \ \mathtt{while} \ (\mathit{cnd}) \ \mathtt{prg} \ \omega \rangle \phi} \tag{1}$$

$$\frac{\Gamma \vdash \mathit{cnd} \doteq \mathtt{false} \quad \Gamma \vdash \langle \pi \omega \rangle \phi}{\Gamma \vdash \langle \pi \ \mathtt{while} \ (\mathit{cnd}) \ \mathtt{prg} \ \omega \rangle \phi} \tag{2}$$

$$\frac{\Gamma \vdash \mathit{instanceof}(\mathit{exc}, T) \quad \Gamma \vdash \langle \pi \ \mathtt{try}\{e=\mathit{exc}; \ q\}\mathtt{finally}\{r\} \ \omega \rangle \phi}{\Gamma \vdash \langle \pi \ \mathtt{try}\{\mathtt{throw} \ \mathit{exc}; \ p\}\mathtt{catch}(T \ e)\{q\}\mathtt{finally}\{r\} \ \omega \rangle \phi} \tag{3}$$

$$\frac{\Gamma \vdash \neg\mathit{instanceof}(\mathit{exc}, T) \quad \Gamma \vdash \langle \pi \ r; \ \mathtt{throw} \ \mathit{exc}; \ \omega \rangle \phi}{\Gamma \vdash \langle \pi \ \mathtt{try}\{\mathtt{throw} \ \mathit{exc}; \ p\}\mathtt{catch}(T \ e)\{q\}\mathtt{finally}\{r\} \ \omega \rangle \phi} \tag{4}$$

$$\frac{\Gamma \vdash \langle \pi \ r \ \omega \rangle \phi}{\Gamma \vdash \langle \pi \ \mathtt{try}\{\}\mathtt{catch}(T \ e)\{q\}\mathtt{finally}\{r\} \ \omega \rangle \phi} \tag{5}$$

Table 1. Some of the rules of our calculus for Java Card DL.

the first *active* command p of a program $\pi p \omega$. The non-active prefix π consists of an arbitrary sequence of opening braces "{", labels, beginnings "try{" of try-catch blocks, etc. The prefix is needed to keep track of the blocks that the (first) active command is part of, such that the commands throw, return, break, and continue that abruptly change the control flow are handled correctly. (In classical DL, where no prefixes are needed, any formula of the form $\langle p \ q \rangle \phi$ can be replaced by $\langle p \rangle \langle q \rangle \phi$. In our calculus, splitting of $\langle \pi p q \omega \rangle \phi$ into $\langle \pi p \rangle \langle q \omega \rangle \phi$ is not possible (unless the prefix π is empty) because πp is not a valid program; and the formula $\langle \pi p \omega \rangle \langle \pi q \omega \rangle \phi$ cannot be used either because its semantics is in general different from that of $\langle \pi p q \omega \rangle \phi$.)

As examples, we present the rules for while loops and for exception handling. The rules operate on sequents $\Gamma \vdash \phi$. The semantics of a sequent is that the conjunction of the DL formulas in Γ implies the DL formula ϕ. Sequents are used to represent proof obligations, proof (sub-)goals, and lemmata.

Rules (1) and (2) in Table 1 allow to "unwind" while loops. They are simplified versions that only work if (a) the condition cnd is a logical term (i.e., has side effects), and (b) the program prg does not contain a continue statement. These rules allow to handle loops if used in combination with induction schemata. Similar rules are defined for do-while and for loops.

Rules (3)–(5) handle try-catch-finally blocks and the throw statement. Again, these are simplified versions of the actual rules; they are only applicable if (a) exc is a logical term (e.g., a program variable), and (b) the statements break, continue, return do not occur. Rule (3) applies, if an exception exc is thrown that is an instance of exception class T, i.e., the exception is caught; otherwise, if the exception is not caught, rule (4) applies. Rule (5) applies if the try block is empty and terminates normally.

3.5   The KeY Deduction Component

The KeY system comprises a deductive component, that can handle KeY-DL. This KeY prover combines interactive and automated theorem proving techniques. Experience with the KIV system [27] has shown how to cope with DL proof obligations. The original goal is reduced to first-order predicate logic using such DL rules as shown in the previous subsections. First-order goals can be proven using theory specific knowledge about the used data types.

We developed a language for expressing knowledge of specific theories—we are thinking here mainly of theories of abstract data types—in the form of proof rules. We believe that this format, stressing the operational aspect, is easier to understand and simpler to use than alternative approaches coding the same knowledge in declarative axioms, higher-order logic, or fixed sets of special proof rules. This format, called *schematic theory specific rules*, is explained in detail in [16] and has been implemented in the interactive proof system IBIJa (illwww.ira.uka.de/~ibija). In particular, a schematic theory specific rule contains: (a) Pure logical knowledge, (b) information on how this knowledge is to be used, and (c) information on when and where this knowledge should be presented for interactive use.

Nearly all potential rule applications are triggered by the occurrence of certain terms or formulas in the proof context. The easy-to-use graphical user interface of IBIJa supports invocation of rule applications by mouse clicks on the relevant terms and formulas. The rule schema language is expressive enough to describe even complex induction rules. The rule schema language is carefully designed in such a way that for every new schematic theory specific rule, IBIJa automatically generates proof obligations in first-order logic. Once these obligations are shown to be true the soundness of all applications of this rule is guaranteed. Hence, during each state of a proof, soundness-preserving new rules can be introduced.

To be practically useful, interactive proving must be enhanced by automating intermediate proof steps as much as possible. Therefore, the KeY prover combines IBIJa with automated proof search in the style of analytic tableaux. This integration is based on the concepts described in [12,13]. A screen shot of a typical situation as it may arise during proof construction with our prototype is shown below. The user may either interactively apply a rule (button "Apply Selected Rule") or invoke the automated deduction component (button "Start PRINS").

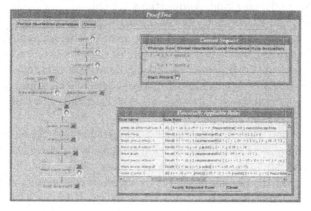

In a real development process, resulting programs often are bug-ridden, there-fore, the ability of *disproving* correctness is as important as the ability of *proving* it. The interesting and common case is that neither correctness nor its negation are deducible from given assumptions. A typical reason is that data structures are underspecified. We may, for example, not have any knowledge about the be-havior of, say, pop(s:Stack):Stack if s is empty. To recognize such situations, which often lead to bugs in the implementation, we develop special deductive techniques. They are based on automatically constructing *interpretations* (of data type operations) that fulfill all assumptions but falsify the hypothesis.

4   Related Work

There are many projects dealing with formal methods in software engineering including several ones aimed at JAVA as a target language. There is also work on security of JAVA CARD and ACTIVEX applications as well as on secure smart card applications in general. We are, however, not aware of any project quite like ours. We mention some of the more closely related projects.

A thorough mathematical analysis of Java using Abstract State Machines has been given in [6]. Following another approach, a precise semantics of a Java sublanguage was obtained by embedding it into Isabelle/HOL [23]; there, an axiomatic semantics is used in a similar spirit as in the present paper.

The COGITO project [30] resulted in an integrated formal software develop-ment methodology and support system based on extended Z as specification language and Ada as target language. It is not integrated into a CASE tool, but stand-alone.

The FuZE project [10] realized CASE tool support for integrating the Fu-SION OOAD process with the formal specification language Z. The aim was to formalize OOAD methods and notations such as the UML, whereas we are interested to derive formal specifications with the help of an OOAD process extension.

The goal of the QUEST project [28] is to enrich the CASE tool AUToFo-CUS for description of distributed systems with means for formal specification and support by model checking. Applications are embedded systems, description formalisms are state charts, activity diagrams, and temporal logic.

Aim of the SYSLAB project is the development of a scientifically founded ap-proach for software and systems development. At the core is a precise and formal notion of hierarchical "documents" consisting of informal text, message sequence charts, state transition systems, object models, specifications, and programs. All documents have a "mathematical system model" that allows to precisely describe dependencies or transformations [7].

The goal of the PROSPER project was to provide the means to deliver the benefits of mechanized formal specification and verification to system designers in industry (**www.dcs.gla.ac.uk/prosper/index.html**). The difference to the KeY project is that the dominant goal is hardware verification; and the software part involves only specification.

5   Conclusion and the Future of K͡e̲Y

In this paper we described the current state of the KeY project and its ultimate
goal: To facilitate and promote the use of formal verification in an industrial
context for real-world applications. It remains to be seen to which degree this
goal can be achieved.

Our vision is to make the logical formalisms transparent for the user with respect to OO modeling. That is, whenever user interaction is required, the current
state of the verification task is presented in terms of the environment the user
has created so far and not in terms of the underlying deduction machinery. The
situation is comparable to a symbolic debugger that lets the user step through
the source code of a program while it actually executes compiled machine code.

Acknowledgements

Thanks to S. Klingenbeck and J. Posegga for valuable comments on earlier versions of this paper. We also thank our former group members T. Fuchß, R. Preiß,
and A. Schönegge for their input during the preparation of the project. KeY is
supported by the Deutsche Forschungsgemeinschaft (grant no. Ha 2617/2-1).

References

1. J.-R. Abrial. *The B Book - Assigning Programs to Meanings.* Cambridge University
   Press, Aug. 1996.
2. K. R. Apt. Ten years of Hoare logic: A survey – part I. *ACM Transactions on
   Programming Languages and Systems*, 1981.
3. T. Baar. Experiences with the UML/OCL-approach to precise software modeling:
   A report from practice. Submitted, see i12www.ira.uka.de/~key, 2000.
4. T. Baar, R. Hähnle, T. Sattler, and P. H. Schmitt. Entwurfsmustergesteuerte
   Erzeugung von OCL-Constraints. In G. Snelting, editor, *Softwaretechnik-Trends*,
   Informatik Aktuell. Springer, 2000.
5. B. Beckert. A dynamic logic for Java Card. In *Proc. 2nd ECOOP Workshop on Formal Techniques for Java Programs, Cannes, France*, 2000. See
   i12www.ira.uka.de/~key/doc/2000/beckert00.pdf.gz.
6. E. Börger and W. Schulte. A programmer friendly modular definition of the semantics of Java. In J. Alves-Foss, editor, *Formal Syntax and Semantics of Java*,
   LNCS 1523, pages 353–404. Springer, 1999.
7. R. Breu, R. Grosu, F. Huber, B. Rumpe, and W. Schwerin. Towards a precise
   semantics for object-oriented modeling techniques. In H. Kilov and B. Rumpe,
   editors, *Proc Workshop on Precise Semantics for Object-Oriented Modeling Techniques at ECOOP'97*. Techn. Univ. of Munich, Tech. Rep. TUM-I9725, 1997.
8. E. Clarke and J. M. Wing. Formal methods: State of the art and future directions.
   ACM Computing Surveys, 28(4):626–643, 1996.
9. D. L. Dill and J. Rushby. Acceptance of formal methods: Lessons from hardware
   design. *IEEE Computer*, 29(4):23–24, 1996. Part of: Hossein Saiedian (ed.). *An
   Invitation to Formal Methods.* Pages 16–30.

10. R. B. France, J.-M. Bruel, M. M. Larrondo-Petrie, and E. Grant. Rigorous object-oriented modeling: Integrating formal and informal notations. In M. Johnson, editor, *Proc. Algebraic Methodology and Software Technology (AMAST)*, Berlin, Germany, LNCS 1349. Springer, 1997.

11. E. Gamma, R. Helm, R. Johnson, and J. Vlissides. *Design Patterns: Elements of Reusable Object-Oriented Software*. Addison-Wesley, 1995.

12. M. Giese. Integriertes automatisches und interaktives Beweisen: Die Kalkülebene. Diploma Thesis, Fakultät für Informatik, Universität Karlsruhe, June 1998.

13. M. Giese. A first-order simplification rule with constraints. In *Proc. Int. Workshop on First-Order Theorem Proving, St. Andrews, Scotland*, 2000. See i12www.ira.uka.de/~key/doc/2000/giese00a.ps.gz.

14. M. Grand. *Patterns in Java*, volume 2. John Wiley & Sons, 1999.

15. S. B. Guthery. Java Card: Internet computing on a smart card. *IEEE Internet Computing*, 1(1):57–59, 1997.

16. E. Habermalz. Interactive theorem proving with schematic theory specific rules. Technical Report 19/00, Fakultät für Informatik, Universität Karlsruhe, 2000. See i12www.ira.uka.de/~key/doc/2000/stsr.ps.gz.

17. U. Hansmann, M. S. Nicklous, T. Schäck, and F. Seliger. *Smart Card Application Development Using Java*. Springer, 2000.

18. M. G. Hinchey and J. P. Bowen, editors. *Applications of Formal Methods*. Prentice Hall, 1995.

19. I. Jacobson, G. Booch, and J. Rumbaugh. *The Unified Software Development Process*. Object Technology Series. Addison-Wesley, 1999.

20. D. Kozen and J. Tiuryn. Logic of programs. In J. van Leeuwen, editor, *Handbook of Theoretical Computer Science*, volume B: Formal Models and Semantics, chapter 14, pages 789–840. Elsevier, Amsterdam, 1990.

21. K. Lano. *The B Language and Method: A guide to Practical Formal Development*. Springer Verlag London Ltd., 1996.

22. J. Martin and J. J. Odell. *Object-Oriented Methods: A Foundation, UML Edition*. Prentice-Hall, 1997.

23. T. Nipkow and D. von Oheimb. Machine-checking the Java specification: Proving type safety. In J. Alves-Foss, editor, *Formal Syntax and Semantics of Java*, LNCS 1523, pages 119–156. Springer, 1999.

24. Object Management Group, Inc., Framingham/MA, USA, www.omg.org. *OMG Unified Modeling Language Specification, Version 1.3*, June 1999.

25. L. C. Paulson. *Isabelle: A Generic Theorem Prover*. Springer, Berlin, 1994.

26. A. Poetzsch-Heffter and P. Müller. A programming logic for sequential Java. In S. D. Swierstra, editor, *Proc. Programming Languages and Systems (ESOP)*, Amsterdam, The Netherlands, LNCS 1576, pages 162–176. Springer, 1999.

27. W. Reif. The KIV-approach to software verification. In M. Broy and S. Jähnichen, editors, *KORSO: Methods, Languages, and Tools for the Construction of Correct Software – Final Report*, LNCS 1009. Springer, 1995.

28. O. Slotosch. Overview over the project QUEST. In *Applied Formal Methods, Proc. FM-Trends 98, Boppard, Germany*, LNCS 1641, pages 346–350. Springer, 1999.

29. Sun Microsystems, Inc., Palo Alto/CA, USA. *Java Card 2.1 Application Programming Interfaces, Draft 2, Release 1.3*, 1998.

30. O. Traynor, D. Hazel, P. Kearney, A. Martin, R. Nickson, and L. Wildman. The Cogito development system. In M. Johnson, editor, *Proc. Algebraic Methodology and Software Technology*, Berlin, LNCS 1349, pages 586–591. Springer, 1997.

31. J. Warmer and A. Kleppe. *The Object Constraint Language: Precise Modelling with UML*. Object Technology Series. Addison-Wesley, 1999.

Semi-qualitative Reasoning about Distances: A Preliminary Report

Holger Sturm[1], Nobu-Yuki Suzuki[2], Frank Wolter[1], and
Michael Zakharyaschev[3]

[1] Institut für Informatik, Universität Leipzig,
Augustus-Platz 10-11, 04109 Leipzig, Germany
[2] Department of Mathematics, Faculty of Science
Shizuoka University, Ohya 836, Shizuoka 422–8529, Japan
[3] Division of Artificial Intelligence, School of Computer Studies
University of Leeds, Leeds LS2 9JT, UK.

Abstract We introduce a family of languages intended for representing knowledge and reasoning about metric (and more general distance) spaces. While the simplest language can speak only about distances between individual objects and Boolean relations between sets, the more expressive ones are capable of capturing notions such as 'somewhere in (or somewhere out of) the sphere of a certain radius', 'everywhere in a certain ring', etc. The computational complexity of the satisfiability problem for formulas in our languages ranges from NP-completeness to undecidability and depends on the class of distance spaces in which they are interpreted. Besides the class of all metric spaces, we consider, for example, the spaces $\mathbb{R} \times \mathbb{R}$ and $\mathbb{N} \times \mathbb{N}$ with their natural metrics.

1 Introduction

The concept of 'distance between objects' is one of the most fundamental abstractions both in science and in everyday life. Imagine for instance (only imagine) that you are going to buy a house in London. You then inform your estate agent about your intention and provide her with a number of constraints:

(A) The house should not be too far from your college, say, not more than 10 miles.

(B) The house should be close to shops, restaurants, and a movie theatre; all this should be reachable, say, within 1 mile.

(C) There should be a 'green zone' around the house, at least within 2 miles in each direction.

(D) Factories and motorways must be far from the house, not closer than 5 miles.

(E) There must be a sports center around, and moreover, all sports centers of the district should be reachable on foot, i.e., they should be within, say, 3 miles.

(F) And of course there must be a tube station around, not too close, but not too far either—somewhere between 0.5 and 1 mile.

M. Ojeda-Aciego et al. (Eds.): JELIA 2000, LNAI 1919, pp. 37–56, 2000.

'Distances' can be induced by different measures. We may be interested in the physical distance between two cities a and b, i.e., in the length of the straight (or geodesic) line between a and b. More pragmatic would be to bother about the length of the railroad connecting a and b, or even better the time it takes to go from a to b by train (plane, ship, etc.). But we can also define the distance as the number of cities (stations, friends to visit, etc.) on the way from a to b, as the difference in altitude between a and b, and so forth.

The standard mathematical models capturing common features of various notions of distance are known as metric spaces (see e.g. [4]). We define a *metric space* as a pair $\mathfrak{D} = \langle W, d \rangle$, where W is a set (of points) and d a function from $W \times W$ into $\mathbb{R}$, the *metric* on W, satisfying the following conditions, for all $x, y, z \in W$:

$$d(x, y) = 0 \text{ iff } x = y, \tag{1}$$
$$d(x, z) \le d(x, y) + d(y, z), \tag{2}$$
$$d(x, y) = d(y, x). \tag{3}$$

The value $d(x, y)$ is called the *distance* from the point x to the point y.[1]

It is to be noted, however, that although quite acceptable in many cases, the defined concept of metric space is not universally applicable to all interesting measures of distances between points, especially those used in everyday life. Here are some examples:

(i) Suppose that W consists of the villages in a certain district and $d(x, y)$ denotes the time it takes to go from x to y by train. Then the function d is not necessarily total, since there may be villages without stations.

(ii) If $d(x, y)$ is the flight-time from x to y then, as we know it too well, d is not necessarily symmetric, even approximately (just go from Malaga to Tokyo and back).

(iii) Often we do not measure distances by means of real numbers but rather using more fuzzy notions such as 'short', 'medium', 'long'. To represent these measures we can, of course, take functions d from $W \times W$ into the set $\{1, 2, 3\} \subseteq \mathbb{R}$ and define *short* $:= 1$, *medium* $:= 2$, and *long* $:= 3$. So we can still regard these distances as real numbers. However, for measures of this type the triangle inequality (2) does not make sense (short plus short can still be short, but it can be also medium or long).

In this paper we assume first that distance functions are total and satisfy (1)–(3), i.e., we deal with standard metric spaces. But then, in Section 6, we discuss how far our results can be extended if we consider more general *distance spaces*.

Our main aim in the paper is to

> *design formal languages of metric (or more general distance) spaces that can be used to represent and reason about (a substantial part of) our*

---

[1] Usually axioms (2) and (3) are combined into one axiom $d(y, z) \le d(x, y) + d(x, z)$ which implies the symmetry property (3); cf. [4]. In our case symmetry does not follow from the triangle inequality (2). We will use this fact in Section 6.

everyday knowledge of distances, and that are at the same time as computationally tractable as possible.

The next step will be to integrate the developed languages with formalisms intended for qualitative spatial reasoning (e.g. RCC-8), temporal reasoning, and maybe even combined spatio-temporal reasoning (e.g. [19]).

The requirement of computational effectiveness imposes rather severe limitations on possible languages of metric spaces. For instance, we can hardly use the full power of the common mathematical formalism which allows arithmetic operations and quantification over distances as in the usual definition of a continuous function f from $\mathfrak{D}$ to $\mathbb{R}$:

$$\forall x \in W \; \forall \epsilon > 0 \; \exists \delta > 0 \; \forall y \in W \;\; (d(x,y) < \epsilon \to |f(x) - f(y)| < \delta).$$

On the other hand, in everyday life a great deal of assertions about distances can be (and are) made without such operations and quantification. Although we operate quantitative information about distances, as in examples (A)–(F) above, the reasoning is quite often rather qualitative, with numerical data being involved only in comparisons ('everywhere within 7 m distance', 'in more than 3 hours', etc.), which as we observed above can also encode such vague concepts as 'short', 'medium', 'long'. As travelling scientists, we don't care about the precise location of Malaga, being content with the (qualitative) information that it is in Spain, Spain is disconnected from Germany and the U.K., and the flight-time to any place in Spain from Germany or the U.K. is certainly less than 4 hours. That is why we call our formalisms *semi-qualitative*, following a suggestion of A. Cohn.

In the next section we propose a hierarchy of 'semi-qualitative' propositional languages intended for reasoning about distances. We illustrate their expressive power and formulate the results on the finite model property, decidability, and computational complexity we have managed to obtain so far. (The closest 'relatives' of our logics in the literature are the logics of place from [14,18,15,11,12] and metric temporal logics from [13]; see also [5].) Sections 3–5 show how some of these results can be proved. And in Section 6 we discuss briefly more general notions of 'distance spaces.'

The paper is a preliminary report on our ongoing research; that is why it contains more questions than answers (some of them will certainly be solved by the time of publication).

2   The Logics of Metric Spaces

All the logics of metric spaces to be introduced in this section are based on the following *Boolean logic of space* $\mathcal{BS}$. The alphabet of $\mathcal{BS}$ contains

- an infinite list of *set* (or *region*) *variables* $X_1, X_2, \ldots$;
- an infinite list of *location variables* $x_1, x_2, \ldots$;
- the Boolean operators $\wedge$ and $\neg$.

Boolean combinations of set variables are called *set* (or *region*) *terms*. *Atomic formulas* in $\mathcal{BS}$ are of two types:

- $x \mathbin{\mathsf{E}} t$, where x is a location variable and t a set term,
- $t_1 = t_2$, where t_1 and t_2 are set terms.

The intended meaning of these formulas should be clear from their syntax: $x \mathbin{\mathsf{E}} t$ means that x belongs to t, and $t_1 = t_2$ says that t_1 and t_2 have the same extensions.

$\mathcal{BS}$-*formulas* are just arbitrary Boolean combinations of atoms.

The language $\mathcal{BS}$, as well as all other languages to be introduced below, is interpreted in metric spaces $\mathfrak{D} = \langle W, d \rangle$ by means of *assignments* $\mathfrak{a}$ associating with every set variable X a subset $\mathfrak{a}(X)$ of W and with every location variable x an element $\mathfrak{a}(x)$ of W. The *value* $t^{\mathfrak{a}}$ of a set term t in the model $\mathfrak{M} = \langle \mathfrak{D}, \mathfrak{a} \rangle$ is defined inductively:

$$X_i^{\mathfrak{a}} = \mathfrak{a}(X_i), \ X_i \text{ a set variable,}$$
$$(t_1 \wedge t_2)^{\mathfrak{a}} = t_1^{\mathfrak{a}} \cap t_2^{\mathfrak{a}},$$
$$(\neg t)^{\mathfrak{a}} = W - t^{\mathfrak{a}}.$$

(If the space $\mathfrak{D}$ is not clear from the context, we write $t^{\mathfrak{M}}$ instead of $t^{\mathfrak{a}}$.)

The *truth-relation* for $\mathcal{BS}$-formulas reflects the intended meaning:

$$\mathfrak{M} \models x \mathbin{\mathsf{E}} t \quad \text{iff} \quad \mathfrak{a}(x) \in t^{\mathfrak{a}},$$
$$\mathfrak{M} \models t_1 = t_2 \ \text{iff} \ t_1^{\mathfrak{a}} = t_2^{\mathfrak{a}},$$

plus the standard clauses for the Booleans.

We write $\top$ instead of $\neg(X \wedge \neg X)$, $\emptyset$ instead of $X \wedge \neg X$, and $t_1 \sqsubseteq t_2$ instead of $\neg(t_1 \wedge \neg t_2) = \top$. It should be clear that $\mathfrak{M} \models t_1 \sqsubseteq t_2$ iff $t_1^{\mathfrak{a}} \subseteq t_2^{\mathfrak{a}}$.

$\mathcal{BS}$ can only talk about relations between sets, about their members, but not about distances. For instance, we can construct the following knowledge base in $\mathcal{BS}$:

$$Leipzig \mathbin{\mathsf{E}} Germany, \qquad Malaga \mathbin{\mathsf{E}} Spain,$$
$$Germany \sqsubseteq Europe, \qquad Spain \sqsubseteq Europe,$$
$$Spain \wedge Germany = \emptyset.$$

The metric d in $\mathfrak{D}$ is irrelevant for $\mathcal{BS}$. 'Real' metric logics are defined by extending $\mathcal{BS}$ with a number of set term and formula constructs which involve distances. We define five such logics and call them $\mathcal{MS}_0, \ldots, \mathcal{MS}_4$.

$\mathcal{MS}_0$. To begin with, let us introduce constructs which allow us to speak about distances between locations. Denote by $\mathcal{MS}_0$ the language extending $\mathcal{BS}$ with the possibility of constructing atomic formulas of the form

- $\delta(x, y) = a$,
- $\delta(x, y) < a$,

- $\delta(x, y) = \delta(x', y')$,
- $\delta(x, y) < \delta(x', y')$,

where x, y, x', y' are location variables and $a \in \mathbb{R}_+$ (i.e., a is a non-negative real number). The truth-conditions for such formulas are obvious:

$$
\begin{aligned}
\mathfrak{M} &\models \delta(x, y) = a & &\text{iff } d(\mathfrak{a}(x), \mathfrak{a}(y)) = a, \\
\mathfrak{M} &\models \delta(x, y) < a & &\text{iff } d(\mathfrak{a}(x), \mathfrak{a}(y)) < a, \\
\mathfrak{M} &\models \delta(x, y) = \delta(x', y') & &\text{iff } d(\mathfrak{a}(x), \mathfrak{a}(y)) = d(\mathfrak{a}(x'), \mathfrak{a}(y')), \\
\mathfrak{M} &\models \delta(x, y) < \delta(x', y') & &\text{iff } d(\mathfrak{a}(x), \mathfrak{a}(y)) < d(\mathfrak{a}(x'), \mathfrak{a}(y')).
\end{aligned}
$$

$\mathcal{MS}_0$ provides us with some primitive means for basic reasoning about regions and distances between locations. For example, constraint (A) from Section 1 can be represented as

$$(\delta(house, college) < 10) \vee (\delta(house, college) = 10). \tag{4}$$

The main reasoning problem we are interested in is satisfiability of finite sets of formulas in arbitrary metric spaces or in some special classes of metric spaces, say, finite ones, the Euclidean n-dimensional space $\langle \mathbb{R}^n, d_n \rangle$ with the standard metric

$$d_n(\boldsymbol{x}, \boldsymbol{y}) = \sqrt{\sum_{i=1}^{n} (x_i - y_i)^2},$$

the subspace $\langle \mathbb{N}^n, d'_n \rangle$ of $\langle \mathbb{R}^n, d_n \rangle$ (with the induced metric), etc. The choice of metric spaces depends on applications. For instance, if we deal with time constraints then the intended space can be one-dimensional $\langle \mathbb{R}, d_1 \rangle$ or its subspaces based on $\mathbb{Q}$ or $\mathbb{N}$. If we consider a railway system, then the metric space is finite.

It is to be noted from the very beginning that the language $\mathcal{MS}_0$ as well as other languages $\mathcal{MS}_i$ are *uncountable* because all of them contain uncountably many formulas of the form $\delta(x, y) = a$, for $a \in \mathbb{R}_+$. So in general it does not make sense to ask whether the satisfiability problem for such languages is decidable.

To make the satisfiability problem sensible we have to restrict the languages $\mathcal{MS}_i$ to at least recursive (under some coding) subsets of $\mathbb{R}_+$. Natural examples of such subsets are the non-negative rational numbers $\mathbb{Q}_+$ or the natural numbers $\mathbb{N}$.

Given a set $\mathbb{S} \subseteq \mathbb{R}_+$, we denote by $\mathcal{MS}_i[\mathbb{S}]$ the fragment of $\mathcal{MS}_i$ consisting of only those $\mathcal{MS}_i$-formulas all real numbers in which belong to $\mathbb{S}$.

For the logic $\mathcal{MS}_0$ we have the following:

Theorem 1. (i) *The satisfiability problem for $\mathcal{MS}_0[\mathbb{Q}]$-formulas in arbitrary metric spaces is decidable.*

(ii) *Every finite satisfiable set of $\mathcal{MS}_0$-formulas is satisfiable in a finite metric space, or in other words, $\mathcal{MS}_0$ has the finite model property.*

This theorem follows immediately from the proof of the finite model property of $\mathcal{MS}_2$ in Section 5. We don't know whether satisfiability of $\mathcal{MS}_0[\mathbb{Q}]$-formulas in $\mathbb{R}^n$ is decidable. We conjecture that it is and that the complexity of the satisfiability problem for both arbitrary metric spaces and $\mathbb{R}^n$ is in NP.

In $\mathcal{MS}_0$ we can talk about distances between *points* in metric spaces. Now we extend the language by providing constructs capable of saying that a point is within a certain distance from a set, which is required to represent constraint (B) from Section 1.

$\mathcal{MS}_1$. Denote by $\mathcal{MS}_1$ the language that is obtained by extending $\mathcal{MS}_0$ with the following set term constructs:

- if t is a set term and $a \in \mathbb{R}_+$, then $\exists_{\leq a} t$ and $\forall_{\leq a} t$ are set terms as well.

The semantical meaning of the new set terms is defined by

$$(\exists_{\leq a} t)^a = \{x \in W : \exists y \in W \ (d(x,y) \leq a \wedge y \in t^a)\},$$
$$(\forall_{\leq a} t)^a = \{x \in W : \forall y \in W \ (d(x,y) \leq a \rightarrow y \in t^a)\}.$$

Thus $x \in \exists_{\leq a} t$ means that 'somewhere in or on the sphere with center x and radius a there is a point from t'; $x \in \forall_{\leq a} t$ says that 'the whole sphere with center x and radius a, including its surface, belongs to t.'

Constraints (B)–(D) are now expressible by the formulas:

$$house \in \exists_{\leq 1} shops \wedge \exists_{\leq 1} restaurants \wedge \exists_{\leq 1} cinemas, \tag{5}$$

$$house \in \forall_{\leq 2} green\_zone, \tag{6}$$

$$house \in \neg \exists_{\leq 5}(factories \vee motorways). \tag{7}$$

Here is what we know about this language:

Theorem 2. (i) *The satisfiability problem for $\mathcal{MS}_1[\mathbb{Q}]$-formulas in arbitrary metric spaces is decidable.*

(ii) *$\mathcal{MS}_1$ has the finite model property.*

(iii) *The satisfiability problem for $\mathcal{MS}_1[\{1\}]$-formulas in $\langle \mathbb{N}^2, d_2' \rangle$ is undecidable.*

Claims (i) and (ii) follow from the proof of the finite model property in Section 5. The proof of (iii) is omitted. It can be conducted similarly to the undecidability proof in Section 3. Note that at the moment we don't know whether the satisfiability in $\mathbb{R}^2$ is decidable and what is the complexity of satisfiability of $\mathcal{MS}_1[\mathbb{Q}]$-formulas.

$\mathcal{MS}_2$. In the same manner we can enrich the language $\mathcal{MS}_1$ with the constructs for expressing 'somewhere outside the sphere with center x and radius a' and 'everywhere outside the sphere with center x and radius a'. To this end we add to $\mathcal{MS}_1$ two term-formation constructs:

- if t is a set term and $a \in \mathbb{R}_+$, then $\exists_{>a} t$ and $\forall_{>a} t$ are set terms.

The resulting language is denoted by $\mathcal{MS}_2$. The intended semantical meaning of the new constructs is as follows:

$$(\exists_{>a}t)^{\mathfrak{a}} = \{x \in W : \exists y \in W \ (d(x,y) > a \wedge y \in t^{\mathfrak{a}})\},$$
$$(\forall_{>a}t)^{\mathfrak{a}} = \{x \in W : \forall y \in W \ (d(x,y) > a \rightarrow y \in t^{\mathfrak{a}})\}.$$

Constraint (E) can be represented now as the formula

$$house \sqsubseteq \exists_{\leq 3} district\_sports\_center \wedge \forall_{>3}\neg \ district\_sports\_center. \qquad (8)$$

The language $\mathcal{MS}_2$ is quite expressive. First, it contains an analogue of the difference operator from modal logic (see [6]), because using $\forall_{>0}$ we can say 'everywhere but here':

$$\mathfrak{M} \models x \in \forall_{>0}t \text{ iff } \mathfrak{M} \models y \in t \text{ for all } y \neq x.$$

We also have the universal modalities of [9]: the operators $\forall$ and $\exists$ can be defined by taking

$$\forall t = t \wedge \forall_{>0}t, \qquad \text{i.e., } \forall t \text{ is } \emptyset \text{ if } t \neq \top \text{ and } \top \text{ otherwise,}$$
$$\exists t = t \vee \exists_{>0}t, \qquad \text{i.e., } \forall t \text{ is } \top \text{ if } t \neq \emptyset \text{ and } \emptyset \text{ otherwise.}$$

Second, we can simulate the nominals of [1]. Denote by $\mathcal{MS}_2'$ the language that results from $\mathcal{MS}_2$ by allowing set terms of the form $\{x\}$, for every location variable x, with the obvious interpretation:

– $\mathfrak{a}(\{x\}) = \{\mathfrak{a}(x)\}$.

In $\mathcal{MS}_2'$ we can say, for example, that

$$(\exists_{\leq 1100}\{Leipzig\} \wedge \exists_{\leq 1100}\{Malaga\}) \sqsubseteq France,$$

i.e., 'if you are not more than 1100 km away from Leipzig and not more than 1100 km away from Malaga, then you are in France'.

As far as the satisfiability problem is concerned, $\mathcal{MS}_2'$ is not more expressive than $\mathcal{MS}_2$. To see this, consider a finite set of $\mathcal{MS}_2'$-formulas Γ and suppose that $x_1, \ldots, x_n$ are all location variables which occur in Γ as set terms $\{x_i\}$. Take fresh set variables $X_1, \ldots, X_n$ and let Γ' be the result of replacing all $\{x_i\}$ in Γ with X_i. It is readily checked that Γ is satisfiable in a model based on a metric space $\mathfrak{D}$ iff the set of $\mathcal{MS}_2$-formulas

$$\Gamma' \cup \{(X_i \wedge \neg\exists_{>0}X_i) \neq \emptyset : i \leq n\}$$

is satisfiable in $\mathfrak{D}$.

It is worth noting that, as will become obvious in the next section, the relation between the operators $\forall_{\leq a}$ and $\forall_{>a}$ corresponds to the relation between modal operators $\square$ and $\square^-$ interpreted in Kripke frames by an accessibility relation R and its complement $\overline{R}$, respectively; see [8] for a study of modal logics with such boxes.

Theorem 3. (i) *The satisfiability problem for* $\mathcal{MS}_2[\mathbb{Q}]$*-formulas in arbitrary metric spaces is decidable.*
(ii) $\mathcal{MS}_2$ *has the finite model property.*

This result will be proved in Section 5. We don't know, however, what is the complexity of the satisfiability problem from (i).

$\mathcal{MS}_3$. To be able to express the last constraint (F) from Section 1, we need two more constructs:

- if t is a set term and $a < b$, then $\exists^{\geq a}_{\leq b} t$ and $\forall^{\geq a}_{\leq b} t$ are set terms.

The extended language will be denoted by $\mathcal{MS}_3$. The truth-conditions for these operators are as follows:

$$(\exists^{\geq a}_{\leq b} t)^{\mathfrak{a}} = \{x \in W : \exists y \in W \ (a < d(x,y) \leq b \wedge y \in t^{\mathfrak{a}})\},$$
$$(\forall^{\geq a}_{\leq b} t)^{\mathfrak{a}} = \{x \in W : \forall y \in W \ (a < d(x,y) \leq b \rightarrow y \in t^{\mathfrak{a}})\}.$$

In other words, $x \in \exists^{\geq a}_{\leq b} t$ iff 'somewhere in the ring with center x, the inner radius a and the outer radius b, including the outer circle, there is a point from t'.
Constraint (F) is represented then by the formula:

$$house \in \exists^{\geq 0.5}_{\leq 1} tube\_station. \tag{9}$$

(By the way, the end of the imaginary story about buying a house in London was not satisfactory. Having checked her knowledge base, the estate agent said: "Unfortunately, your constraints (4)–(9) are not satisfiable in London, where we have

$$tube\_station \sqsubseteq \exists_{\leq 3.5}(factory \vee motorway).$$

In view of the triangle inequality, this contradicts constraints (7) and (9).")

Unfortunately, the language $\mathcal{MS}_3$ is too expressive for many important classes of metric spaces.

Theorem 4. *Let* $\mathcal{K}$ *be a class of metric spaces containing* $\mathbb{R}^2$*. Then the satisfiability problem for* $\mathcal{MS}_3[\{0,\dots,100\}]$*-formulas in* $\mathcal{K}$ *is undecidable.*

This result will be proved in the next section (even for a small fragment of $\mathcal{MS}_3$).

$\mathcal{MS}_4$. The most expressive language $\mathcal{MS}_4$ we have in mind is an extension of $\mathcal{MS}_3$ with the operators $\exists_{<a} t$, $\forall_{<a} t$, $\exists_{\geq a} t$, $\forall_{\geq a} t$, $\exists^{\geq a}_{<b} t$, $\forall^{\geq a}_{<b} t$.
Here is what we know about these operators: the satisfiability problem for the full language in the class of all metric spaces is of course undecidable—it contains $\mathcal{MS}_3$. Moreover, the operators $\forall^{\geq a}_{<b}$ alone determine an undecidable language for the class of arbitrary metric spaces (this can be proved similarly to the undecidability proof in Section 3). Also, a similar proof shows that the language with the operators $\forall_{<a}$ only is undecidable both in $\langle \mathbb{R}^2, d_2 \rangle$ and in $\langle \mathbb{N}^2, d'_2 \rangle$. Still, various questions are open, however: for example, whether the language with the operators $\forall_{<a}$ only is decidable in arbitrary metric spaces or whether there are interesting classes of metric spaces in which $\mathcal{MS}_4$ is decidable.

3   Undecidability

In this section we prove a rather general undecidability result. In particular, Theorem 4 is its immediate consequence.

Theorem 5. *Let $\mathcal{K}$ be a class of metric spaces containing $\mathbb{R}^2$. Then the satisfiability problem for $\mathcal{MS}_3[\{0, 9, 10, 20, 80\}]$-formulas (even for those with the operators $\forall_{\leq a}^{\geq 0}$ and $\exists_{\leq a}$ only) in $\mathcal{K}$ is undecidable.*

Proof. To prove this result, we reduce the undecidable $\mathbb{N} \times \mathbb{N}$-tiling problem (see [17,2] and references therein) to the satisfiability problem in $\mathcal{K}$. We remind the reader that the tiling problem for $\mathbb{N} \times \mathbb{N}$ is formulated as follows: given a finite set $\mathcal{T} = \{T_1, \ldots, T_l\}$ of tiles (i.e., squares T_i with colors $left(T_i)$, $right(T_i)$, $up(T_i)$, and $down(T_i)$ on their edges), determine whether tiles in $\mathcal{T}$ can cover the grid $\mathbb{N} \times \mathbb{N}$ in such a way that the colors of adjacent edges on adjacent tiles match, or more precisely, whether there exists a function $\tau : \mathbb{N} \times \mathbb{N} \to \mathcal{T}$ such that for all $n, m \in \mathbb{N}$:

(a) $right(\tau(n, m)) = left(\tau(n + 1, m))$,
(b) $up(\tau(n, m)) = down(\tau(n, m + 1))$.

So, suppose a set of tiles $\mathcal{T} = \{T_1, \ldots, T_l\}$ is given. Our aim is to construct a finite set of $\mathcal{MS}_3[\{0, 9, 10, 20, 80\}]$-formulas which is satisfiable in $\mathcal{K}$ iff $\mathcal{T}$ can tile $\mathbb{N} \times \mathbb{N}$.

Take set variables $Z_1, \ldots, Z_l, X_0, \ldots, X_4, Y_0, \ldots, Y_4$. Let $\chi_{ij} = \forall_{\leq 9}(X_i \wedge Y_j)$, for $i, j \leq 4$, and let Γ be the set of the following formulas, where $i, j \leq 4$ and $k \leq l$:

$$X_i \wedge Y_j \sqsubseteq \exists_{\leq 9}\chi_{ij}, \quad \chi_{ij} \sqsubseteq \forall_{\leq 80}^{\geq 0}\neg\chi_{ij}, \quad \chi_{ij} \sqsubseteq \neg\chi_{mn} \; ((i, j) \neq (m, n)), \tag{10}$$

$$\chi_{ij} \sqsubseteq \bigvee_{k \leq l} \forall_{\leq 9}Z_k, \quad Z_m \sqsubseteq \neg Z_n \; (n \neq m), \tag{11}$$

$$\chi_{ij} \wedge Z_k \sqsubseteq \exists_{\leq 20}\big(\chi_{i+51j} \wedge \bigvee_{right(T_k)=left(T_m)} Z_m\big), \tag{12}$$

$$\chi_{ij} \wedge Z_k \sqsubseteq \exists_{\leq 20}\big(\chi_{ij+51} \wedge \bigvee_{up(T_k)=down(T_m)} Z_m\big), \tag{13}$$

where $+_5$ denotes addition modulo 5.

The first formula in (10) is satisfied in a model $\mathfrak{M} = \langle W, d, \mathfrak{a} \rangle$ iff $\mathfrak{a}(X_i \wedge Y_j)$ is the union of a set of spheres of radius 9. The second one is satisfied in $\mathfrak{M}$ iff the distance between any two distinct centers of spheres, all points in which belong to $\mathfrak{a}(X_i \wedge Y_j)$, is more than 80.

We are going to show that the set $\{x \sqsubseteq \chi_{00}\} \cup \Gamma$ is satisfiable in $\mathcal{K}$ iff $\mathcal{T}$ can tile $\mathbb{N} \times \mathbb{N}$.

Lemma 1. *If $\mathcal{T}$ can tile $\mathbb{N} \times \mathbb{N}$, then $\{x \sqsubseteq \chi_{00}\} \cup \Gamma$ is satisfiable in $\mathbb{R}^2$.*

Proof. Suppose $\tau : \mathbb{N} \times \mathbb{N} \to \mathcal{T}$ is a tiling. For $r \in \mathbb{R}^2$, put

$$S(r) = \{y \in \mathbb{R}^2 : d_2(r, y) \leq 9\}.$$

Define an assignment $\mathfrak{a}$ into $\mathbb{R}^2$ by taking, for $i, j \leq 4$ and $k \leq l$:

- $\mathfrak{a}(X_i) = \bigcup\{S(50m + 10i, 20n) : m, n \in \mathbb{N}\}$,
- $\mathfrak{a}(Y_j) = \bigcup\{S(20n, 50m + 10j) : m, n \in \mathbb{N}\}$,
- $\mathfrak{a}(Z_k) = \bigcup\{S(n, m) : \tau(n, m) = T_k\}$.

It is not difficult to see that $\langle \mathbb{R}^2, \mathfrak{a} \rangle$ satisfies $\{x \mathbb{E} \chi_{00}\} \cup \Gamma$.

Lemma 2. *Suppose a model* $\mathfrak{M} = \langle W, d, \mathfrak{a} \rangle$ *satisfies* $\{x \mathbb{E} \chi_{00}\} \cup \Gamma$. *Then there exists a function* $f : \mathbb{N} \times \mathbb{N} \to W$ *such that, for all* $i, j \leq 4$ *and* $k_1, k_2 \in \mathbb{N}$,

- $f(5k_1 + i, 5k_2 + j) \in \chi_{ij}^{\mathfrak{a}}$,
- $d(f(k_1, k_2), f(k_1 + 1, k_2)) \leq 20$,
- $d(f(k_1, k_2), f(k_1, k_2 + 1)) \leq 20$.

The map $\tau : \mathbb{N} \times \mathbb{N} \to \mathcal{T}$ *defined by taking* $\tau(n, m) = T_k$ *iff* $f(n, m) \in Z_k^{\mathfrak{a}}$, *for all* $k \leq l$ *and all* $n, m \in \mathbb{N}$, *is a tiling.*

Proof. We define f inductively. Put $f(0, 0) = \mathfrak{a}(x)$. By (12), we find a sequence $w_n \in W$, $n \in \mathbb{N}$, such that

- $w_0 = f(0, 0)$,
- $w_{5k+i} \in \chi_{i0}^{\mathfrak{a}}$, for all $i \leq 4$ and $k \in \mathbb{N}$,
- $d(w_n, w_{n+1}) \leq 20$.

We put $f(n, 0) = w_n$ for all $n \in \mathbb{N}$. Similarly, by (13) we find a sequence v_n, $n \in \mathbb{N}$, such that

- $v_0 = f(0, 0)$,
- $v_{5k+j} \in \chi_{0j}^{\mathfrak{a}}$, for all $j \leq 4$ and $k \in \mathbb{N}$,
- $d(v_n, v_{n+1}) \leq 20$.

Put $f(0, m) = v_m$ for all $m \in \mathbb{N}$. Suppose now that f satisfies the conditions listed in the formulation of the lemma (on its defined domain), that it has been defined for all (m', n') with $m' + n' < m + n$, but not for (m, n). Without loss of generality we can assume that $n = 5k_1$, $m = 5k_2 + 1$, for some $k_1, k_2 \in \mathbb{N}$. Then $f(n, m - 1) \in \chi_{00}^{\mathfrak{a}}$, and so $f(n, m - 1) \in (\exists_{\leq 20} \chi_{01})^{\mathfrak{a}}$. So we can find a $w' \in W$ with $d(f(n, m - 1), w') \leq 20$ such that $w' \in \chi_{01}^{\mathfrak{a}}$. We then put $f(n, m) = w'$. It remains to prove that f still has the required properties. To this end it suffices to show that $d(f(n - 1, m), w') \leq 20$. We have $f(n - 1, m) \in \chi_{41}^{\mathfrak{a}}$, and so there exists a w'' such that $w'' \in \chi_{01}^{\mathfrak{a}}$ and $d(f(n - 1, m), w'') \leq 20$. So it is enough to show that $w' = w''$. Suppose otherwise. Then

- $d(w'', f(n - 1, m)) \leq 20$,
- $d(f(n - 1), m), f(n - 1, m - 1)) \leq 20$,
- $d(f(n - 1, m - 1), f(n, m - 1)) \leq 20$,
- $d(f(n, m - 1), w') \leq 20$.

By the triangle inequality, we then have $d(w'', w') \leq 80$, contrary to the second formula in (10).

The reader can readily check that τ is a tiling.

4 Relational Semantics

To prove the finite model property of $\mathcal{MS}_2$, we require a relational representation of metric space models defined in Section 2. Let $M \subseteq \mathbb{R}_+$.

A *relational metric M-model* is a quadruple of the form

$$\mathfrak{S} = \langle W, (R_a)_{a \in M}, (R_{\overline{a}})_{a \in M}, \mathfrak{a} \rangle,$$

where W is a non-empty set, $(R_a)_{a \in M}$ and $(R_{\overline{a}})_{a \in M}$ are families of binary relations on W, and $\mathfrak{a}$ is an assignment in W. The *value* $t^{\mathfrak{S}}$ of a set term t in $\mathfrak{S}$ is defined inductively. The basis of induction and the case of Booleans are the same as in metric space models. And for set terms of the form $\forall_{\leq a} t$ and $\forall_{> a} t$ we put

- $(\forall_{\leq a} t)^{\mathfrak{S}} = \{w \in W : \forall v \in W \ (wR_a v \to v \in t^{\mathfrak{S}})\}$,
- $(\forall_{> a} t)^{\mathfrak{S}} = \{w \in W : \forall v \in W \ (wR_{\overline{a}} v \to v \in t^{\mathfrak{S}})\}$.

The values of $\exists_{\leq a} t$ and $\exists_{> a} t$ are defined dually.

Say that the model $\mathfrak{S}$ is *M-standard* if the following conditions are satisfied for all $a, b \in M$ and $w, u, v \in W$:

(i) $R_a \cup R_{\overline{a}} = W \times W$,

(ii) $R_a \cap R_{\overline{a}} = \emptyset$,

(iii) if $uR_a v$ and $a \leq b$, then $uR_b v$,

(iv) if $uR_{\overline{a}} v$ and $a \geq b$, then $uR_{\overline{b}} v$,

(v) $uR_0 v$ iff $u = v$,

(vi) if $uR_a v$ and $vR_b w$, then $uR_{a+b} w$ whenever $a + b \in M$,

(vii) $uR_a v$ iff $vR_a u$.

Note that as a consequence of (i), (ii) and (vi) we have:

(viii) if $uR_a v$ and $uR_{\overline{a+b}} w$ then $vR_{\overline{b}} w$.

With every metric space model $\mathfrak{M} = \langle W, d, \mathfrak{a} \rangle$ we can associate the relational metric M-model

$$\mathfrak{S}(\mathfrak{M}) = \langle W, (R_a)_{a \in M}, (R_{\overline{a}})_{a \in M}, \mathfrak{a} \rangle,$$

in which the relations R_a and $R_{\overline{a}}$ are defined as follows:

$$\forall w, v \in W \ (wR_a v \leftrightarrow d(w, v) \leq a),$$
$$\forall w, v \in W \ (wR_{\overline{a}} v \leftrightarrow d(w, v) > a).$$

It is easy to see that $\mathfrak{S}(\mathfrak{M})$ is M-standard. Note that (v), (vi) and (vii) reflect axioms (1)–(3) of metric spaces.

The model $\mathfrak{S}(\mathfrak{M})$ can be regarded as a relational representation of $\mathfrak{M}$. For we clearly have the following:

Lemma 3. *For every metric space model $\mathfrak{M}$ and every set term $t \in \mathcal{MS}_2[M]$, the value of t in $\mathfrak{M}$ coincides with the value of t in $\mathfrak{S}(\mathfrak{M})$.*

5   The Finite Model Property of $\mathcal{MS}_2$

In this section we prove that $\mathcal{MS}_2$ has the finite model property. The idea of the proof is as follows.

Let φ be an $\mathcal{MS}_2$-formula and let $\mathfrak{M} \models \varphi$ for some metric space model $\mathfrak{M} = \langle W, d, \mathfrak{a} \rangle$. Depending on $\mathfrak{M}$, we transform φ into a set Φ, containing only formulas of the form $x \in t$, $s = t$, $s \neq t$, and $\delta(x, y) = a$, in such a way that φ is satisfiable in a finite model whenever Φ is finitely satisfiable. Starting from Φ, we compute a finite set $M[\Phi]$ of real numbers containing, in particular, all the numbers occurring in Φ. Then we replace the metric d by a new metric d' with (finite) range $M[\Phi]$. The new model $\mathfrak{M}_1$ still satisfies Φ. The next step is to filtrate (as in modal logic; see e.g. [3]) the relational metric model $\mathfrak{S} = \mathfrak{S}(\mathfrak{M}_1)$ through some suitable set of terms $cl(\Phi)$. To define $cl(\Phi)$, we first transform Φ into a set Φ' which, roughly speaking, is obtained from Φ by replacing every formula of the form $\delta(y, z) = a$ with two formulas $z \in X^z$ and $y \in \exists_{\leq a} X^z$, where the X^z are fresh set variables. $cl(\Phi)$ will be the closure of the terms in Φ' under syntactical rules that are similar to the rules of the Fischer–Ladner closure for **PDL**-formulas (cf. [10]). (Note, however, that in contrast to the Fischer–Ladner closure the closure considered here results in an exponential blow up.)

As a result of the filtration we get a *finite* relational metric model $\mathfrak{S}^f$. But unlike $\mathfrak{S}$, in general $\mathfrak{S}^f$ is not $M[\Phi]$-standard, which means that we cannot directly transform it into a finite metric space model. However, $\mathfrak{S}^f$ still has all the properties of $M[\Phi]$-standard models save (ii): there may exist $v \in W^f$ such that $w R_a v$ and $w R_{\overline{a}} v$, for some $w \in W^f$, and $a \in M[\Phi]$. To 'cure' these defects, we make copies of such 'bad' points v and modify the relations R_a and $R_{\overline{a}}$ in $\mathfrak{S}^f$ obtaining a finite standard relational metric model $\mathfrak{S}^*$. (The 'copying-method' was developed by the Bulgarian school of modal logic; see [7,16]. Our technique follows [8]). The final step is to transform $\mathfrak{S}^*$ into a metric space model $\mathfrak{M}^*$.

Let us now turn to details. Denote by $term(\varphi)$ the set of all set terms occurring in φ; $sub(\varphi)$ stands for the set of all subformulas of φ. Define a set $\Phi = \Phi_1 \cup \Phi_2 \cup \Phi_3$ by taking:

$$\Phi_1 = \{x \in t : (x \in t) \in sub(\varphi),\ \mathfrak{M} \models x \in t\} \cup$$
$$\{x \in \neg t : (x \in t) \in sub(\varphi),\ \mathfrak{M} \not\models x \in t\},$$
$$\Phi_2 = \{s = t : (s = t) \in sub(\varphi),\ \mathfrak{M} \models s = t\} \cup$$
$$\{s \neq t : (s = t) \in sub(\varphi),\ \mathfrak{M} \models s \neq t\},$$
$$\Phi_3 = \{\delta(y, z) = a : \delta(y, z) \in term(\varphi),\ a = d(\mathfrak{a}(y), \mathfrak{a}(z))\}.$$

It should be clear from the definition that we have

Lemma 4. (1) $\mathfrak{M} \models \Phi$.

(2) *For every metric space model* $\mathfrak{M}'$, *if* $\mathfrak{M}' \models \Phi$ *then* $\mathfrak{M}' \models \varphi$.

Next we construct $M[\Phi]$ and Φ'. Let

$$M(\Phi) = \{a \in \mathbb{R} : a \text{ occurs in } \Phi\}.$$

Denote by γ the smallest natural number that is greater than all numbers in $M(\Phi) \cup \{0\}$ and define $M[\Phi]$ as

$$M[\Phi] = \{a_1 + \cdots + a_n < \gamma : a_1, \ldots, a_n \in M(\Phi), \ n < \omega\} \cup \{\gamma\} \cup \{0\}.$$

Let $\mu = \min\{M(\Phi) - \{0\}\}$ and let χ be the least natural number such that $\chi \geq \gamma/\mu$. An easy (but tedious) computation yields:

Lemma 5. $|M[\Phi]| \leq |M(\Phi)|^{\chi}$, whenever $|M(\Phi)| \geq 2$.

For each location variable x occurring in Φ_3 we pick a new set variable X^x and define Φ'_3, Φ', and $t(\Phi)$ by taking

$$\Phi'_3 = \{y \in \exists_{\leq a} X^z : \delta(y, z) = a \in \Phi_3\} \cup$$
$$\{z \in X^z : \delta(y, z) = a \in \Phi_3\} \cup$$
$$\{y \in \forall_{\leq b} \neg X^z : \delta(y, z) = a \in \Phi_3, \ b < a, \ b \in M[\Phi]\},$$
$$\Phi' = \Phi_1 \cup \Phi_2 \cup \Phi'_3,$$
$$t(\Phi) = \{t : t \in term(\Phi')\}.$$

The *closure* $cl(\Phi)$ of $t(\Phi)$ is the smallest set of terms T such that $t(\Phi) \subseteq T$ and

1. T is closed under subterms;
2. if $t \in T$, then $\forall_{\leq 0} t \in T$ whenever t is not of the form $\forall_{\leq 0} s$;
3. if $\forall_{\leq a} t \in T$ and $a \geq a_1 + \cdots + a_n$, for $a_i \in M[\Phi] - \{0\}$, then $\forall_{\leq a_1} \ldots \forall_{\leq a_n} t \in T$;
4. if $\forall_{>a} t \in T$ and $b \in M[\Phi]$, then $\neg \forall_{<b} \neg \forall_{>a} t \in T$;
5. if $\forall_{>a} t \in T$ and $b > a$, for $b \in M[\Phi]$, then $\forall_{>b} t \in T$ and $\neg \forall_{>b} \neg \forall_{>a} t \in T$.

By an easy but tedious computation the reader can check that we have:

Lemma 6. If $|M(\Phi)| \geq 4$ and $\chi \geq 3$, then

$$|cl(\Phi)| \leq S(\Phi) = |t(\Phi)| \cdot |M[\Phi]|^{(\chi+1) \cdot (|M(\Phi)|+1)}.$$

We are in a position now to prove the following:

Theorem 6. Φ is satisfied in a metric space model $\mathfrak{M}^* = \langle W^*, d^*, \mathfrak{b}^* \rangle$ such that $|W^*| \leq 2 \cdot 2^{S(\Phi)}$ and the range of d^* is a subset of $M[\Phi]$.

Proof. We first show that Φ is satisfied in a metric space model $\langle W, d', \mathfrak{a} \rangle$ with the range of d' being a subset of $M = M[\Phi]$. Indeed, define d' by taking

$$d'(w, v) = \min\{\gamma, a \in M : d(w, v) \leq a\},$$

for all $w, v \in W$, and let $\mathfrak{M}_1 = \langle W, d', \mathfrak{a} \rangle$. Clearly, the range of d' is a subset of M. We check that d' is a metric. It satisfies (1) because $0 \in M$. That d' is symmetric follows from the symmetry of d. To show (2), suppose $d'(w, v) + d'(v, u) \leq a$, for $a \in M$. By the definition of d', we then have $d(w, v) + d(v, u) \leq a$, and so $d(w, u) \leq a$. Hence $d'(w, u) \leq a$. Thus we have shown that

$$\{a \in M : d'(w, v) + d'(v, u) \leq a\} \subseteq \{a \in M : d'(w, u) \leq a\},$$

from which one easily concludes that $d'(w, u) \leq d'(w, v) + d'(v, u)$.

Lemma 7. *The set Φ is satisfied in $\mathfrak{M}_1$.*

Proof. Clearly, for each $(\delta(y,z) = a) \in \Phi_3$, $d(\mathfrak{a}(y), \mathfrak{a}(z)) = d'(\mathfrak{a}(y), \mathfrak{a}(z)) = a$. So $\mathfrak{M}_1 \models \Phi_3$. To show $\mathfrak{M}_1 \models \Phi_1 \cup \Phi_2$ it suffices to prove that

$$\forall w \in W \forall t \in t(\Phi) \; (w \in t^{\mathfrak{M}} \leftrightarrow w \in t^{\mathfrak{M}_1}).$$

This can be done by a straightforward induction on the construction of t. The basis of induction and the case of Booleans are trivial. So suppose t is $\forall_{\leq a} s$ (then $a \in M$). Then we have:

$$w \in t^{\mathfrak{M}} \leftrightarrow_1 \forall v \in W \; (d(w,v) \leq a \rightarrow v \in s^{\mathfrak{M}})$$
$$\leftrightarrow_2 \forall v \in W \; (d'(w,v) \leq a \rightarrow v \in s^{\mathfrak{M}_1})$$
$$\leftrightarrow_3 w \in t^{\mathfrak{M}_1}.$$

The equivalences $\leftrightarrow_1$ and $\leftrightarrow_3$ are obvious. $\leftrightarrow_2$ holds by the induction hypothesis and the fact that, for all $w, v \in W$ and every $a \in M$, $d(x,y) \leq a$ iff $d'(x,y) \leq a$. The case $\forall_{>a} s$ is considered in a similar way.

Before filtrating $\mathfrak{M}_1$ through $\Theta = cl(\Phi)$, we slightly change its assignment. Recall that Θ contains the new set variables X^z which function as nominals and which will help to fix the distances between the points occurring in Φ_3. Define $\mathfrak{b}$ to be the assignment that acts as $\mathfrak{a}$ on all variables save the X^z, where

- $\mathfrak{b}(X^z) = \{\mathfrak{a}(z)\}$.

Let $\mathfrak{M}_2 = \langle W, d', \mathfrak{b} \rangle$. It should be clear from the definition and Lemma 7 that we have:

(a) $t^{\mathfrak{M}_1} = t^{\mathfrak{M}_2}$, for all set terms $t \in t(\Phi)$;
(b) $\mathfrak{M}_1 \models \psi$ iff $\mathfrak{M}_2 \models \psi$, for all formulas $\psi \in MS_2(\Phi)$;
(c) $\mathfrak{M}_2 \models \Phi$;
(d) $\mathfrak{M}_2 \models \Phi'$.

Consider the relational counterpart of $\mathfrak{M}_2$, i.e., the model

$$\mathfrak{S}(\mathfrak{M}_2) = \langle W, (R_a)_{a \in M}, (R_{\bar{a}})_{a \in M}, \mathfrak{b} \rangle$$

which, for brevity, will be denoted by $\mathfrak{S}$. Define an equivalence relation $\equiv$ on W by taking $u \equiv v$ when $u \in t^{\mathfrak{S}}$ iff $v \in t^{\mathfrak{S}}$ for all $t \in \Theta$. Let $[u] = \{v \in W : u \equiv v\}$. Note that if $(z \in X^z) \in \Phi_3$ then $[\mathfrak{b}(z)] = \{\mathfrak{b}(z)\}$, since $X^z \in \Theta$.

Construct a filtration $\mathfrak{S}^f = \left\langle W^f, (R_a^f)_{a \in M}, (R_{\bar{a}}^f)_{a \in M}, \mathfrak{b}^f \right\rangle$ of $\mathfrak{S}$ through Θ by taking

- $W^f = \{[u] : u \in W\}$;
- $\mathfrak{b}^f(x) = [\mathfrak{b}(x)]$;
- $\mathfrak{b}^f(X) = \{[u] : u \in \mathfrak{b}(X)\}$;
- $[u]R_a^f[v]$ iff for all terms $\forall_{\leq a} t \in \Theta$,
  - $u \in (\forall_{\leq a} t)^{\mathfrak{S}}$ implies $v \in t^{\mathfrak{S}}$ and
  - $v \in (\forall_{\leq a} t)^{\mathfrak{S}}$ implies $u \in t^{\mathfrak{S}}$;

- $[u]R_{\bar{a}}^f[v]$ iff for all terms $\forall_{>a}t \in \Theta$,
  - $u \in (\forall_{>a}t)^{\mathfrak{S}}$ implies $v \in t^{\mathfrak{S}}$ and
  - $v \in (\forall_{>a}t)^{\mathfrak{S}}$ implies $u \in t^{\mathfrak{S}}$.

Since Θ is finite, W^f is finite as well. Note also that $\mathfrak{b}^f(X^z) = \{\mathfrak{b}^f(z)\}$ whenever $(z \in X^z) \in \Phi_3'$.

Lemma 8. (1) *For every $t \in \Theta$ and every $u \in W$, $u \in t^{\mathfrak{S}}$ iff $[u] \in t^{\mathfrak{S}^f}$.*
(2) *For all $(\delta(y,z) = a) \in \Phi_3$, $a = \min\{b \in M : \mathfrak{b}^f(y)R_b^f \mathfrak{b}^f(z)\}$.*
(3) $\mathfrak{S}^f$ *satisfies* (i), (iii)–(vii) *in Section 4.*

Proof. (1) is proved by an easy induction on the construction of t. To prove (2), take $(\delta(y,z) = a) \in \Phi_3$. We must show that $\mathfrak{b}^f(y)R_a^f \mathfrak{b}^f(z)$ and $\neg \mathfrak{b}^f(y)R_b^f \mathfrak{b}^f(z)$, for all $a > b \in M$. Notice first that $uR_a v$ implies $[u]R_{\bar{a}}^f[v]$ and $uR_{\bar{a}}v$ implies $[u]R_{\bar{a}}^f[v]$. Since $\mathfrak{M}_2 \models \Phi$, we have $\mathfrak{M}_2 \models \delta(y,z) = a$, and so $d'(\mathfrak{b}(y), \mathfrak{b}(z)) = a$. Hence $\mathfrak{b}(y)R_a \mathfrak{b}(z)$ and $\mathfrak{b}^f(y)R_a^f \mathfrak{b}^f(z)$. Suppose now that $b < a$ and consider $\forall_{\leq b}\neg X^z$. By definition, $\mathfrak{b}(X^z) = \{\mathfrak{b}(z)\}$. Hence $\mathfrak{b}(z) \notin (\neg X^z)^{\mathfrak{S}}$. On the other hand, we have $b < d'(\mathfrak{b}(y), \mathfrak{b}(z))$, from which $\mathfrak{b}(y) \in (\forall_{\leq b}\neg X^z)^{\mathfrak{S}}$. Since $(\forall_{\leq b}\neg X^z) \in \Theta$, we then obtain $\neg \mathfrak{b}^f(y)R_b^f \mathfrak{b}^f(z)$.

Now let us prove (3). Condition (vii), i.e., $[w]R_a^f[u]$ iff $[u]R_a^f[w]$, holds by definition.

(i), i.e, $R_a^f \cup R_{\bar{a}}^f = W^f \times W^f$. If $\neg[u]R_a^f[v]$ then $\neg uR_a v$, and so $uR_{\bar{a}}v$, since $\mathfrak{S}$ satisfies (i). Thus $[u]R_{\bar{a}}^f[v]$.

(iii), i.e., if $[u]R_a^f[v]$ and $a \leq b$ then $[u]R_b^f[v]$. Let $[u]R_a^f[v]$ and $a < b$, for $b \in M$. Suppose $u \in (\forall_{\leq b}t)^{\mathfrak{S}}$. By the definition of $\Theta = cl(\Phi)$, $\forall_{\leq a}t \in \Theta$, and so $u \in (\forall_{\leq a}t)^{\mathfrak{S}}$. Hence $v \in t^{\mathfrak{S}}$. The other direction is considered in the same way.

(iv), i.e., if $[u]R_{\bar{a}}^f[v]$ and $a \geq b$ then $[u]R_{\bar{b}}^f[v]$. Let $[u]R_{\bar{a}}^f[v]$ and $a > b$, and suppose that $u \in (\forall_{>b}t)^{\mathfrak{S}}$. Then $\forall_{>a}t \in \Theta$, $u \in (\forall_{>a}t)^{\mathfrak{S}}$, and so $v \in t^{\mathfrak{S}}$. Again, the other direction is treated analogously.

(v), i.e., $[u]R_0^f[v]$ iff $[u] = [v]$. The implication ($\Leftarrow$) is obvious. So suppose $[u]R_0^f[v]$. Take some $t \in \Theta$ with $u \in t^{\mathfrak{S}}$. Without loss of generality we may assume that t is not of the form $\forall_{\leq 0}s$. Then, by the definition of Θ, $u \in (\forall_{\leq 0}t)^{\mathfrak{S}}$ and $\forall_{\leq 0}t \in \Theta$. Hence $v \in t^{\mathfrak{S}}$. In precisely the same way one can show that for all $t \in \Theta$, $v \in t^{\mathfrak{S}}$ implies $u \in t^{\mathfrak{S}}$. Therefore, $[u] = [v]$.

(vi), i.e., if $[u]R_a^f[v]$ and $[v]R_b^f[w]$, then $[u]R_{a+b}^f[w]$, for $(a+b) \in M$. Suppose $u \in (\forall_{\leq a+b}t)^{\mathfrak{S}}$. Then $\forall_{\leq a}\forall_{\leq b}t \in \Theta$ and $u \in (\forall_{\leq a}\forall_{\leq b}t)^{\mathfrak{S}}$. Hence $w \in t^{\mathfrak{S}}$. For the other direction, assume $w \in (\forall_{\leq a+b}t)^{\mathfrak{S}}$. Again, we have $\forall_{\leq a}\forall_{\leq b}t \in \Theta$ and $w \in (\forall_{\leq a}\forall_{\leq b}t)^{\mathfrak{S}}$. In view of (vii) we then obtain $u \in t^{\mathfrak{S}}$.

(viii), i.e., if $[u]R_a^f[v]$ and $[u]R_{\overline{a+b}}^f[w]$ then $[v]R_{\bar{b}}^f[w]$, for $(a+b) \in M$. Suppose $v \in (\forall_{>b}t)^{\mathfrak{S}}$. Then $\neg\forall_{\leq a}\neg\forall_{>b}t \in \Theta$ and $u \in (\neg\forall_{\leq a}\neg\forall_{>b}t)^{\mathfrak{S}}$. Hence $u \in (\forall_{>(a+b)}t)^{\mathfrak{S}}$ and so $w \in t^{\mathfrak{S}}$. For the other direction, suppose $w \in (\forall_{>b}t)^{\mathfrak{S}}$. Then $u \in (\neg\forall_{>(a+b)}\neg\forall_{>b}t)^{\mathfrak{S}}$ and $\neg\forall_{>(a+b)}\neg\forall_{>b}t \in \Theta$. Hence $u \in (\forall_{\leq a}t)^{\mathfrak{S}}$ and so $v \in t^{\mathfrak{S}}$.

Unfortunately, $\mathfrak{S}^f$ does not necessarily satisfy (ii) which is required to construct the model $\mathfrak{M}^*$ we need: it may happen that for some points $[u], [v]$ in W^f and $a \in M$, we have both $[u]R_a^f[v]$ and $[u]R_{\bar{a}}^f[v]$. To 'cure' these defects, we have to perform some surgery. The defects form the set

$$D(W^f) = \{d \in W^f : \exists a \in M \exists x \in W^f \, (xR_a^f d \,\&\, xR_{\bar{a}}^f d)\}.$$

Let

$$W^* = \{\langle d, i \rangle : d \in D(W^f), i \in \{0,1\}\} \cup \{\langle c, 0 \rangle : c \in W^f - D(W^f)\}.$$

So for each $d \in D(W^f)$ we have now two copies $\langle d, 0 \rangle$ and $\langle d, 1 \rangle$. Define an assignment $\mathfrak{b}^*$ in W^* by taking

- $\mathfrak{b}^*(x) = \langle \mathfrak{b}^f(x), 0 \rangle$ and
- $\mathfrak{b}^*(X) = \{\langle c, i \rangle \in W^* : c \in \mathfrak{b}^f(X)\}$.

Finally, we define accessibility relations R_a^* and $R_{\bar{a}}^*$ as follows:

- if $a > 0$ then $\langle c, i \rangle \, R_a^* \, \langle d, j \rangle$ iff either
  - $cR_a^f d$ and $\neg cR_{\bar{a}}^f d$, or
  - $cR_a^f d$ and $i = j$;
- if $a = 0$ then $\langle c, i \rangle \, R_a^* \, \langle d, j \rangle$ iff $\langle c, i \rangle = \langle d, j \rangle$;
- $R_{\bar{a}}^*$ is defined as the complement of R_a^*, i.e., $\langle c, i \rangle \, R_{\bar{a}}^* \, \langle d, j \rangle$ iff $\neg \langle c, i \rangle \, R_a^* \, \langle d, j \rangle$.

Lemma 9. $\mathfrak{S}^* = \langle W^*, (R_a^*)_{a \in M}, (R_{\bar{a}}^*)_{a \in M}, \mathfrak{b}^* \rangle$ *is an M-standard relational metric model.*

Proof. That $\mathfrak{S}^*$ satisfies (i), (ii), and (v) follows immediately from the definition. Let us check the remaining conditions.

(iii) Suppose $\langle c, i \rangle \, R_a^* \, \langle d, j \rangle$ and $a < b \in M$. If $i = j$ then clearly $\langle c, i \rangle \, R_b^* \, \langle d, j \rangle$. So assume $i \neq j$. Then, by definition, $cR_a^f d$ and $\neg cR_{\bar{a}}^f d$. Since $\mathfrak{S}^f$ satisfies (iii) and (iv), we obtain $cR_b^f d$ and $\neg cR_{\bar{b}}^f d$. Thus $\langle c, i \rangle \, R_b^* \, \langle d, j \rangle$.

(iv) Suppose that $\langle c, i \rangle \, R_{\bar{a}}^* \, \langle d, j \rangle$ and $a > b \in M$, but $\neg \langle c, i \rangle \, R_{\bar{b}}^* \, \langle d, j \rangle$. By (i), $\langle c, i \rangle \, R_b^* \, \langle d, j \rangle$. And by (iii), $\langle c, i \rangle \, R_a^* \, \langle d, j \rangle$. Finally, (ii) yields $\neg \langle c, i \rangle \, R_{\bar{a}}^* \, \langle d, j \rangle$, which is a contradiction.

(vi) Suppose $\langle c, i \rangle \, R_a^* \, \langle d, j \rangle$, $\langle d, j \rangle \, R_b^* \, \langle e, k \rangle$ and $a + b \in M$. Then $cR_a^f d$ and $dR_b^f e$. As $\mathfrak{S}^f$ satisfies (vii), we have $cR_{a+b}^f e$. If $i = k$ then clearly $\langle c, i \rangle \, R_{a+b}^* \, \langle e, k \rangle$. So assume $i \neq k$. If $i = j \neq k$ then $\neg cR_{\overline{a+b}}^f e$, since $cR_a^f d$ and $\neg dR_{\bar{b}}^f e$. The case $i \neq j = k$ is considered analogously using the fact that the relations in $\mathfrak{S}^f$ are symmetric.

(vii) follows from the symmetry of R_a^f and $R_{\bar{a}}^f$.

Lemma 10. *For all $\langle d, i \rangle \in W^*$ and $t \in \Theta$, we have $\langle d, i \rangle \in t^{\mathfrak{S}^*}$ iff $d \in t^{\mathfrak{S}^f}$.*

Proof. The proof is by induction on t. The basis of induction and the case of Booleans are trivial. The cases $t = (\forall_{\leq a} s)$ and $t = (\forall_{>a} s)$ are consequences of the following claims:

Claim 1: if $cR_a^f d$ and $i \in \{0, 1\}$, then there exists j such that $\langle c, i \rangle R_a^* \langle d, j \rangle$. Indeed, this is clear for $i = 0$. Suppose $i = 1$. If d was duplicated, then $\langle d, 1 \rangle$ is as required. If d was not duplicated, then $\neg cR_{\bar{a}}^f d$, and so $\langle d, 0 \rangle$ is as required.

Claim 2: if $\langle c, i \rangle R_a^* \langle d, j \rangle$ then $cR_a^f d$. This is obvious.

Claim 3: if $cR_{\bar{a}}^f d$ and $i \in \{0, 1\}$ then there exists j such that $\neg \langle c, i \rangle R_a^* \langle d, j \rangle$. Suppose $i = 0$. If d was not duplicated, then $\neg cR_a^f d$. Hence $\neg \langle c, 0 \rangle R_a^* \langle d, 0 \rangle$. If d was duplicated, then $\neg \langle c, 0 \rangle R_a^* \langle d, 1 \rangle$. In the case $i = 1$ we have $\neg \langle c, 1 \rangle R_a^* \langle d, 0 \rangle$.

Claim 4: if $\neg \langle c, i \rangle R_a^* \langle d, j \rangle$ then $cR_{\bar{a}}^f d$. Indeed, if $i = j$ then $\neg cR_a^f d$ and so $cR_{\bar{a}}^f d$. And if $i \neq j$ then $cR_{\bar{a}}^f d$.

To complete the proof of Theorem 6, we transform $\mathfrak{S}^*$ into a finite metric space model and show that this model satisfies Φ. Put $\mathfrak{M}^* = \langle W^*, d^*, \mathfrak{b}^* \rangle$, where for all $w, v \in W^*$,

$$d^*(w, v) = min\{\gamma, a \in M : wR_a^* v\}.$$

As M is finite, d^* is well-defined. Using (v)–(vii), it is easy to see that d^* is a metric. So $\mathfrak{M}^*$ is a finite metric space model. It remains to show that $\mathfrak{M}^*$ satisfies Φ. Note first that

(†) for all $w \in W^*$ and $t \in t(\Phi)$, we have $w \in t^{\mathfrak{S}^*}$ iff $w \in t^{\mathfrak{M}^*}$.

This claim is proved by induction on t. The basis and the Boolean cases are clear. So let $t = (\forall_{\leq a} s)$ for some $a \in M$. Then

$$
\begin{aligned}
w \in (\forall_{\leq a} s)^{\mathfrak{S}^*} &\Leftrightarrow_1 \forall v \, (wR_a^* v \rightarrow v \in s^{\mathfrak{S}^*}) \\
&\Leftrightarrow_2 \forall v \, (wR_a^* v \rightarrow v \in s^{\mathfrak{M}^*}) \\
&\Leftrightarrow_3 \forall v \, (d^*(w, v) \leq a \rightarrow v \in s^{\mathfrak{M}^*}) \\
&\Leftrightarrow_4 w \in (\forall_{\leq a} s)^{\mathfrak{M}^*}.
\end{aligned}
$$

Equivalences $\Leftrightarrow_1$ and $\Leftrightarrow_4$ are obvious; $\Leftrightarrow_2$ holds by the induction hypothesis; $\Leftarrow_3$ is an immediate consequence of the definition of d^*, and $\Rightarrow_3$ follows from (iii). The case $t = (\forall_{>a} s)$ is proved analogously.

We can now show that $\mathfrak{M}^* \models \Phi$. Let $(x \sqsubseteq t) \in \Phi_1$. Then we have:

$$\mathfrak{M}^* \models x \sqsubseteq t \Leftrightarrow_1 \mathfrak{b}^*(x) \in t^{\mathfrak{M}^*} \Leftrightarrow_2 \mathfrak{b}^*(x) \in t^{\mathfrak{S}^*} \Leftrightarrow_3 \langle \mathfrak{b}^f(x), 0 \rangle \in t^{\mathfrak{S}^*} \Leftrightarrow_4$$
$$\mathfrak{b}^f(x) \in t^{\mathfrak{S}^f} \Leftrightarrow_5 [\mathfrak{b}(x)] \in t^{\mathfrak{S}^f} \Leftrightarrow_6 \mathfrak{b}(x) \in t^{\mathfrak{S}} \Leftrightarrow_7 \mathfrak{b}(x) \in t^{\mathfrak{M}_2} \Leftrightarrow_8 \mathfrak{M}_2 \models x \sqsubseteq t.$$

Equivalences $\Leftrightarrow_1$ and $\Leftrightarrow_8$ are obvious; $\Leftrightarrow_2$ follows from (†); $\Leftrightarrow_3$ and $\Leftrightarrow_5$ hold by definition; $\Leftrightarrow_4$ follows from Lemma 10, $\Leftrightarrow_6$ from Lemma 8, and $\Leftrightarrow_7$ from Lemma 3.

Since $\mathfrak{M}_2 \models \Phi$, we have $\mathfrak{M}^* \models \Phi_1$. That $\mathfrak{M}^* \models \Phi_2$ is proved analogously using (†).

It remains to show that $\mathfrak{M}^* \models \Phi_3$. Take any $\delta(y, z) = a$ from Φ_3. We must show that $d^*(\mathfrak{b}^*(y), \mathfrak{b}^*(z)) = a$. By Lemma 8 (2),

$$a = min\{b \in M : \mathfrak{b}^f(y)R_b^f\mathfrak{b}^f(z)\}.$$

So $a = min\{b \in M : \langle \mathfrak{b}^f(y), 0 \rangle R_b^* \langle \mathfrak{b}^f(z), 0 \rangle\}$. By the definition of $\mathfrak{b}^*$ we have $a = min\{b \in M : \mathfrak{b}^*(y)R_b^*\mathfrak{b}^*(z)\}$, which means that $d^*(\mathfrak{b}^*(y), \mathfrak{b}^*(z)) = a$.

This completes the proof of Theorem 6.

Thus, by Theorem 6 and Lemma 4 (2), φ is satisfied in the finite model $\mathfrak{M}^*$.

Yet this is not enough to prove the decidability of $\mathcal{MS}_2[\mathbb{Q}]$: we still do not know an effectively computable upper bound for the size of a finite model satisfying φ. Indeed, the set $M(\Phi)$ depends not only on φ, but also on the initial model $\mathfrak{M}$ satisfying φ. Note, however, that by Lemmas 5 and 6 the size of $\mathfrak{M}^*$ can be computed from the maximum of $M(\Phi)$, the minimum of $M(\Phi) - \{0\}$, and φ. Hence, to obtain an effective upper bound we need, it suffices to start the construction with a model satisfying φ for which both the maximum of $M(\Phi)$ and the minimum of $M(\Phi) - \{0\}$ are known. The next lemma shows how to obtain such a model.

Lemma 11. *Suppose a formula $\varphi \in \mathcal{MS}_2[\mathbb{Q}]$ is satisfied in a metric space model $\langle W, d, \mathfrak{a} \rangle$. Denote by $\mathcal{D}$ the set of all $\delta(x, y)$ occurring in φ, and let a and b be the minimal positive number and the maximal number occurring in φ, respectively (if no such number exists, then put $a = b = 1$). Then there is a metric d' on W such that φ is satisfied in $\langle W, d', \mathfrak{a} \rangle$ and*

$$min\{d'(\mathfrak{a}(x), \mathfrak{a}(y)) > 0 : \delta(x, y) \in \mathcal{D}\} \geq a/2,$$
$$max\{d'(\mathfrak{a}(x), \mathfrak{a}(y)) : \delta(x, y) \in \mathcal{D}\} \leq 2b.$$

Proof. Let

$$a' = min\{d(\mathfrak{a}(x), \mathfrak{a}(y)) > 0 : \delta(x, y) \in \mathcal{D}\},$$
$$b' = max\{d(\mathfrak{a}(x), \mathfrak{a}(y)) : \delta(x, y) \in \mathcal{D}\}.$$

We consider here the case when $a' < a/2$ and $2b < b'$. The case when this is not so is easy; we leave it to the reader. Define d' by taking

$$d'(x, y) = \begin{cases} d(x, y) & \text{if } a \leq d(x, y) \leq b \text{ or } d(x, y) = 0, \\ b + (b/(b' - b)) \cdot (d(x, y) - b) & \text{if } d(x, y) > b, \\ a + (a/2(a - a')) \cdot (d(x, y) - a) & \text{if } 0 < d(x, y) < a. \end{cases}$$

One can readily show now that d' is a metric and $\langle W, d', \mathfrak{a} \rangle$ satisfies φ.

6   Weaker Distance Spaces

As was mentioned in Section 1, our everyday life experience gives interesting measures of distances which lack some of the features characteristic to metric spaces. Not trying to cover all possible cases, we list here some possible ways of defining such alternative measures by modifying the axioms of standard metric spaces:

- we can omit either the symmetry axiom or the triangular inequality;
- we can omit both of them;
- we can allow d to be a partial function satisfying the following conditions for all $w, v, u \in W$, where $dom(d)$ is the domain of d:
  - $\langle w, w \rangle \in dom(d)$ and $d(w, w) = 0$,
  - if $\langle w, v \rangle \in dom(d)$ and $d(w, v) = 0$, then $w = v$,
  - if $\langle w, v \rangle \in dom(d)$ and $\langle v, u \rangle \in dom(d)$, then $\langle w, u \rangle \in dom(d)$ and $d(w, u) \leq d(w, v) + d(v, u)$,
  - if $\langle w, v \rangle \in dom(d)$, then $\langle v, w \rangle \in dom(d)$ and $d(w, v) = d(v, w)$.

Using almost the same techniques as above one can generalize the obtained results on the decidability and finite model property of $\mathcal{MS}_2$ to these weaker metric spaces as well.

Acknowledgments:

We are grateful to A. Cohn for stimulating discussions and comments. The first author was supported by the Deutsche Forschungsgemeinschaft (DFG), the second author was partially supported by the Sumitomo Foundation, and the fourth author was partially supported by grant no. 99–01–0986 from the Russian Foundation for Basic Research.

References

1. P. Blackburn. Nominal tense logic. *Notre Dame Journal of Formal Logic*, 34:56–83, 1993.
2. E. Börger, E. Grädel, and Yu. Gurevich. *The Classical Decision Problem*. Perspectives in Mathematical Logic. Springer, 1997.
3. A. Chagrov and M. Zakharyaschev. *Modal Logic*. Oxford University Press, Oxford, 1997.
4. E.T. Copson. *Metric Spaces*. Number 57 in Cambridge Tracts in Mathematics and Mathematical Physics. Cambridge University Press, 1968.
5. T. de Laguna. Point, line and surface as sets of solids. *The Journal of Philosophy*, 19:449–461, 1922.
6. M. de Rijke. The modal logic of inequality. *Journal of Symbolic Logic*, 57:566–584, 1990.
7. G. Gargov, S. Passy, and T. Tinchev. Modal environment for boolean speculations. In D. Scordev, editor, *Mathematical Logic*. Plenum Press, New York, 1988.
8. V. Goranko. Completeness and incompleteness in the bimodal base $\mathcal{L}(R, -R)$. In P. Petkov, editor, *Mathematical Logic*, pages 311–326. Plenum Press, New York, 1990.
9. V. Goranko and S. Passy. Using the universal modality. *Journal of Logic and Computation*, 2:203–233, 1992.
10. D. Harel. Dynamic logic. In D. Gabbay and F. Guenthner, editors, *Handbook of Philosophical Logic*, pages 605–714. Reidel, Dordrecht, 1984.
11. R. Jansana. Some logics related to von Wright's logic of place. *Notre Dame Journal of Formal Logic*, 35:88–98, 1994.

12. O. Lemon and I. Pratt. On the incompleteness of modal logics of space: advancing complete modal logics of place. In M. Kracht, M. de Rijke, H. Wansing, and M. Zakharyaschev, editors, *Advances in Modal Logic*, pages 115–132. CSLI, 1998.
13. A. Montanari. *Metric and layered temporal logic for time granularity*. PhD thesis, Amsterdam, 1996.
14. N. Rescher and J. Garson. Topological logic. *Journal of Symbolic Logic*, 33:537–548, 1968.
15. K. Segerberg. A note on the logic of elsewhere. *Theoria*, 46:183–187, 1980.
16. D. Vakarelov. Modal logics for knowledge representation. *Theoretical Computer Science*, 90:433–456, 1991.
17. P. van Emde Boas. The convenience of tilings. In A. Sorbi, editor, *Complexity, Logic and Recursion Theory*, volume 187 of *Lecture Notes in Pure and Applied Mathematics*, pages 331–363. Marcel Dekker Inc., 1997.
18. G.H. von Wright. A modal logic of place. In E. Sosa, editor, *The Philosophy of Nicholas Rescher*, pages 65–73. D. Reidel, Dordrecht, 1979.
19. F. Wolter and M. Zakharyaschev. Spatio-temporal representation and reasoning based on RCC-8. In *Proceedings of the seventh Conference on Principles of Knowledge Representation and Reasoning, KR2000, Breckenridge, USA*, pages 3–14, Montreal, Canada, 2000. Morgan Kaufman.

Hybrid Probabilistic Logic Programs as Residuated Logic Programs

Carlos Viegas Damásio and Luís Moniz Pereira

Centro de Inteligéncia Artificial (CENTRIA)
Departamento de Informática, Faculdade de Ciéncias e Tecnologia
Universidade Nova de Lisboa
2825-114 Caparica, Portugal.
cd|lmp@di.fct.unl.pt

Abstract. In this paper we show the embedding of Hybrid Probabilistic Logic Programs into the rather general framework of Residuated Logic Programs, where the main results of (definite) logic programming are validly extrapolated, namely the extension of the immediate consequences operator of van Emden and Kowalski. The importance of this result is that for the first time a framework encompassing several quite distinct logic programming semantics is described, namely Generalized Annotated Logic Programs, Fuzzy Logic Programming, Hybrid Probabilistic Logic Programs, and Possibilistic Logic Programming. Moreover, the embedding provides a more general semantical structure paving the way for defining paraconsistent probabilistic reasoning logic programming semantics.

1 Introduction

The literature on logic programming theory is brimming with proposals of languages and semantics for extensions of definite logic programs (e.g. [7,15,4,10]), i.e. without non-monotonic or default negation. Usually, the authors characterize their programs with a model theoretic semantics, where a minimum model is guaranteed to exist, and a corresponding monotonic fixpoint operator (continuous or not). In many cases these semantics are many-valued.

In this paper we start by defining a rather general framework of Residuated Logic Programs. We were inspired by the deep theoretical results of many-valued logics and fuzzy logic (see [1,9] for excellent accounts) and applied these ideas to logic programming. In fact, a preliminary work in this direction is [15], but the authors restrict themselves to a linearly ordered set of truth-values (the real closed interval $[0, 1]$) and to a very limited syntax: the head of rules is a literal and the body is a multiplication (t-norm) of literals. Our main semantical structures are residuated (or residual) lattices (c.f. [1,9]), where a generalized modus ponens rule is defined. This characterizes the essence of logic programming: from the truth-value of bodies for rules for an atom we can determine the truth-value of that atom, depending on the confidence in the rules.

M. Ojeda-Aciego et al. (Eds.): JELIA 2000, LNAI 1919, pp. 57–72, 2000.

Besides fuzzy reasoning, probabilistic reasoning forms are essential for knowledge representation in real-world applications. However, a major difficulty is that there are several logical ways of determining the probabilities of complex events (conjunctions or disjunctions) from primitive ones. To address this issue, a model theory, fixpoint theory and proof theory for hybrid probabilistic logic programs were recently introduced [4,3]. The generality of Residuated Logic Programming is illustrated in practice by presenting an embedding of Hybrid Probabilistc Logic Programs [4,3] into our framework.

Our paper proceeds as follows. In the next section we present the residuated logic programs. Afterwards, we overview the hybrid probabilistic logic programming setting and subsequently provide the embedding. We finally draw some conclusions and point out future directions. We included the main proofs for the sake of completeness.

2   Residuated Logic Programs

The theoretical foundations of logic programming were clearly established in [11,14] for definite logic programs (see also [12]), i.e. programs made up of rules of the form $A_0 \subset A_1 \wedge \ldots \wedge A_n (n \geq 0)$ where each $A_i (0 \leq i \leq n)$ is a propositional symbol (an atom), $\subset$ is classical implication, and $\wedge$ the usual Boolean conjunction[1]. In this section we generalize the language and semantics of definite logic programs in order to encompass more complex bodies and heads and, evidently, multi-valued logics. For simplicity, we consider only the propositional (ground) case.

In general, a logic programming semantics requires a notion of consequence (implication) which satisfies a generalization of *Modus Ponens* to a multi-valued setting. The generalization of *Modus Ponens* to multi-valued logics is very well understood, namely in Fuzzy Propositional Logics [13,1,9]. Since one of our initial goals was to capture Fuzzy Logic Programming [6,15], it was natural to adopt as semantical basis the *residuated lattices* (see [5,1]). This section summarizes the results fully presented and proved in [2]. We first require some definitions.

Definition 1 (Adjoint pair). *Let* $< P, \preceq_P >$ *be a partially ordered set and* $(\leftarrow, \otimes)$ *a pair of binary operations in* P *such that:*

(a_1) *Operation* $\otimes$ *is isotonic, i.e. if* $x_1, x_2, y \in P$ *such that* $x_1 \preceq_P x_2$ *then* $(x_1 \otimes y) \preceq_P (x_2 \otimes y)$ *and* $(y \otimes x_1) \preceq_P (y \otimes x_2)$;

(a_2) *Operation* $\leftarrow$ *is isotonic in the first argument (the consequent) and antitonic in the second argument (the antecedent), i.e. if* $x_1, x_2, y \in P$ *such that* $x_1 \preceq_P x_2$ *then* $(x_1 \leftarrow y) \preceq_P (x_2 \leftarrow y)$ *and* $(y \leftarrow x_2) \preceq_P (y \leftarrow x_1)$;

(a_3) *For any* $x, y, z \in P$, *we have that* $x \preceq_P (y \leftarrow z)$ *holds if and only if* $(x \otimes z) \preceq_P y$ *holds.*

Then we say that $(\leftarrow, \otimes)$ *forms an adjoint pair in* $< P, \preceq_P >$.

---

[1] We remove the parentheses to simplify the reading of the rule.

The intuition of the two above properties is immediate, the third one may be more difficult to grasp. In one direction, it is simply asserting that the following *Fuzzy Modus Ponens* rule is valid (cf. [9]):

If x is a lower bound of $\psi \leftarrow \varphi$, and z is a lower bound of φ then a lower bound y of ψ is $x \otimes z$.

The other direction is ensuring that the truth-value of $y \leftarrow x$ is the maximal z satisfying $x \otimes z \preceq_P y$.

Besides (a_1)–(a_3) it is necessary to impose extra conditions on the multiplication operation $(\otimes)$, namely associativity, commutativity and existence of a unit element. It is also indispensable to assume the existence of a bottom element in the lattice of truth-values (the zero element). Formally:

Definition 2 (Residuated Lattice). *Consider the lattice $< L, \preceq_L >$. We say that $(L^{\preceq}, \leftarrow, \otimes)$ is a residuated lattice whenever the following three conditions are met:*

(l_1) $< L, \preceq_L >$ *is a bounded lattice, i.e. it has bottom $(\perp)$ and top $(\top)$ elements;*
(l_2) $(\leftarrow, \otimes)$ *is an adjoint pair in $< L, \preceq_L >$;*
(l_3) $(L, \otimes, \top)$ *is a commutative monoid.*

We say that the residuated lattice is complete whenever $< L, \preceq_L >$ is complete. In this case, condition (l_1) is immediately satisfied.

Our main semantical structure is a residuated algebra, an algebra where a multiplication operation is defined, the corresponding residuum operation (or implication), and a constant representing the top element of the lattice of truth-values (whose set is the carrier of the algebra). They must define a complete residuated lattice, since we intend to deal with infinite programs (theories). Obviously, a residuated algebra may have additional operators. Formally:

Definition 3 (Residuated Algebra). *Consider a algebra $\mathfrak{R}$ defining operators $\leftarrow, \otimes$ and $\top$ on carrier set $T_{\mathfrak{R}}$ such that $\preceq$ is a partial order on $T_{\mathfrak{R}}$. We say that $\mathfrak{R}$ is a residuated algebra with respect to $(\leftarrow, \otimes)$ if $(T_{\mathfrak{R}}^{\preceq}, \leftarrow, \otimes)$ is a complete residuated lattice. Furthermore, operator $\top$ is a constant mapped to the top element of $T_{\mathfrak{R}}$.*

Our Residuated Logic Programs will be constructed from the abstract syntax induced by a residuated algebra and a set of propositional symbols. The way of relating syntax and semantics in such algebraic setting is well-known and we refer to [8] for more details.

Definition 4 (Residuated Logic Programs). *Let $\mathfrak{R}$ be a residuated algebra with respect to $(\leftarrow, \otimes, \top)$. Let Π be a set of propositional symbols and the corresponding algebra of formulae $\mathfrak{F}$ freely generated from Π. A residuated logic program is a set of weighted rules of the form $\langle (A \leftarrow \Psi), \vartheta \rangle$ such that:*

1. *The rule $(A \leftarrow \Psi)$ is a formula of $\mathfrak{F}$;*
2. *The confidence factor ϑ is a truth-value of $\mathfrak{R}$ belonging to $T_{\mathfrak{R}}$;*

3. The head of the rule A is a propositional symbol of Π.

4. The body formula Ψ corresponds to an isotonic function with propositional symbols $B_1, \ldots, B_n$ $(n \geq 0)$ as arguments.

To simplify the notation, we represent the above pair as $A \xleftarrow{\vartheta} \Psi[B_1, \ldots, B_n]$, where $B_1, \ldots, B_n$ are the propositional variables occurring in Ψ. Facts are rules of the form $A \xleftarrow{\top} \top$.

A rule of a residuated logic program expresses a (monotonic) computation rule of the truth-value of the head propositional symbol from the truth-values of the symbols in the body. The monotonicity of the rule is guaranteed by isotonicity of formula Ψ: if an argument of Ψ is monotonically increased then the truth-value of Ψ also monotonically increases.

As usual, an interpretation is simply an assignment of truth-values to every propositional symbol in the language. To simplify the presentation we assume, throughout the rest of this section, that a residuated algebra $\mathfrak{R}$ is given with respect to $(\leftarrow, \otimes, \top)$.

Definition 5 (Interpretation). *An interpretation is a mapping $I : \Pi \to \mathcal{T}_{\mathfrak{R}}$. It is well known that an interpretation extends uniquely to a valuation function $\hat{I}$ from the set of formulas to the set of truth values. The set of all interpretations with respect to the residuated algebra $\mathfrak{R}$ is denoted by $\mathcal{I}_{\mathfrak{R}}$.*

The ordering $\preceq$ of the truth-values $\mathcal{T}_{\mathfrak{R}}$ is extended to the set of interpretations as usual:

Definition 6 (Lattice of interpretations). *Consider the set of all interpretations with respect to the residuated algebra $\mathfrak{R}$ and the two interpretations $I_1, I_2 \in \mathcal{I}_{\mathfrak{R}}$. Then, $< \mathcal{I}_{\mathfrak{R}}, \sqsubseteq >$ is a complete lattice where $I_1 \sqsubseteq I_2$ iff $\forall_{p \in \Pi} I_1(p) \preceq I_2(p)$. The least interpretation Δ maps every propositional symbol to the least element of $\mathcal{T}_{\mathfrak{R}}$.*

A rule of a residuated logic program is satisfied whenever the truth-value of the rule is greater or equal than the confidence factor associated with the rule. Formally:

Definition 7. *Consider an interpretation $I \in \mathcal{I}_{\mathfrak{R}}$. A weighted rule $\langle (A \leftarrow \Psi), \vartheta \rangle$ is satisfied by I iff $\hat{I}((A \leftarrow \Psi)) \succeq \vartheta$. An interpretation $I \in \mathcal{I}_{\mathfrak{R}}$ is a model of a residuated logic program P iff all weighted rules in P are satisfied by I.*

Mark that we used $\hat{I}$ instead of I in the evaluation of the truth-value of a rule, since a complex formula is being evaluated instead of a propositional symbol. If $\leftarrow_{\mathfrak{R}}$ is the function in $\mathfrak{R}$ defining the truth-table for the implication operator, the expression $\hat{I}((A \leftarrow \Psi))$ is equal to

$$\hat{I}(A) \leftarrow_{\mathfrak{R}} \hat{I}(\Psi) = I(A) \leftarrow_{\mathfrak{R}} \hat{I}(\Psi)$$

The evaluation of $\hat{I}(\Psi)$ proceeds inductively as usual, till all propositional symbols in Ψ are reached and evaluated in I.

The immediate consequences operator of van Emden and Kowalski [14] is extended to the very general theoretical setting of residuated logic programs as follows:

Definition 8. *Let P be a residuated logic program. The monotonic immediate consequences operator $T_P^\mathfrak{R} : \mathcal{I}_\mathfrak{R} \to \mathcal{I}_\mathfrak{R}$, mapping interpretations to interpretations, is defined by:*

$$T_P^\mathfrak{R}(I)(A) = lub \left\{ \vartheta \otimes \hat{I}(\Psi) \text{ such that } A \xleftarrow{\vartheta} \Psi[B_1, \dots, B_n] \in P \right\}$$

As remarked before, the monotonicity of the operator $T_P^\mathfrak{R}$ has been shown in [2]. The semantics of a residuated logic program is characterized by the post-fixpoints of $T_P^\mathfrak{R}$:

Theorem 1. *An interpretation I of $\mathcal{I}_\mathfrak{R}$ is a model of a residuated logic program P iff $T_P^\mathfrak{R}(I) \sqsubseteq I$. Moreover, the semantics of P is given by its least model which is exactly the least fixpoint of $T_P^\mathfrak{R}$. The least model of P and can be obtained by trasfinitely iterating $T_P^\mathfrak{R}$ from the least interpretation Δ.*

The major difference from classical logic programming is that our $T_P^\mathfrak{R}$ may not be continuous, and therefore more than ω iterations may be necessary to reach the least fixpoint. This is unavoidable if of one wants to keep generality. All the other important results carry over to our general framework.

3   Hybrid Probabilistic Logic Programs

In this section we provide an overview of the main definitions and results in [4,3]. We do not address any of the aspects of the proof theory present in these works. A major motivation for the Hybrid Probabilistic Logic Programs is the need for combining several probabilistic reasoning forms within a general framework. To capture this generality, the authors introduced the new notion of probabilistic strategies.

A first important remark is that the probabilites of compound events may be closed intervals in $[0, 1]$, and not simply real-valued probability assignments. The set of all closed intervals of $[0,1]$ is denoted by $\mathcal{C}[0, 1]$. Recall that the empty set $\varnothing$ is a closed interval. In $\mathcal{C}[0, 1]$ two partial-orders are defined. Let $[a, b] \in \mathcal{C}[0, 1]$ and $[c, d] \in \mathcal{C}[0, 1]$, then:

- $[a, b] \leq_t [c, d]$ if $a \leq c$ and $b \leq d$, meaning that $[c, d]$ is closer to 1 than $[a, b]$.
- $[a, b] \subseteq [c, d]$ if $c \leq a$ and $b \leq d$, meaning that $[a, b]$ is more precise than $[c, d]$.

The probabilistic strategies must obey the following natural properties:

Definition 9 (Probabilistic strategy). *A p-strategy is a pair of functions $\rho = < c, md >$ such that:*

1. $c : C[0,1] \times C[0,1] \rightarrow C[0,1]$ *is called a probabilistic composition function satisfying the following axioms:*
   Commutativity: $c([a_1, b_1], [a_2, b_2]) = c([a_2, b_2], [a_1, b_1])$
   Associativity: $c(c([a_1, b_1], [a_2, b_2]), [a_3, b_3]) = c([a_1, b_1], c([a_2, b_2], [a_3, b_3]))$
   Inclusion Monotonicity: *If* $[a_1, b_1] \subseteq [a_3, b_3]$ *then* $c([a_1, b_1], [a_2, b_2]) \subseteq c([a_3, b_3], [a_2, b_2])$
   Separation: *There exist two functions* $c^1, c^2 : [0,1] \times [0,1] \rightarrow [0,1]$ *such that* $c([a, b], [c, d]) = [c^1(a, c), c^2(b, d)]$.
2. $md : C[0,1] \rightarrow C[0,1]$ *is called a maximal interval function.*

The strategies are either conjuntive or disjunctive:

Definition 10. *A p-strategy* $< c, md >$ *is called a conjunctive (disjunctive) p-strategy if it satisfies the following axioms:*

	Conjunctive p-strategy	Disjunctive p-strategy
Bottomline	$c([a_1, b_1], [a_2, b_2]) \leq_t$ $[min(a_1, a_2), min(b_1, b_2)]$	$[max(a_1, a_2), max(b_1, b_2)]$ $\leq_t c([a_1, b_1], [a_2, b_2])$
Identity	$c([a, b], [1, 1]) = [a, b]$	$c([a, b], [0, 0]) = [a, b]$
Annihilator	$c([a, b], [0, 0]) = [0, 0]$	$c([a, b], [1, 1]) = [1, 1]$
Max. Interval	$md([a, b]) = [a, 1])$	$md([a, b]) = [0, b]$

The syntax of hybrid probabilistic logic programs (hp-programs) is built on a first-order language L generated from finitely many constants and predicate symbols. Thus, the Herbrand base B_L of L is finite. Without loss of generality, we restrict the syntax to a propositional language: variables are not admitted in atoms. This simplifies the embedding into residuated logic programs.

In a hp-program one can use arbitrary p-strategies. By definition, for each conjunctive p-strategy the existence of a corresponding disjunctive p-strategy is assumed, and vice-versa. Formally:

Definition 11. *Let* $CONJ$ *be a finite set of coherent conjunctive p-strategies and* $DISJ$ *be a finite set of coherent disjunctive p-strategies. Let* $\mathcal{L}$ *denote* $CONJ \cup DISJ$. *If* $\rho \in CONJ$ *then connective* $\wedge_\rho$ *is called a* ρ-*annotated conjunction. If* $\rho \in DISJ$ *then* $\vee_\rho$ *is called a* ρ-*annotated disjunction.*

The elementary syntactic elements of hp-programs are basic formulas:

Definition 12. *Let* ρ *be a conjunctive p-strategy,* ρ' *be a disjunctive p-strategy and* $A_1, \ldots, A_k$ *be atoms. Then* $A_1 \wedge_\rho A_2 \wedge_\rho \ldots \wedge_\rho A_k$ *and* $A_1 \vee_{\rho'} A_2 \vee_{\rho'} \ldots \vee_{\rho'} A_k$ *are hybrid basic formulas. Let* $bf_\rho(B_L)$ *denote the set of all ground hybrid basic formulas for a connective. The set of ground hybrid basic formulas is* $bf_{\mathcal{L}} = \cup_{\rho \in \mathcal{L}} bf_\rho(B_L)$.

Basic formulas are annotated with probability intervals. Here we differ from [4] where basic formulas can be additionally annotated with variables and functions.

Definition 13. *A hybrid probabilistic annotated basic formula is an expression of the form $B : \mu$ where B is a hybrid basic formula and $\mu \in C[0, 1]$.*

Finally, we can present the syntax of hybrid rules and hp-programs:

Definition 14. *A hybrid probabilistic program over the set $\mathcal{L}$ of p-strategies is a finite set of hp-clauses of the form $B_0 : \mu_0 \leftarrow B_1 : \mu_1 \wedge \ldots \wedge B_k : \mu_k$ where each $B_i : \mu_i$ is a hp-annotated basic formula over $\mathcal{L}$.*

Intuitively, an hp-clause means that "if the probability of B_1 falls in the interval μ_1 and ... and the probability of B_k falls within the interval μ_k, then the probability of B_0 lies in the interval μ_0". Mark that the conjunction symbol $\wedge$ in the antecedent of hp-clauses should be interpreted as logical conjunction and should not be confused with a conjunctive p-strategy.

The semantics of hp-programs is given by a fixpoint operator. Atomic functions are akin to our notion of interpretation and are functions $f : B_L \to C[0, 1]$. They may be extended to hybrid basic formulas. For this the notion of splitting a formula into two disjoint parts is necessary::

Definition 15. *Let $F = F_1 *_\rho \ldots *_\rho F_n$, $G = G_1 *_\rho \ldots *_\rho G_k$, $H = H_1 *_\rho \ldots *_\rho H_m$ where $* \in \{\wedge, \vee\}$. We write $G \oplus_\rho H = F$ iff*

1. $\{G_1, \ldots, G_k\} \cup \{H_1, \ldots, H_m\} = \{F_1, \ldots, F_n\}$,
2. $\{G_1, \ldots, G_k\} \cap \{H_1, \ldots, H_m\} = \varnothing$,
3. $k > 0$ and $m > 0$.

The extension to atomic formulas is as follows:

Definition 16. *A hybrid formula function is a function $h : bf_\mathcal{L}(B_L) \to C[0, 1]$ which satisfies the following properties:*

1. **Commutativity.** *If $F = G_1 \oplus_\rho G_2$ then $h(F) = h(G_1 *_\rho G_2)$.*
2. **Composition.** *If $F = G_1 \oplus_\rho G_2$ then $h(F) \subseteq c_\rho(h(G_1), h(G_2))$.*
3. **Decomposition.** *For any basic formula F, $h(F) \subseteq md_\rho(h(F *_\rho G)$ for all $\rho \in \mathcal{L}$ and $G \in bf_\mathcal{L}(B_L)$.*

Let h_1 and h_2 be two hybrid formula functions. We say that $h_1 \leq h_2$ iff $(\forall F \in bf_\mathcal{L}(B_L))$ $h_1(F) \supseteq h_2(F)$. In particular, this means that there is a minimum element of $\mathcal{HFF}$ mapping every hybrid basic formula to $[0, 1]$.

The immediate consequences operator for hp-programs resorts to the following auxiliary operator. Again, we consider the ground case only:

Definition 17. *Let P be a hp-program. Operator $S_P : \mathcal{HFF} \to \mathcal{HFF}$ is defined as follows, where F is a basic formula. $S_P(h)(F) = \cap M$ where $M = \{\mu | F : \mu \leftarrow F_1 : \mu_1 \wedge \ldots \wedge F_n : \mu_n$ is an instance of some hp-clause in P and $(\forall j \leq n)$ $h(F_j) \subseteq \mu_j\}$. Obviously, if $M = \varnothing$ then $S_P(h)(F) = [0, 1]$.*

Definition 18. *Let P be a hp-program. Operator $T_P : \mathcal{HFF} \to \mathcal{HFF}$ is defined as follows:*

1. *Let F be an atomic formula.*
   (a) *if $S_P(h)(F) = \varnothing$ then $T_P(h)(F) = \varnothing$.*
   (b) *if $S_P(h)(F) \neq \varnothing$ then let $M = \{< \mu, \rho > |(F \oplus_\rho G) : \mu \leftarrow F_1 : \mu_1 \wedge \ldots \wedge$*
   $F_n : \mu_n$ where $* \in \{\vee, \wedge\}$, $\rho \in \mathcal{L}$ and $(\forall j \leq n)\ h(F_j) \subseteq \mu_j\}$. We define

   $$T_P(h)(F) = (\cap \{md_\rho(\mu)| < \mu, \rho >\in M\}) \cap S_P(h)(F)$$

2. *If F is not atomic, then*

   $$T_P(h)(F) = S_P(h)(F) \cap (\cap \{c_\rho(T_P(h)(G), T_P(h)(H))\ |\ G \oplus H = F\}) \cap$$
   $$(\{md_\rho(\mu)| < \mu, \rho >\in M\})$$

   where $M = \{< \mu, \rho >|\ D_1 *_\rho \ldots *_\rho D_k : \mu \leftarrow E_1 : \mu_1 \wedge \ldots \wedge E_m :$
   μ_m such that $(\forall j \leq n)\ h(E_j) \subseteq \mu_j$ and $\exists_H F \oplus_\rho H = \{D_1, \ldots, D_k\}\}$

A full explanation and intuition of the above operators can be found in [4]. Mark that the interval intersection operator $\cap$ in operators S_P and T_P corresponds to the join operation in lattice $C[0, 1]$ ordered by containment relation $\supseteq$. For the continuation of our work it is enough to recall that the T_P operator is monotonic (on the containment relation) and that it has a least fixpoint. Furthermore, the least model of a hp-program is given by the least fixpoint of T_P. We will base our results in these properties of the T_P operator. We end this section with a small example from [4], adapted to the ground case.

Example 1. Assume that if the CEO of a company sells the stock, retires with the probability over 85% and we are ignorant about the relationship between the two events, then the probability that the stock of the company drops is 40-90%. However, if the CEO retires and sells the stock, but we know that the former entails the latter, then the probability that the stock of the company will drop is only 5-20%. This situation is formalized with the following two rules:

   price-drop:[0.4,0.9] $\leftarrow$ (ch-sells-stock $\wedge_{igc}$ ch-retires):[0.85,1]
   price-drop:[0.05,0.2] $\leftarrow$ (ch-sells-stock $\wedge_{pcc}$ ch-retires):[1,1]

Where $\wedge_{igc}$ is a conjunctive ignorance p-strategy with $c_{igc}([a_1, b_1], [a_2, b_2]) = [max(0, a_1 + a_2 - 1), min(b_1, b_2)]$, and $\wedge_{pcc}$ is the positive correlation conjunctive p-strategy such that $c_{pcc}([a_1, b_1], [a_2, b_2]) = [min(a_1, a_2), min(b_1, b_2)]$.

Now assume we have the two facts *ch-sells-stock:[1,1]* and *ch-retires:[0.9,1]*. In this case, we obtain in the model of P that the probability of *price-drop* is in [0.4,0.9] since the first rule will fire and the second won't. If instead of the above two facts we have *(ch-sells-stock $\wedge_{igc}$ ch-retires):[1,1]* then in the least fixpoint of T_P *price-drop* will be assigned $\varnothing$.

4   Embedding of Hybrid Probabilistic Logic Programs into Residuated Logic Programs

In this section we present the embedment result. This will require some effort. First, we need to define our underlying residuated lattice. We will not restrict ourselves to closed intervals of $[0, 1]$. We require additional truth-values:

Definition 19. *Let $\mathcal{INT}$ be the set of pairs formed from values in $[0,1]$. We represent a value $< a,b >\in \mathcal{INT}$ by $[a,b]$. We say that $[a_1,b_1] \leq [a_2,b_2]$ iff $a_1 \leq a_2$ and $b_2 \leq b_1$.*

A pair $[a,b]$ in $\mathcal{INT}$ (with $a \leq b$) represents a non-empty closed interval of $C[0,1]$. The intuition for the remaining "intervals" of the form $[c,d]$ with $c > d$ will be provided later on, but we can advance now that they represent a form of inconsistent probability intervals. They correspond to $\varnothing$ in $C[0,1]$. The relation $\leq$ on $\mathcal{INT}$ forms a partial order, and extends the containment relation of $C[0,1]$ to $\mathcal{INT}$. In particular, $[0,1]$ and $[1,0]$ are, respectively, the least and greatest elements of $\mathcal{INT}$. These remarks are justified by the following two results:

Proposition 1. *The set $\mathcal{INT}$ with the partial order forms a complete lattice with the following meet and join operators:*

$$[a_1,b_1] \sqcap [a_2,b_2] = [min(a_1,a_2), max(b_1,b_2)]$$
$$[a_1,b_1] \sqcup [a_2,b_2] = [max(a_1,a_2), min(b_1,b_2)]$$

In general, consider the family $\{[a_i,b_i]\}_{i\in I}$ then

$$\sqcap_{i\in I}[a_i,b_i] = [inf\{a_i \mid i \in I\}, sup\{b_i \mid i \in I\}]$$
$$\sqcup_{i\in I}[a_i,b_i] = [sup\{a_i \mid i \in I\}, inf\{b_i \mid i \in I\}]$$

Proposition 2. *Consider the mapping $\overline{}$ from $\mathcal{INT}$ to $C[0,1]$ such that $\overline{[a,b]} = [a,b]$ if $a \leq b$, otherwise it is $\varnothing$. Let $[a_1,b_1]$ and $[a_2,b_2]$ belong to $C[0,1]$. Then,*

$$[a_1,b_1] \cap [a_2,b_2] = \overline{[a_1,b_1] \sqcup [a_2,b_2]}$$

Example 2. Consider the intervals $[0.5,0.7]$ and $[0.6,0.9]$. Their intersection is $[0.6,0.7]$ which is identical to their join in lattice $\mathcal{INT}$. Now, the intervals $[0.5,0.7]$ and $[0.8,0.9]$ have empty intersection. However their join is $[0.8,0.7]$. This will mean that there is some inconsistency in the assignment of probability intervals. In fact, we know that there is a gap from $[0.7,0.8]$. Thus, $\overline{[0.8,0.7]}$ is $\varnothing$.

The interpretation is a little more complex when more than two intervals are involved in the join operation. The intersection of $[0.1,0.2]$, $[0.4,0.6]$ and $[0.7,0.9]$ is empty again. Their join is $[0.7,0.2]$, meaning that the leftmost interval ends at 0.2 while the rightmost begins at 0.7. Again, $\overline{[0.7,0.2]} = \varnothing$.

We have seen that the meet and join operations perform the union and intersection of "intervals" of $\mathcal{INT}$, respectively. Our objective is to construct a residuated lattice from $\mathcal{INT}$ and the meet operation, which will be the multiplication operation. The adjoint residuum operation (implication) is defined as follows:

Definition 20. *Let $[a_1, b_1]$ and $[a_2, b_2]$ belong to $\mathcal{INT}$. Then:*

$$[a_1, b_1] \hookleftarrow [a_2, b_2] = \begin{cases} [1, 0] & \text{if } a_2 \leq a_1 \text{ and } b_2 \geq b_1 \\ [1, b_1] & \text{if } a_2 \leq a_1 \text{ and } b_2 < b_1 \\ [a_1, 0] & \text{if } a_2 > a_1 \text{ and } b_2 \geq b_1 \\ [a_1, b_1] & \text{if } a_2 > a_1 \text{ and } b_2 < b_1 \end{cases}$$

The result of the residuum operation is not obvious but still intuitive. In fact, we are testing whether $[a_2, b_2]$ contains $[a_1, b_1]$ (i.e. if $[a_1, b_1] \geq [a_2, b_2]$) and how $[a_2, b_2]$ should be extended in order to satisfy the inclusion. If the first (second) component of $[a_1, b_1] \hookleftarrow [a_2, b_2]$ is 1 (respectively 0) we do not have to do anything to $[a_2, b_2]$. Otherwise, a_2 (resp. b_2) should be reduced (increased) to a_1 (b_1). Notice again that $[1, 0]$ is our top element in lattice $\mathcal{INT}$.

Theorem 2. *The operations $(\hookleftarrow, \sqcap)$ form an adjoint pair in the partially ordered set $< \mathcal{INT}, \leq >$.*

Clearly, the structure $< \mathcal{INT}, \hookleftarrow, \sqcap >$ is a complete residuated lattice, with top element $[1, 0]$. A corresponding residuated algebra is easily constructed. We proceed by presenting a result which will enable the embedding of hybrid probabilistic logic programs into residuated logic programs:

Theorem 3. *Consider the operator T'_P which is identical to T_P except for when its argument formula F is not atomic; then:*

$$T'_P(h)(F) = S_P(h)(F) \sqcap (\sqcap \{c_\rho(h(G), h(H)) \mid G \oplus H = F\}) \sqcap$$
$$(\{md_\rho(\mu) \mid < \mu, \rho > \in M\})$$

with M defined as before. Then h is a fixpoint of T_P iff h is a fixpoint of T'_P.

Proof: The only difference between the operators is that we have replaced $c_\rho(T_P(h)(G), T_P(h)(H))$ in T_P by $c_\rho(h(G), h(H))$ in T'_P. Clearly, if h is a fixpoint of T_P then it is also a fixpoint of T'_P, since $h = T_P(h)$ we can substitute h by $T_P(h)$ in the definition of T'_P getting T_P. For the other direction, we prove the result by induction on the number of atoms in F. If F is atomic then $T_P(h) = T'_P(h)$, by definition. Otherwise, F is not an atomic formula. Since h is a fixpoint of T'_P we have:

$$T'_P(h)(F) = S_P(h)(F) \sqcap (\sqcap \{c_\rho(h(G), h(H)) \mid G \oplus H = F\}) \sqcap$$
$$(\{md_\rho(\mu) \mid < \mu, \rho > \in M\})$$
$$= S_P(h)(F) \sqcap (\sqcap \{c_\rho(T'_P(h)(G), T'_P(h)(H)) \mid G \oplus H = F\}) \sqcap$$
$$(\{md_\rho(\mu) \mid < \mu, \rho > \in M\})$$

But clearly G and H have a smaller number of atoms. So, from the induction hypothesis we know that $T'_P(h)(G) = T_P(h)(G)$ and $T'_P(h)(H) = T_P(h)(H)$. Substituting these equalities into the above equation we get $T'_P(h)(F) = S_P(h)(F) \sqcap (\sqcap \{c_\rho(T_P(h)(G), T_P(h)(H)) \mid G \oplus H = F\}) \sqcap (\{md_\rho(\mu) \mid < \mu, \rho > \in M\})$ which is $T_P(h)(F)$. $\qquad \square$

Before we present the embedding, we need some auxiliary functions in $\mathcal{INT}$:

Definition 21. *The double bar function $\overline{\overline{\cdot}}$ from $\mathcal{INT}$ to $\mathcal{INT}$ and the functions $s_\mu : \mathcal{INT} \to \mathcal{INT}$ where μ in $\mathcal{INT}$ are defined as follows:*

$$\overline{\overline{[a,b]}} = \begin{cases} [1,0], & \text{if } a > b \\ [a,b], & \text{otherwise.} \end{cases} \qquad s_\mu(\vartheta) = \begin{cases} [1,0], & \text{if } \mu \le \vartheta \\ [0,1], & \text{otherwise.} \end{cases}$$

The above functions are clearly monotonic. Furthermore, the s_μ functions are "two-valued" and will be used to perform the comparisons in the rule bodies of a probabilistic logic program. Now, the embedding is immediate:

Definition 22. *Consider the hp-program P on the set of p-strategies $\mathcal{L}$. First, we construct the residual algebra $\mathfrak{I}$ from the carrier set $\mathcal{INT}$, and operations $\hookleftarrow$, $\sqcap$, $c_\rho(\rho \in \mathcal{L})$, $s_\mu(\mu \in \mathcal{INT})$, the double bar function, and the top constant $[1,0]$. Next, we build the residuated logic program P_{hp} from P as follows, where every ground hybrid basic formula in $bf_\mathcal{L}$ is viewed as a new propositional symbol[2] in the language of P_{hp}.*

1. *For each rule in P of the form $F : \mu \leftarrow F_1 : \mu_1 \wedge \ldots \wedge F_k : \mu_k$ we add to P_{hp} the rule[3] $F \overset{\mu}{\hookleftarrow} s_{\mu_1}\left(\overline{\overline{F_1}}\right) \sqcap \ldots \sqcap s_{\mu_k}\left(\overline{\overline{F_k}}\right).$*

2. *For every, F, G, and H in $bf_\mathcal{L}$ such that $H = F \oplus_\rho G$, and ρ is a conjunctive p-strategy, then for every rule $H : [a,b] \leftarrow E_1 : \mu_1 \wedge \ldots \wedge E_m : \mu_m$ in P we add to P_{hp} the rule $F \overset{[a,1]}{\hookleftarrow} s_{\mu_1}\left(\overline{\overline{E_1}}\right) \sqcap \ldots \sqcap s_{\mu_m}\left(\overline{\overline{E_m}}\right).$*

3. *For every, F, G, and H in $bf_\mathcal{L}$ such that $H = F \oplus_\rho G$, and ρ is a disjunctive p-strategy, then for every rule $H : [a,b] \leftarrow E_1 : \mu_1 \wedge \ldots \wedge E_m : \mu_m$ in P we add to P_{hp} the rule $F \overset{[0,b]}{\hookleftarrow} s_{\mu_1}\left(\overline{\overline{E_1}}\right) \sqcap \ldots \sqcap s_{\mu_m}\left(\overline{\overline{E_m}}\right).$*

4. *Finally, for every F, G, and H in $bf_\mathcal{L}$ such that $F = G \oplus_\rho H$ then include in P_{hp} the rule $F \overset{[1,0]}{\hookleftarrow} c_\rho\left(\overline{\overline{G}}, \overline{\overline{H}}\right).$*

Some remarks are necessary to fully clarify the above translation. First, the c_ρ functions were previously defined on domain $\mathcal{C}[0,1]$. It is required to extend them to $\mathcal{INT}$. For elements of $\mathcal{INT}$ isomorphic to elements of $\mathcal{C}[0,1]$ the functions should coincide. For values in $\mathcal{INT}$ not in $\mathcal{C}[0,1]$ the functions c_ρ can take arbitrary values, since in the embedding the arguments of these functions always take values from $\mathcal{C}[0,1]$.

Also, the above translation produces a residuated logic program. The rules belong to the algebra of formulae freely generated from the set of propositional

---

[2] Without loss of generality, we assume that the ocurrences of atoms in each hybrid basic formula are ordered according to some total order in the set of all atoms.

[3] We assign to the body of translated facts the top constant $[1,0]$.

symbols and operators in the corresponding residual algebra. Thus, when evaluating $F \hookleftarrow s_{\mu_1}\left(\overline{\overline{F_1}}\right) \sqcap \ldots \sqcap s_{\mu_k}\left(\overline{\overline{F_k}}\right)$ with respect to interpretation I we really mean $I(F) \hookleftarrow s_{\mu_1}\left(\overline{I(F_1)}\right) \sqcap \ldots \sqcap s_{\mu_k}\left(\overline{I(F_k)}\right)$, as usual. It should be clear that every body formula is isotonic on its arguments: for the first three types of rules the body is the composition of isotonic functions and therefore the resulting function is also isotonic. The probabilistic composition functions are isotonic by definition (check Definition 9).

The rules introduced in the fourth step are exponential in the number of atoms (width) in F. This is expected since it is known that the computation of the least fixpoint of an HPP is exponential in the width of the largest formula of interest, as shown in [3]. The complexity of the entailment and consistency problems for HPPs are more subtle and the reader is referred again to [3] for these profound results.

Theorem 4. *Let P be a hybrid probabilistic logic program and P_{hp} the corresponding residuated logic program over $\mathfrak{I}$. Let h be the least fixpoint of $T^{\mathfrak{I}}_{P_{hp}}$ and h' be the least fixpoint of T'_P. Then, for every F in $bf_{\mathcal{L}}$, we have $h'(F) = \overline{h(F)}$.*

Proof: We will prove that for every F in $bf_{\mathcal{L}}$ we have $T'_P \uparrow^\alpha (F) = \overline{T^{\mathfrak{I}}_{P_{hp}} \uparrow^\alpha (F)}$. To simplify notation we drop the subscripts in the operators. The proof is by transfinite induction on α:

$\alpha = 0$: Trivial since every hybrid basic formula is mapped to $[0,1]$ in both operators.

Sucessor ordinal $\alpha = \beta + 1$: Let $h' = T' \uparrow^\beta$ and $h = T^{\mathfrak{I}} \uparrow^\beta$. By induction hypothesis we know that for every F in $bf_{\mathcal{L}}$ we have $h'(F) = \overline{h(F)}$. The essential point is that $h'(F) \subseteq \mu$ iff $s_\mu\left(\overline{h(F)}\right) = [1,0]$. Therefore, we have the body of a rule in P satisfied by h' iff the body of the corresponding rule in P_{hp} evaluates to $[1,0]$. Otherwise, the body of the rule in P_{hp} has truth-value $[0,1]$.

Rules of the first kind in the embedding implement the S_P operator because

$$T^{\mathfrak{I}}(h)(F) = \bigsqcup \left\{ \mu \sqcap \hat{h}\left(s_{\mu_1}\left(\overline{\overline{F_1}}\right) \sqcap \ldots \sqcap s_{\mu_k}\left(\overline{\overline{F_k}}\right)\right)\right.$$
$$\left. \text{such that } F \overset{\mu}{\hookleftarrow} s_{\mu_1}\left(\overline{\overline{F_1}}\right) \sqcap \ldots \sqcap s_{\mu_k}\left(\overline{\overline{F_k}}\right) \in P_{hp}\right\}$$
$$= \bigsqcup \left\{ \mu \sqcap [1,0] \text{ such that } F \overset{\mu}{\hookleftarrow} s_{\mu_1}\left(\overline{\overline{F_1}}\right) \sqcap \ldots \sqcap s_{\mu_k}\left(\overline{\overline{F_k}}\right) \in P_{hp}\right.$$
$$\left. \text{and } \hat{h}\left(s_{\mu_1}\left(\overline{\overline{F_1}}\right) \sqcap \ldots \sqcap s_{\mu_k}\left(\overline{\overline{F_k}}\right)\right) = [1,0]\right\}$$
$$= \bigsqcup \{\mu \text{ where } F : \mu \leftarrow F_1 : \mu_1 \wedge \ldots \wedge F_k : \mu_k \text{ is satisfied by } h'\}$$

Rules of the second and third kind extract the maximal interval associated with F with respect to connective ρ. By definition, we know that the maximal interval $md_\rho([a,b])$ is $[a,1]$ for a conjunctive p-strategy ρ, or $[0,b]$ if ρ is a disjunctive p-strategy. Therefore the rules of the second and third kind implement $\bigsqcup \{md_\rho(\mu)| < \mu, \rho >\in M\}$ for both cases 1b) and 2 of Definition 18. Finally, the remaining rules compute $\bigsqcup \{c_\rho(h'(G), h'(H)) \mid G \oplus H = F\}$. The result immediately follows from Proposition 2.

Limit ordinal α other than 0: We have to show that

$$\bigcap_{\beta < \alpha} T' \uparrow^\beta (F) = \bigsqcup_{\beta < \alpha} T^{\mathfrak{I}} \uparrow^\beta (F)$$

Suppose for every $\beta < \alpha$ we have $T' \uparrow^\beta (F) \neq \varnothing$. This means that $T' \uparrow^\beta (F) = T^{\mathfrak{I}} \uparrow^\beta (F)$. The result then follows again from Proposition 2.
If for some β it is the case that $T' \uparrow^\beta (F) = \varnothing$ this means $T' \uparrow^\alpha (F) = \varnothing$. By induction hypothesis, $T^{\mathfrak{I}} \uparrow^\beta (F) = [a_\beta, b_\beta]$ with $a_\beta > b_\beta$. Let $T^{\mathfrak{I}} \uparrow^\alpha (F) = [a_\alpha, b_\alpha]$. We conclude $[a_\beta, b_\beta] \leq [a_\alpha, b_\alpha]$ by monotonicity of $T^{\mathfrak{I}}$, i.e. $a_\alpha \geq a_\beta$ and $b_\beta \geq b_\alpha$. Obviously, $a_\alpha > b_\alpha$ and the theorem holds.

$\square$

By Theorem 3 we conclude immediately that $\overline{lfpT^{\mathfrak{I}}}$ is the least fixpoint of Dekhtyar and Subrahmanian's T_P operator, and the embedding is proved. The convergence of the process is guaranteed both by the properties of the T_P operator and the fixpoint results for residuated logic programs. We now return to Example 1 to illustrate the embedding. For simplicity, we ignore the rules generated in the fourth step for annotated disjunctions since they will not be required.

Example 3. The first two rules will be encoded as follows:

$$\text{price-drop} \overset{[0.4,0.9]}{\longleftarrow} s_{[0.85,1]} \left(\overline{\text{ch-sells-stock} \wedge_{igc} \text{ch-retires}} \right)$$
$$\text{price-drop} \overset{[0.05,0.2]}{\longleftarrow} s_{[1,1]} \left(\overline{\text{ch-sells-stock} \wedge_{pcc} \text{ch-retires}} \right)$$

Additionally, the following two rules will be introduced by the fourth step in the transformation:

$$\text{ch-sells-stock} \wedge_{igc} \text{ch-retires} \overset{[1,0]}{\longleftarrow} c_{igc} \left(\overline{\text{ch-sells-stock}}, \overline{\text{ch-retires}} \right)$$
$$\text{ch-sells-stock} \wedge_{pcc} \text{ch-retires} \overset{[1,0]}{\longleftarrow} c_{pcc} \left(\overline{\text{ch-sells-stock}}, \overline{\text{ch-retires}} \right)$$

In the first situation, the two facts will be translated to

$$\text{ch-sells-stock} \overset{[1,1]}{\longleftarrow} [1,0]$$
$$\text{ch-retires} \overset{[0.9,1]}{\longleftarrow} [1,0]$$

In the least fixpoint of $T^{\mathfrak{I}}$ the literals *ch-sells-stock* and *ch-retires* have truth-value $[1,1]$ and $[0.9,1]$, respectively. From this we obtain for the literals representing hybrid basic formulas *ch-sells-stock* $\wedge_{igc}$ *ch-retires* and *ch-sells-stock* $\wedge_{pcc}$ *ch-retires* the same truth-value of $[0.9,1]$. Finally, we obtain the interval $[0.4, 0.9]$ by application of the first rule.

The fact (ch-sells-stock $\wedge_{igc}$ ch-retires):[1,1] will be encoded instead as follows, by application of the first and second rules:

$$\text{ch-sells-stock } \wedge_{igc} \text{ ch-retires} \overset{[1,1]}{\hookleftarrow} [1,0]$$
$$\text{ch-sells-stock} \overset{[1,1]}{\hookleftarrow} [1,0]$$
$$\text{ch-retires} \overset{[1,1]}{\hookleftarrow} [1,0]$$

From the above facts we conclude that *ch-sells-stock* $\wedge_{pcc}$ *ch-retires* gets truth-value [1,1], and by application of the rules for *price-drop* we obtain for this literal the assignment [0.4, 0.2], and as expected $\overline{[0.4, 0.2]} = \varnothing$.

We conclude by remarking that the substitution of $F : \mu$ by $s_\mu \left(\overline{\overline{F}} \right)$ instead of by $s_\mu (F)$ in the transformed program is of the essence. Otherwise, we could get different semantics when some literal is mapped to $\varnothing$. However, it is not clear what is the better semantics in that case, and further work is necessary. We illustrate the distinction in the next example:

Example 4. Consider the hp-program:

$$a : [0.5, 0.7] \leftarrow \qquad a : [0.8, 0.9] \leftarrow \qquad b : [1,1] \leftarrow a : [0.9, 0.95]$$

According to the transformation of Definition 22 we have:

$$a \overset{[0.5,0.7]}{\hookleftarrow} [1,0] \qquad a \overset{[0.8,0.9]}{\hookleftarrow} [1,0] \qquad b \overset{[1,1]}{\hookleftarrow} s_{[0.9,0.95]} \left(\overline{\overline{a}} \right)$$

In the model of the program a is mapped to [0.8, 0.7] and b to [1, 1]. Now, if we translate the rule for b as $b \overset{[1,1]}{\hookleftarrow} s_{[0.9,0.95]}(a)$, literal a is still mapped to [0.8, 0.7]. However, the body of the rule for b has truth-value [0, 1], and b also has this value, since $[1, 1] \sqcap [0, 1] = [0, 1]$.

5 Conclusions and Further Work

The major contribution of this paper is the generality of our setting, both at the language and the semantic level. We presented an algebraic characterization of Residuated Logic Programs. Program rules have arbitrary monotonic body functions and our semantical structures are residuated lattices, where a generalized form of *Modus Ponens Rule* is valid. After having defined an implication (or residuum operator) and the associated multiplication (t-norm in the fuzzy logic setting) we obtain a logic programming semantics with corresponding model and fixpoint theory.

The embedding of hybrid probabilistic logic programs into residuated logic programs relies on a generalization of the complete lattice of closed intervals in [0, 1]. The extra truth-values capture invalid probability interval assignments, not used in [4]. The program transformation capturing the hp-semantics is a direct translation of the fixpoint conditions on a logic program. This aspect illustrates the generality and potential of our approach. Besides hp-programs we have shown that Generalized Annotated Logic Programs, Fuzzy Logic Programming, and Possibilistic Logic Programming are all captured by Residuated Logic Programs. These results could not be included for lack of space.

Our work paves the way to combine and integrate several forms of reasoning into a single framework, namely fuzzy, probabilistic, uncertain, and paraconsistent. We have also defined another class of logic programs, extending the Residuated one, where rule bodies can be anti-monotonic functions, with Well-Founded and Stable Model like semantics. This brings together non-monotonic and incomplete forms of reasoning to those listed before. It will be the subject of a forthcoming paper.

Acknowledgements

Work partially supported by PRAXIS project MENTAL. We thank V. S. Subrahmanian, J. Dix, T. Eiter, J. Alferes and L. Caires for helpful discussions. We also thank the anonoymous referees for their valuable comments.

References

1. L. Bolc and P. Borowik. *Many-Valued Logics. Theoretical Foundations.* Springer–Verlag, 1992.
2. C. V. Dam·sio and L. M. Pereira. Residuated logic programs. Technical report, Universidade Nova de Lisboa, 2000. Available at http://centria.di.fct.unl.pt/~cd.
3. A. Dekhtyar, M. Dekhtyar, and V. S. Subrahmanian. Hybrid probabilistic programs: Algorithms and complexity. In *Proc. of 1999 Conference on Uncertainty in AI*, 1999.
4. A. Dekhtyar and V. S. Subrahmanian. Hybrid probabilistic programs. *J. of Logic Programming*, 43:187–250, 2000.
5. R. P. Dilworth and M. Ward. Residuated lattices. *Trans. A.M.S.*, 45:335–354, 1939.
6. D. Dubois, J. Lang, and H. Prade. Fuzzy sets in approximate reasoning, part 2: Logical approaches. *Fuzzy Sets and Systems*, 40:203–244, 1991.
7. D. Dubois, J. Lang, and H. Prade. Towards possibilistic logic programming. In *International Conference on Logic Programming 1991*, pages 581–598. MIT Press, 1991.
8. J. H. Gallier. *Logic for Computer Science.* John Wiley & Sons, 1987.
9. P. H·jek. *Metamathematics of Fuzzy Logic.* Trends in Logic. Studia Logica Library. Kluwer Academic Publishers, 1998.
10. M. Kifer and V. S. Subrahmanian. Theory of generalized annotated logic programming and its applications. *J. of Logic Programming*, 12:335–367, 1992.
11. R. Kowalski. Predicate logic as a programming language. In *Proceedings of IFIP'74*, pages 569–574. North Holland Publishing Company, 1974.
12. J. W. Lloyd. *Foundations of Logic Programming.* Springer–Verlag, 1987. 2nd. edition.

13. J. Pavelka. On fuzzy logic I, II, III. *Zeitschr. f. Math. Logik und Grundl. der Math.*, 25, 1979.
14. M. van Emden and R. Kowalski. The semantics of predicate logic as a programming language. *Journal of ACM*, 4(23):733–742, 1976.
15. P. Vojt·s and L. Paulìk. Soundness and completeness of non-classical extended SLD-resolution. In *Proc. of the Ws. on Extensions of Logic Programming (ELP'96)*, pages 289–301. LNCS 1050, Springer–Verlag., 1996.

Topo-distance: Measuring the Difference between Spatial Patterns

Marco Aiello

Institute for Logic, Language and Information, and
Intelligent Sensory and Information Systems,
University of Amsterdam, the Netherlands
aiellom@wins.uva.nl

Abstract. A framework to deal with spatial patterns at the qualitative level of mereotopology is proposed. The main contribution is to provide formal tools for issues of model equivalence and model similarity. The framework uses a multi-modal language S4$_u$ interpreted on topological spaces (rather than Kripke semantics) to describe the spatial patterns. Model theoretic notions such as topological bisimulations and topological model comparison games are introduced to define a distance on the space of all topological models for the language S4$_u$. In the process, a new take on mereotopology is given, prompting for a comparison with prominent systems, such as RCC.

Keywords: qualitative spatial reasoning, RCC, mereotopology, model comparison games

1 Introduction

There are various ways to take space qualitatively. Topology, orientation or distance have been investigated in a non-quantitative manner. The literature especially is abundant in mereotopological theories, i.e. theories of parthood P and connection C. Even though the two primitives can be axiomatized independently, the definition of part in terms of connection suffices for AI applications. Usually, some fragment of topology is axiomatized and set inclusion is used to interpret parthood (see the first four chapters of [9] for a complete overview).

Most of the efforts in mereotopology have gone into the axiomatization of the specific theories, disregarding important model theoretic questions. Issues such as model equivalence are seldom (if ever) addressed. Seeing an old friend from high-school yields an immediate comparison with the image one had from the school days. Most often, one immediately notices how many aesthetic features have changed. Recognizing a place as one already visited involves comparing the present sensory input against memories of the past sensory inputs. "Are these trees the same as I saw six hours ago, or are they arranged differently?" An image retrieval system seldom yields an exact match, more often it yields a series of 'close' matches. In computer vision, object occlusion cannot be disregarded. One 'sees' a number of features of an object and compares them with other sets of

M. Ojeda-Aciego et al. (Eds.): JELIA 2000, LNAI 1919, pp. 73–86, 2000.
© Springer-Verlag Berlin Heidelberg 2000

features to perform object recognition. Vision is not a matter of precise matching, it is more closely related to similarity. The core of the problem lies in the precise definition of 'close' match, thus the question shall be: *How similar are two spatial patterns?*

In this paper, a general framework for mereotopology is presented, providing a language that subsumes many of the previously proposed ones, and then model theoretic questions are addressed. Not only a notion of model equivalence is provided, but also a precise definition of distance between models.

2   A General Framework for Mereotopology

2.1   The Language S4$_u$

The proposed framework takes the beaten road of mereotopology by extending topology with a mereological theory based on the interpretation of set inclusion as parthood. Hence, a brief recall here of the basic topological definitions is in order.

A *topological space* is a couple $\langle X, O \rangle$, where X is a set and $O \subseteq \mathcal{P}(X)$ such that: $\emptyset \in O$, $X \in O$, O is closed under arbitrary union, O is closed under finite intersection. An element of O is called an *open*. A subset A of X is called *closed* if $X - A$ is open. The *interior* of a set $A \subseteq X$ is the union of all open sets contained in A. The *closure* of a set $A \subseteq X$ is the intersection of all closed sets containing A.

To capture a considerable fragment of topological notions a multi-modal language S4$_u$ interpreted on topological spaces (à la Tarski [17]) is used. A *topological model* $M = \langle X, O, \nu \rangle$ is a topological space $\langle X, O \rangle$ equipped with a valuation function $\nu : P \to \mathcal{P}(X)$, where P is the set of proposition letters of the language.

The definition and interpretation of S4$_u$ follows that given in [2]. In that paper though, emphasis is given to the topological expressivity of the language rather than the mereotopological implications. Every formula of S4$_u$ represents a region. Two modalities are available. $\Box\varphi$ to be interpreted as "interior of the region φ", and $U\varphi$ to be interpreted as "it is the case everywhere that φ." The truth definition can now be given. Consider a topological model $M = \langle X, O, \nu \rangle$ and a point $x \in X$:

$$
\begin{array}{lll}
M, x \models p & \text{iff} & x \in \nu(p) \, (\text{with } p \in P) \\
M, x \models \neg\varphi & \text{iff} & \text{not } M, x \models \varphi \\
M, x \models \varphi \to \psi & \text{iff} & \text{not } M, x \models \varphi \text{ or } M, x \models \psi \\
M, x \models \Box\varphi & \text{iff} & \exists o \in O : x \in o \land \\
& & \forall y \in o : M, y \models \varphi \\
M, x \models U\varphi & \text{iff} & \forall y \in X : M, y \models \varphi
\end{array}
$$

Since $\square$ is interpreted as interior and $\diamond$ (defined dually as $\diamond\varphi \leftrightarrow \neg\square\neg\varphi$, for all φ) as closure, it is not a surprise that these modalities obey the following axioms[1], [17]:

$$\square A \to A \qquad\qquad\qquad\text{(T)}$$
$$\square A \to \square\square A \qquad\qquad\qquad\text{(4)}$$
$$\square\top \qquad\qquad\qquad\text{(N)}$$
$$\square A \wedge \square B \leftrightarrow \square(A \wedge B) \qquad\qquad\qquad\text{(R)}$$

(4) is idempotence, while (N) and (R) are immediately identifiable in the definition of topological space. For the universal—existential modalities U and E (defined dually: $E\varphi \leftrightarrow \neg U\neg\varphi$) the axioms are those of S5:

$$U(\varphi \to \psi) \to (U\varphi \to U\psi) \qquad\qquad\qquad\text{(K)}$$
$$U\varphi \to \varphi \qquad\qquad\qquad\text{(T)}$$
$$U\varphi \to UU\varphi \qquad\qquad\qquad\text{(4)}$$
$$\varphi \to UE\varphi \qquad\qquad\qquad\text{(B)}$$

In addition, the following 'connecting' principle is part of the axioms:

$$\diamond\varphi \to E\varphi$$

The language $S4_u$ is thus a multi-modal S4*S5 logic interpreted on topological spaces. Extending S4 with universal and existential operators to get rid of its intrinsic 'locality' is a known technique used in modal logic, [12]. In the spatial context, similar settings have been used initially in [7] to encode decidable fragments of the region connection calculus RCC (the fundamental and most widely used qualitative spatial reasoning calculi in the field of AI, [14]), then by [15] to identify maximal tractable fragments of RCC and, recently, by [16]. Even though the logical technique is similar to that of [7,15], there are two important differences. First, in the proposed use of $S4_u$ there is no commitment to a specific definition of connection (as RCC does by forcing the intersection of two regions to be non-empty). Second, the stress is on model equivalence and model comparison issues, not only spatial representation. On the other hand, there is no treatment here of consistency checking problems, leaving them for future investigation.

2.2 Expressivity

The language $S4_u$ is perfectly suited to express mereotopological concepts. Parthood P: a region A is part of another region B if it is the case everywhere that A implies B:

$$P(A, B) := U(A \to B)$$

---

[1] The axiomatization of $\square$ given is known as S4. Usually thought S4's axiomatization is given replacing axioms (N) and (R) by (K), see [7].

This captures exactly the set-inclusion relation of the models. As for connection C, two regions A and B are connected if there exists a point where both A and B are true:

$$C(A, B) := E(A \wedge B)$$

From here it is immediate to define all the usual mereotopological predicates such as proper part, tangential part, overlap, external connection, and so on. Notice that the choice made in defining P and C is arbitrary. So, why not take a more restrictive definition of parthood? Say, A is part of B whenever the closure of A is contained in the interior of B?

$$P(A, B) := U(\Diamond A \rightarrow \Box B)$$

As this formula shows, $S4_u$ is expressive enough to capture also this definition of parthood. In [10], the logical space of mereotopological theories is systematized. Based on the intended interpretation of the connection predicate C, and the consequent interpretation of P (and fusion operation), a type is assigned to mereotopological theories. More precisely, a *type* is a triple $\tau = \langle i, j, k \rangle$, where the first i refers to the adopted definition of C_i, j to that of P_j and k to the sort of fusion. The index i, referring to the connection predicate C, accounts for the different definition of connection at the topological level. Using $S4_u$ one can repeat here the three types of connection:

$$C_1(A, B) := E(A \wedge B)$$
$$C_2(A, B) := E(A \wedge \Diamond B) \vee E(\Diamond A \wedge B)$$
$$C_3(A, B) := E(\Diamond A \wedge \Diamond B)$$

Looking at previous mereotopological literature, one remarks that RCC uses a C_3 definition, while the system proposed in [4] uses a C_1. Similarly to connectedness, one can distinguish the various types of parthood, again in terms of $S4_u$:

$$P_1(A, B) := U(A \rightarrow B)$$
$$P_2(A, B) := U(A \rightarrow \Diamond B)$$
$$P_3(A, B) := U(\Diamond A \rightarrow \Diamond B)$$

In [10], the definitions of the C_i are given directly in terms of topology, and the definitions of P_j in terms of a first order language with the addition of a predicate C_i. Finally, a general fusion ϕ_k is defined in terms of a first order language with a C_i predicate. Fusion operations are like algebraic operations on regions, such as adding two regions (product), or subtracting two regions. One cannot repeat the general definition given in [10] at the $S4_u$ level. Though, one can show that various instances of fusion operations are expressible in $S4_u$. For example, the product $A \times_k B$:

$$A \times_1 B := A \wedge B$$
$$A \times_2 B := (\Diamond A \wedge B) \vee (A \wedge \Diamond B)$$
$$A \times_3 B := (\Diamond A \wedge \Diamond B)$$

The above discussion has shown that $S4_u$ is a general language for mereotopology. All the different types $\tau = \langle i, j, k \rangle$ of mereotopological theories are expressible within $S4_u$.

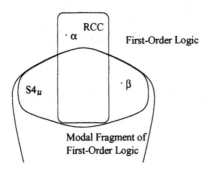

Fig. 1. The positioning of $S4_u$ and RCC with respect to well-known logics.

Before diving into the similarity results of this paper a remark is in order. The language $S4_u$ is a multi-modal language with nice computational properties. It is complete with respect to topological models, it is decidable, it has the finite model property (see [3] for the proofs of these facts). It captures a large and "well-behaved" fragment of mereotopology, though it is not a first-order language. In other words, it is not possible to quantify over regions. A comparison with the best-known RCC is in order.

Comparison with RCC RCC is a first order language with a distinguished connection predicate C_3. The driving idea behind this qualitative theory of space is that regions of space are primitive objects and connection is the basic predicate. This reflects in the main difference between RCC and the proposed system, which instead builds on traditional point-based topology.

RCC and S4u capture different portions of mereotopology.

To show this, two formulas are given: an RCC formula which is not expressible in $S4_u$ and, vice-versa, one expressible in $S4_u$, but not in RCC. The situation is depicted in Figure 1. In RCC, one can write:

$$\forall A \exists B : P(A, B) \tag{α}$$

meaning that every region is part of another one (think of the entire space). On the other hand, one can write a $S4_u$ formula such as:

$$\neg E(p \wedge \Diamond \Box \neg p) \tag{β}$$

which expresses the regularity of the region p. It is easy to see that α is not expressible in $S4_u$ and that β is not in RCC.

This fact may though be misleading. It is not the motivations, nor the core philosophical intuitions that draw the line between RCC and $S4_u$. Rather, it is the logical apparatus which makes the difference. To boost the similarities, next it is shown how the main predicates of RCC can be expressed within $S4_u$. Consider the case of RCC8:

RCC8	$S4_u$	Interpretation
DC(A, B)	$\neg E(A \wedge B)$	A is DisConnected from B
EC(A, B)	$E(\Diamond A \wedge \Diamond B) \wedge$ $\neg E(\Box A \wedge \Box B)$	A and B are Externally Connected
PO(A, B)	$E(A \wedge B) \wedge E(A \wedge \neg B) \wedge$ $E(\neg A \wedge B)$	A and B Properly Overlap
TPP(A, B)	$U(A \rightarrow B) \wedge$ $E(\Diamond A \wedge \Diamond B \wedge \Diamond \neg A \wedge \Diamond \neg B)$	A is a Tangential Proper Part of B
NTPP(A, B)	$U(\Diamond A \rightarrow \Box B)$	A is a Non Tangential Proper Part of B
TPPi(A, B)	$U(B \rightarrow A) \wedge$ $E(\Diamond B \wedge \Diamond A \wedge \Diamond \neg B \wedge \Diamond \neg A)$	The inverse of the TTP predicate
NTPPi(A, B)	$U(\Diamond B \rightarrow \Box A)$	The inverse of the NTTP predicate
EQ(A, B)	$U(A \leftrightarrow B)$	A and B are EQual

Indeed one can define the same predicates as RCC8, but as remarked before the nature of the approach is quite different. Take for instance the non tangential part predicate. In RCC it is defined by means of the non existence of a third entity C:

$$\text{NTTP}(A, B) \text{ iff } P(A, B) \wedge \neg P(B, A) \wedge \neg \exists C[\text{EC}(C, A) \wedge \text{EC}(C, B)]$$

On the other hand, in $S4_u$ it is simply a matter of topological operations. As in the previous table, for NTTP(A, B) it is sufficient to take the interior of the containing region $\Box B$, the closure of the contained region $\Diamond A$ and check if all points that satisfy the latter $\Diamond A$ also satisfy the former $\Box B$.

The RCC and $S4_u$ are even more similar if one takes the perspective of looking at RCC's modal decidable encoding of Bennett, [7]. Bennett's approach is to start from Tarski's original interpretation of modal logic in terms of topological spaces (Tarski proves S4 to be the complete logic of all topological spaces) and then to increase the expressive power of the language by means of a universal modality. The positive side effect is that the languages obtained in this manner usually maintain nice computational properties. The road to $S4_u$ has followed the same path and was inspired by Bennett's original work.

Here is the most important difference of the two approaches: the motivation for the work of Bennett comes from RCC, the one for the proposed framework from topology. $S4_u$ keeps a general topological view on spatial reasoning, it gives means to express more of the topological intricacy of the regions in comparison with RCC. For example regularity is not enforced by axioms (like in RCC), but it is expressible directly by a $S4_u$ formula (β). More on the 'topological expressive power' of S4 and its universal extension can be found in [2].

3   When Are Two Spatial Patterns the Same?

One is now ready to address questions such as: *When are two spatial patterns the same?* or *When is a pattern a sub-pattern of another one?* More formally, one wants to define a notion of equivalence adequate for $S4_u$ and the topological models. In first-order logic the notion of 'partial isomorphism' is the building block of model equivalence. Since $S4_u$ is multi-modal language, one resorts to bisimulation, which is the modal analogue of partial isomorphism. Bisimulations compare models in a structured sense, 'just enough' to ensure the truth of the same modal formulas [8,13].

Definition 1 (Topological bisimulation).   Given two topological models $\langle X, O, \nu \rangle$, $\langle X', O', \nu' \rangle$, a *total topological bisimulation* is a non-empty relation $\leftrightarrows$ $\subseteq X \times X'$ defined for all $x \in X$ and for all $x' \in X'$ such that if $x \leftrightarrows x'$:

(base): $x \in \nu(p)$ iff $x' \in \nu'(p)$ (for any proposition letter p)

(forth condition):   if $x \in o \in O$ then
                    $\exists o' \in O' : x' \in o'$ and $\forall y' \in o' : \exists y \in o : y \leftrightarrows y'$

(back condition):   if $x' \in o' \in O'$ then
                    $\exists o \in O : x \in o$ and $\forall y \in o : \exists y' \in o' : y \leftrightarrows y'$

If only conditions (i) and (ii) hold, the second model *simulates* the first one.

The notion of bisimulation is used to answer questions of 'sameness' of models, while simulation will serve the purpose of identifying sub-patterns. Though, one must show that the above definition is adequate with respect to the mereotopological framework provided in this paper.

Theorem 1. *Let $M = \langle X, O, \nu \rangle$, $M' = \langle X', O', \nu' \rangle$ be two models, $x \in X$, and $x' \in X'$ bisimilar points. Then, for any modal formula φ in $S4_u$, $M, x \models \varphi$ iff $M', x' \models \varphi$.*

Theorem 2. *Let $M = \langle X, O, \nu \rangle$, $M' = \langle X', O', \nu' \rangle$ be two models with finite O, O', $x \in X$, and $x' \in X'$ such that for every φ in $S4_u$, $M, x \models \varphi$ iff $M', x' \models \varphi$. Then there exists a total bisimulation between M and M' connecting x and x'.*

In words, extended modal formulas are invariant under total bisimulations, while finite modally equivalent models are totally bisimilar. The proofs are straightforward extensions of those of Theorem 1 and Theorem 2 in [2], respectively. In the case of Theorem 1, the inductive step must be extended also to consider the universal and existential modalities; while for Theorem 2, one needs to add an universal quantification over all points of the two equivalent models. One may notice, that in Theorem 2 a finiteness restriction is posed on the open sets. This will not surprise the modal logician, since the same kind of restriction holds for Kripke semantics and does not affect the proposed use for bisimulations in the mereotopological framework.

4   How Different Are Two Spatial Patterns?

If topological bisimulation is satisfactory from the formal point of view, one needs more to address qualitative spatial reasoning problems and computer vision issues. If two models are not bisimilar, or one does not simulate the other, one must be able to quantify the difference between the two models. Furthermore, this difference should behave in a coherent manner across the class of all models. Informally, one needs to answer questions like: *How different are two spatial patterns?*

To this end, the game theoretic definition of topo-games as in [2] is recalled, and the prove of the main result of this paper follows, namely the fact that topo-games induce a distance on the space of all topological models for S4$_u$. First, the definition and the theorem that ties together the topo-games, S4$_u$ and topological models is given.

Definition 2 (Topo-game). Consider two topological models $\langle X, O, \nu \rangle$, $\langle X', O', \nu' \rangle$ and a natural number n. A *topo-game* of length n, notation $TG(X, X', n)$, consists of n rounds between two players, Spoiler and Duplicator, who move alternatively. Spoiler is granted the first move and always the choice of which type of round to engage, either global or local. The two sorts of rounds are defined as follows:

- **global**
    (i) Spoiler chooses a model X_s and picks a point $\bar{x}_s$ anywhere in X_s
    (ii) Duplicator chooses a point $\bar{x}_d$ anywhere in the other model X_d

- **local**
    (i) Spoiler chooses a model X_s and an open o_s containing the current point x_s of that model
    (ii) Duplicator chooses an open o_d in the other model X_d containing the current point x_d of that model
    (iii) Spoiler picks a point $\bar{x}_d$ in Duplicator's open o_d in the X_d model
    (iv) Duplicator replies by picking a point $\bar{x}_s$ in Spoiler's open o_s in X_s

The points $\bar{x}_s$ and $\bar{x}_d$ become the new current points. A game always starts by a global round. By this succession of actions, two sequences are built. The form after n rounds is:

$$\{x_1, x_2, x_3, \ldots, x_n\}$$

$$\{x'_1, x'_2, x'_3, \ldots, x'_n\}$$

After n rounds, if x_i and x'_i (with $i \in [1, n]$) satisfy the same propositional atoms, Duplicator *wins*, otherwise, Spoiler wins. A *winning strategy (w.s.)* for Duplicator is a function from any sequence of moves by Spoiler to appropriate responses which always end in a win for him. Spoiler's winning strategies are defined dually.

The *multi-modal rank* of a $S4_u$ formula is the maximum number of nested modal operators appearing in it (i.e. $\Box$, $\Diamond$, U and E modalities). The following adequacy of the games with respect to the mereotopological language holds.

Theorem 3 (Adequacy). Duplicator has a winning strategy for n rounds in $TG(X, X', n)$ iff X and X' satisfy the same formulas of multi-modal rank at most n.

The reader is referred to [2] for a proof, various examples of plays and a discussion of winning strategies.

The interesting result is that of having a game theoretic tool to compare topological models. Given any two models, they can be played upon. If Spoiler has a winning strategy in a certain number of rounds, then the two models are different up to a certain degree. The degree is exactly the minimal number of rounds needed by Spoiler to win. On the other hand, one knows (see [2]) that if Spoiler has no w.s. in any number of rounds, and therefore Duplicator has in all games, including the infinite round game, then the two models are bisimilar.

A way of comparing any two given models is not of great use by itself. It is essential instead to have some kind of measure. It turns out that topo-games can be used to define a distance measure.

Definition 3 (isosceles topo-distance). Consider the space of all topological models T. *Spoiler's shortest possible win* is the function $spw : T \times T \to \mathbb{N} \cup \{\infty\}$, defined as:

$$spw(X_1, X_2) = \begin{cases} n & \text{if Spoiler has a winning} \\ & \text{strategy in } TG(X_1, X_2, n), \\ & \text{but not in } TG(X_1, X_2, n-1) \\ \\ \infty & \text{if Spoiler does not have a} \\ & \text{winning strategy in} \\ & TG(X_1, X_2, \infty) \end{cases}$$

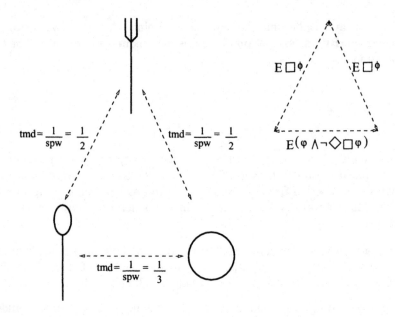

Fig. 2. On the left, three models and their relative distance. On the right, the distinguishing formulas.

The *isosceles topo-model distance (topo-distance,* for short*)* between X_1 and X_2 is the function $tmd : T \times T \to [0,1]$ defined as:

$$tmd(X_1, X_2) = \frac{1}{spw(X_1, X_2)}$$

The distance was named 'isosceles' since it satisfies the triangular property in a peculiar manner. Given three models, two of the distances among them (two sides of the triangle) are always the same and the remaining distance (the other side of the triangle) is smaller or equal. On the left of Figure 2, three models are displayed: a spoon, a fork and a plate. Think these cutlery objects as subsets of a dense space, such as the real plane, which evaluate to ϕ, while the background of the items evaluates to $\neg\phi$. The isosceles topo-distance is displayed on the left next to the arrow connecting two models. For instance, the distance between the fork and the spoon is $\frac{1}{2}$ since the minimum number of rounds that Spoiler needs to win the game is 2. To see this, consider the formula $E\Box\phi$, which is true on the spoon (there exists an interior point of the region ϕ associated with the spoon) but not on the fork (which has no interior points). On the right of the figure, the formulas used by spoiler to win the three games between the fork, the spoon and the plate are shown. Next the proof that *tmd* is really a distance, in particular the triangular property, exemplified in Figure 2, is always satisfied by any three topological models.

Theorem 4 (isosceles topo-model distance). *tmd* is a distance measure on the space of all topological models.

Proof. tmd satisfies the three properties of distances; i.e., for all X_1, $X_2 \in T$:

   (i) $tmd(X_1, X_2) \geq 0$ and $tmd(X_1, X_2) = 0$ iff $X_1 = X_2$
  (ii) $tmd(X_1, X_2) = tmd(X_2, X_1)$
 (iii) $tmd(X_1, X_2) + tmd(X_2, X_3) \geq tmd(X_1, X_3)$

As for (i), from the definition of topo-games it follows that the amount of rounds that can be played is a positive quantity. Furthermore, the interpretation of $X_1 = X_2$ is that the spaces X_1, X_2 satisfy the same modal formulas. If Spoiler does not have a w.s. in $\lim_{n \to \infty} TG(X_1, X_2, n)$ then X_1, X_2 satisfy the same modal formulas. Thus, one correctly gets

$$tmd(X_1, X_2) = \lim_{n \to \infty} \frac{1}{n} = 0.$$

Equation (ii) is immediate by noting that, for all X_1, X_2, $TG(X_1, X_2, n) = TG(X_2, X_1, n)$.

As for (iii), the triangular property, consider any three models X_1, X_2, X_3 and the three games playable on them,

$$TG(X_1, X_2, n), \; TG(X_2, X_3, n), \; TG(X_1, X_3, n) \tag{1}$$

Two cases are possible. Either Spoiler does not have a winning strategy in all three games (1) for any amount of rounds, or he has a winning strategy in at least one of them.

    If Spoiler does not have a winning strategy in all the games (1) for any number of rounds n, then Duplicator has a winning strategy in all games (1). Therefore, the three models satisfy the same modal formulas, $spw \to \infty$, and $tmd \to 0$. Trivially, the triangular property (iii) is satisfied.

    Suppose Spoiler has a winning strategy in one of the games (1). Via Theorem 3 (adequacy), one can shift the reasoning from games to formulas: there exists a modal formula γ of multi-modal rank m such that $X_i \models \gamma$ and $X_j \models \neg\gamma$. Without loss of generality, one can think of γ as being in normal form:

$$\gamma = \bigvee \bigwedge [\neg] U(\varphi_{S4}) \tag{2}$$

This last step is granted by the fact that every formula φ of S4$_u$ has an equivalent one in normal form whose modal rank is equivalent or smaller to that of φ.[2] Let γ^* be the formula with minimal multi-modal depth m^* with the property: $X_i \models \gamma^*$ and $X_j \models \neg\gamma^*$. Now, the other model X_k either satisfies γ^* or its

---

[2] In the proof, the availability of the normal form is not strictly necessary, but it gives gives a better impression of the behavior of the language and it has important implementation consequences, [2].

negation. Without loss of generality, $X_k \models \gamma^*$ and therefore X_j and X_k are distinguished by a formula of depth m^*. Suppose X_j and X_k to be distinguished by a formula β of multi-modal rank $h < m^*$: $X_j \models \beta$ and $X_k \models \neg\beta$. By the minimality of m^*, one has that $X_i \models \beta$, and hence, X_i and X_k can be distinguished at depth h. As this argument is symmetric, it shows that either

- one model is at distance $\frac{1}{m*}$ from the other two models, which are at distance $\frac{1}{l}$ $(\leq \frac{1}{m*})$, or
- one model is at distance $\frac{1}{h}$ from the other two models, which are at distance $\frac{1}{m*}$ $(\leq \frac{1}{h})$ one from the other.

It is a simple matter of algebraic manipulation to check that m^*, l and h, m^* (as in the two cases above), always satisfy the triangular inequality.

The nature of the isosceles topo-distance triggers a question. Why, given three spatial models, the distance between two couples of them is always the same?

First an example, consider a spoon, a chop-stick and a sculpture from Henry Moore. It is immediate to distinguish the Moore's sculpture from the spoon and from the chop-stick. The distance between them is high and the same. On the other hand, the spoon and the chop-stick look much more similar, thus, their distance is much smaller. Mereotopologically, it may even be impossible to distinguish them, i.e., the distance may be null.

In fact one is dealing with models of a qualitative spatial reasoning language of mereotopology. Given three models, via the isosceles topo-distance, one can easily distinguish the very different patterns. In some sense they are far apart as if they were belonging to different equivalence classes. Then, to distinguish the remaining two can only be harder, or equivalently, the distance can only be smaller.

5   Concluding Remarks

In this paper, a new perspective on mereotopology is taken, addressing issues of model equivalence and especially of model comparison. Defining a distance that encodes the mereotopological difference between spatial models has important theoretical and application implications. In addition, the use of model comparison games is novel. Model comparison games have been used only to compare two given models, but the issue of setting a distance among a whole class of models has not been addressed. The technique employed in Theorem 4 for the language $S4_u$ is more general, as it can be used for all Ehrenfeucht-Fraïssé style model comparison games[3] adequate for modal and first-order languages equipped with negation. A question interesting *per se*, but out of the scope of the present paper, is: which is the class of games (over which languages) for which a notion of isosceles distance holds? (E.g. are pebble games suited too?)

Another question open for further investigation is the computability of the topo-distance. First, there is a general issue on how to calculate the distance

---

[3] For an introduction to Ehrenfeucht-Fraïssé games see, for instance, [11].

for any topological space. One may be pessimistic at a first glance, since the definition and the proof of the Theorem 4 are not constructive, but actually the proof of the adequacy theorem for topo-games given in [2] is. Furthermore, decidability results for the logic $S4_u$ on the usual Kripke semantics (cf. [12]) should extend to the topological interpretation. Second, in usual applications the topological spaces at hand are much more structured and tractable. For example in a typical geographical information system, regions are represented as a finite number of open and/or closed polygons. With these structures, it is known that finiteness results apply (cf. [3]) and one should be able to compute the topo-distance by checking a finite number of points of the topological spaces. Currently, an image retrieval system based on spatial relationships where the indexing parameter is the topo-distance is being built, [1]. The aim is twofold, on the one hand one wants to build a system effectively computing the topo-distance, on the other one wants to check with the average user whether and how much the topo-distance is an intuitive and meaningful notion.

Broadening the view, another important issue is that of increasing the expressive power of the spatial language, then considering how and if the notion of isosceles distance extends. The most useful extensions are those capturing geometrical properties of regions, e.g. orientation, distance or shape. Again one can start by Tarski's ideas, who fell for the fascinating topic of axiomatizing geometry, [18], but can also follow different paths. For example, staying on the ground of modal logics, one can look at languages for incidence geometries. In this approach, one distinguishes the sorts of elements that populate space and considers the incidence relation between elements of the different sorts (see [6,5,19]).

Acknowledgments

The author is thankful to Johan van Benthem for fruitful discussions on the topic and to Kees Doets for feedback on a previous version of the proof of Theorem 4. The author would also like to extend his gratitude to the anonymous referees for their constructive comments. This work was supported in part by CNR grant 203.15.10.

References

1. M. Aiello. IRIS. An Image Retrieval System based on Spatial Relationship. Manuscript, 2000.
2. M. Aiello and J. van Bentham. Logical Patterns in Space. In D. Barker-Plummer, D. Beaver, J. van Benthem, and P. Scotto di Luzio, editors, *First CSLI Workshop on Visual Reasoning*, Stanford, 2000. CSLI. To appear.
3. M. Aiello, J. van Benthem, and G. Bezhanishvili. Reasoning about Space: the Modal Way. Manuscirpt, 2000.
4. N. Asher and L. Vieu. Toward a Geometry of Common Sense: a semantics and a complete axiomatization of mereotopology. In *IJCAI95*, pages 846–852. International Joint Conference on Artificial Itelligence, 1995.

5. Ph. Balbiani. The modal multilogic of geometry. *Journal of Applied Non-Classical Logics*, 8:259–281, 1998.
6. Ph. Balbiani, L. Fariñas del Cerro, T. Tinchev, and D. Vakarelov. Modal logics for incidence geometries. *Journal of Logic and Computation*, 7:59–78, 1997.
7. B. Bennett. Modal Logics for Qualitative Spatial Reasoning. *Bulletin of the IGPL*, 3:1 – 22, 1995.
8. J. van Benthem. *Modal Correspondence Theory*. PhD thesis, University of Amsterdam, 1976.
9. R. Casati and A. Varzi. *Parts and Places*. MIT Press, 1999.
10. A. Cohn and A. Varzi. Connection Relations in Mereotopology. In H. Prade, editor, *Proc. 13th European Conf. on AI (ECAI98)*, pages 150–154. John Wiley, 1998.
11. K. Doets. *Basic Model Theory*. CSLI Publications, Stanford, 1996.
12. V. Goranko and S. Pasy. Using the universal modality: gains and questions. *Journal of Logic and Computation*, 2:5–30, 1992.
13. D. Park. Concurrency and Automata on Infinite Sequences. In *Proceedings of the 5th GI Conference*, pages 167–183, Berlin, 1981. Springer Verlag.
14. D. Randell, Z. Cui, and A Cohn. A Spatial Logic Based on Regions and Connection. In *Proceedings of the Third International Conference on Principles of Knowledge Representation and Reasoning (KR'92)*, pages 165–176. San Mateo, 1992.
15. J. Renz and B. Nebel. On the Complexity of Qualitative Spatial Reasoning: A Maximal Tractable Fragment of the Region Connection Calculus. *Artificial Intelligence*, 108(1-2):69–123, 1999.
16. V. Shehtman. "Everywhere" and "Here". *Journal of Applied Non-Classical Logics*, 9(2-3):369–379, 1999.
17. A. Tarski. Der Aussagenkalkül und die Topologie. *Fund. Math.*, 31:103–134, 1938.
18. A. Tarski. What is Elementary Geometry? In L. Henkin and P. Suppes and A. Tarski, editor, *The Axiomatic Method, with Special Reference to Geometry ad Physics*, pages 16–29. North-Holland, 1959.
19. Y. Venema. Points, Lines and Diamonds: a Two-Sorted Modal Logic for Projective Planes. *Journal of Logic and Computation*, 9(5):601–621, 1999.

An Abductive Mechanism for Natural Language Processing Based on Lambek Calculus*

Antonio Frias Delgado[1] and Jose Antonio Jimenez Millan[2]

[1] Departamento de Filosofia
[2] Escuela Superior de Ingenieria de Cadiz
Universidad de Cadiz, Spain
{antonio.frias,joseantonio.jimenez}@uca.es

Abstract. We present an abductive mechanism that works as a robust parser in realistic tasks of Natural Language Processing involving incomplete information in the *lexicon*, whether it lacks lexical items or the items are partially and/or wrongly tagged. The abductive mechanism is based on an algorithm for automated deduction in Lambek Calculus for Categorial Grammar. Most relevant features, from the Artificial Intelligence point of view, lie in the ability for handling incomplete information input, and for increasing and reorganizing automatically lexical data from large scale *corpora*.

1 Introduction

1.1 Logic and Natural Language Processing

Natural Language Processing (NLP) is an interdisciplinary field where lots of research communities meet. Out of all NLP objectives, parsing is among the basic tasks on which other treatments of natural language can be founded. Development of efficient and robust parsing methods is a pressing need for computational linguistics; some of these methods are also relevant to Logic in AI whether they are founded on Logic or they use AI characteristic techniques.

Lambek Calculus (LC) for Categorial Grammar (CG) is a good candidate for developing parsing techniques in a logic framework. Some of the major advantages of CG lie in: (a) its ability for treating incomplete subphrases; (b) it is (weakly) equivalent to context free grammars, but (c) CG is radically lexicalist, it owns no (production) rule except logical ones; therefore, (d) syntactic revisions are reduced to type reassignments of lexical data of a given *lexicon*.

On the other hand, the Gentzen-style sequent formulation of LC for CG also presents several attractive features: (a) a well-known logical behaviour —LC corresponds to intuitionistic non-commutative multiplicative linear logic with non empty antecedent; (b) the cut-rule elimination, and hence the subformula property that is desirable with regard to its implementation.

---

* Partially supported by grant no. PB98-0590 of the Comisión Interministerial de Ciencia y Tecnología. We would like to thank two anonymous referees for their valuable comments.

M. Ojeda-Aciego et al. (Eds.): JELIA 2000, LNAI 1919, pp. 87–101, 2000.

When it comes to using LC in realistic tasks of NLP, one must admit that LC has two possible disadvantages: (a) its complexity is unknown; (b) in so far as it is equivalent to context free grammars, LC cannot account for several linguistic phaenomena. These limitations accepted, we encounter another kind of dificulties: the realistic tasks of NLP involve characteristic problems that cannot be solved by the sole use of deductive systems. A deduction is always something closed, in accordance with immovable rules; however our language understanding is robust enough and it succeeds even if partial information is lacking.

1.2   Learning and Revising Data

The AI researches intend to enlarge the logical machinery from the precise mathematical reasoning to the real situations in the real world. That means, for example: to learn from experience, to reorganize the knowledge, to operate even if the information is incomplete. The task of building robust parsers comes right into the goals of AI in a natural way.

The (informal) notion of robustness refers to the indifference of a system to a wide range of external disruptive factors [Ste92], [Men95]. Out of all desirable properties of a robust parser we focus on two ones chiefly: (a) a robust parser has to work in absence of information (hence it must learn from data); (b) a robust parser has to revise and to update the information.

In the last years, the idea that systematic and reliable acquisition on a large scale of linguistic information is the real challenge to NLP has been actually stressed. Moreover, currently available *corpora* make it is possible to build the core of a grammar and to increase the grammatical knowledge automatically from *corpora*. Two strategies vie with each other when it comes to approaching the specific problems of NLP we refer before: statistical versus rule-based strategies. From an engineering point of view, statistical extensions of linguistic theories have gained a vast popularity in the field of NLP: purely rule-based methods suffer from a lack of robustness in solving uncertainty due to overgeneration (if too many analyses are generated for a sentence) and undergeneration (if no analysis is generated for a sentence) [Bod98]. We think this 'lack of robustness' can be filled in the AI intention using abductive mechanisms that enlarge the deductive systems.

1.3   Abductive Mechanisms

We use the terms 'abductive mechanism' in a sense that may require a deeper explanation.

A deductive logical system typically offers a 'yes/no' answer to a closed question stated in the language of this logic. The two situations pointed out above can be found whenever we try to use a deduction system in realistic tasks of NLP:

(a) Lack of information in the *lexicon*. Thus, we have to use variables that do not belong to the logical language —$\alpha(X)$— for unknown values . An equivalent problem in classical logic would be the following task: $p \vee q, p \rightarrow r, X \vdash r$. Stated in this way, it is not a deduction problem properly.

(b) A negative answer merely: $\nvdash \alpha$.

In both cases we could consider we have a theory (here, the *lexicon, L*) and a problem to solve: how the *lexicon* has to be modified and/or increased in order to obtain a deduction:

(a') $L \vdash Subs_X^A \alpha(X)$, where A belongs to the used logical language;

(b') $L \vdash \beta$, where β is obtained from α according to some constraints.

That is precisely what we have called 'abductive' problems (inasmuch as it is not a new rule, but new data that have to be searched for), and 'abductive mechanism' (as the method for its solution). One matter is the logical system on whose rules we justify a concrete yes/no answer to a closed question, and another matter is the procedure of searching for some answer, that admits to be labelled as abductive.

Our purpose is to introduce an abductive mechanism that enlarges LC in order to obtain a robust parser that can be fruitfully employed in realistic applications of NLP.[1]

1.4   State-of-the-Art in Categorial Grammar Learning

Large electronic *corpora* make the induction of linguistic knowledge a challenge. Most of the work in this field falls within the paradigm of classical automata and formal language theory [HU79], whether it uses symbolic methods, or statistical methods, or both.[2] As formal automata and language theory does not use the mechanisms of deductive logics, the used methods for learning a language from a set of data are not abductive or inductive mechanisms. Instead, they build an infinite sequence of grammars that converges in the limit.

This being the background, much of the work about learning Categorial Grammars deals with the problem of what classes of categorial grammars may be built from positive or negative examples in the limit.[3] This approach manages *corpora* that hold no tags at all, or that are tagged with the information of which item acts as functor and which item acts as argument.

The difference between those works and ours is that the former ones (a) have a wider goal—that of learning a whole class of categorial grammars from tagged *corpora*—, and (b) that they do not make use of any abductive mechanism, but follow the steps made in the field of formal language theory.

---

[1] Currently, LC seems to be relegated to an honourable *logical* place. It is far from constituting an indispensable methodology in NLP. Let us use the TMR Project *Learning Computational Grammars* as an illustration. This project "will apply several of the currently interesting techniques for machine learning of natural language to a common problem, that of learning noun-phrase syntax." Eight techniques are used. None is related to LC.

[2] Cfr. Gold [Gol67], Angluin [Ang80], [AS83], Bod [Bod98] and references therein.

[3] For this approach, cfr. Buszkowski [Bus87a], [Bus87b], Buszkowski and Penn [BP90], Marciniec [Mar94], Kanazawa [Kan98].

On the other hand, our work is (i) of a narrower scope—we are only interested in filling some gaps that the *lexicon* may have, or we want to change the category assigned by the *lexicon* to some lexical item when it does not lead to success —, and (ii) we use an abductive mechanism.

Finding the right category to assign to a lexical item is possible because we make use of a goal directed parsing algorithm that avoids infinite ramifications of the search tree trying only those categories that are consistent with the context.

2   A Parsing Algorithm Based on Lambek Calculus

2.1   Lambek Calculus

First, we introduce the Gentzen-style sequent formulation of LC. The underlying basic idea in the application of LC to natural language parsing is to assign a syntactic type (or category) to a lexical item. A concrete sequence of lexical items (words in some natural language) is grammatically acceptable if the sequent with these types as antecedent and the type s (sentence) as succedent is provable in LC.

The *language* of the (product-free) LC for CG is defined by a set of basic or atomic categories ($BASCAT$) -also called primitive types-, from which we form complex categories -also called types- with the set of right and left division operators $\{/, \backslash\}$:

If A and B are categories, then A/B, and $B\backslash A$ are categories.

We define a formula as being a category or a type.

In the following we shall use lower case latin letters for basic categories, upper case latin letters for whatever categories, lower case greek letters for non-empty sequences of categories, and upper case greek letters for, possible empty, sequences of categories.

The rules of LC are [Lam58]:

1. *Axioms:*

$$\frac{}{A \Rightarrow A}(Ax)$$

2. *Right Introduction: /R, \R*

$$\frac{\gamma, B \Rightarrow A}{\gamma \Rightarrow A/B}(/R) \qquad \frac{B, \gamma \Rightarrow A}{\gamma \Rightarrow B\backslash A}(\backslash R)$$

3. *Left Introduction: /L, \L*

$$\frac{\gamma \Rightarrow B \quad \Gamma, A, \Delta \Rightarrow C}{\Gamma, A/B, \gamma, \Delta \Rightarrow C}(/L) \qquad \frac{\gamma \Rightarrow B \quad \Gamma, A, \Delta \Rightarrow C}{\Gamma, \gamma, B\backslash A, \Delta \Rightarrow C}(\backslash L)$$

4. *Cut*

$$\frac{\gamma \Rightarrow A \quad \Gamma, A, \Delta \Rightarrow C}{\Gamma, \gamma, \Delta \Rightarrow C}(Cut)$$

It is required that each sequent has a non-empty antecedent and precisely one succedent category. The cut-rule is eliminable.

2.2   Automated Deduction in Lambek Calculus

Given a *lexicon* for a natural language, the problem of determining the grammatical correctness of a concrete sequence of lexical items (a parsing problem) becomes into a deductive problem in LC. Therefore, a parsing algorithm is just an LC theorem prover.

LC-theoremhood is decidable. However, LC typically allows many distinct proofs of a given sequent that assign the same meaning; this problem is called 'spurious ambiguity'. An efficient theorem prover has to search for (all) non-equivalent proofs only. There are in the literature two approaches to this problem, based on a normal form of proofs (Hepple [Hep90], König, Moortgat [Moo90], Hendriks [Hen93]) or on proof nets (Roorda [Roo91]). LC theorem prover we present is related to König's method [Kön89], but it solves problems which are proper to König's algorithm.

First, we introduce some definitions.

1. *Value and Argument Formulae*

   1.1. If $F = a$, then a is the value formula of F;

   1.2. If (i) $F = G/H$ or (ii) $F = H\backslash G$, then G is the value formula of F and H is the argument formula of F. In the case (i), H is the right argument formula; in the case (ii), H is the left argument formula.

2. *Value Path*

   The value path of a complex formula F is the ordered set of formulae $\langle A_1, \ldots, A_n \rangle$ such that A_1 is the value formula of F and A_j is the value formula of A_{j-1} for $2 \leq j \leq n$.

3. *Argument Path*

   The argument path of a complex formula F is the ordered set of formulae $\langle B_1, \ldots, B_n \rangle$ such that B_1 is the argument formula of F and B_j is the argument formula of A_{j-1}, for $2 \leq j \leq n$, and $\langle A_1, \ldots, A_n \rangle$ being the value path of F.

   The right (resp. left) argument path of a complex formula F is the ordered subset of its argument path owning right (resp. left) argument formulae only.

4. *Main Value Formula*

   A is the main value formula of a complex formula F whose value path is $\langle A_1, \ldots, A_n \rangle$ if and only if $A = A_n$.

   It follows that: (i) if A is a main value formula, then $A \in BASCAT$; (ii) every complex formula has exactly one main value formula.

2.3   The Algorithm

We now sketch the algorithm implemented in both C language and Prolog. We present the algorithm in a pseudo-Prolog fashion in order to provide an easier understanding. This is not Prolog, as we have simplified the management of data structures and other practical problems of the language. At the same time we assume a "try or fail" strategy of control like that of Prolog, as well as mechanisms of unification to build data structures. Self-evident procedures (`search_value`, etc.) are not included.

procedure proof
input : $data \Rightarrow target$
output :   Proof tree if $\{\vdash data \Rightarrow target\}$, otherwise FAIL.
process :
      CASE data = target: RETURN $\{\frac{}{target \Rightarrow target}(Ax)\}$
      CASE target = A/B: RETURN $\{\frac{proof(data, B \Rightarrow A)}{data \Rightarrow A/B}(/R)\}$
      CASE target = $B \backslash A$: RETURN $\{\frac{proof(B, data \Rightarrow A)}{data \Rightarrow B \backslash A}(\backslash R)\}$
      CASE atomic(target):
          LET $c := target$
          LET $[list_1, list_2, \ldots, list_n] :=$ search_value(c in $data$)
          FOREACH $list_i \in [list_1, list_2, \ldots, list_n]$ DO
              LET $[\alpha, F, \beta] := list_i$
              LET $[A_1, \ldots, A_k] :=$ left_argument_path(c in F)
              LET $[B_1, \ldots, B_m] :=$ right_argument_path(c in F)
              LET $tree_i :=$ STACK reduce($[\,], \alpha, [A_k, \ldots, A_1]$)
                      WITH reduce($[\,], \beta, [B_1, \ldots, B_m]$)
              IF $tree_i =$ FAIL
                  THEN CONTINUE
                  ELSE RETURN $\{\frac{tree_i \quad c \Rightarrow c}{data \Rightarrow c}(|L)\}$
          END FOR
END **procedure proof**

procedure reduce
input :   ($[acums], [data], [targets]$)
output :   proof tree if $\{\Vdash_{LC} acums, data \Rightarrow targets\}$, otherwise FAIL.
process :
      CASE $acums = data = targets = [\,]$: RETURN $\{\frac{}{}(empty)\}$
      CASE $targets = [A]$:
          RETURN proof($acums, data \Rightarrow A$)
      OTHERWISE:
          CASE $acums \neq [\,]$ AND length($data$) $\geq$ length(tail($targets$)):
              LET $tree :=$ STACK proof($acums \Rightarrow$ head($targets$))
                     WITH reduce(head($data$),tail($data$),tail($targets$))
              IF $tree \neq$ FAIL
                  THEN RETURN $tree$
                  ELSE try next case
          CASE length(tail($data$)) $\geq$ length($targets$) - 1:
              RETURN reduce($acums$+head($data$),tail($data$),tail($targets$))
          OTHERWISE RETURN FAIL
END **procedure reduce**

2.4   Remarks on the Algorithm

(i) The proof procedure behaves as expected when input is an axiom.
(ii) The algorithm decomposes any target complex formula until it has to prove an atomic one, $c \in BASCAT$.

(iii) The reduce procedure is the main charasteristic of our algorithm. When we have to prove an atomic target, (i) we search for the formulae in the antecedent whose main value formula is the same as the atomic target $(F_1, \ldots, F_n)$; (ii) for each $F_i, 1 \leq i \leq n$, the left-hand side (resp. right-hand side) of the antecedent (with respect to F_i) and the left argument path of F_i (resp. right argument path) have to be cancelled out. The algorithm speeds up the deduction trying to satisfy the argument paths of F_i. The major advantages are obtained when the length of the sequence of data is long enough (note that a sentence in natural language may be up to 40 to 50 words long), and argument paths of the formulae are high. This property lies in the fact that the reduce procedure cares for still- not-consumed data and target formulae remaining to be proved. Efficient implementation for this algorithm has to avoid unnecessary calls to proof procedure from the reduce procedure, memorizing the proofs already tried.

(iv) FAIL may be regarded as an error propagating value. If any of the arguments of the proof-tree constructors —such as STACK, $(|L)$, $(/R)$, etc.— is FAIL, then resultant proof-tree is FAIL. A sensible implementation should be aware of this feature to stop the computational current step and to continue with the next one.

2.5   Properties of the Algorithm

(1) *The algorithm is correct*: If the output of proof procedure is not FAIL, then the proof tree constructed is a deduction of the input in LC.

Proof. Every rule we employ is a direct LC rule: axiom, $/R$, $\backslash R$. Note that the symbol $|L$ stands for successive applications of $/L$ and/or $\backslash L$. The conditions needed for applying each rule are exactly the same as they are required in LC. Hence, we can construct a proof tree in LC from the output of the proof procedure. $\square$

(2) *The algorithm is complete*: If $\vdash_{LC} data \Rightarrow target$, then the output of the proof procedure is a proof tree.

The proof follows from (2.1) and (2.2) below:

(2.1) If there is no deduction in LC for $\gamma, B \Rightarrow A$, then there is no deduction in LC for $\gamma \Rightarrow A/B$. (Similarly for $B, \gamma \Rightarrow A$, and $\gamma \Rightarrow B \backslash A$)

Proof: Let us suppose that there is a proof tree, Π, in LC for $\gamma \Rightarrow A/B$.

Case 1: If every rule in Π is either a L-rule either an axiom, then we follow the deduction tree in a bottom-up fashion and we reach the sequent $A/B \Rightarrow A/B$. We can construct a proof Π' from Π in this way:

$$\frac{B \Rightarrow B \quad A \Rightarrow A}{A/B, B \Rightarrow A}(/L)$$

Next we apply the rules of Π over A/B that yield $\gamma \Rightarrow A/B$ in Π, and we obtain in Π': $\gamma, B \Rightarrow A$.

Case 2: If there is an application of $/R$ in Π that yields

$$\frac{\delta, B \Rightarrow A}{\delta \Rightarrow A/B}(/R)$$

but it is not at the bottom of Π, we can postpone the application of the /R rule in Π' till remaining rules of Π have beeing applied, and so we have in Π' the sequent $\gamma, B \Rightarrow A$. $\square$

We use these properties to decompose any complex succedent until we reach an atomic one.

(2.2) Let $c \in BASCAT$, and $\gamma = \gamma_1, \ldots, \gamma_n$ $(n > 0)$.
If $\vdash_{LC} \gamma \Rightarrow c$, then it exists some γ_j, $(1 \leq j \leq n)$ such that:
(i) c is the main value formula of γ_j;
(ii) $\Vdash_{LC} \gamma_1, \ldots, \gamma_{j-1} \Rightarrow \Phi$;
(iii) $\Vdash_{LC} \gamma_{j+1}, \ldots, \gamma_n \Rightarrow \Delta$;
(iv) A deduction tree for $\gamma \Rightarrow c$ can be reconstructed from (ii), (iii), and from the axiom $c \Rightarrow c$.
(Where $\langle A_1, \ldots, A_k \rangle$ is the left argument path of γ_j, $\Phi = \langle A_k, \ldots, A_1 \rangle$, and $\Delta = \langle B_1, \ldots, B_m \rangle$ is the right argument path of γ_j).

The symbol $\Vdash_{LC}$ stands for the fact that a sequence of formulae (data) proves a sequence of target formulae *keeping the order*. If we consider the Lambek Calculus with the product operator, $\bullet$, Φ and Δ can be constructed as the product of all A_i and all B_i respectively, and $\Vdash_{LC}$ can be substituted for $\vdash_{LC}$ in (ii), (iii).
Note that (ii) and (iii) state that $\gamma_1, \ldots, \gamma_{j-1}$ can be split up in k sequences of categories $(\alpha_k, \ldots, \alpha_1)$, and $\gamma_{j+1}, \ldots, \gamma_n$ can be split up in m sequences of categories $(\beta_1, \ldots, \beta_m)$ such that
(ii') $\vdash_{LC} \alpha_n \Rightarrow A_n$, for $1 \leq n \leq k$;
(iii') $\vdash_{LC} \beta_n \Rightarrow B_n$, for $1 \leq n \leq m$.
Proof:
Ad (i) No rule except an axiom allows to introduce c in the succedent. Following the deduction tree in a bottom-up fashion, successive applications of /L and \L are such that (a) the argument formulae in the conclusion turn into the succedent of the premise on the left; (b) the value formula remains as part of the antecedent of the premise on the right; (c) the succedent of the conclusion remains as the succedent of the premise on the right — note that this ordering of the premises is always possible. Therefore we will reach the sequent $c \Rightarrow c$ eventually, being c the main value formula of γ_j. $\square$
This property allows us to restrict, without loss of completeness, the application of the L-rules to complex formulae whose main value formula is the same as the (atomic) target succedent.
Ad (ii) Let $\vdash_{LC} \gamma \Rightarrow c$. The only possibility of introducing A_n as a left argument formula of γ_j is from a L-rule. Hence, it exists some α_n such that $\vdash_{LC} \alpha_n \Rightarrow A_n$, because of $\alpha_n \Rightarrow A_n$ is the left-hand side premise of the L-rule. Otherwise, A_n together with c have to be introduced as an axiom, but the succedent is supposed to be an atomic type.
Note that we can first apply all L-rules for (/), followed by all L-rules for (\) —or vice versa—, whatever the formula may be. That follows from the theorems:
(a) $\vdash_{LC} (A\backslash(B/D))/C \Rightarrow ((A\backslash B)/D)/C$
(b) $\vdash_{LC} ((A\backslash B)/D)/C \Rightarrow (A\backslash(B/D))/C$

(c) $\vdash_{LC} C\backslash((D\backslash B)/A) \Rightarrow C\backslash(D\backslash(B/A))$
(d) $\vdash_{LC} C\backslash(D\backslash(B/A)) \Rightarrow C\backslash((D\backslash B)/A)$ □

Ad (iii) Similar to (ii). □

Ad (iv) Immediate from successive applications of /L and \L. □

(3) *The algorithm stops.* For whatever sequence of data and target, the number of tasks is finite, and every step simplifies the complexity of the data and/or the target. □

(4) *The algorithm finds all different deduction and only once.*

If there are several formulae in γ such that (i)–(iii) hold, each case corresponds to a non equivalent deduction of $\gamma \Rightarrow c$.

The proof is based upon the fact that property (2.2) may be regarded as the construction of a proof-net for $\gamma \Rightarrow c$ (in the equivalent fragment of non-commutative linear logic). The axiom $c \Rightarrow c$ becomes the construction of an *axiom-link*, and the points (ii) and (iii) become the construction of the corresponding sub-proof-nets with no overlap. Different axiom-links produce different proof-nets. □

3   An Abductive Mechanism for NLP

We say a sequent is *open* if it has any unknown category instance in the antecedent and/or in the succedent; otherwise we say the sequent is *closed*. We use upper case latin letters from the end of the alphabet (X, Y, Z) for non-optional unknown categories, and X^*, Y^*, Z^* for optional unknown categories.

3.1   Learning and Discovery Processes

We would consider two abductive mechanisms that we shall call *learning* and *discovery* processes, depending on the form of the target sequent. Discovery processes are related to tasks involving open sequents; learning processes are related to tasks involving closed sequents.

1. Given a closed sequent, we may subdivide the possible tasks into:
   (a) Grammatical correctness: to check either or not a sequence of data yields a target, merely. This is the normal use of LC.
   (b) If a closed sequent is not provable, we can introduce a procedure for learning in two ways: according to data priority or according to target priority.
      i. If we have certainty about data, and a closed target is not provable from them, we remove the given target and we search for a (minimum) new target that may be provable from data. We need the target to be a minimum in order to avoid the infinite solutions produced by the type-raising rule.
      ii. If we have certainty about target, and the set of closed data does not prove it, we remove data, by means of re-typing the necessary lexical items, in such a way that the target becomes provable from these new data.

iii. If we have certainty about data and about target, we could consider
the sequence as a linguistic phaenomenon that falls beyond a context
free grammar, ellipsis, etc.

In both cases (b.i) and (b.ii) we can appropriately say that we learn new
syntactic uses. Moreover, in case (b.ii) we carry out a revision of the
lexicon.

2. An open sequent is related to discovery tasks. In a sense, every discovery
task is also susceptible of being considered as a learning one (or vice versa).
However, we would rather prefer to differentiate them by pointing out that
they are based on formal features of the sequents.

3.2   The Abductive Mechanism

The objectives we pointed out above need the parsing algorithm —hereafter,
$\mathcal{LC}$— to be enlarged using an abductive mechanism —hereafter, $\mathcal{ACG}$, Abductive Categorial Grammar— for handling open sequents and removing types if
necessary. $\mathcal{ACG}$ manages:

(i) input sequences either from *corpora* or users;

(ii) information contained in the *lexicon*;

(iii) data transfer to $\mathcal{LC}$;

(iv) input adaptation and/or modification, if necessary;

(v) output of $\mathcal{LC}$;

(vi) request for a choice to the user;

(vii) addition of new types to the *lexicon* —its update.

What we have called an abductive mechanism has to do with the point (iv)
most of all. We sketch only its main steps for taking into account the learning
and discovery processes. Similarly to the parsing algorithm (2.3.), we present the
procedure in a pseudo-prolog fashion.

procedure learning
input: $(data \Rightarrow target)(A)$
    such that $\nvdash_{LC} data \Rightarrow target$, $closed(data)$, $closed(target)$
output: substitution $\{A := B\}$
    such that $\vdash_{LC} (data \Rightarrow target)\{A := B\}$
process:
    CASE certainty_about_target:
      LET $[A_1, \dots, A_n] := data$
      FOREACH $A_i \in [A_1, \dots, A_n]$ DO
        LET $new\_data := [\dots, A_{i-1}, X_i, A_{i+1}, \dots]$
        $\{X_i := B_i\} :=$ discovering $new\_data \Rightarrow target$
      END FOR
      RETURN $\{A_1 := B_1, \dots, A_n := B_n\}$
    CASE certainty_about_data:
      $\{X := B\} :=$ discovering $data \Rightarrow X$
      RETURN $\{A := B\}$
END **procedure learning**

procedure discovering
input: $(data \Rightarrow target)(X)$
output: $\{X := B\}$
          such that $\vdash_{LC} (data \Rightarrow target)\{X := B\}$
process:
          CASE open_target: $data \Rightarrow X$
                IF $data = [B]$
                THEN RETURN $\{X := B\}$
                IF $data = [F_1, \dots, F_n]$
                THEN FOREACH $F_i(1 \leq i \leq n)$, $F_i \notin BASCAT$, DO
                      LET $c_i := search\_value(F_i)$
                      $\{Y_i^* := B_i, Z_i^* := C_i\} := $ **new_proof** $Y_i^*, data, Z_i^* \Rightarrow c_i$
                END FOR
                RETURN $\{X_1 := B_1 \backslash c_1/C_1, \dots, X_n := B_n \backslash c_n/C_n\}$
          CASE open_data: $data(X_1, \dots, X_n) \Rightarrow target$
                IF $data = [X]$
                THEN RETURN $\{X := target\}$
                FOREACH $X_i(1 \leq i \leq n)$ DO
                      LET $[F_1, \dots, F_{i-1}, X_i, F_{i+1}, \dots, F_n] := data$
                      LET $c := target$
                      LET $new\_data := [F_1, \dots, Y^* \backslash c/Z^*, \dots, F_n]$
                      $\{X_i := B_i \backslash c/C_i\} := $ **new_proof** $new\_data \Rightarrow target$
                END FOR
                RETURN $\{X_1 := B_1 \backslash c/C_1, \dots, X_n := B_n \backslash c/C_n\}$
END **procedure discovering**

3.3   Remarks on $\mathcal{ACG}$

The old **proof** procedure (2.3) has to be adapted to a **new_proof** one. To achieve
this goal, we make two main changes: (a) the old **proof** procedure was built to
work with closed sequents and now it should be able to deal with open ones;
(b) the old **proof** procedure was initially designed to return a *proof tree* but it
should now return the substitution that makes the open sequent provable.

The old **proof** algorithm may work with open sequents, behaving as an abductive mechanism, if we consider the (=) operator as *unification*. It is well
known that the unification algorithm produces the substitution we are looking
for.

Two major changes come (a) from the **search_value**(c in $data$) procedure,
and (b) from the **reduce** procedure.

(a) The **search_value** procedure was considered to be self-evident, but now
it needs further explanations inasmuch as unknown data or targets are present.
What does it mean a value occurrence of X in Y? We will discuss the change in
the process that considers a formula to be the main value of another one.

procedure search_value
input: (Formula from data, target formula)

output: ([right_argument_path],target formula,[left_argument_path]) or FAIL
    CASE closed data (F) and closed target (c):
        CASE $F = c$: RETURN $([\,],c,[\,])$
        CASE $F = B\backslash A$: RETURN $([B] + \gamma, c, \delta)$
                where $(\gamma, c, \delta) :=$ search_value(A, c)
        CASE $F = A/B$: RETURN $(\gamma, c, [B] + \delta)$
                where $(\gamma, c, \delta) :=$ search_value(A, c)
        OTHERWISE RETURN FAIL
    CASE closed data (F) and open target (X):
        CASE $F = c$: RETURN $X := c$
        CASE $F = B\backslash A$: RETURN STACK F
                WITH search_value(A, c)
        CASE $F = A/B$: RETURN STACK F
                WITH search_value(A, c)
    CASE open data (Y) and closed target (c): RETURN $([\,], Y := c, [\,])$
    OTHERWISE RETURN FAIL
end procedure search_value

(b) Unknown categories may be either basic or complex ones. A treatment of the second case is rather difficult and it forces us to introduce constraints for bounding the search. We have to decide the upper bound of the complexity; i.e. X may be $A\backslash c/B$, or $A_1\backslash A_2 \backslash c/B_1/B_2$, etc. The reduce procedure requires some adaptations for working with optional categories. Optional categories are matched only if they are needed in the proof.

CASE X^* in target:
    IF $data = [\,]$
        THEN $X^* := [\,]$
    ELSE $X^* := X$
CASE X^* in data
    IF $target = [\,]$
        THEN $X^* := [\,]$
    ELSE
        LET $[F_1, \ldots, X^*, \ldots, F_n] := data$
        IF proof $[F_1, \ldots, F_n] \Rightarrow target \neq$ FAIL
            THEN $X^* := [\,]$
        ELSE $X^* :=$ new_proof $[F_1, \ldots, X, \ldots, F_n] \Rightarrow target$

Finally, let us note that type-raising rules yield sequents like following: $A \Rightarrow X/(A\backslash X)$ or $A \Rightarrow (X/A)\backslash X$ —where A and X are whichever formulae— that are provable in LC. The basic (deductive) proof algorithm is complete and has no problem with the proof of such sequents, although some LC parsing algorithms in the literature (mainly natural deduction based ones) are not complete because of the type-raising rules are not provable in them. Regarding our new_proof algorithm, the problem arises when it works as an abductive process in which X, the target consequent, is unknown; then it may be regarded as atomic or as a complex one. To regard it as atomic —our choice— causes no trouble but makes the type rising rule not provable (if the consequent is unknown). If we consider the possibility of an unknown consequent to be complex, then it may yield an endless loop. In fact, the type raising rule allows us to infer an endless number of more and more complex types.

3.4   Running $\mathcal{ACG}$

Example 1:
Data: "John loves".
Initial state of the *lexicon*:
John $= np$
loves $= np\backslash s/np$
Sketch of the abductive process:
(1) **proof** $(np, np\backslash s/np \Rightarrow s) = $ FAIL
(2) Certainty about data:
(2.1) $np, np\backslash s/np \Rightarrow X$
(2.2) $X := Y^*\backslash s/Z^*$
(2.3) $Y^*, np, np\backslash s/np, Z^* \Rightarrow s$
(2.4) $Y^*, np \Rightarrow np$; $Z^* \Rightarrow np$
(2.5) $Y^* := [\,]$; $Z^* := np$; $X := s/np$
Output:
• John loves $= s/np$
(3) Certainty about target:
(3.1) $X, np\backslash s/np \Rightarrow s$
(3.2) $X := s/Y^*$
(3.3) $np\backslash s/np \Rightarrow Y^*$
(3.4) $Y^* := np\backslash s/np$; $X := s/(np\backslash s/np)$
Output:
• John $= s/(np\backslash s/np)$
(3.5) $np, X \Rightarrow s$
(3.6) $X := Y^*\backslash s/Z^*$
(3.7) $np, Y^*\backslash s/Z^* \Rightarrow s$
(3.8) $np \Rightarrow Y^*$;
(3.9) $Y^* := np$; $Z^* := [\,]$
(3.10) $X := np\backslash s$
Output:
• loves $= np\backslash s$
(4) Certainty about data and target:
Output:
• John loves $\boxed{X} = np, np\backslash s/np, \boxed{np} \Rightarrow s$.
Example 2:
Data: "someone bores everyone".
Initial state of the *lexicon*:
someone $= ?$ (unknown)
bores $= np\backslash s/np$

everyone = ? (unknown)

$X, np\backslash s/np, Z \Rightarrow s$

Sketch of the abductive process:

(1)$X := s/Y^*$; $s/Y^*, np\backslash s/np, Z \Rightarrow s$

(1.1) $np\backslash s/np, Z \Rightarrow Y^*$

(1.2) $Y^* := Y_1^*\backslash s/Y_2^*$

(1.3) $Y_1^*, np\backslash s/np, Z, Y_2^* \Rightarrow s$

(1.4) $Y_1^* \Rightarrow np$

(1.5) $Z, Y_2^* \Rightarrow np$

(1.6) $Y_1^* := np$; $Z := np$; $Y_2^* := [\,]$

Output:

• someone = $s/(np\backslash s)$

• everyone = np

(2) $X \Rightarrow np$; $Z \Rightarrow np$

(2.1) $X := np$; $Z := np$

Output:

• someone = np

(3) $Z := Y^*\backslash s$; $X, np\backslash s/np, Y^*\backslash s \Rightarrow s$

(3.1) $X, np\backslash s/np \Rightarrow Y^*$

(3.2) $Y^* := Y_1^*\backslash s/Y_2^*$

(3.3) $Y_1^*, X, np\backslash s/np, Y_2^* \Rightarrow s$

(3.4) $Y_1^*, X \Rightarrow np$

(3.5) $Y_2^* \Rightarrow np$

(3.6) $Y_1^* := [\,]$; $X := np$; $Y_2^* := np$

Output:

• everyone = $(s/np)\backslash s$

State of the *lexicon* after runing $\mathcal{ACG}$:

someone = np, $s/(np\backslash s)$

bores = $np\backslash s/np$

everyone = np, $(s/np)\backslash s$

References

[Ang80]   Angluin, D. Inductive Inference of Formal Languges from Positive Data. *Information and Control* **45** (1980), 117–135.

[AS83]   Angluin, Dana and Smith, Carl H. Inductive Inference: Theory and Methods. *Computing Surveys* **15** (1983), 237–269.

[Bod98]   Bod, R. *Beyond Grammar. An Experience-Based Theory of Language.* CSLI Publications, 1998.

[Bus87a] Buszkowski, W. Solvable Problems for Classical Categorial Grammars. *Bulletin of the Polish Academy of Sciences: Mathematics* **35**, (1987) 507–516.

[Bus87b] Buszkowski, W Discovery Procedures for Categorial Grammars. In E. Klein and J. van Benthem, eds. *Polimorphisms and Unifications.* University of Amsterdam, 1987.

[Bus97]   Buszkowski, W. Mathematical linguistics and proof theory. In Benthem and Meulen [VBTM97], pages 683–736.

[BP90]   Buszkowski, W. and Penn, G. Categorial Grammars Determined from Lin-
           guistics data by unification. *Studia Logica* **49**, (1990) 431–454.
[Gol67]   Gold, E. M. Language Identification in the Limit. *Information and Control*
           10, (1967) 447–474.
[Hen93]   Hendriks, H. *Studied Flexibility. Categories and Types in Sybtax and Seman-
           tics*. PhD Thesis, University of Amsterdam, 1993.
[Hep90]   Hepple, M. *The Grammar and Processing of Order and Dependency: a Cat-
           egorical Approach*. PhD. Thesis, University of Edinburgh, 1990
[HU79]   Hopcroft, John E. and Ullman, Jeffrey D. *Introduction to Automata Theory,
           Languages, and Computation*. Addison-Wesley, 1979.
[Kan98]   Kanazawa, M. *Learnable Classes of Categorial Grammars*. CSLI Publica-
           tions, 1998.
[Kön89]   König, E. Parsing as Natural Deduction. In *Proceedings of the Annual
           Meeting of the Association for Computational Linguistics*. Vancouver, 1989.
[Lam58]   Lambek, J. The Mathematics of Sentence Structure. *American Mathematical
           Monthly* **65** (1958), 154–169.
[Mar94]   Marciniec, J. Learning Categorial Grammars by Unification with negative
           constraints. *Journal of Applications of Non-Classical Logics* **4** (1994), 181–
           200.
[Men95]   Menzel, W. Robust Processing of Natural Language. cmp-lg/9507003.
[Moo90]   Moortgat, M. Unambiguous Proof Representation for the Lambek Calculus.
           In *Proceedings of the Seventh Amsterdam Colloquium*, 1990.
[Moo97]   Moortgat, M. Categorial type logics. In Benthem and Meulen [VBTM97],
           pages 93–178.
[Plot71]   Plotkin, G.D. *Automatic Methods of Inductive Inference*. PhD Thesis, Ed-
           imburgh University, August 1971.
[Roo91]   Roorda, D. *Resource Logics: Proof-theoretical Investigation*. PhD Thesis,
           University of Amsterdam, 1991.
[Ste92]   Stede, M. The search for robustness in natural language understanding.
           Artificial Intelligence Review, **6** (1992) 383-414
[VBTM97]   Van Benthem, J. and Ter Meulen, A., Editors. *Handbook of Logic and
           Language*. Elsevier Science B.V. and MIT Press, 1997.

Capturing Stationary and Regular Extensions with Reiter's Extensions

Tomi Janhunen

Helsinki University of Technology
Laboratory for Theoretical Computer Science
P.O.Box 5400, FIN-02015 HUT, Finland
Tomi.Janhunen@hut.fi

Abstract. Janhunen et al. [14] have proposed a translation technique for normal logic programs in order to capture the alternating fix-points of a program with the stable models of the translation. The same technique is also applicable in the disjunctive case so that partial stable models can be captured. In this paper, the aim is to capture Przymusinska and Przymusinski's stationary extensions with Reiter's extensions using the same translational idea. The resulting translation function is polynomial, but only weakly modular and not perfectly faithful. Fortunately, another technique leads to a polynomial, faithful and modular (PFM) translation function. As a result, stationary default logic (STDL) is ranked in the expressive power hierarchy (EPH) of non-monotonic logics [13]. Moreover, reasoning with stationary extensions as well as brave reasoning with regular extensions (i.e., maximal stationary extensions) can be implemented using an inference engine for reasoning with Reiter's extensions.

1   Introduction

Quite recently, Janhunen et al. [14] have proposed a translation for *normal logic programs*. Using this translation the *alternating fix-points* of a program P [23] can be captured with the *stable models* [5] of the translation $\mathrm{Tr}_{\mathrm{AFP}}(P)$. This is interesting, since the alternating fix-points of P include the *well-founded model* of P [25], the *stable models* of P [5] as well as the *regular models* of P [26]. Formally speaking, an alternating fix-point M of P satisfies (i) $M = \Gamma_P^2(M)$ and (ii) $M \subseteq \Gamma_P(M)$ where Γ_P is the famous Gelfond-Lifschitz operator [5] and Γ_P^2 corresponds to applying Γ_P twice. Such a fix-point M can be understood as follows: M and $M' = \Gamma_P(M)$ specify *true* and *possibly true* atoms, respectively. Thus M induces a *partial* (or *three-valued*) model of P in which an atom a can be *true* (a $\in M$), *undefined* (a $\in M' - M$) or *false* (a $\notin M'$). Note that M becomes a (total) stable model of P if $M = M'$. These observations justify the view that the translation function $\mathrm{Tr}_{\mathrm{AFP}}$ lets us to unfold partiality under the stable model semantics [14]. A similar setting arises in conjunction with *disjunctive logic programs*: *partial stable models* [20] can be captured with total ones [6].

Since normal and disjunctive logic programs can be seen as special cases of Reiter's *default theories* [22] one could expect the same translational idea can

M. Ojeda-Aciego et al. (Eds.): JELIA 2000, LNAI 1919, pp. 102–117, 2000.

be applied to Reiter's *default logic* (DL). In this context, Przymusinska and Przymusinski have proposed a partial semantics for default logic [19]: *stationary extensions* of default theories are analogous to alternating fix-points of normal logic programs (an equivalent notion is used by Dix [4]). One of the main goals of this paper is to analyze the possibilities of generalizing the translation $\text{Tr}_{\text{AFP}}(P)$ for default theories under stationary extensions. Moreover, the author [13] has used *polynomial, faithful,* and *modular* (PFM) translation functions in order to classify non-monotonic logics by their expressive powers. As a result of this analysis, the expressive power hierarchy of non-monotonic logics (EPH) was obtained. Further refinements to EPH are given in [12]. From the perspective of EPH, it would be important to find out the exact position of *stationary default logic* (STDL) in EPH. A crucial step in this respect is that we succeed to embed STDL to conventional DL using a PFM translation function.

The rest of the paper is organized as follows. Basic notions of DL and STDL are reviewed in Sections 2 and 3, respectively. Then the classification method based on polynomial, faithful and modular (PFM) translation functions is introduced in Section 4. These properties of translation functions play an important role in the subsequent analysis. Starting from the translation function proposed for normal and disjunctive logic programs by Janhunen et al. [14], a preliminary translation function Tr_{ST1} for default theories is worked out in Section 5. Unfortunately, this translation function turns out to be unsatisfactory: it is not perfectly faithful and it is only weakly modular. These problems are addressed in Section 6 where another translational technique is applied successfully: a PFM translation function Tr_{ST2} is obtained. In addition, comparisons with other logics in EPH are made in order to classify STDL properly in EPH. Brave reasoning with *regular extensions* turns also to be manageable via Tr_{ST2}. Finally, the conclusions of the paper are presented in Section 7. Future work is also sketched.

2 Default Logic

In this section, we review the basic definitions of Reiter's default logic [22] in the propositional case. The reader is assumed to be familiar with classical propositional logic (CL). We write $\mathcal{L}(\mathcal{A})$ to declare a *propositional language* $\mathcal{L}$ based on propositional *connectives* ($\neg$, $\wedge$, $\vee$, $\rightarrow$, $\leftrightarrow$) and *constants* (truth $\top$ and falsity $\perp$) and a set of propositional *atoms* $\mathcal{A}$. On the semantical side, propositional *interpretations* $I \subseteq \mathcal{A}$ and *models* $M \subseteq \mathcal{A}$ are defined in the standard way. The same applies to conditions when a sentence $\phi \in \mathcal{L}$ is *valid* (denoted by $\models \phi$) and a *propositional consequence* of a theory $T \subseteq \mathcal{L}$ (denoted by $T \models \phi$). The theory $\text{Cn}(T) = \{\phi \in \mathcal{L} \mid T \models \phi\}$ is the closure of a theory $T \subseteq \mathcal{L}$ under propositional consequence. A sentence $\phi \in \mathcal{L}$ is *consistent* with a theory $T \subseteq \mathcal{L}$ (denoted by $T * \phi$) whenever $T \cup \{\phi\}$ is *propositionally consistent*, i.e. $T \cup \{\phi\}$ has at least one model. Note that $T * \phi \Leftrightarrow T \not\models \neg\phi$ holds in general. Moreover, $T * \top$ expresses that a theory $T \subseteq \mathcal{L}$ is propositionally consistent, i.e. $T \not\models \perp$.

In Reiter's default logic [22], basic syntactic elements are *default rules* (or simply *defaults*) which are expressions of the form $\frac{\alpha : \beta_1, \dots, \beta_n}{\gamma}$ where $\alpha, \beta_1, \dots, \beta_n$, and γ are sentences of $\mathcal{L}$. The intuition behind such a rule is that if the *prerequisite* α has been inferred and each of the *justifications* β_i is (separately) consistent with our beliefs, then the *consequent* γ can be inferred. A *default theory* in a propositional language $\mathcal{L}(\mathcal{A})$ is a pair $\langle D, T \rangle$ where D is a set of defaults in $\mathcal{L}$ and $T \subseteq \mathcal{L}$ is a propositional theory. For a set of defaults D, we let $\mathrm{Cseq}(D)$ denote the set of consequents $\{\gamma \mid \frac{\alpha:\beta_1,\ldots,\beta_n}{\gamma} \in D\}$ that appear in D.

The semantics of a default theory $\langle D, T \rangle$ is determined by its *extensions*, i.e. *sets of conclusions* that are propositionally closed theories associated with $\langle D, T \rangle$. Rather than presenting Reiter's definition of extensions [22] we resort to one by Marek and Truszczyński [16]. The justifications of a set of defaults are interpreted as follows. For any $E \subseteq \mathcal{L}$, the reduct D_E contains an (ordinary) inference rule $\frac{\alpha}{\gamma}$ whenever there is a default $\frac{\alpha:\beta_1,\ldots,\beta_n}{\gamma} \in D$ such that $E * \beta_i$ for all $i \in \{1, \ldots, n\}$. Given $T \subseteq \mathcal{L}$ and a set of inference rules R in $\mathcal{L}$, we let $\mathrm{Cn}^R(T)$ denote the closure of T under R and propositional consequence. More precisely, the closure $\mathrm{Cn}^R(T)$ is the least theory $T' \subseteq \mathcal{L}$ satisfying (i) $T \subseteq T'$, (ii) for every rule $\frac{\alpha}{\gamma} \in R$, $\alpha \in T'$ implies $\gamma \in T'$, and (iii) $\mathrm{Cn}(T') \subseteq T'$. The closure $\mathrm{Cn}^R(T)$ can be characterized using a proof system [11,16]. A sentence ϕ is R-provable from T if there is a sequence $\frac{\alpha_1}{\gamma_1}, \ldots, \frac{\alpha_n}{\gamma_n}$ of rules from R such that $T \cup \{\gamma_1, \ldots, \gamma_{i-1}\} \models \alpha_i$ for all $i \in \{1, \ldots, n\}$ and $T \cup \{\gamma_1, \ldots, \gamma_n\} \models \phi$. Then $\phi \in \mathcal{L}$ is R-provable from $T \Leftrightarrow \phi \in \mathrm{Cn}^R(T)$. The definition of extensions follows.

Definition 1 (Marek and Truszczyński [16]). *A theory $E \subseteq \mathcal{L}$ is an extension of a default theory $\langle D, T \rangle$ in $\mathcal{L}$ if and only if $E = \mathrm{Cn}^{D_E}(T)$.*

By *default logic* (DL) we mean default theories under Reiter's extensions. It is not necessary that a default theory $\langle D, T \rangle$ has a unique extension nor extensions at all. Typically two approaches are used. In the *brave* approach, it is sufficient to find one extension E containing the query $\phi \in \mathcal{L}$. In the *cautious* approach, the query $\phi \in \mathcal{L}$ should belong to every extension, i.e. the intersection of extensions.

3   Stationary Default Logic

As already stated, the existence of Reiter's extensions is not guaranteed in general. Motivated by the well-founded semantics [24] and alternating fix-points [23] of normal logic programs, Przymusinska and Przymusinski [19] propose a weaker notion of extensions as a solution to the problem. Dix [4] considers an equivalent semantics in order to establish a *cumulative* variant of DL.

Definition 2 (Przymusinska and Przymusinski [19]). *A theory $E \subseteq \mathcal{L}$ is a stationary extension of a default theory $\langle D, T \rangle$ in $\mathcal{L}$ if and only if $E = \mathrm{Cn}^{D_{E'}}(T)$ holds for the theory $E' = \mathrm{Cn}^{D_E}(T)$ and $E \subseteq E'$.*

The intuition is that the theory E provides the set of *actual* conclusions associated with $\langle D, T \rangle$ while E' can be understood as the set of *potential* conclusions (cf. the alternating fix-points of normal logic programs described in the introduction). This explains why the requirement $E \subseteq E'$ is reasonable, i.e. actual conclusions must also be potential conclusions. Note that if (in addition) $E' \subseteq E$ holds, then $E = E'$ is a Reiter-style extension of $\langle D, T \rangle$. By *stationary default logic* (STDL) we mean default theories under stationary extensions.

Every default theory $\langle D, T \rangle$ is guaranteed to have at least one stationary extension E known as the *least stationary* extension of $\langle D, T \rangle$. It serves as an *approximation* of any other stationary extension F of $\langle D, T \rangle$ in the sense that $E \subseteq F$. This applies equally to any Reiter-style extension E of $\langle D, T \rangle$ which is also a stationary extension of $\langle D, T \rangle$. Complexity results on DL [7] and STDL [8] support the approximative view: cautious reasoning with Reiter's extensions (a $\mathbf{\Pi}_2^p$-complete decision problem) is strictly more complex than cautious reasoning with stationary extensions (a $\mathbf{\Delta}_2^p$-complete decision problem). The least stationary extension of a finite default theory $\langle D, T \rangle$ can be iteratively constructed [4,19]. Initially, let $E_0 = \emptyset$ and $E_0' = \mathrm{Cn}^{D_0}(T)$. Then compute $E_i = \mathrm{Cn}^{D_{E_{i-1}'}}(T)$ and $E_i' = \mathrm{Cn}^{D_{E_i}}(T)$ for $i = 1, 2, \dots$ until $E_i = E_{i-1}$ holds. For instance, the set of defaults $D = \{ \frac{\mathsf{T}:\neg\mathsf{a}}{\mathsf{b}\vee\mathsf{c}}, \frac{\mathsf{T}:\neg\mathsf{p}\wedge\neg\mathsf{q}}{\mathsf{q}}, \frac{\mathsf{T}:\neg\mathsf{b},\neg\mathsf{c}}{\mathsf{s}}, \frac{\mathsf{T}:\neg\mathsf{q}}{\mathsf{r}} \}$ and the theory $T = \{\mathsf{b} \to \mathsf{p}, \mathsf{c} \to \mathsf{p}\}$ (adopted from [11, Example 10.18]) give rise to the following iteration sequence: $E_0 = \emptyset$, $E_1 = \mathrm{Cn}(\{\mathsf{b} \vee \mathsf{c}, \mathsf{p}, \mathsf{s}\})$, $E_2 = \mathrm{Cn}(\{\mathsf{b} \vee \mathsf{c}, \mathsf{p}, \mathsf{s}, \mathsf{r}\})$ and $E_3 = E_2$. Consequently, the theory E_2 is the least stationary extension of $\langle D, T \rangle$. In fact, E_2 is the unique (Reiter-style) extension of $\langle D, T \rangle$, as $E_2 = E_2'$.

There are two ways to distinguish *propositionally consistent* stationary extensions of a default theory $\langle D, T \rangle$. The first one is simply to require that E is propositionally consistent. The other demands that the set of potential conclusions $E' = \mathrm{Cn}^{D_E}(T)$ is propositionally consistent, too. In the latter case, we say that E is *strongly* propositionally consistent. Let us highlight the difference of these notions of consistency by a set of defaults $D = \{ \frac{\mathsf{T}:\mathsf{a}}{\mathsf{a}}, \frac{\mathsf{T}:\neg\mathsf{a}}{\neg\mathsf{a}}, \frac{\mathsf{T}:}{\mathsf{b}} \}$ in $\mathcal{L}(\{\mathsf{a}, \mathsf{b}\})$. Now $\langle D, \emptyset \rangle$ has three stationary extensions: $E_1 = \mathrm{Cn}(\{\mathsf{b}\})$, $E_2 = \mathrm{Cn}(\{\mathsf{a}, \mathsf{b}\})$, and $E_3 = \mathrm{Cn}(\{\neg\mathsf{a}, \mathsf{b}\})$. The respective sets of potential conclusions are $E_1' = \mathcal{L}$, $E_2' = E_2$, and $E_3' = E_3$. Thus E_1 is (only) propositionally consistent while E_2 and E_3 are strongly propositionally consistent.

4   PFM Translations Functions and EPH

In this section, we recall the classification method [12,13] which has been designed for comparing the expressive powers of non-monotonic logics. In the sequel, we assume that non-monotonic logics under consideration use a propositional language $\mathcal{L}$ as a sublanguage. Therefore, we let $\langle X, T \rangle$ stand for a non-monotonic theory in general. Here $T \subseteq \mathcal{L}$ is a propositional theory and X is a set of *parameters* specific to the non-monotonic logic L in question. For instance, the set of parameters X is a set of defaults in default logic. We let $||\langle X, T \rangle||$ stand for the the *length* of $\langle X, T \rangle$ in symbols.

Generally speaking, a translation function $\mathrm{Tr} : L_1 \to L_2$ transforms a theory $\langle X, T \rangle$ of one non-monotonic logic L_1 into a theory of another non-monotonic logic L_2. Both logics are assumed to have a notion of extensions available. Our requirements for Tr are the following. A translation function Tr is (i) **polynomial**, if for all X and T, the time required to compute $\mathrm{Tr}(\langle X, T \rangle)$ is polynomial in $||\langle X,T\rangle||$, (ii) **faithful**, if for all X and T, the propositionally consistent extensions of $\langle X,T\rangle$ and $\mathrm{Tr}(\langle X,T\rangle)$ are in one-to-one correspondence and coincide up to $\mathcal{L}$, and (iii) **modular**, if for all X and T, the translation $\mathrm{Tr}(\langle X,T\rangle) = \langle X',T'\cup T\rangle$ where $\langle X',T'\rangle = \mathrm{Tr}(\langle X,\emptyset\rangle)$. A translation function $\mathrm{Tr}: L_1 \to L_2$ is called **PFM** if it satisfies all the three criteria.

Note that a modular translation function translates the set of parameters X independently of T which remains untouched in the translation. For the purposes of this paper, we distinguish also *weakly* modular translation functions considered by Gottlob [9]. A translation function Tr is **weakly modular**, if for all X and T, $\mathrm{Tr}(\langle X,T\rangle) = \langle X',T'\cup t(T)\rangle$ where $\langle X',T'\rangle = \mathrm{Tr}(\langle X,\emptyset\rangle)$ and t is a separate translation function for T. Note that the translation of X remains independent of the translation of T even in this setting.

Given two non-monotonic logics L_1 and L_2, we write $L_1 \xrightarrow{\text{PFM}} L_2$, if there exists a PFM translation function $\mathrm{Tr}: L_1 \to L_2$. Then L_2 is considered to be as expressive as L_1. In certain cases, we are able to construct a counter-example which shows that a translation function satisfying our criteria does not exist. We use the notation $L_1 \xslashedrightarrow{\text{PFM}} L_2$ in such cases and we may also drop any of the three letters (referring to the three criteria) given that the corresponding criterion is not needed in the counter-example (note that $L_1 \xslashedrightarrow{\text{FM}} L_2$ implies $L_1 \xslashedrightarrow{\text{PFM}} L_2$, for instance). Further relations are definable for non-monotonic logics in terms of the base relations $\xrightarrow{\text{PFM}}$ and $\xslashedrightarrow{\text{PFM}}$: (i) L_1 is *less expressive* than L_2 (denoted by $L_1 \xrightarrow{\text{PFM}} L_2$) if $L_1 \xrightarrow{\text{PFM}} L_2$ and $L_2 \xslashedrightarrow{\text{PFM}} L_1$, (ii) L_1 and L_2 are *equally expressive* (denoted by $L_1 \xleftrightarrow{\text{PFM}} L_2$) if $L_1 \xrightarrow{\text{PFM}} L_2$ and $L_2 \xrightarrow{\text{PFM}} L_1$, and (iii) L_1 and L_2 are *mutually incomparable* (denoted by $L_1 \xslashedleftrightarrow{\text{PFM}} L_2$) if $L_1 \xslashedrightarrow{\text{PFM}} L_2$ and $L_2 \xslashedrightarrow{\text{PFM}} L_1$.

In Fig. 1, we have depicted the current EPH using only single representatives of the classes that have been obtained from DL via *syntactic restrictions*. *Normal* DL (NDL) is based on defaults of the form $\frac{\alpha:\beta}{\beta}$. In *prerequisite-free* DL (PDL) only defaults of the form $\frac{\top:\beta_1,\ldots,\beta_n}{\gamma}$ are allowed. The third variant (PNDL) is a hybrid of NDL and PDL with defaults of the form $\frac{\top:\beta}{\beta}$. The semantics of these syntactic variants is determined by Reiter's extensions. Recall that CL stands for propositional logic. The reader is referred to [12,13] for the complete EPH with 11 non-monotonic logics.

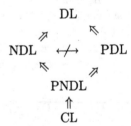

Fig. 1: Classes of EPH Represented by Syntactic Variants of DL

5   A Weakly Modular Translation

The goal of this section is to generalize the translation proposed by Janhunen et al. [14] so that the stationary extensions of a default theory $\langle D,T\rangle$ can be captured with the (Reiter-style) extensions of the translation. For a while, we restrict ourselves to the case of normal logic programs in order to explain the ideas behind the translation function $\mathrm{Tr}_{\mathrm{APF}}$ discussed in the introduction. The way to represent partial models of a normal logic program P is to introduce a new atom $a^{\bullet}$ for each atom a that appears in P. The intuitive reading of $a^{\bullet}$ is that a is *potentially* true. Then undefined atoms are captured with a total model N as follows: an atom a is undefined in a partial model if and only if a is false in N and $a^{\bullet}$ is true in N. The translation $\mathrm{Tr}_{\mathrm{AFP}}(P)$ is obtained as follows: a rule of the form $a \leftarrow b_1, \ldots, b_n, {\sim}c_1, \ldots, {\sim}c_m$ is translated into two rules $a \leftarrow b_1, \ldots, b_n, {\sim}c_1^{\bullet}, \ldots, {\sim}c_m^{\bullet}$ and $a^{\bullet} \leftarrow b_1^{\bullet}, \ldots, b_n^{\bullet}, {\sim}c_1, \ldots, {\sim}c_m$. In addition, a rule of the form $a^{\bullet} \leftarrow a$ is introduced for each atom a that appears in P. Rules of the latter type make sure that any atom that is true is also potentially true (cf. Section 1). As a result of this transformation on the rules of P, the stable models of $\mathrm{Tr}_{\mathrm{AFP}}(P)$ capture the alternating fix-points of P exactly.

Let us now devise an analogous translation for a default theory $\langle D, T \rangle$ in a propositional language $\mathcal{L}(\mathcal{A})$. A new atom $a^{\bullet}$ is introduced for each atom $a \in \mathcal{A}$ and we define $A^{\bullet} = \{a^{\bullet} \mid a \in A\}$ for any set of atoms $A \subseteq \mathcal{A}$. Since propositional logic is based on a much richer syntax than bare atoms, we have to find a way to express that an arbitrary sentence $\phi \in \mathcal{L}$ is a *potential* conclusion (i.e. a member of E' in Definition 2). As a solution, we introduce a sentence $\phi^{\bullet}$ for each $\phi \in \mathcal{L}$.

Definition 3. *The sentences ϕ of $\mathcal{L}(\mathcal{A})$ are translated by the following rules: (i) $(\top)^{\bullet} = \top$, (ii) $(\bot)^{\bullet} = \bot$, (iii) $(a)^{\bullet} = a^{\bullet}$ for an atom $a \in \mathcal{A}$, (iv) $(\neg\psi)^{\bullet} = \neg(\psi)^{\bullet}$, and (v) $(\psi_1 \circ \psi_2)^{\bullet} = (\psi_1)^{\bullet} \circ (\psi_2)^{\bullet}$ for any connective $\circ \in \{\wedge, \vee, \rightarrow, \leftrightarrow\}$.*

By this definition, any sentence $\phi \in \mathcal{L}(\mathcal{A})$ is translated into a sentence $\phi^{\bullet}$ in the propositional language $\mathcal{L}^{\bullet}$ based on $\mathcal{A}^{\bullet}$. For instance, $(\neg a \rightarrow (b \vee \bot))^{\bullet}$ is rewritten as $\neg a^{\bullet} \rightarrow (b^{\bullet} \vee \bot)$. For a theory $T \subseteq \mathcal{L}$ and a set of inference rules R in $\mathcal{L}$, we let $T^{\bullet}$ and $R^{\bullet}$ stand for the theory $\{\phi^{\bullet} \mid \phi \in T\} \subseteq \mathcal{L}^{\bullet}$ and the set of inference rules $\{\frac{\alpha^{\bullet}}{\gamma^{\bullet}} \mid \frac{\alpha}{\gamma} \in R\}$ in $\mathcal{L}^{\bullet}$, respectively. The following lemmas state some useful properties of theories involving sentences from $\mathcal{L}$ and $\mathcal{L}^{\bullet}$.

Lemma 1. *Let $T \subseteq \mathcal{L}(\mathcal{A})$ and $S \subseteq \mathcal{L}(\mathcal{A})$ be theories so that $S^{\bullet} \subseteq \mathcal{L}^{\bullet}(\mathcal{A}^{\bullet})$ and $T \cup S^{\bullet} \subseteq \mathcal{L}'(\mathcal{A} \cup \mathcal{A}^{\bullet})$. Consider any $\phi \in \mathcal{L}(\mathcal{A})$. Then (i) $(T \cup S^{\bullet}) * T \Leftrightarrow T * T$ and $S^{\bullet} * T$, (ii) if $S^{\bullet} * T$, then $(T \cup S^{\bullet}) * \phi \Leftrightarrow T * \phi$ and $T \cup S^{\bullet} \models \phi \Leftrightarrow T \models \phi$, and (iii) if $T * T$, then $(T \cup S^{\bullet}) * \phi^{\bullet} \Leftrightarrow S^{\bullet} * \phi^{\bullet}$ and $T \cup S^{\bullet} \models \phi^{\bullet} \Leftrightarrow T' \models \phi^{\bullet}$.*

Lemma 2. *Let T be a propositional theory in $\mathcal{L}(\mathcal{A})$ and $\phi \in \mathcal{L}$ any sentence. Then it holds that (i) $T * \phi \Leftrightarrow T^{\bullet} * \phi^{\bullet}$, (ii) $T \models \phi \Leftrightarrow T^{\bullet} \models \phi^{\bullet}$, (iii) $[\mathrm{Cn}(T)]^{\bullet} = \mathrm{Cn}(T^{\bullet})$, and (iv) $[\mathrm{Cn}^R(T)]^{\bullet} = \mathrm{Cn}^{R^{\bullet}}(T^{\bullet})$.*

The generalization of $\mathrm{Tr}_{\mathrm{APF}}$ for default theories follows.

Definition 4. *For any default theory $\langle D, T \rangle$ in $\mathcal{L}(\mathcal{A})$, let $\mathrm{Tr}_{\mathrm{ST1}}(\langle D, T \rangle) =$*

$$\langle \{\frac{\alpha^{\bullet} : \beta_1, \ldots, \beta_n}{\gamma^{\bullet}}, \frac{\alpha : \beta_1^{\bullet}, \ldots, \beta_n^{\bullet}}{\gamma} \mid \frac{\alpha : \beta_1, \ldots, \beta_n}{\gamma} \in D\} \cup \{\frac{\gamma}{\gamma^{\bullet}} \mid \gamma \in \mathit{Cseq}(D)\}, T \cup T^{\bullet} \rangle.$$

The intuition behind the translation is to capture a stationary extension E of $\langle D, T \rangle$ as well as the associated set of potential conclusions E' with an extension $\mathrm{Cn}(E \cup (E')^{\bullet})$ of the translation $\mathrm{Tr}_{\mathrm{ST1}}(\langle D, T \rangle)$. The defaults of the forms

$\frac{\alpha:\beta_1{}^{\bullet},\ldots,\beta_n{}^{\bullet}}{\gamma}$ and $\frac{\alpha^{\bullet}:\beta_1,\ldots,\beta_n}{\gamma^{\bullet}}$ capture the closures $\mathrm{Cn}^{D_{E'}}(T)$ and $(\mathrm{Cn}^{D_E}(T))^{\bullet}$, respectively. The latter closure (i.e. $(E')^{\bullet}$) is encoded in $\mathcal{L}^{\bullet}$ rather than $\mathcal{L}$. The defaults of the form $\frac{\alpha:}{\gamma^{\bullet}}$ enforce the relationship $E \subseteq E'$, i.e. actual conclusions have to be potential as well. Using Lemmas 1 and 2, we may compute the reduct of the set of defaults D' involved in the translation $\mathrm{Tr}_{\mathrm{ST1}}(\langle D, T\rangle)$.

Proposition 1. *Let $\langle D, T\rangle$ be a default theory in $\mathcal{L}(\mathcal{A})$ and $\langle D', T \cup T^{\bullet}\rangle$ the translation $\mathrm{Tr}_{\mathrm{ST1}}(\langle D, T\rangle)$ in $\mathcal{L}'(\mathcal{A} \cup \mathcal{A}^{\bullet})$. Moreover, let $E = \mathrm{Cn}(E_1 \cup E_2{}^{\bullet})$ hold for propositionally consistent theories E_1 and E_2 in $\mathcal{L}$. Then for $\frac{\alpha:\beta_1,\ldots,\beta_n}{\gamma} \in D$,*
(i) $\frac{\alpha:}{\gamma^{\bullet}} \in D'_E \Leftrightarrow \frac{\alpha}{\gamma} \in D_{E_1}$, (ii) $\frac{\alpha}{\gamma} \in D'_E \Leftrightarrow \frac{\alpha}{\gamma} \in D_{E_2}$, and (iii) $\frac{\gamma}{\gamma^{\bullet}} \in D'_E$.

Using these relationships of D'_E, D_{E_1} and D_{E_2} as well as Lemmas 1 and 2, it can be shown that $\mathrm{Tr}_{\mathrm{ST1}}$ captures stationary extensions in the following way.

Theorem 1. *Let $\langle D, T\rangle$ be a default theory in $\mathcal{L}(\mathcal{A})$ and $\langle D', T \cup T^{\bullet}\rangle$ the translation $\mathrm{Tr}_{\mathrm{ST1}}(\langle D, T\rangle)$ in $\mathcal{L}'(\mathcal{A} \cup \mathcal{A}^{\bullet})$. If $E_1 \subseteq \mathcal{L}$ is a strongly propositionally consistent stationary extension of $\langle D, T\rangle$ and $E_2 = \mathrm{Cn}^{D_{E_1}}(T)$, then $E = \mathrm{Cn}(E_1 \cup E_2{}^{\bullet}) \subseteq \mathcal{L}'$ is a propositionally consistent extension of $\langle D', T \cup T^{\bullet}\rangle$.*

Theorem 2. *Let $\langle D, T\rangle$ be a default theory in $\mathcal{L}(\mathcal{A})$ and $\langle D', T \cup T^{\bullet}\rangle$ the translation $\mathrm{Tr}_{\mathrm{ST1}}(\langle D, T\rangle)$ in $\mathcal{L}'(\mathcal{A} \cup \mathcal{A}^{\bullet})$. If $E \subseteq \mathcal{L}'$ is a propositionally consistent extension of $\langle D', T \cup T^{\bullet}\rangle$, then $E_1 = E \cap \mathcal{L}$ is a strongly propositionally consistent stationary extension of $\langle D, T\rangle$ such that $E_2 = \{\phi \in \mathcal{L} \mid \phi^{\bullet} \in E\}$ satisfies $E_2 = \mathrm{Cn}^{D_{E_1}}(T)$.*

A shortcoming of the translation function $\mathrm{Tr}_{\mathrm{ST1}}$ is that it is unable to capture stationary extensions of a default theory $\langle D, T\rangle$ that are propositionally consistent but not strongly propositionally consistent. In other words, $\mathrm{Tr}_{\mathrm{ST1}}$ is not faithful in the sense it is required in Section 4. Let us recall the set of defaults $D = \{\frac{T:a}{a}, \frac{T:\neg a}{\neg a}, \frac{T:}{b}\}$ from Section 3 in order to demonstrate this feature of $\mathrm{Tr}_{\mathrm{ST1}}$. The translation $\mathrm{Tr}_{\mathrm{ST1}}(\langle D, \emptyset\rangle) = \langle D', \emptyset\rangle$ where the set of defaults $D' = \{\frac{T:a}{a^{\bullet}}, \frac{T:a^{\bullet}}{a}, \frac{T:\neg a}{\neg a^{\bullet}}, \frac{T:\neg a^{\bullet}}{\neg a}, \frac{T:}{b}, \frac{T:}{b^{\bullet}}, \frac{b:}{b^{\bullet}}, \frac{a:}{a^{\bullet}}, \frac{\neg a:}{\neg a^{\bullet}}\}$. The default theory $\langle D', \emptyset\rangle$ has two extensions $E'_2 = \mathrm{Cn}(\{a, a^{\bullet}, b, b^{\bullet}\})$ and $E'_3 = \mathrm{Cn}(\{\neg a, \neg a^{\bullet}, b, b^{\bullet}\})$ corresponding to the stationary extensions $E_2 = \mathrm{Cn}(\{a, b\})$ and $E_3 = \mathrm{Cn}(\{\neg a, b\})$ of $\langle D, \emptyset\rangle$. However, there is no extension corresponding to the stationary extension $E_1 = \mathrm{Cn}(\{b\})$ of $\langle D, \emptyset\rangle$, since E_1 is not strongly propositionally consistent. Nevertheless, the translation function $\mathrm{Tr}_{\mathrm{ST1}}$ is very close to being faithful.

Theorem 3. *Let $\langle D, T\rangle$ be a default theory in $\mathcal{L}(\mathcal{A})$ and $\langle D', T \cup T^{\bullet}\rangle$ the translation $\mathrm{Tr}_{\mathrm{ST1}}(\langle D, T\rangle)$ in $\mathcal{L}'(\mathcal{A} \cup \mathcal{A}^{\bullet})$. Then the strongly propositionally consistent stationary extensions of $\langle D, T\rangle$ and the propositionally consistent extensions of $\langle D', T \cup T^{\bullet}\rangle$ are in one-to-one correspondence and coincide up to $\mathcal{L}$.*

There is a further reason to consider $\mathrm{Tr}_{\mathrm{ST1}}$ as an unsatisfactory translation function: it is only weakly modular. This is because $\mathrm{Tr}_{\mathrm{ST1}}$ duplicates the propositional subtheory T in $\mathcal{L}^{\bullet}$, i.e. it forms the theory $T \cup T^{\bullet}$. To enforce full modularity, we should generate $T^{\bullet}$ in terms of defaults. It is shown in the following that this is not possible if we wish to keep $\mathrm{Tr}_{\mathrm{ST1}}$ polynomial.

Proposition 2. *It is impossible to translate a finite set of atoms $\mathcal{A}$ into a fixed set of defaults D in $\mathcal{L}'(\mathcal{A} \cup \mathcal{A}^{\bullet})$ such that (i) the time needed to translate $\mathcal{A}$ is polynomial in $|\mathcal{A}|$ and (ii) for all $T \subseteq \mathcal{L}(\mathcal{A})$, the theory $\langle D, T \rangle$ has a unique extension $E = \mathrm{Cn}(T \cup T^{\bullet}) \subseteq \mathcal{L}'$.*

Proof. It is worth stating some relevant properties of propositional logic. Consider a fixed propositional language $\mathcal{L}(\mathcal{A})$ based on a finite set of atoms $\mathcal{A}$. Any two propositional theories $T_1 \subseteq \mathcal{L}$ and $T_2 \subseteq \mathcal{L}$ are considered to be $\mathcal{L}$-equivalent if $\mathrm{Cn}(T_1) = \mathrm{Cn}(T_2)$. Consequently, there are $2^{2^{|\mathcal{A}|}}$ different propositional theories T in $\mathcal{L}$ up to $\mathcal{L}$-equivalence. This is because the models of any propositional theory $T \subseteq \mathcal{L}$ form a subset of the set of all interpretations $\{I \mid I \subseteq \mathcal{A}\}$ which has the cardinality $2^{|\mathcal{A}|}$. Of course, the number of different theories $T \subseteq \mathcal{L}$ becomes infinite if $\mathcal{L}$-equivalence of theories is not taken into account. Let us also recall that it is possible to represent any theory $T \subseteq \mathcal{L}$ in a disjunctive normal form $\phi_1 \vee \ldots \vee \phi_n$ based on the models $M_i \subseteq \mathcal{A}$ of T such that each disjunct ϕ_i is a conjunction of the literals in $\{a \mid a \in M_i\} \cup \{\neg a \mid a \in \mathcal{A} - M_i\}$.

Let us then assume that $\mathcal{A}$ can be translated into a fixed set of defaults D in $\mathcal{L}'(\mathcal{A} \cup \mathcal{A}^{\bullet})$ such that (i) and (ii) hold. Consequently, the length $||D||$ is also majored by a polynomial $p(|\mathcal{A}|)$. Moreover, the unique extension of $\langle D, T \rangle$ is of the form $\mathrm{Cn}(T \cup \Gamma) \subseteq \mathcal{L}'$ where $\Gamma \subseteq \mathrm{Cseq}(D)$ [16] regardless of the choice for T. It is clear that $p(|\mathcal{A}|)$ provides also an upper limit for $|\mathrm{Cseq}(D)|$. Since $T \subseteq \mathcal{L}$, the theory $E = \mathrm{Cn}(T \cup \Gamma)$ has at most $2^{p(|\mathcal{A}|)}$ different projections with respect to $\mathcal{L}^{\bullet}$ up to $\mathcal{L}^{\bullet}$-equivalence. Let us then consider a sufficiently large set of atoms $\mathcal{A}$ such that $p(|\mathcal{A}|) < 2^{|\mathcal{A}|}$ (this is possible regardless of the polynomial $p(|\mathcal{A}|)$) and the set of defaults D obtained as a translation. Now the number of different propositional theories in $\mathcal{L}^{\bullet}$ (up to $\mathcal{L}^{\bullet}$-equivalence) exceeds that of projections $\mathrm{Cn}(T \cup \Gamma) \cap \mathcal{L}^{\bullet}$ (up to $\mathcal{L}^{\bullet}$-equivalence). Consequently, there is a theory $S^{\bullet} \subseteq \mathcal{L}^{\bullet}$ which is not propositionally equivalent to any of the projections $\mathrm{Cn}(T \cup \Gamma) \cap \mathcal{L}^{\bullet}$ where $T \subseteq \mathcal{L}$ and $\Gamma \subseteq \mathrm{Cseq}(D)$. This means that $\langle D, S \rangle$ cannot have an extension E such that $E \cap \mathcal{L}^{\bullet} = S^{\bullet}$. But this would be the case if $\langle D, S \rangle$ had a unique extension $E = \mathrm{Cn}(S \cup S^{\bullet})$, a contradiction with (ii). $\square$

However, there is a modular but exponential translation of $\mathcal{A}$ into a set of defaults that satisfies the second criteria of Proposition 2. Given a finite set of literals $L = \{l_1, \ldots, l_n\}$, we write $\bigvee L$ to denote the sentence $l_1 \vee \ldots \vee l_n$. A set of atoms $\mathcal{A}$ is translated into a set of defaults $D = \{\frac{\bigvee L}{\bigvee L^{\bullet}} \mid L \subseteq \mathcal{A} \cup \{\neg a \mid a \in \mathcal{A}\}\}$. The length of D grows exponentially in $|\mathcal{A}|$. Since each finite $T \subseteq \mathcal{L}(\mathcal{A})$ is equivalent to a sentence $(\bigvee L_1) \wedge \ldots \wedge (\bigvee L_n)$ in a conjunctive normal form where each $L_i \subseteq \mathcal{A} \cup \{\neg a \mid a \in \mathcal{A}\}$, it is clear that the unique extension E of $\langle D, T \rangle$ contains exactly the logical consequences of $(\bigvee L_1) \wedge \ldots \wedge (\bigvee L_n)$ and $(\bigvee L_1^{\bullet}) \wedge \ldots \wedge (\bigvee L_n^{\bullet})$. Thus $E = \mathrm{Cn}(T \cup T^{\bullet})$ results for all $T \subseteq \mathcal{L}(\mathcal{A})$.

6   A Fully Modular Translation

The analysis in Section 5 reveals two weaknesses of the translation function $\mathrm{Tr}_{\mathrm{ST1}}$, as it is only weakly modular and it is not faithful, i.e. it does not capture all propositionally consistent stationary extensions. In this section, we shall consider another technique in order to overcome these shortcomings of Tr_{ST1}. The technique is adopted from [1] where Bonatti and Eiter embed DL into PDL (such a translation cannot be PFM, as indicated by the classes of EPH [13]). In their approach, new atoms are introduced as *guards* (i.e., as antecedents of implications) in order to encode several propositional theories in one.

Before demonstrating guarding atoms in practice, let us introduce some notation. Given $T \subseteq \mathcal{L}(\mathcal{A})$ and a new atom $\mathsf{g} \notin \mathcal{A}$, we let T^{g} denote the theory $\{\mathsf{g} \to \phi \mid \phi \in T\}$ where the sentences of T are guarded by g. Similarly for a set of inference rules R in $\mathcal{L}(\mathcal{A})$ and a new atom $\mathsf{g} \notin \mathcal{A}$, we define $R^{\mathsf{g}} = \{\frac{\mathsf{g} \to \alpha}{\mathsf{g} \to \gamma} \mid \frac{\alpha}{\gamma} \in R\}$. Then consider propositional theories $T_1 = \{\mathsf{a}, \mathsf{a} \to \mathsf{b}\}$ and $T_2 = \{\neg \mathsf{b}\}$. Using guards g_1 and g_2, we define a theory $T = T^{\mathsf{g}_1} \cup T^{\mathsf{g}_2} = \{\mathsf{g}_1 \to \mathsf{a}, \mathsf{g}_1 \to (\mathsf{a} \to \mathsf{b}), \mathsf{g}_2 \to \neg \mathsf{b}\}$. The guards g_1 and g_2 let us distinguish the two subtheories within T. For instance, $T \models \mathsf{g}_1 \to \mathsf{b}$ holds, since $T_1 \models \mathsf{b}$ holds. Moreover, we have that $T \models \mathsf{g}_2 \to \neg \mathsf{b}$, because $T_2 \models \neg \mathsf{b}$ holds. It is also possible to combine guards: $T \models \mathsf{g}_1 \wedge \mathsf{g}_2 \to \bot$ holds, since $T_1 \cup T_2$ is propositionally inconsistent. Note that T remains propositionally consistent although this is the case. Let us then state some useful properties of theories and sentences involving one guarding atom (a generalization for multiple guards is also possible).

Lemma 3. *Let T_1 and T_2 be propositional theories in $\mathcal{L}(\mathcal{A})$ and $\mathsf{g} \notin \mathcal{A}$ a new atom. Then it holds for any $\phi \in \mathcal{L}$ that (i) $(T_1 \cup (T_2)^{\mathsf{g}}) * \phi \Leftrightarrow T_1 * \phi$, (ii) $(T_1 \cup (T_2)^{\mathsf{g}}) * (\mathsf{g} \wedge \phi) \Leftrightarrow (T_1 \cup T_2) * \phi$, (iii) $T_1 \cup (T_2)^{\mathsf{g}} \models \phi \Leftrightarrow T_1 \models \phi$ and (iv) $T_1 \cup (T_2)^{\mathsf{g}} \models \mathsf{g} \to \phi \Leftrightarrow T_1 \cup T_2 \models \phi$.*

Our forthcoming translation will use only one guarding atom, namely p, which refers to any "potential" conclusion associated with a stationary extension. This resembles our previous approach in which a potential conclusion ϕ is encoded as $\phi^{\bullet}$. Given a stationary extension E_1 and $E_2 = \text{Cn}^{D_{E_1}}(T)$, our idea is (i) to include E_1 (i.e. the set of conclusions) without guards and (ii) to represent E_2 (i.e. the set of potential conclusions) using p as a guard. This approach provides an implicit encoding of the inclusion $E_1 \subseteq E_2$, since $\models \phi \to (\mathsf{p} \to \phi)$ holds for any propositional sentence $\phi \in \mathcal{L}$. This is the key observation that lets us to define a fully modular translation: there is no need to provide a separate translation for the propositional subtheory T (in contrast to Tr_{ST1}).

Definition 5. *For any default theory $\langle D, T \rangle$ in $\mathcal{L}(\mathcal{A})$, let $\text{Tr}_{\text{ST2}}(\langle D, T \rangle) =$*

$$\langle \{ \frac{\alpha : \mathsf{p} \wedge \beta_1, \ldots, \mathsf{p} \wedge \beta_n}{\gamma}, \frac{\mathsf{p} \to \alpha : \beta_1, \ldots, \beta_n}{\mathsf{p} \to \gamma} \mid \frac{\alpha : \beta_1, \ldots, \beta_n}{\gamma} \in D\}, T \rangle$$

where p is a new atom not appearing in $\mathcal{A}$.

Using the first two items of Lemma 3, the reduct of the set of defaults introduced by Tr_{ST2} may be computed.

Proposition 3. *Let $\langle D, T \rangle$ be a default theory in $\mathcal{L}(\mathcal{A})$ and $\langle D', T \rangle$ the translation $\text{Tr}_{\text{ST2}}(\langle D, T \rangle)$ in $\mathcal{L}'(\mathcal{A} \cup \{\mathsf{p}\})$. Moreover, let $E = \text{Cn}(E_1 \cup (E_2)^{\mathsf{p}})$ and $E_1 \subseteq E_2$ hold for theories E_1 and E_2 in $\mathcal{L}$. Then it holds for any default $\frac{\alpha : \beta_1, \ldots, \beta_n}{\gamma} \in D$ that (i) $\frac{\mathsf{p} \to \alpha}{\mathsf{p} \to \gamma} \in D'_E \Leftrightarrow \frac{\alpha}{\gamma} \in D_{E_1}$ and (ii) $\frac{\alpha}{\gamma} \in D'_E \Leftrightarrow \frac{\alpha}{\gamma} \in D_{E_2}$.*

By the theorems that follow, we shall establish that Tr_{ST2} fulfills the requirements set up in Section 4. In contrast to Tr_{ST1}, the translation function Tr_{ST2} is modular and it captures also propositionally consistent stationary extensions which are not strongly propositionally consistent. As a matter of fact, even propositionally inconsistent stationary extensions are captured by Tr_{ST2}.

Theorem 4. *Let $\langle D, T \rangle$ be a default theory in $\mathcal{L}(\mathcal{A})$ and $\langle D', T \rangle$ the translation $\text{Tr}_{\text{ST2}}(\langle D, T \rangle)$ in $\mathcal{L}'(\mathcal{A} \cup \{p\})$. If $E_1 \subseteq \mathcal{L}$ is a stationary extension of $\langle D, T \rangle$ and $E_2 = \text{Cn}^{D_{E_2}}(T)$, then $E = \text{Cn}(E_1 \cup (E_2)^p) \subseteq \mathcal{L}'$ is an extension of $\langle D', T \rangle$.*

Proof sketch. Let E_1 be a stationary extension of $\langle D, T \rangle$ and $E_2 = \text{Cn}^{D_{E_1}}(T)$. Then define the theory $E = \text{Cn}(E_1 \cup (E_2)^p) \subseteq \mathcal{L}'$. Since $E_1 \subseteq E_2$, it follows by Proposition 3 that $D'_E = D_{E_2} \cup (D_{E_1})^p$. It remains to be established that $\text{Cn}^{D_{E_2} \cup (D_{E_1})^p}(T) = \text{Cn}(E_1 \cup (E_2)^p)$. $(\subseteq)$ It can be shown by Lemma 3 that $\text{Cn}(E_1 \cup (E_2)^p)$ has sufficient closure properties: (i) $T \subseteq \text{Cn}(E_1 \cup (E_2)^p)$, (ii) if $\frac{\alpha}{\gamma} \in D_{E_2}$ and $\alpha \in \text{Cn}(E_1 \cup (E_2)^p)$, then also $\gamma \in \text{Cn}(E_1 \cup (E_2)^p)$, (iii) if $(\frac{p \rightarrow \alpha}{p \rightarrow \gamma}) \in (D_{E_1})^p$ and $(p \rightarrow \alpha) \in \text{Cn}(E_1 \cup (E_2)^p)$, then also $p \rightarrow \gamma \in \text{Cn}(E_1 \cup (E_2)^p)$, and (iv) $\text{Cn}(E_1 \cup (E_2)^p)$ is propositionally closed in $\mathcal{L}'$. $(\supseteq)$ It can be shown that $T' = \text{Cn}^{D_{E_2} \cup (D_{E_1})^p}(T)$ shares the essential closure properties of $E_1 = \text{Cn}^{D_{E_2}}(T) \subseteq \mathcal{L}$ and $(E_2)^p = \text{Cn}^{(D_{E_1})^p}(T^p) \subseteq \mathcal{L}^p$: (i) $T \subseteq T'$ and $T^p \subseteq T'$, (ii) T' is closed under the rules of D_{E_2} and the rules of $(D_{E_1})^p$, and (iii) T' is propositionally closed in $\mathcal{L}$ and $\mathcal{L}^p$. Thus $E_1 \subseteq T'$ and $(E_2)^p \subseteq T'$ so that $\text{Cn}(E_1 \cup (E_2)^p) \subseteq T'$ holds. □

Theorem 5. *Let $\langle D, T \rangle$ be a default theory in $\mathcal{L}(\mathcal{A})$ and $\langle D', T \rangle$ the translation $\text{Tr}_{\text{ST2}}(\langle D, T \rangle)$ in $\mathcal{L}'(\mathcal{A} \cup \{p\})$. If $E \subseteq \mathcal{L}'$ is an extension of the translation $\langle D', T \rangle$, then $E_1 = E \cap \mathcal{L}$ is a stationary extension of $\langle D, T \rangle$ such that $E_2 = \{\phi \in \mathcal{L} \mid p \rightarrow \phi \in E\}$ satisfies $E_2 = \text{Cn}^{D_{E_1}}(T)$.*

Proof sketch. Let $E = \text{Cn}^{D'_E}(T)$ be an extension of $\langle D', T \rangle$ and let E_1 and E_2 be defined as above. Moreover, define $\Gamma_1 = \{\gamma \mid \frac{\alpha}{\gamma} \in D'_E$ and $\alpha \in E\}$ and $\Gamma_2 = \{\gamma \mid \frac{p \rightarrow \alpha}{p \rightarrow \gamma} \in D'_E$ and $p \rightarrow \alpha \in E\}$. It follows by a characterization of extensions [16] that $E = \text{Cn}(T \cup \Gamma_1 \cup (\Gamma_2)^p)$. Thus $E_1 = \text{Cn}(T \cup \Gamma_1)$, $E_2 = \text{Cn}(T \cup \Gamma_1 \cup \Gamma_2)$ and $E = \text{Cn}(E_1 \cup (E_2)^p)$ hold by Lemma 3. It follows that $E_1 \subseteq E_2$.

(A) It is established that $E_1 = E \cap \mathcal{L}$ equals to $\text{Cn}^{D_{E_2}}(T)$. $(\subseteq)$ Consider any $\phi \in E_1$ so that $\phi \in \mathcal{L}$, $\phi \in E$ and ϕ is D'_E-provable from T in $i \geq 0$ steps. It can be proved by induction on i that $\phi \in \text{Cn}^{D_{E_2}}(T)$ holds using Lemma 3 and Proposition 3. $(\supseteq)$ It can be shown using Proposition 3 that E_1 has the closure properties of $\text{Cn}^{D_{E_2}}(T)$: (i) $T \subseteq E_1$, (ii) if $\frac{\alpha}{\gamma} \in D_{E_2}$ and $\alpha \in E_1$, then also $\gamma \in E_1$, and (iii) $E_1 = E \cap \mathcal{L}$ is propositionally closed in $\mathcal{L} \subset \mathcal{L}'$.

(B) It remains to be shown that E_2 equals to $\text{Cn}^{D_{E_1}}(T)$. $(\subseteq)$ Consider any $\phi \in E_2$. It follows that $\phi \in \mathcal{L}$ and $p \rightarrow \phi \in E$, i.e. $p \rightarrow \phi$ is D'_E-provable from T in $i \geq 0$ steps. Then it can be proved by induction on i that $\phi \in \text{Cn}^{D_{E_1}}(T)$ holds using Lemma 3 and Proposition 3. In particular, note that $E_1 \subseteq E_2$ implies $D_{E_2} \subseteq D_{E_1}$. $(\supseteq)$ It can be shown by Proposition 3 that E_2 shares the closure properties of $\text{Cn}^{D_{E_1}}(T)$: (i) $T \subseteq E_2$, (ii) if $\frac{\alpha}{\gamma} \in D_{E_1}$ and $\alpha \in E_2$, then also $\gamma \in E_2$, and (iii) $E_2 = \{\phi \in \mathcal{L} \mid p \rightarrow \phi \in E\}$ is propositionally closed in $\mathcal{L}$. □

Theorem 6. *Let $\langle D,T\rangle$ be a default theory in $\mathcal{L}(\mathcal{A})$ and $\langle D',T\rangle$ the translation $\mathrm{Tr}_{\mathrm{ST2}}(\langle D,T\rangle)$ in $\mathcal{L}'(\mathcal{A}\cup\{p\})$. Then the stationary extensions of $\langle D,T\rangle$ and the extensions of $\langle D',T\rangle$ are in one-to-one correspondence and coincide up to $\mathcal{L}$.*

Proof sketch. Theorems 4 and 5 provide us two implicit mappings. The first one maps a stationary extension $E \subseteq \mathcal{L}$ of $\langle D,T\rangle$ to an extension $m_1(E) = \mathrm{Cn}(E \cup (E')^p)$ of $\langle D',T\rangle$ where $E' = \mathrm{Cn}^{D_E}(T)$. The second one maps an extension $E \subseteq \mathcal{L}'$ of $\langle D',T\rangle$ to a stationary extension $m_2(E) = E \cap \mathcal{L}$ of $\langle D,T\rangle$. Using Lemma 3 and the proof of Theorem 5, it can be shown that m_1 and m_2 are injective and inverses of each other. Moreover, the extensions involved in the one-to-one correspondence coincide up to $\mathcal{L}$ by the definition of m_2. $\square$

From now on, our goal is to to locate the exact position of STDL in EPH [12,13]. The results established so far let us draw the first conclusion in this respect. The translation function $\mathrm{Tr}_{\mathrm{ST2}}$ is PFM by Definition 5 and Theorems 4–6 (restricted to propositionally consistent extensions). We conclude the following.

Corollary 1. STDL $\xrightarrow{\text{PFM}}$ DL.

Theorems 7 and 8 establish that STDL resides between CL and DL in EPH.

Theorem 7. STDL $\overrightarrow{\underset{\text{PFM}}{\Rightarrow}}$ DL.

Proof. Consider a set of defaults $D = \{\frac{:a}{a}, \frac{:\neg a}{\neg a}\}$ in $\mathcal{L}$ based on $\mathcal{A} = \{a\}$. The default theory $\langle D,\emptyset\rangle$ has two propositionally consistent extensions: $E_1 = \mathrm{Cn}(\{a\})$ and $E_2 = \mathrm{Cn}(\{\neg a\})$. Suppose there is a PFM translation function Tr that maps $\langle D,\emptyset\rangle$ to a default theory $\langle D',T'\rangle$ in $\mathcal{L}'$ based on $\mathcal{A}' \supseteq \mathcal{A}$ such that the propositionally consistent extensions of the former and the propositionally consistent stationary extensions of the latter are in one-to-one correspondence and coincide up to $\mathcal{L}$. Then the translation $\langle D',T'\rangle$ has at least one propositionally consistent stationary extension E by the one-to-one correspondence of extensions. Consequently, the least stationary extension F of $\langle D',T'\rangle$ is also propositionally consistent, since F is contained in E which is propositionally consistent.

Then consider the extension of $\langle D,\emptyset\rangle$ corresponding to F which is either E_1 or E_2. Let us analyze the case that E_1 corresponds to F (the case that E_2 corresponds to F is covered by symmetry). Since $a \in E_1$, it follows that $a \in F$ by the faithfulness of Tr. Then let E' be the stationary extension of $\langle D',T'\rangle$ corresponding to E_2. Since $F \subseteq E'$ it follows that $a \in E'$. Thus $a \in E_2$ by the faithfulness of Tr, a contradiction. Hence DL $\xrightarrow{\;\;}{}\!\!\!\!\!/_{\,F}$ STDL and DL $\xrightarrow{}\!\!\!\!\!/_{\,\text{PFM}}$ STDL. $\square$

Theorem 8. CL $\overrightarrow{\underset{\text{PFM}}{\Rightarrow}}$ STDL.

Proof. The unique extension associated with a classical propositional theory $T \subseteq \mathcal{L}(\mathcal{A})$ is $\mathrm{Cn}(T)$. Consider the translation function $\mathrm{Tr}(T) = \langle\emptyset,T\rangle$. It is clear that the default theory $\mathrm{Tr}(T)$ has a unique stationary extension $E = \mathrm{Cn}^{\emptyset_E}(T) = \mathrm{Cn}^{\emptyset}(T) = \mathrm{Cn}(T)$ regardless of T. Thus CL $\xrightarrow{\text{PFM}}$ STDL. $\square$

Then consider the set of defaults $D = \{\frac{:\neg b}{a}, \frac{:\neg a}{b}\}$ and the possibilities of translating the default theory $\langle D, \emptyset \rangle$ under stationary extensions into a classical propositional theory T'. Now $\langle D, \emptyset \rangle$ has three stationary extensions, namely $E_1 = \mathrm{Cn}(\emptyset)$, $E_2 = \mathrm{Cn}(\{a\})$ and $E_3 = \mathrm{Cn}(\{b\})$. However, the translation has only one extension $\mathrm{Cn}(T')$. Hence STDL $\overset{\not\longrightarrow}{\mathrm{F}}$ CL and STDL $\overset{\not\longrightarrow}{\mathrm{PFM}}$ CL.   □

The set of defaults D involved in the proof of Theorem 7 is normal and prerequisite-free. We may conclude the following by the same counter-example.

Corollary 2. NDL $\overset{\not\longrightarrow}{\mathrm{PFM}}$ STDL, PDL $\overset{\not\longrightarrow}{\mathrm{PFM}}$ STDL *and* PNDL $\overset{\not\longrightarrow}{\mathrm{PFM}}$ STDL.

It remains to explore whether STDL is captured by PDL, PNDL and NDL.

Theorem 9. STDL $\overset{\not\longrightarrow}{\mathrm{PFM}}$ PDL *and* STDL $\overset{\not\longrightarrow}{\mathrm{PFM}}$ PNDL.

Proof. Consider the set of defaults $D = \{\frac{a:}{b}, \frac{a\to b:}{a}\}$ (adopted from [9, Theorem 3.2]) and the theories $T_1 = \{a\}$, $T_2 = \{a \to b\}$ and $T_3 = \{a, a \to b\}$ in $\mathcal{L}(\mathcal{A})$ where $\mathcal{A} = \{a, b\}$. Each default theory $\langle D, T_i \rangle$ where $i \in \{1, 2, 3\}$ has a unique propositionally consistent stationary extension $E = \mathrm{Cn}(\{a, b\})$. Then suppose that there is a PFM translation function $\mathrm{Tr}_{\mathrm{PDL}}$ from STDL to PDL. Let $\langle D', T' \rangle$ be the translation $\mathrm{Tr}_{\mathrm{PDL}}(\langle D, \emptyset \rangle)$ in $\mathcal{L}'(\mathcal{A}')$ where $\mathcal{A}' \supseteq \mathcal{A}$. Since $\mathrm{Tr}_{\mathrm{PDL}}$ is modular, we know that $\mathrm{Tr}_{\mathrm{PDL}}(\langle D, T_i \rangle) = \langle D', T' \cup T_i \rangle$ holds for every $i \in \{1, 2, 3\}$. By the faithfulness of $\mathrm{Tr}_{\mathrm{PDL}}$, each default theory $\langle D', T' \cup T_i \rangle$ with $i \in \{1, 2, 3\}$ has a unique propositionally consistent extension E_i' such that $E = E_i' \cap \mathcal{L}$. Since D' is prerequisite-free, each extension E_i' is of the form $\mathrm{Cn}(T' \cup T_i \cup \Gamma_i)$ where Γ_i is the set of consequents $\{\gamma \mid \frac{T:\beta_1,\dots,\beta_n}{\gamma} \in D'$ and $\forall j \in \{1, \dots, n\} : E_i' * \beta_j\}$.

Since $a \to b \in E \cap \mathcal{L}$, it follows that $a \to b \in E_1'$ holds for $E_1' = \mathrm{Cn}(T' \cup T_1 \cup \Gamma_1)$. Thus $E_1' = \mathrm{Cn}(T' \cup T_3 \cup \Gamma_1)$ so that E_1' is also a propositionally consistent extension of $\langle D', T' \cup T_3 \rangle$. On the other hand, it holds that $a \in E \cap \mathcal{L}$. Thus $a \in E_2'$ holds for $E_2' = \mathrm{Cn}(T' \cup T_2 \cup \Gamma_2)$. It follows that $E_2' = \mathrm{Cn}(T' \cup T_3 \cup \Gamma_2)$, i.e. E_2' is also a propositionally consistent extension of $\langle D', T' \cup T_3 \rangle$.

Then $E_1' = E_2' = E_3'$ is the case, as E_3' is the unique propositionally consistent extension of $\langle D', T' \cup T_3 \rangle$. It follows that $\Gamma_1 = \Gamma_2 = \Gamma_3$ as well. Thus we let E' denote any of E_1', E_2' and E_3', as well as Γ any of Γ_1, Γ_2 and Γ_3. Recall that E' is a propositionally consistent extension of $\langle D', T' \cup T_1 \rangle$ and $b \in E'$, since $b \in E$. It follows that $T' \cup \{a\} \cup \Gamma \models b$ as well as that $T' \cup \Gamma \models a \to b$. Thus $E' = \mathrm{Cn}(T' \cup T_2 \cup \Gamma) = \mathrm{Cn}(T' \cup \Gamma)$ holds, indicating that E' is also an extension of $\langle D', T' \rangle$. A contradiction, since $a \in E'$ and $b \in E'$, but the unique propositionally consistent stationary extension of $\langle D, \emptyset \rangle$ is $\mathrm{Cn}(\emptyset)$. Hence STDL $\overset{\not\longrightarrow}{\mathrm{PFM}}$ PDL.

Let us then assume that STDL $\overset{\longrightarrow}{\mathrm{PFM}}$ PNDL. Since PNDL $\overset{\longrightarrow}{\mathrm{PFM}}$ PDL holds by the classes of EPH, we obtain STDL $\overset{\longrightarrow}{\mathrm{PFM}}$ PDL by the *compositionality* of PFM translation functions [13], a contradiction. Hence STDL $\overset{\not\longrightarrow}{\mathrm{PFM}}$ PNDL.   □

Theorem 10. STDL $\not\xrightarrow[\text{PFM}]{}$ NDL.

Proof. Consider a set of defaults $D = \{\frac{a:}{\neg a}\}$ and a theory $T = \{a\}$ in $\mathcal{L}(\{a\})$. Note that the default theory $\langle D, T \rangle$ has no propositionally consistent stationary extensions. Suppose there is a PFM translation function Tr such that $\text{Tr}(\langle D, T \rangle) = \langle D', T' \cup T \rangle$ is a normal default theory which guaranteed to have an extension E' [22]. Since Tr is faithful, E' must be propositionally inconsistent. Thus $T' \cup T$ must be propositionally inconsistent [22]. It follows that $T' \models \neg a$.

On the other hand, the default theory $\langle D, \emptyset \rangle$ has a propositionally consistent stationary extension $E = \text{Cn}(\emptyset)$. By modularity, the translation $\text{Tr}(\langle D, \emptyset \rangle)$ is $\langle D', T' \rangle$. By faithfulness, the translation $\langle D', T' \rangle$ has a corresponding propositionally consistent extension $F = \text{Cn}^{D'F}(T')$ such that $E = F \cap \mathcal{L}$. Since $T' \models \neg a$, it follows that $\neg a \in F$. A contradiction, since $\neg a \notin E = \text{Cn}(\emptyset)$. $\square$

By the theorems presented, STDL is incomparable with PDL, PNDL and NDL. Thus STDL is located in its own class of EPH (not present in Fig. 1).

6.1   Regular Extensions

Let us address a further semantics for default logic which is obtained as a generalization of *regular models* proposed for normal logic programs by You and Yuan [26]. An alternating fix-point M of a normal logic program P is a regular model of P if there is no alternating fix-point M' of P such that $M \subset M'$. In this way, regular models minimize undefinedness. Stable models of P are also regular models of P but in general, a normal logic program may possess more regular models than stable models. *Regular extensions* are definable for default theories in an analogous fashion as maximal stationary extensions.

Definition 6. *A stationary extension E of a default theory $\langle D, T \rangle$ is a regular extension of $\langle D, T \rangle$ iff $\langle D, T \rangle$ has no stationary extension E' such that $E \subset E'$.*

Despite this maximization principle, stationary and regular extensions behave very similarly under the brave reasoning approach. More precisely, a query ϕ belongs to some *regular extension E* of a default theory $\langle D, T \rangle$ if and only if ϕ belongs to some *stationary extension* of $\langle D, T \rangle$. By this tight interconnection of decision problems, Gottlob's complexity results [7,8] imply that brave reasoning with regular extensions forms a Σ_2^p-complete decision problem in analogy to brave reasoning with stationary extensions. The results of this paper enable implementing brave reasoning with stationary and regular extensions. In addition to an inference engine for brave reasoning with Reiter's extensions (such as the system DeReS [2]) we need a program that computes the translation $\text{Tr}_{\text{ST2}}(\langle D, T \rangle)$ for a default theory $\langle D, T \rangle$ given as input.

7   Conclusions and Future Work

In this paper, we have analyzed the possibilities of reducing *stationary default logic* (i.e., default theories under stationary extensions) to Reiter's default logic (i.e., default theories under Reiter's extensions). It turned out that the translation function proposed for normal and disjunctive logic programs [14] does not generalize for default theories in a satisfactory way. In fact, it is established in Section 5 that a PFM translation function cannot be obtained using a similar technique. Fortunately, guarding atoms provide an alternative technique that leads to a PFM translation function in Section 6. This is how we obtain further evidence for the adequacy of PFM translations, because even non-monotonic logics with a partial semantics can be classified using the existence of a PFM translation function as the criterion. It is also interesting to note that Tr_{ST2} does not specialize for normal nor disjunctive logic programs, since conditional inference with guards is not supported by them. However, the situation could be different if *nested logic programs* [15] are taken into consideration. Moreover, the properties of stationary and regular extensions and the translation function Tr_{ST2} enable implementing brave reasoning with stationary and regular extensions simply by using existing implementations of DL (such as DeReS [2]).

By the theorems presented, the stationary default logic (STDL) is strictly less expressive than default logic (DL), but strictly more expressive than classical propositional logic (CL). Moreover, STDL is incomparable with the other representatives of the classes of EPH: NDL (normal DL), PDL (prerequisite-free DL) and PNDL (prerequisite-free and normal DL). Thus STDL determines a class of its own between CL and DL. This is quite understandable, since STDL is the only non-monotonic logic based on a partial semantics and located in EPH. Nevertheless, the results of this paper indicate that EPH can be extended further with semantic variants of default logic. Only weak default logic (WDL) has been considered earlier while a number of syntactic variants have been already classified. One obvious way to extend EPH is to analyze syntactic variants of default logic under stationary extensions. Moreover, analogs of stationary extensions [10,3] have been proposed for Moore's autoepistemic logic [18] and Reiter's closed world assumption (CWA) [21] can be understood as the "stationary counterpart" of McCarthy's circumscription [17] as shown in [11]. It seems that a partial fragment of EPH can be established by comparing STDL with these logics such that STDL links this fragment to the rest of EPH.

References

1. P.A. Bonatti and T. Eiter. Querying disjunctive database through nonmonotonic logics. *Theoretical Computer Science*, 160:321–363, 1996.
2. P. Cholewiński, V.W. Marek, A. Mikitiuk, and M. Truszczyński. Computing with default logic. *Artificial Intelligence*, 112:105–146, 1999.
3. M. Denecker, W. Marek, and M. Truszczyński. Uniform semantic treatment of default and autoepistemic logic. In *Principles of Knowledge Representation and Reasoning: Proceedings of the 7th International Conference*, pages 74–84, Breckenridge, Colorado, April 2000. Morgan Kaufmann.
4. J. Dix. Default theories of Poole-type and a method for constructing cumulative versions of default logic. In B. Neumann, editor, *Proceedings of the 10th European Conference on AI*, pages 289–293, Vienna, Austria, August 1992. Wiley.

5. M. Gelfond and V. Lifschitz. The stable model semantics for logic programming. In *Proceedings of the 5th International Conference on Logic Programming*, pages 1070–1080, Seattle, USA, August 1988. The MIT Press.

6. M. Gelfond and V. Lifschitz. Classical negation in logic programs and disjunctive databases. *New Generation Computing*, 9:365–385, 1991.

7. G. Gottlob. Complexity results for nonmonotonic logics. *Journal of Logic and Computation*, 2(3):397–425, June 1992.

8. G. Gottlob. The complexity of default reasoning under the stationary fixed point semantics. *Information and Computation*, 121:81–92, 1995.

9. G. Gottlob. Translating default logic into standard autoepistemic logic. *Journal of the Association for Computing Machinery*, 42(2):711–740, 1995.

10. T. Janhunen. Separating disbeliefs from beliefs in autoepistemic reasoning. In J. Dix, U. Furbach, and A. Nerode, editors, *Proceedings of the 4th International Conference on Logic Programming and Non-Monotonic Reasoning*, pages 132–151, Dagstuhl, Germany, July 1997. Springer-Verlag. LNAI 1265.

11. T. Janhunen. Non-monotonic systems: A framework for analyzing semantics and structural properties of non-monotonic reasoning. Doctoral dissertation. Research report A49, Helsinki University of Technology, Digital Systems Laboratory, Espoo, Finland, March 1998. 211 p.

12. T. Janhunen. Classifying semi-normal default logic on the basis of its expressive power. In M. Gelfond, N. Leone, and G. Pfeifer, editors, *Proceedings of the 5th International Conference on Logic Programming and Non-Monotonic Reasoning, LPNMR'99*, pages 19–33, El Paso, Texas, December 1999. Springer-Verlag. LNAI.

13. T. Janhunen. On the intertranslatability of non-monotonic logics. *Annals of Mathematics in Artificial Intelligence*, 27(1-4):79–128, 1999.

14. T. Janhunen, I. Niemel, P. Simons, and J.-H. You. Unfolding partiality and disjunctions in stable model semantics. In *Principles of Knowledge Representation and Reasoning: Proceedings of the 7th International Conference*, pages 411–422, Breckenridge, Colorado, April 2000. Morgan Kaufmann.

15. V. Lifschitz, L.R. Tang, and H. Turner. Nested expressions in logic programs. *Annals of Mathematics in Artificial Intelligence*, 25:369–389, 1999.

16. W. Marek and M. Truszczyński. *Nonmonotonic Logic: Context-Dependent Reasoning*. Springer-Verlag, Berlin, 1993.

17. J. McCarthy. Circumscription—a form of non-monotonic reasoning. *Artificial Intelligence*, 13:27–39, 1980.

18. R.C. Moore. Semantical considerations on nonmonotonic logic. *Artificial Intelligence*, 25:75–94, 1985.

19. H. Przymusinska and T.C. Przymusinski. Stationary default extensions. In *Working Notes of the 4th International Workshop on on Nonmonotonic Reasoning*, pages 179–193, Plymouth, Vermont, USA, May 1992.

20. T. Przymusinski. Extended stable semantics for normal and disjunctive logic programs. In *Proceedings of the 7th International Conference on Logic Programming*, pages 459–477. MIT Press, 1990.

21. R. Reiter. On closed world data bases. In H. Gallaire and J. Minker, editors, *Logic and Data Bases*, pages 55–76. Plenum Press, New York, 1978.

22. R. Reiter. A logic for default reasoning. *Artificial Intelligence*, 13:81–132, 1980.

23. A. Van Gelder. The alternating fixpoints of logic programs with negation. In *ACM Symposium on Principles of Database Systems*, pages 1–10, 1989.

24. A. Van Gelder, K.A. Ross, and J.S. Schlipf. Unfounded sets and the well-founded semantics for general logic programs, extended abstract. In *Proceedings of the*

7th Symposium on Principles of Database Systems, pages 221–230, Austin, Texas, March 1988. ACM Press.

25. A. Van Gelder, K.A. Ross, and J.S. Schlipf. The well-founded semantics for general logic programs. *Journal of the ACM*, 38(3):620–650, July 1991.

26. J.-H. You and L. Yuan. A three-valued semantics for deductive databases and logic programs. *Journal of Computer and System Sciences*, 49:334–361, 1994.

Representing the Process Semantics in the Event Calculus

Chunping Li

Department of Computer Science,
Hong Kong University of Science and Technology,
Clear Water Bay, Kowloon, Hong Kong
cli@cs.ust.hk

Abstract. In this paper we shall present a translation of the process semantics [5] to the event calculus. The aim is to realize a method of integrating high-level semantics with logical calculi to reason about continuous change. The general translation rules and the soundness and completeness theorem of the event calculus with respect to the process semantics are main technical results of this paper.

1 Introduction

In the real world a vast variety of applications need logical reasoning about physical properties in dynamic, continuous systems, e.g., specifying and describing physical systems with continuous actions and changes.

The early research work on this aspect was encouraged to address the problem of representing continuous change in a temporal reasoning formalism [1]. The standard approach is equidistant, discrete time points, namely to quantify the whole scenario into a finite number of points in time at which all system parameters are presented as variables. If there were infinitely many points at infinitely small distance, this might be sufficient. But, since discretization is always finite, a problem arises when an action or event happens in between two of these points.

Some work has been done to extend specific action calculi in order to deal with continuous change. The event calculus [7] is one formalism reasoning about time and change. It uses general rules to derive that a new property holds as the result of the event. In [9, 11, 12, 2], the attempts based on the logical formalisms of the event calculus have been exploited for representing continuous change. However, these ideas have not yet been exploited to define a high level action semantics serving as basis for a formal justification of such calculi, their comparison, and an assessment of the range of their applicability [5].

Whereas these previously described formalisms have directly focused on creating new or extending already existing specialized logical formalisms, the other research direction consists in the development of an appropriate semantics [4, 10, 14] as the basis for a general theory of action and change, and successfully applied to concrete calculi [6, 3, 13]. In [4], the *Action Description Language*

M. Ojeda-Aciego et al. (Eds.): JELIA 2000, LNAI 1919, pp. 118–132, 2000.
© Springer-Verlag Berlin Heidelberg 2000

was developed which is based on the concept of single-step actions, and does not include the notion of time. In [10], the duration of actions is not fixed, but an equidistant discretization of time is assumed and state transitions only occur when actions are executed. In [14], it is allowed for user-independent events to cause state transitions. Again equidistant discretization is assumed. But these formalisms are not suitable for calculi dealing with continuous change.

In 1996, Herrmann and Thielscher [5] proposed a logic of processes for reasoning about continuous change which allows for varying temporal distances between state transitions, and a more general notion of a *process* is proposed as the underlying concept for constructing state descriptions. In the process semantics, a state transition may cause existing processes to disappear and new processes to arise. State transitions are either triggered by the execution of actions or by interactions between processes, which both are specified by transition laws.

In this paper we shall present a translation of the process semantics to the event calculus. The aim is to realize a method of integrating high-level semantics with logical calculi to reason about continuous change. In the following, we first review the event calculus and the logic of processes, and then show how the process semantics can be represented in the event calculus. On this basis, we prove the soundness and completeness of the event calculus with respect to the process semantics.

2   Event Calculus

The event calculus [7] was developed as a theory for reasoning about time and events in a logic programming framework. In the event calculus, the ontological primitives are *events*, which initiate periods during which *properties* hold. A property which has been initiated continues to hold by default until some event occurs which terminates it. Time periods are identified by giving their start and end times which are named by terms of the form $after(e, p)$ or $before(e, p)$ where the first argument is the name of the event which starts or ends the time period and the second argument the name of the property itself. A general, one-argument predicate $hold$ is used to express that a property p holds for a period.

The occurrence of an event e at time t is denoted by $Happens(e, t)$. The formula $Initiates(e, p)$ ($Terminates(e, p)$) means that event e initiates (terminates) the property p.

The reasoning can be formalized by employing a predicate $HoldsAt(p, t)$ where p denotes a property and t a time point:

$$HoldsAt(p, t) \leftarrow Holds(after(e, p)), \quad time(e, t_0),$$
$$In(t, after(e, p)), \quad t_0 \leq t.$$

$$Holds(after(e, p)) \leftarrow Happens(e, t), \quad Initiates(e, p).$$

It means that a property p holds at the time t if p holds for the period after an event e happens at time t_0, and there exists no such an event which happens between t_0 and t and terminates the property p.

The further domain dependent axioms are needed to define the predicates *Happens, Initiates* and *Terminates*.

For example, we express an assertion that the property of *possess*(*Antje, Book*) holds after the event *E(Tom give the book to Antje)* happens. In this case, the predicates *Initiates* and *Terminates* can be defined as:

$$Initiates\,(e, possess\,(x, y)) \leftarrow Act\,(e, Give), \ Recipient\,(e, x), \ Object\,(e, y).$$

$$Terminates\,(e, possess\,(x, y)) \leftarrow Act\,(e, Give), \ Donor\,(e, x), \ Object\,(e, y).$$

where predicates *Act* represents the type of event (action), *Recipient* and *Donor* represent the recipient and the donor of this event (action), and *Object* the object acted be this event (action).

Thereafter, the assertion *HoldsAt*(*possess*(*Antje, Book*), t) can be derived from the predicates defined above for the event description.

3   Logic of Processes

In this section, we introduce a formal, high-level semantics proposed by Herrmann and Thielscher [5], for reasoning about continuous processes, their interaction in the course of time, and their manipulation.

Definition 1. *A* process scheme *is a pair* $\langle C, F \rangle$ *where C is a finite, ordered set of symbols of size $l > 0$ and F is a finite set functions f: $\mathbb{R}^{l+2} \to \mathbb{R}$.*

Example 1. Let $\langle C, F \rangle$ be a process scheme describing continuous movement of an object on a line as follows: $C = \{l_0, v\}$ and $F = \{f(l_0, v, t_0, t) = l_0 + v \cdot (t - t_0)\}$, where l_0 denotes the initial location coordinate, v the velocity, t_0 and t the initial and the actual time, and we denote $l = f(l_0, v, t_0, t)$ as the actual location of the object at time t.

Definition 2. *Let N be a set of symbols (called names). A* process *is a 4-tuple* $\langle n, \tau, t_0, \boldsymbol{p} \rangle$ *where*

1. $n \in N$;
2. $\tau = \langle C, F \rangle$ *is a process scheme where C is of size m;*
3. $t_0 \in \mathbb{R}$; and
4. $\boldsymbol{p} = (p_1, \ldots, p_m) \in \mathbb{R}^m$ *is an m-dimensional vector over $\mathbb{R}$.*

Example 2. Let τ_{move} denote the example scheme from above then

$\langle TrainA, \tau_{move}, 1{:}00pm, (0mi, \; 25mph)\rangle$
$\langle TrainB, \tau_{move}, 1{:}30pm, (80mi, \; -20mph)\rangle$

are two processes describing two trains moving toward each other with different speeds at different starting times.

Definition 3. *A* situation *is a pair* $\langle S, t_s\rangle$ *where* S *is a set of processes and* t_s *is a time-point which denotes the time when* S *started.*

Definition 4. *An* event *is a triple* $\langle P_1, t, P_2\rangle$ *where* P_1 *(the precondition) and* P_2 *(the effect) are finite sets of processes and* $t \in \mathbb{R}$ *is the time at which the event is expected to occur.*

Definition 5. *An event* $\langle P_1, t, P_2\rangle$ *is potentially applicable in a situation* $\langle S, t_s\rangle$ *iff* $P_1 \subseteq S$ *and* $t > t_s$. *If* ε *is a set of events then an event* $\langle P_1, t, P_2\rangle \in \varepsilon$ *is applicable to* $\langle S, t_s\rangle$ *iff it is potentially applicable and for each potentially applicable* $\langle P_1', t', P_2'\rangle \in \varepsilon$ *we have* $t \le t'$.

Example 3. Let S denote the two processes of Example 2. Further, let $t_s = 3{:}00pm$, then the following event, which describes an inelastic collision which is interpreted as a coupling of trains, is applicable to $\langle S, t_s\rangle$:

$\langle P_1 \;\; = \;\; \{\langle TrainA, \tau_{move}, 1{:}00pm, (0mi, 25mph)\rangle,$
$\qquad\qquad \langle TrainB, \tau_{move}, 1{:}30pm, (80mi, -20mph)\rangle\}$

$\quad t \;\; = \;\; 3{:}00pm$

$\quad P_2 \;\; = \;\; \{\langle TrainA, \tau_{move}, 3{:}00pm, (50mi, \; 5mph)\rangle,$
$\qquad\qquad \langle TrainB, \tau_{move}, 3{:}00pm, (50mi, \; 5mph)\rangle\}\rangle$

In fact, concrete events are instances of general *transition laws* which contain variables and constraints to guide the process of instantiation, and the event's time is usually determined by the instances of other variables. We can describe the transition law for inelastic collisions of two continuously moving objects as follows.

$\langle P_1 \;\; = \;\; \{\langle N_A, \tau_{move}, T_{A0}, (X_{A0}, V_A)\rangle,$
$\qquad\qquad \langle N_B, \tau_{move}, T_{B0}, (X_{B0}, V_B)\rangle\}$
$\quad t \;\; = \;\; T$ $\qquad\qquad\qquad\qquad\qquad\qquad\qquad\qquad$ (T1)
$\quad P_2 \;\; = \;\; \{\langle N_A, \tau_{move}, T, (X_{new}, V_A + V_B)\rangle,$
$\qquad\qquad \langle N_B, \tau_{move}, T, (X_{new}, V_A + V_B)\rangle\}\rangle$

where it is required that $N_A \neq N_B$, $V_A - V_B \neq 0$, and $x_A = x_B = X_{new}$ at time T. x_A and x_B represent the actual location of *TrainA* and *TrainB* respectively when the collision occurs. Suppose that the two movement differentials are $x_A = X_{A0} + V_A \cdot (T - T_{A0})$ and $x_B = X_{B0} + V_B \cdot (T - T_{B0})$; then the result is:

$$T = \frac{X_{A0} - X_{B0} - V_A \cdot T_{A0} + V_B \cdot T_{B0}}{V_B - V_A}, \quad X_{new} = X_{A0} + V_A \cdot (T - T_{A0}) \tag{T2}$$

Definition 6. *Let ε be a set of events and $\langle S, t_s \rangle$ a situation, then the* successor *situation $\Phi(\langle S, t_s \rangle)$ is defined as follows.*

1. *If no applicable event exists in ε then $\Phi(\langle S, t_s \rangle) = \langle S, \infty \rangle$;*
2. *if $\langle P_1, t, P_2 \rangle \in \varepsilon$ is the only applicable event then $\Phi(\langle S, t_s \rangle) = \langle S', t_s \rangle$ where $S' = (S \setminus P_1) \cup P_2$ and $t_{s'} = t$;*
3. *Otherwise $\Phi(\langle S, t_s \rangle)$ is undefined, i.e., events here are not allowed to occur simultaneously.*

Definition 7. *An* observation *is an expression of the form $[t] \propto (n) = r$ where*

1. *$t \in \mathbb{R}$ is the time of the observation;*
2. *$\propto$ is either a symbol in C or the name of a function in F for some process scheme $\langle C, F \rangle$;*
3. *n is a symbol denoting a process name; and*
4. *$r \in \mathbb{R}$ is the observed value.*

Given an initial situation and a set of events, such an observation is *true* iff the following holds. Let S be the collection of processes describing the system at time t, then S contains a process $\langle n, (C, F), t_0, (r_1, \ldots, r_n, t_0) \rangle$ such that

1. either $C = (c_0, \ldots, c_{k-1}, \propto, c_{k+1}, \ldots, c_{m-1})$ and $r_k = r$;
2. or $\propto \in F$ and $\propto (r_1, \ldots, r_n, t_0, t) = r$.

Example 4. The observation $[2{:}15\text{pm}]l(TrainB) = 65mi$ is true in Example 3, while the observation $[3{:}15\text{pm}]l(TrainB) = 45mi$ is not true since the latter does not take into account the train collision.

Definition 8. *A* model *for a set of observations Ψ (under given sets of names $\mathcal{N}$ and events $\mathcal{E}$) is a system development $\langle S_0, t_0 \rangle$, $\Phi(\langle S_0, t_0 \rangle)$, $\Phi^2(\langle S_0, t_0 \rangle)$, $\ldots$ which satisfies all elements of Ψ. Such a set Ψ* entails *an (additional) observation ψ iff ψ is true in all models of Ψ.*

All definitions concerning successor situations, developments, and observations carry over to the case where a set of actions, which are to be executed (external events), and interactions between processes (internal events) are given.

4   Translation of the Process Semantics to the Event Calculus

In order to represent the process semantics in the event calculus, we here adopt the formalisms of the event calculus of Kowalski [7] and a variant presented by Shanahan [12].

Let $D = (P, E_{proc})$ be a domain description in the process semantics. D consists of a set of processes P and a sequence of events E_{proc}. The corresponding formalism of the event calculus uses variables of two sorts: event variables $e_1, e_2, \ldots$, time variables $t_1, t_2, \ldots$, and a *process* is represented as a relation $P(n, R, C)$ where n denotes the process name, and R and C the sets of dynamic and static parameters respectively defined in the process semantics. The relation $Q(n, F, R, C)$ expresses the property of the process, which holds true during the period of continuous change. F denotes a finite set of functions describing the relationship between the dynamic and static parameters. In fact, the content of the process scheme in the process semantics is specified by the function $Q(n, F, R, C)$. There are also some predicate symbols whose meaning will be clear from their use in the rules below.

Processes and events defined in the process semantics can be formalized as the following general rules by the event calculus.

$$
\begin{aligned}
&HoldsAt(P(n, R, C), t) \leftarrow \\
&\quad Holds\,(after\,(e, Q(n, F, R, C))),\ time\,(e, t_0), \\
&\quad In\,(t, after\,(e, Q(n, F, R, C))),\ t_0 \leq t, \\
&\quad State\,(t, s),\ HoldsIn\,(P(n, R, C), s), \\
&\quad ContinuousProperty\,(Q(n, F, R, C), t_0, P(n, R, C), t).
\end{aligned}
\tag{G1}
$$

$$
\neg HoldsAt\,(P(n, R, C), t) \leftarrow State\,(t, s),\ \neg HoldsIn\,(P(n, R, C), s).
\tag{G2}
$$

$$
\begin{aligned}
&Holds\,(after\,(e, Q(n, F, R, C))) \leftarrow \\
&\quad EventTrigger\,(e, t),\ Initiates(e, Q(n, F, R, C)).
\end{aligned}
\tag{G3}
$$

$$
EventTrigger\,(e, t) \leftarrow Happens\,(e, t).
\tag{G4}
$$

$$
EventTrigger\,(e, t) \leftarrow ImplicitHappens\,(e, t).
\tag{G5}
$$

$$
\begin{aligned}
&In\,(t, p) \leftarrow Start\,(p, e_1),\ End\,(p, e_2),\ Time\,(e_1, t_1), \\
&\quad Time\,(e_2,\ t_2), t_1 < t < t_2.
\end{aligned}
\tag{G6}
$$

$$
\begin{aligned}
&ContinuousProperty\,(Q(n, F, R, C), t_0, P(n, R, C), t) \leftarrow \\
&\quad R = F(C, t, t_0).
\end{aligned}
\tag{G7}
$$

In (G1) the predicate *ContinuousProperty* in the event calculus treats continuous change in correspondence with the process semantics. It means that property $P(n, R, C)$ holds during the period of continuous change $Q(n, F, R,$

C) which starts at time t_0 and varies with time t. The rule (G7) specifies the premise condition required for the predicate *ContinuousProperty* to hold.

In addition, there are two cases for the occurrence of an event: an event is triggered by an external action and initial condition, or implicitly by the transition between processes (defined in the process semantics). In (G4) and (G5), the trigger of an event is formalized by the predicates *EventTrigger, Happens, ImplicitHappens*. In (G6), it is represented by the predicate $In(t, p)$ that t is a time point in the time period p.

An event in the process semantics is defined as a triple $\langle P, t, P' \rangle$. P and P' are finite sets of processes. The event is expected to occur at time t. The result of occurrence of the event is that each process in P is transformed into the corresponding new process in P'. It is assumed that the set P (resp. P') includes k processes $P = (p_1, \ldots, p_k)$ (resp. $P' = (p'_1, \ldots, p'_k)$). The transition of processes from P into P' happens by the event implicitly. For that we can define the event of the process semantics in the event calculus as follows.

$$
\begin{aligned}
&ImplicitHappens\,(e, t) \leftarrow \\
&\quad Start\,(after\,(e, Q(n_i, F_i, R_i, C_i)), e), \\
&\quad End\,(after\,(e', Q(n'_i, F'_i, R'_i, C'_i)), e),\; e' < e, \\
&\quad ConstraintRelation\,(R_1, R'_1, \ldots, R_k, R'_k, t).
\end{aligned}
\tag{G8}
$$

$$
\begin{aligned}
&ConstraintRelation\,(R_1, R'_1, \ldots, R_k, R'_k, t) \leftarrow \\
&\quad g(F_1(C_1, t), F'_1(C'_1, t), \ldots, F_k(C_k, t), F'_k(C'_k, t)) = Constant.
\end{aligned}
\tag{G9}
$$

Here the predicate *ConstraintRelation* is conditioned by a constraint equation. The dynamic and static parameters $(R_1, \ldots, R_k)$, $(C_1, \ldots, C_k)$ of the processes in the sets P and $(R'_1, \ldots, R'_k)$, $(C'_1, \ldots, C'_k)$ in the sets P' meet the equation at a specific time t. With this equation we can calculate the value of the time at which the event occurs.

To avoid the concurrent events which can not be represented in the process semantics, we give the following rule.

$$
\begin{aligned}
&e = e' \leftarrow Happens\,(e, t),\, Happens\,(e', t), \\
&\quad after\,(e, Q(n, F, R, C)) = after\,(e', Q(n, F, R, C))
\end{aligned}
\tag{G10}
$$

In order to formalize properties of processes and continuous change in the event calculus, we furthermore introduce the following basic axioms (ES1) – (ES6) partly based on the Shanahan's work [12]. In the Shanahan's variant version of event calculus, a many-sorted language of the first-order predicate calculus with equality is used, including variables for time points $(t, t_1, t_2, \ldots)$, properties $(p, p_1, p_2, q, q_1, q_2, \ldots)$, states $(s, s_1, s_2, \ldots)$, truth values $(v, v_1, v_2, \ldots)$, and truth elements $(f, f_1, f_2, \ldots)$. The domain of truth values has two members, denoted by the constants *True* and *False*. A pair $\langle p, v \rangle$ is a truth element. A state is represented as a set of truth elements.

$$
s_1 = s_2 \leftrightarrow (\forall f)\,[f \in s_1 \leftrightarrow f \in s_2].
\tag{ES1}
$$

$$
(\forall s_1, f_1)(\exists s_2 \forall f_2)\,[f_2 \in s_2 \leftrightarrow [f_2 \in s_1 \vee f_2 = f_1]].
\tag{ES2}
$$

$$(\exists s)(\forall f)\ [\neg f \in s]. \tag{ES3}$$

$$HoldsIn\,(p, s) \leftarrow [\langle p, True\rangle \in s \wedge \neg Abstate\,(s)]. \tag{ES4}$$

$$\neg HoldsIn\,(p, s) \leftarrow [\langle p, False\rangle \in s \wedge \neg Abstate\,(s)]. \tag{ES5}$$

$$\begin{aligned} &State\,(t, s) \leftrightarrow \\ &\quad (\forall e, p)\ [[\langle p, True\rangle \in s \leftrightarrow (Initiates\,(e, p) \wedge Happens\,(e, t))] \wedge \\ &\quad [\langle p, False\rangle \in s \leftrightarrow (Terminates\,(e, p) \wedge Happens\,(e, t))]]. \end{aligned} \tag{ES6}$$

In the rest of this paper, the set of axioms (ES1) – (ES6) and rules (G1) – (G10) will simply be denoted by ES and G.

5 An Example

Consider two trains $TrainA$ and $TrainB$ starting at different times and moving towards each other with different speed. At the time t_s a collision happens after which they continue to move as a couple with a common speed together.

In the process semantics we may describe this scenario by the definition of processes as follows:

$$\langle TrainA, \tau_{move}, T_{A0}, (X_{A0}, V_A)\rangle$$
$$\langle TrainB, \tau_{move}, T_{B0}, (X_{B0}, V_B)\rangle$$

where T_{A0} and T_{B0} denote the start times of the trains $TrainA$, $TrainB$, X_{A0}, V_A and X_{B0}, V_B initial locations and velocities, respectively. τ_{move} is a symbol which denotes the process scheme describing the continuous movement of the trains $TrainA$ and $TrainB$.

In Section 4 we have defined two relations $P(n, R, C)$ and $Q(n, F, R, C)$ to represent the processes in the event calculus. For instance, we instantiate these as the relations $moving(N, x_N, (l_N, v_N, t_N))$ and $engine(N, \mathcal{F}, x_N, (l_N, v_N, t_N))$ to formalize the two processes above in the event calculus. Here N represents a variable of process name $N \in (TrainA$ and $TrainB)$. The static parameters $l_N, v_N, t_N \in C$ correspond to the initial location, the velocity and the starting time of the train N. The dynamic parameter $x_N \in R$ corresponds to the actual location of the train N, which varies with time. $\mathcal{F}$ corresponds to the process scheme τ_{move} of the continuous movement of the trains $TrainA$ and $TrainB$.

The description of the two processes can be translated into rules in the event calculus:

$$\begin{aligned} &HoldsAt\,(moving\,(TrainA, x_A, (l_A, v_A, t_A)), t) \leftarrow \\ &\quad Holds\,(after\,(e, engine\,(TrainA, \mathcal{F}, x_A, (l_A, v_A, t_A)))), time\,(e, t_0), \\ &\quad In\,(t, after\,(e, engine\,(TrainA, \mathcal{F}, x_A, (l_A, v_A, t_A)))), t_0 \leq t, \\ &\quad State\,(t, s),\ HoldsIn\,(moving\,(TrainA, x_A, (l_A, v_A, t_A)), s), \\ &\quad ContinuousProperty\,(engine\,(TrainA, \mathcal{F}, x_A, (l_A, v_A, t_A)), \\ &\quad\quad t_0, moving\,(TrainA, x_A, (l_A, v_A, t_A)), t). \end{aligned} \tag{S1}$$

$$HoldsAt\,(moving\,(TrainB, x_B, (l_B, v_B, t_B)), t) \leftarrow$$
$$\quad Holds\,(after\,(e, engine\,(TrainB, \mathcal{F}, x_B, (l_B, v_B, t_B)))), time\,(e, t_0),$$
$$\quad In\,(t, after\,(e, engine\,(TrainB, \mathcal{F}, x_B, (l_B, v_B, t_B)))), t_0 \leq t, \tag{S2}$$
$$\quad State\,(t, s),\ HoldsIn\,(moving\,(TrainB, x_B, (l_B, v_B, t_B)), s),$$
$$\quad ContinuousProperty\,(engine\,(TrainB, \mathcal{F}, x_B, (l_B, v_B, t_B)),$$
$$\quad\quad t_0, moving\,(TrainB, x_B, (l_B, v_B, t_B)), t).$$

$$ContinuousProperty\,(engine\,(N, \mathcal{F}, x, (l, v, t_0)),$$
$$\quad t_0, moving\,(N, x, (l, v, t_0)), t) \leftarrow x = l + v \cdot (t - t_0). \tag{S3}$$

By using *moving* and *engine* as the general properties we describe a process in which a train N moves continuously. t and x denote the actual time and location of the train which satisfies the equation $x = l + v \cdot (t - t_0)$. l and t_0 denote the initial location and time of the occurrence of event e which initiates the property *engine* (engine of train is on) so that the process happens in which the train starts to move continuously from the initial location l with velocity v till a new event terminates this process.

In the process semantics, an event is represented as a triple $\langle P, t, P' \rangle$ whereby each concrete event is viewed as an instance of the general translation laws. The occurrence of an event at time t terminates the former processes P and results in new processes P' to occur. We can describe the transition law for inelastic collisions of two continuously moving objects by (T1) and (T2).

The event for an inelastic collision which is interpreted as a couple of trains can be formalized in the event calculus as the following rules.

$$ImplicitHappens\,(e, t) \leftarrow$$
$$\quad Start\,(after\,(e, engine\,(TrainA, \mathcal{F}, x_A, (l_{newA}, v_{newA}, t))), e),$$
$$\quad End\,(after\,(e', engine\,(TrainA, \mathcal{F}, x_A, (l_{oldA}, v_{oldA}, t_{oldA}))), e),$$
$$\quad Start\,(after\,(e, engine\,(TrainB, \mathcal{F}, x_B, (l_{newB}, v_{newB}, t))), e), \tag{S4}$$
$$\quad End\,(after\,(e'', engine\,(TrainB, \mathcal{F}, x_B, (l_{oldB}, v_{oldB}, t_{oldB}))), e),$$
$$\quad e' < e, e'' < e, ConstraintRelation\,(l_{newA}, v_{newA}, l_{newB}, v_{newB},$$
$$\quad\quad l_{oldA}, v_{oldA}, l_{oldB}, v_{oldB}, t_{oldA}, t_{oldB}, t).$$

$$ConstraintRelation\,(l_{newA}, v_{newA}, l_{newB}, v_{newB}, l_{oldA}, v_{oldA},$$
$$\quad l_{oldB}, v_{oldB}, t_{oldA}, t_{oldB}, t) \leftarrow$$
$$\quad l_{oldA} + v_{oldA} \cdot (t - t_{oldA}) = l_{oldB} + v_{oldB} \cdot (t - t_{oldB}), \tag{S5}$$
$$\quad l_{newA} = l_{newB} = l_{oldA} + v_{oldA} \cdot (t - t_{oldA}),$$
$$\quad v_{newA} = v_{newB} = v_{oldA} + v_{oldB}.$$

We suppose that *TrainA* (initial location is 0mi) starts to move at time 1:00 pm with the velocity 25mph, while *TrainB* at time 1:30 pm with the velocity -20mph. We describe two events *MoveA* and *MoveB* and have the domain-dependent formulae as follows.

$Happens\,(MoveA,\ 1{:}00pm).$ \hfill (H1)

$Happens\,(MoveB,\ 1{:}30pm).$ \hfill (H2)

In the following, we show that it holds that a collision occurs between *TrainA* and *TrainB* at 3:00pm and then they move as a couple with a common speed together. Here we use circumscription [8] to minimise the extensions of certain predicates.

Let χ be the conjunction of the axioms *ES*, *G*, *S* and *H* without *ES6* and *G1*. $CIRC_{ec}[\chi]$ is defined as the conjunction of

$$CIRC[\chi;\ Happens,\ Initiates,\ Terminates;\ State,\ HoldsAt]$$

with

$$CIRC[\chi;\ AbState;\ Happens,\ Initiates,\ Terminates,\ State,\ HoldsAt].$$

We take the first conjunct of $CIRC_{ec}[\chi]$. Since all occurrence of *Happens*, *Initiates*, *Terminates* in χ are positive,

$$CIRC[\chi;\ Happens,\ Initiates,\ Terminates]$$

is equivalent to

$$CIRC[\chi;\ Happens]\ \wedge\ CIRC[\chi;\ Initiates]\ \wedge\ CIRC[\chi;\ Terminates]$$

(See Theorem 3 in the next section). It can be seen that the *Happens*, *Initiates*, *Terminates* are true in all of its models, and we have

$$Happens\,(e,t) \leftrightarrow$$
$$[e = MoveA \wedge t = 1{:}00pm]\ \vee\ [e = MoveB \wedge t = 1{:}30pm] \tag{1}$$

$$Initiates\,(e,p) \leftrightarrow$$
$$[e = MoveA \wedge p = engine\,(TrainA, \mathcal{F}, x_A, (0mi,\ 25mph))]\ \vee \tag{2}$$
$$[e = MoveB \wedge p = engine\,(TrainB, \mathcal{F}, x_B, (80mi,\ \text{-}20mph))]$$

Since there are no occurrences of *State*, *HoldsAt* in χ, (1) and (2) are also true in all models of $CIRC_{ec}[\chi]$.

We take the second conjunct of $CIRC_{ec}[\chi]$. The only abnormal combinations of true elements are those which include both $\langle p, False\rangle$ and $\langle p, True\rangle$ for some p. So, in all models of

$$CIRC[\chi;\ AbState;\ Happens,\ Initiates,\ Terminates]$$

we have

$$Abstate\,(s) \leftrightarrow (\exists p)\ [\langle p, False\rangle \in s\ \wedge\ \langle p, True\rangle \in s] \tag{3}$$

Since there are no occurrences of *State*, *HoldsAt* in χ, we allow these predicates to vary does not affect the outcome of circumscription. So, (3) is also true in all models of $CIRC_{ec}[\chi]$. Since (G1) and (ES6) are chronological, we can show that (1), (2) and (3) are also true in all models of $CIRC_{ec}[ES \wedge G \wedge S \wedge H]$(See Theorem 2 in the next section).

The combination of (3) with axioms *ES*, *G*, *S* and *H* ensures that every model includes a state in which properties *engine* and *moving* hold.

By the rules (S1) – (S5), we can deduce that an implicit event denoted as e_{ic} occurs at time 3:00pm, since the condition of the constraint equation in (S4)–(S5) is satisfied. It is easy to show from (1), (2), (3) and (S4) – (S5) that in all models under circumscription we have

$$(\exists s) \; [State\,(30, s) \wedge HoldsIn\,(moving\,(TrainA, x_A, (50mi, \; 5mph, \; 3:00pm)), s)$$
$$\wedge HoldsIn\,(moving\,(TrainB, x_B, (50mi, \; 5mph, \; 3:00pm)), s)]$$

Therefore,

$$HoldsAt\,(moving\,(TrainA, x_A, (50mi, \; 5mph, \; 3:00pm)), t).$$
$$HoldsAt\,(moving\,(TrainB, x_B, (50mi, \; 5mph, \; 3:00pm)), t).$$

where $t \geq$ 3:00pm.

6   Soundness and Completeness Theorem

Definition 9. *A* marker set *is a subset S of R such that, for all T_1 in R, the set of T_2 in S such that $T_2 < T_1$ is finite.*

Definition 10. *A formula ψ is* chronological *in argument k with respect to a formula χ and a marker set S if*
  (a) it has the form $\forall x \; q(x) \leftrightarrow \phi(x)$, where q is a predicate whose kth argument is a time point and $\phi(x)$ is a formula in which x is free, and
  (b) all occurrences of q in $\phi(x)$ are in conjunctions of the form $q(z) \wedge z_k < x_k \wedge \theta$, where $\chi \wedge \psi \models \neg\theta$ if $z_k \notin S$.

Theorem 1. *Consider only models in which the time points are interpreted as reals, and in which $<$ is interpreted accordingly. Let P^* and Q^* be sets of predicates such that Q^* includes q. Let $\psi = \forall x \; q(x) \leftrightarrow \phi(x)$ be a formula which is chronological in some argument with respect to a formula χ which does not mention the predicate q, and a marker set S. Then*

$$CIRC[\chi \wedge \psi; P^*; Q^*] \models CIRC[\chi; P^*; Q^*].$$

In order to minimize domains and histories, two other properties of circumscription will be useful.

Theorem 2. *Let λ be any formula and $\delta(x)$ be any formula in which x is free. $CIRC[\lambda \wedge \forall x \ p(x) \leftarrow \delta(x); p]$ is equivalent to $\lambda \wedge \forall x \ p(x) \leftrightarrow \delta(x)$ if λ and $\delta(x)$ are formulae containing no occurrences of the predicate p.*

Theorem 3. *Let λ be any formula and $\delta(x)$ be any formula in which x is free. If all occurrences of the predicates $p_1, p_2, \ldots, p_n$ in a formula λ are positive, then $CIRC[\lambda; P^*]$, where $P^* = p_1, p_2, \ldots, p_n$, is equivalent to*

$$CIRC[\lambda; p_1] \wedge CIRC[\lambda; p_2] \wedge \ldots \wedge CIRC[\lambda; p_n].$$

Here Theorem 1, 2 and 3 are reproduced without proof, but proofs can be found in Shanahan's [12] and Lifschitz's papers [8], respectively.

Let $\mathcal{D} = (\mathcal{P}, \mathcal{E})$ be consistent domain description for process semantics, where $\mathcal{P}$ is a set of initial processes and $\mathcal{E}$ is a set of events. We write $\mathcal{P} = (p_1, p_2, \ldots, p_m)$ and $\mathcal{E} = (e_1, e_2, \ldots, e_n)$.

Let $OBS(P, \alpha, t_s)$ denote an observation of the process with name n at time t_s, where α is a symbol in C or F for some process scheme (C, F) and $\alpha = r$ (where r is an observed value). In the event calculus we describe an observation in the following form: $HoldsAt(P(n, R, C), t_s) \wedge \alpha = r$, where α is a variable name in R or C.

Lemma 1. *Let π denote the defined translation from the process semantics into the event calculus and $\mathcal{D}$ be a consistent domain description for process semantics, for any process P if $CIRC_{ec}[\pi P \wedge ES \wedge G] \models HoldsAt(P(n, R, C), t_s) \wedge \alpha \in (R \cup C) \wedge \alpha = r$, then $\mathcal{D}$ entails $OBS(P, \alpha, t_s) \wedge \alpha = r$.*

Proof. Let λ denote the conjunction of πP, ES and G. Suppose that for any process P from $\mathcal{D}$, $CIRC_{ec}[\lambda] \models HoldsAt(P(n, R, C), t_s) \wedge \alpha \in (R \cup C) \wedge \alpha = r$. Then there must exist a state s and it follows that

$$(\exists s) \ (State(t, s) \ \wedge \ HoldsIn(P(n, R, C), s)).$$

Since all occurrences of *Happens*, *ImplicitHappens*, *Initiates* and *Terminates* in λ are positive, from Theorem 3, we have

$$CIRC_{ec}[\lambda; Happens, ImplicitHappens, Initiates, Terminates]$$

is equivalent to

$$CIRC[\lambda; Happens] \wedge CIRC[\lambda; ImplicitHappens] \wedge CIRC[\lambda; Initiates] \wedge$$

$$CIRC[\lambda; Terminates].$$

Applying Theorem 2 to each conjunct in this formula, it can be seen that *Happens*, *ImplicitHappens*, *Initiates* and *Terminates* are true in all models under circumscription.

Case 1: If $HoldsAt(P(n, R, C), t_s) \ \wedge \ \alpha \in (R \cup C) \ \wedge \ \alpha = r$ is true, and $(\exists e') \ time(e') < t_s \wedge terminates(e', after(Q(n, F, R, C), e))$ is not true, it is clear that in the process semantics any model of $\mathcal{D}$ is also the model of the observation $OBS(P, \alpha, t_s)$, i.e., $\mathcal{D}$ entails $OBS(P, \alpha, t_s) \wedge \alpha = r$.

Case 2: Assume that there exist a set of events $\mathcal{E}$ and for any event $e \in \mathcal{E}$, $time(e) < t_s$. Since $CIRC_{ec}[\lambda] \models HoldsAt(P(n, R, C), t_s) \wedge \alpha = r$, with the rules (G1)–(G10) we can deduce that the event e is applicable and occurs at a certain time $time(e)$. By applying rules (G8) and (G9) we can further deduce a set of processes which are initiated by the event e and meet the rule (G1) such that for one of these processes $P(n, R, C)$, we have $HoldsAt(P(n, R, C), t_s) \wedge \alpha = r$ holds. It follows that the process $P(n, R, C)$, initiated by the event e, with the observed value r is true in all the models of $CIRC_{ec}[\lambda]$. By the Definition 3.8, under given events and processes, the observed value $\alpha = r$ is true in all the system developments for the observation $OBS(P, \alpha, t_s)$. Thus, we have that $\mathcal{D}$ entails the observation $OBS(P, \alpha, t_s) \wedge \alpha = r$.

Theorem 4. [Soundness Theorem] *Let $\mathcal{D}$ be a consistent domain description for process semantics and π denote the translation from the process semantics into the event calculus, for any process P if $\pi\mathcal{D}$ entails πP, then $\mathcal{D}$ entails P.*

Proof. By Lemma 1, an observation $OBS(P, \alpha, t_s) \wedge \alpha = r$ is entailed by $\mathcal{D}$, if $CIRC[\pi P \wedge ES \wedge G] \models \pi OBS$. Suppose $\pi\mathcal{D}$ entails πP. Since the observation is made during a development of the system being modeled and involved in some concrete process at time t_s, this observed process holds under the development of the system (given the set of initial processes and the set of events), if $HoldsAt(P(n, R, C), t_s)$ is true in all the models of $CIRC[\pi P \wedge ES \wedge G]$. It follows that $\mathcal{D}$ entails P.

Theorem 5. [Completeness Theorem] *Let $\mathcal{D}$ be a consistent domain description for process semantics and π denote the translation from the process semantics into the event calculus, for any process P if $\mathcal{D}$ entails P, then $\pi\mathcal{D}$ entails πP.*

Proof. Assume that $\mathcal{D}$ entails P; then since $\mathcal{D}$ is consistent, every system development of the process P satisfies a set of observations for P under $\mathcal{D}$. Let $OBS(P, \alpha, t_s)$ represent an observation for the process P at time t_S with which the observed value is real and we denote it as $\alpha = r$.

For any process P from $\mathcal{D}$, let χ be the conjunction of πP, ES and G without $ES6$ and $G1$.

$CIRC_{ec}[\chi]$ is defined as the conjunction of

$$CIRC[\chi; Happens, Initiates, Terminates; State, HoldsAt]$$

with

$$CIRC[\chi; AbState; Happens, Initiates, Terminates, State, HoldsAt].$$

We take the first conjunct. Since all occurrences of *Happens*, *ImplicitHappens*, *Initiates*, and *Terminates* in χ are positive,

$$CIRC[\chi; Happens, ImplicitHappens, Initiates, Terminates]$$

is equivalent to

$$CIRC[\chi; Happens] \wedge CIRC[\chi; ImplicitHappens] \wedge CIRC[\chi; Initiates] \wedge$$

$$CIRC[\chi; Terminates].$$

Since there are no occurrences of *State*, *HoldsAt* in χ, from Theorem 3, applying Theorem 2 to each conjunct in this formula, it can be seen that *Happens*, *ImplicitHappens*, *Initiates*, *Terminates* and *Holds* are true in all of its models under circumscription.

We take the second conjunct of $CIRC_{ec}[\chi]$. The only abnormal combinations of true elements are those which include both $\langle P, False \rangle$ and $\langle P, True \rangle$ for P. So, in all models of

$$CIRC[\chi; AbState; Happens, Initiates, Terminates]$$

we have

$$Abstate(s) \leftrightarrow (\exists P) \ [\langle P, False \rangle \in s \ \wedge \ \langle P, True \rangle \in s]$$

Since there are no occurrences of *State*, *HoldsAt* in χ, we allow these predicates to vary, which does not affect the outcome of circumscription. So, the formula above is also true in all models of $CIRC_{ec}[\chi]$.

Since (G1) and (ES6) are chronological, by applying Theorem 1, $CIRC[\chi \wedge G1 \wedge ES6)] \models CIRC[\chi]$.

The combination of axioms (ES) with the general rules (G) ensures that for the process P from $\mathcal{D}$, in all models under circumscription we have

$$(\exists s) \ (State(t, s) \ \wedge \ HoldsIn(P(n, R, C), s)).$$

It follows that $HoldsAt(P(n, R, C), t)$ is true in all of models of $CIRC[\chi \wedge G1 \wedge ES6]$.

For every system development of the process P under $\mathcal{D}$, we have the observation $OBS(P, \alpha, t_s)$ with which the observed value $\alpha = r$ (r is a real) at the time t_s. Thus, for $\alpha \in (R \cup C)$ and $\alpha = r$ in $\mathcal{D}$, we have $CIRC_{ec}[\chi \wedge G1 \wedge ES6] \models HoldsAt(P(n, R, C), t_s) \wedge \alpha \in (R \cup C) \wedge \alpha = r$. It follows that $\pi\mathcal{D}$ entails πP.

7 Concluding Remarks

In this paper we have provided a method to represent the process semantics in the event calculus. For specifying the properties of continuous change, the concept of process, event and state transition law of the process semantics are formalized in the event calculus, based on the described general translation rules. We further have proved the soundness and completeness of the event calculus with respect to the process semantics.

Only a handful of other authors have given attention to the problem of using logic to represent continuous change. Based on the event calculus, some techniques were presented for representing continuous change to complement its existing capability for discrete change. For example, Shanahan [11, 12] outlined a framework for representing continuous change based on the ontology of events. Belleghem, Denecker and de Schreye [2] presented an abductive version of event calculus for this purpose. All of these approaches can be embedded in logic programming but are not yet defined in a high-level description semantics for processes and continuous change, which is in contrast to our method.

8   Acknowledgements

I would like to thank Wolfgang Bibel and Michael Thielscher for their comments on an earlier version of this paper. This work was supported by the German Academic Exchange Service (DAAD).

References

[1] Allen, J.: Toward a General Theory of Action and Time. Journal of Artificial Intelligence **23** (1984) 123–154

[2] Belleghem, K., Denecker, M., De Schreye, D.: Representing Continuous Change in the Abductive Event Calculus. Proceedings of the International Conference on Logic Programming (1995) 225–240

[3] Doherty, P., Lukaszewicz, W.: Circumscribing Features and Fluents. Proceedings of the International Conference on Temporal Logic (1994) 82–100

[4] Gelfond, M., Lifschitz, V.: Representing Action and Change by Logic Programs. Journal of Logic Programming **17** (1993) 301–321

[5] Herrmann, C., Thielsche, M.: Reasoning about Continuous Processes. Proceedings of the AAAI National Conference on Artificial Intelligence (1996) 639–644

[6] Kartha, F. G.: Soundness and Completeness Theorems for Three Formalizations of Actions. Proceedings of IJCAI (1993) 724–729

[7] Kowalski, R., Sergot, M.: A Logic Based Calculus of Events. Journal of New Generation Computing **4** (1986) 319–340

[8] Lifschitz, V.: Circumscription. The Handbook of Logic in Artificial Intelligence and Logic Programming, Vol. 3: Nonmonotoic Reasoning and Uncertain Reasoning (1994) 297–352

[9] Miller, R., Shanahan, M.: Reasoning about Discontinuities in the Event Calculus. Proceedings of the 5th International Conference on Principles of Knowledge Representation and Reasoning (1996) 63–74

[10] Sandewall, E.: The Range of Applicability and Nonmonotonic Logics for the Inertia Problem. Proceedings of IJCAI (1993) 738–743

[11] Shanahan, M.: Representing Continuous Change in the Event Calculus. Proceeding of ECAI (1990) 598–603

[12] Shanahan, M.: A Circumscriptive Calculus of Events. Journal of Artificial Intelligence **77** (1995) 249–284

[13] Thielscher, M.: Representing Actions in Equational Logic Programming. Proceedings of the International Joint Conference on Logic Programming (1994) 207–224

[14] Thielscher, M.: The Logic of Dynamic Systems. Proceedings of IJCAI (1995) 639–644

Declarative Formalization of Strategies for Action Selection: Applications to Planning

Josefina Sierra-Santibáñez

Escuela Técnica Superior de Ingeniería Informática
Universidad Autónoma de Madrid
jsierra@ii.uam.es

Abstract. We present a representation scheme for the declarative formalization of strategies for action selection based on the *situation calculus* and *circumscription*. The formalism is applied to represent a number of *heuristics* for moving blocks in order to solve planning problems in the blocks world. The formal model of a *heuristic forward chaining planner*, which can take advantage of *declarative formalizations of strategies for action selection*, is proposed. Experiments showing how the use of declarative representations of strategies for action selection allows a heuristic forward chaining planner to improve the performance of state of the art planning systems are described.

1 Introduction

Interesting research is being done lately on improving the performance of domain independent planners using declarative representations of domain knowledge [1], [8], [24]. Domain knowledge can be represented in a number of different forms, such as task decomposition schemas [29], search control knowledge [1], or heuristics for action selection [25]. This paper builds on previous work on the declarative formalization of strategies for action selection [25], describing its application to improving the performance of a forward chaining planner.

The idea is to use *heuristics for action selection* (such as "if a block can be moved to final position[1], this should be done right away") to circumscribe the set of situations that should be considered by a planner to those situations that are *selectable* according to a *strategy for action selection*. We use a declarative formalization of strategies for action selection that allows refining the action selection strategy used by a planner (and, therefore, to prune its search space) by simple additions of better heuristics [19]. The incorporation of this idea to a forward chaining planner leads to the notion of a *heuristic forward chaining planner*, which can use *declarative representations of action selection strategies* to reduce considerably the size of its search space. We present the declarative formalization of an action selection strategy for the blocks world in section 4,

---

[1] In the blocks world, a block is in *final* position if it is on the table and it should be on the table in the goal configuration, or if it is on a block it should be on in the goal configuration and that block is in final position.

M. Ojeda-Aciego et al. (Eds.): JELIA 2000, LNAI 1919, pp. 133–147, 2000.

and we show how the planner can use this strategy for solving a number of blocks world problems.

The paper is organized as follows. Section 2 presents a formal model of a simple forward chaining planner. Section 3 introduces and formalizes the concept of a heuristic forward chaining planner, which can use declarative representations of action selection strategies. Section 4 describes a representation scheme for the declarative formalization of action selection strategies proposed in [25]. Section 5 compares our approach to related work on the use of the declarative representations of domain knowledge for planning. Section 6 describes some experiments comparing the performance of our heuristic forward chaining planner and TLPlan [1]. Finally, section 7 summarizes our main contributions.

2 Forward Chaining Planner

We begin with the formal description of a forward chaining planner which explores the space of possible situations, i.e., the set of situations generable by applying executable sequences of actions to the initial situation, until it finds a situation that satisfies the goal conditions. The planner uses a bounded depth first search strategy to explore the space of situations.

The formal model of the forward chaining planner, presented below, is based on a formalization of STRIPS [5] in the *situation calculus* described in [21]. Associated with each situation is a *database* of propositions describing the *state* associated with that situation. The predicate $DB(f, s)$ asserts that propositional fluent f is in the database associated with situation s. Each action is described by a *precondition list*, an *add list*, and a *delete list*, which are formally characterized by the following predicates: (1) $Prec(f, a)$ is true provided proposition f is a precondition of action a; (2) $Del(f, a)$ is true if proposition f becomes false when action a is performed; (3) $Add(f, a)$ is true if proposition f becomes true when action a is performed. The function *Result* maps a situation s and an action a into the situation that results when action a is performed in situation s. When an action is considered, it is first determined whether its preconditions are satisfied (axiom 1). If the preconditions are met, then the sentences on the delete list are deleted from the database, and the sentences on the add list are added to it (axiom 2).

We assume uniqueness of names for every function symbol, and every pair of distinct function symbols[2]. The constant symbols S_0 and S_g denote, respectively, the initial and goal situations. The predicate $Goal(s)$ is true provided situation s satisfies all the conditions that are true at the goal situation S_g. The expression $s <_r s_1$ means that s_1 can be *reached* from s performing a nonempty sequence of *executable* actions. We introduce an axiom of induction for situations that allows us to prove that a property holds for all the situations. This axiom also constrains the domain of situations to those that can be reached ($<_r$) from the initial and goal situations [23].

---

[2] The symbols h and g are meta-variables ranging over distinct function symbols; x and y denote tuples of variables.

The expression $s_1 <_{df} s_2$ is true provided situations s_1 and s_2 can be both reached from S_0 (or can be both reached from S_g), and situation s_1 will be found earlier than situation s_2 if the tree of situations reachable from S_0 (respectively, from S_g) is explored using a *depth first* search strategy ($<_{alph}$ denotes the alphabetic order). Finally, an action sequence p is the solution returned by the planner if the situation resulting from performing p in the initial situation satisfies the goal conditions and it is minimal with respect to the search strategy of the planner. The constant K is a natural number that corresponds to the maximum depth explored by the bounded depth first search strategy used by the planner.

$$Poss(a, s) \leftrightarrow \forall f (Prec(f, a) \rightarrow DB(f, s)) \qquad (1)$$

$$DB(f, Result(a, s)) \leftrightarrow Poss(a, s) \wedge (Add(f, a) \vee (DB(f, s) \wedge \neg Del(f, a))) \qquad (2)$$

$$\forall x, y (h(x) = h(y) \rightarrow x = y); \quad \forall x, y (h(x) \neq g(y)) \qquad (3)$$

$$Goal(s) \leftrightarrow \forall f (DB(f, S_g) \rightarrow DB(f, s)) \qquad (4)$$

$$\forall s (\neg s <_r S_0) \wedge \forall s (\neg s <_r S_g) \wedge \forall a, s, s_1 (s <_r Result(a, s_1) \leftrightarrow Poss(a, s_1) \wedge s \leq_r s_1) \qquad (5)$$

$$\forall P(P(S_0) \wedge P(S_g) \wedge \forall s, a (P(s) \wedge Poss(a, s) \rightarrow P(Result(a, s))) \rightarrow \forall s P(s)) \qquad (6)$$

$$s_1 <_{df} s_2 \leftrightarrow s_1 \leq_r s_2 \vee \exists a, b, s (a \prec_{alph} b \wedge Result(a, s) \leq_r s_1 \wedge Result(b, s) \leq_r s_2) \qquad (7)$$

$$Length(S_0) = 0 \wedge Length(S_g) = 0 \wedge Length(Result(a, s)) = 1 + Length(s) \qquad (8)$$

$$\forall s (Result([\,], s) = s) \wedge \forall a, p, s (Result([a|p], s) = Result(p, Result(a, s))) \qquad (9)$$

$$Sol(p) \leftrightarrow \exists s (s = Result(p, S_0) \wedge S_0 \leq_r s \wedge Goal(s) \wedge Length(s) \leq K \wedge \qquad (10)$$
$$\forall s_1 (S_0 \leq_r s_1 \wedge Goal(s_1) \wedge Length(s) \leq K \rightarrow s <_{df} s_1))$$

The axiom set $T_{FC} = \{1, \dots, 10\}$ is our formal model of a forward chaining planner.

2.1   Blocks World Example

We present now a formal model of the sort of information that must be communicated to the forward chaining planner to solve a planning problem. This information can be divided into *domain* dependent information (the precondition, add and delete lists of the available actions), and *problem* dependent information (the states associated with the initial and goal situations).

The variables x, y and z range over blocks. The constants A, B, C, and T (for *Table*) are of the sort *block*. The function symbol On maps a pair of blocks x and y into the propositional fluent $On(x, y)$ describing the fact that block x is on block y. The function symbol $Clear$ maps a block x into the propositional fluent $Clear(x)$ describing the fact that there is space on block x to place another block. We include a domain closure axiom for blocks. The initial and goal configurations are described by axioms 15 and 16. The function symbol $Move$ maps a triple of blocks x, y and z into the action $Move(x, y, z)$ denoting the act of moving block x from y to z. The precondition, delete and add lists of $Move(x, y, z)$ are as follows.

$$Prec(f, Move(x, y, z)) \leftrightarrow f = Clear(x) \vee f = On(x, y) \vee (z \neq T \rightarrow f = Clear(z)) \qquad (11)$$

$$Del(f, Move(x, y, z)) \leftrightarrow f = On(x, y) \vee (z \neq T \rightarrow f = Clear(z)) \qquad (12)$$

$$Add(f, Move(x, y, z)) \leftrightarrow f = On(x, z) \lor (y \neq T \rightarrow f = Clear(y)) \quad (13)$$

$$\forall x(x = A \lor x = B \lor x = C \lor x = T) \quad (14)$$

$$DB(f, S_0) \leftrightarrow \exists x, y((f = On(x, y) \land ((x = A \land y = T) \lor (x = B \land y = T) \lor \quad (15)$$
$$(x = C \land y = A))) \lor (f = Clear(x) \land (x = B \lor x = C)))$$

$$DB(f, S_g) \leftrightarrow \exists x, y(f = On(x, y) \land ((x = A \land y = B) \lor (x = B \land x = C) \lor \quad (16)$$
$$(x = C \land y = T)))$$

The axiom sets $T_{BW1} = \{11, \ldots, 13\}$ and $T_{P1} = \{14, \ldots, 16\}$ constitute our formal models of the blocks world domain and the problem known as Sussman's anomaly, respectively.

3   Heuristic Forward Chaining Planner

A *heuristic forward chaining planner* is a forward chaining planner that explores the space of *selectable situations*, rather than the space of *possible situations*. Selectable situations are those that can be generated by applying sequences of *selectable actions* to the initial situation. A heuristic forward chaining planner needs information that goes beyond the classical specification of a planning problem. In particular, it needs to know what actions are selectable at a particular situation.

In the following section, we address the issue of how a user can specify such information. Let's assume, for a moment, that the user supplies a definition of the predicate $Sel(a, s)$, which is true provided action a can be selected at situation s, along with the specification of a planning problem. Then, the only modification that we need to make to the formal model of the forward chaining planner T_{FC} in order to obtain the formal model of the heuristic forward chaining planner T_{HFC} is to replace the predicate $Poss$ by the predicate Sel in axiom 5[3].

4   Declarative Formalization of Strategies for Action Selection

In [25], we proposed a representation scheme for the declarative formalization of *strategies for action selection* based on the *situation calculus* [18] and *circumscription* [20]. The idea is to represent strategies for action selection as sets of action selection rules [7]. An *action selection rule* is an implication whose antecedent is a formula of the situation calculus, and whose consequent can take one of the following forms: $Good(a, s)$, $Bad(a, s)$ or $Better(a, b, s)$. The intuitive interpretation of these predicates is that performing action a at situation s is *good, bad,* or *better* than performing action b.

The following action selection rules describe some *heuristics* for determining what blocks should be moved in order to solve planning problems in the blocks

---

[3] This replacement redefines the reachability relation $<_r$ as follows: $s_1 \leq_r s_2$ is true provided s_2 can be reached from s_1 by performing a sequence of *selectable* actions.

world: *(1) If a block can be moved to final position, this should be done right away (axiom 17); (2) If a block is not in final position and cannot be moved to final position, it is better to move it to the table than anywhere else (axioms 18); (3) If a block is in final position, do not move it (axiom 19); (4) If a block is above another block it ought to be above but it is not in final position (i.e., it is in tower-deadlock position), put it on the table (axiom 20).*

$$\neg Holds(Final(x), s) \wedge Holds(On(x,y), s) \wedge Holds(On(x,z), S_g) \wedge (z = T \vee \quad (17)$$
$$Holds(Final(z), s)) \wedge Poss(Move(x,y,z), s) \rightarrow Good(Move(x,y,z), s)$$
$$\neg Holds(Final(x), s) \wedge Holds(On(x,y), s) \wedge Holds(On(x,z), S_g) \wedge \quad (18)$$
$$(\neg Holds(Final(z), s) \vee \neg Poss(Move(x,y,z), s)) \wedge w \neq T \rightarrow$$
$$Better(Move(x,y,T), Move(x,y,w), s)$$
$$Holds(On(x,y), s) \wedge Holds(Final(x), s) \rightarrow Bad(Move(x,y,z), s) \quad (19)$$
$$Holds(On(x,y), s) \wedge Holds(TD(x), s) \rightarrow Good(Move(x,y,T), s) \quad (20)$$

The predicate *Holds(f,s)* is true provided propositional fluent f is true at situation s. A block is in *final* position $Holds(Final(x), s)$ if it is on the table and it should be on the table in the goal configuration, or if it is on a block it should be on in the goal configuration and that block is in final position. A block is in *tower-deadlock* position $Holds(TD(x), s)$ if it is above another block it ought to be above but it is not in final position. Section 4.3 contains formal definitions of these symbols.

A *consistent* set of action selection rules (such as $S1 = \{17, 18, 19, 20\}$) defines a *strategy for action selection*.

4.1 Nonmonotonic Interpretation

The formal semantics of a strategy for action selection T_S is given by $INT(T_S)$ [25], the *nested abnormality theory* specified on the right hand side of formula 22. *Nested abnormality theories* [16] extend *simple abnormality theories* [22] by allowing the specification of nested applications of the *circumscription operator* [20]. $INT(T_S)$ characterizes the conditions under which an action is *good* or *bad* for a particular situation, by jumping to the conclusions that: (1) an action is *"not good"* unless the action selection rules in T_S imply that it is *good*; and (2) an action is *"not bad"* unless the action selection rules in T_S, together with axiom 21, imply that it is *bad*. Axiom 21 asserts that an action is *bad* for a particular situation if there exists a *better* action for the same situation.

$$Better(a_1, a_2, s) \rightarrow Bad(a_2, s) \quad (21)$$
$$INT(T_S) \equiv \{Better, min\ Bad : 21, \{min\ Good : T_S\}\} \quad (22)$$

Formally, this is achieved as follows. First, the predicate *Good* is circumscribed with respect to the conjunction of the universal closures of the axioms in T_S. Then, the predicate *Bad* is circumscribed with respect to the result of the circumscription of *Good* in T_S and the universal closure of axiom 21. *Better* is allowed to vary because minimizing the extension of *Bad* may affect (through axiom 21) the extension of *Better*.

This nonmonotonic interpretation of action selection strategies has both representation and computational advantages. It allows describing strategies: (1) *succinctly*, since it is not necessary to specify negative information (i.e., which actions are not good, not bad, or not better than others); (2) according to a *least commitment* policy, in which it is not necessary to assert that an action is good, bad, or better than other unless it is known for sure; and (3) *incrementally*, since it is possible to refine an action selection strategy by simple additions of better heuristics (i.e., consistent action selection rules that may become available later on). In these three cases, circumscription takes care of appropriately adapting its consequences to the lack of information or the availability of new relevant facts.

The following formal result establishes some conditions under which the interpretation $INT(T_S)$ of a strategy for action selection T_S can be computed by a variant of Clark's completion algorithm [4].

Proposition 1 If every axiom of T_S is a first order action selection rule such that its antecedent does not contain the predicates *Good*, *Bad* or *Better*, then $INT(T_S)$ is equivalent to the conjunction of the first order sentences 23 and 24 resulting from the application of the *completion algorithm* described bellow to T_S.

$$\forall a, s(Good(a, s) \leftrightarrow A_3^{good}(a, s)) \tag{23}$$

$$\forall a, s(Bad(a, s) \leftrightarrow A_3^{bad}(a, s) \vee \exists a_1, a_2(a = a_2 \wedge A_3^{better}(a_1, a_2, s))) \tag{24}$$

Completion Algorithm Let T_S be a declarative formalization of a strategy for action selection. The axioms of T_S are all of the form $A \to P(t_a, t_s)$, where A is a first order formula which does not contain the predicates *Good*, *Bad* or *Better*, t_a is a tuple of terms of the sort *action*, t_s is a term of the sort *situation*, and P is one of the predicates *Good*, *Bad* or *Better*.

Step 1 Replace each rule of the form $A \to P(t_a, t_s)$ in T_S by $A \wedge a = t_a \wedge s = t_s \to P(a, s)$, where a is a tuple of new variables of the sort *action*, and s is a new variable of the sort *situation*.

Step 2 Replace each rule $A_1(a, s) \to P(a, s)$ obtained in the previous step by $\exists x A_1(a, s) \to P(a, s)$, where x are the free variables in the original rule.

Step 3 For each P, replace all the rules of the form $A_2^i(a, s) \to P(a, s)$ obtained in step 2 by a single rule of the form $\bigvee_i A_2^i(a, s) \to P(a, s)$.

Step 4 Replace the rule $A_3^{good}(a, s) \to Good(a, s)$ obtained in step 3 by $\forall a, s(Good(a, s) \leftrightarrow A_3^{good}(a, s))$.

Step 5 Replace the rules $A_3^{bad}(a, s) \to Bad(a, s)$ and $A_3^{better}(a_1, a_2, s) \to Better(a_1, a_2, s)$ obtained in step 3 by a single rule[4] of the form $\forall a, s(Bad(a, s) \leftrightarrow A_3^{bad}(a, s) \vee \exists a_1, a_2(a = a_2 \wedge A_3^{better}(a_1, a_2, s)))$.

---

[4] We assume the variables a, a_1 and a_2 of the sort action are distinct from each other.

Proof (Proposition 1) The semantics of nested abnormality theories is characterized by a map φ that translates blocks[5] into sentences of a second order language. Proposition 1 in [16] allows us to describe the semantics of the nested abnormality theory $INT(T_S)$ as the following circumscription formula[6].

$$\varphi(\{Better, \ min \ Bad: \ 21, \ \{min \ Good: \ T_S\}\}) \equiv$$
$$CIRC(21', \ CIRC(T_S; \ Good); \ Bad; \ Better)$$

We use several rules for computing circumscription described in [15]. Formula (19) and proposition 2 in [15] allow us to prove the following equivalence. Formula 23 is the first order characterization of the predicate *Good* obtained in step 4 of the completion algorithm, T_{Bad} is the conjunction of the universal closures of the action selection rules of the form $A \rightarrow Bad(t_a, t_s)$ in T_S, and T_{Better} is the conjunction of the universal closures of the action selection rules of the form $A \rightarrow Better(t_a, t_s)$ in T_S.

$$CIRC(T_S; \ Good) \equiv 23 \wedge T_{Bad} \wedge T_{Better}$$

The equivalence above, together with formula (19) and proposition 3 in [15] allow us to simplify $INT(T_S)$ as follows. T_{better} is the second order formula obtained from T_{Better} by substituting every instance of the predicate constant *Better* by a similar predicate variable *better*.

$$CIRC(21', \ CIRC(T_S; \ Good); \ Bad; \ Better) \equiv$$
$$CIRC(21', 23 \wedge T_{Bad} \wedge T_{Better}; \ Bad; Better) \equiv$$
$$23 \wedge CIRC(T_{Bad}, \ \exists better(21' \wedge T_{better}); \ Bad)$$

Using equivalence (27) in section 3.2 of [15], we can prove that $\exists better(21' \wedge T_{better})$ is equivalent to the following formula which does not depend on *better*. $A_3^{better}(a_1, a_2, s)$, a_1, a_2, a and s are as described in step 5 of the completion algorithm.

$$\forall a, s(\exists a_1, a_2(a = a_2 \wedge A_3^{better}(a_1, a_2, s)) \rightarrow Bad(a, s)) \tag{25}$$

Finally, proposition 1 in [15] allows us to compute the result of circumscribing T_{Bad} and 25 with respect to *Bad*. Formula 24 is the first order characterization of the predicate *Bad* obtained in step 5 of the completion algorithm.

$$23 \wedge CIRC(T_{bad}, \ 25; \ Bad) \ \equiv 23 \wedge 24 \qquad \square$$

For example, the nonmonotonic interpretation $INT(S1)$ of action selection strategy $S1$ (described by action selection rules 17 to 20) can be computed by the completion algorithm. We show the result of the last step of the algorithm.

$\forall a, s(Good(a, s) \leftrightarrow \exists x, y, z(\neg Holds(Final(x), s) \wedge Holds(On(x, y), s) \wedge Holds(On(x, z), S_g) \wedge$
$(z = T \vee Holds(Final(z), s)) \wedge Poss(Move(x, y, z), s) \wedge a = Move(x, y, z)) \vee$
$\exists x, y(Holds(On(x, y), s) \wedge Holds(TD(x), s) \wedge a = Move(x, y, T)))$
$\forall a, s(Bad(a, s) \leftrightarrow \exists x, y, z(Holds(On(x, y), s) \wedge Holds(Final(x), s) \wedge a = Move(x, y, z)) \vee$
$\exists a_1, a_2(a = a_2 \wedge \exists x, y, z, w(\neg Holds(Final(x), s) \wedge Holds(On(x, y), s) \wedge Holds(On(x, z), S_g) \wedge$
$(\neg Holds(Final(z), s) \vee \neg Poss(Move(x, y, z), s)) \wedge w \neq T \wedge a_1 = Move(x, y, T) \wedge$
$a_2 = Move(x, y, w))))$

---

[5] Blocks are the equivalent of axioms in *nested abnormality theories* (see [16]).
[6] In the following, we denote the universal closure of a formula A by A'.

These two formulas characterize the conditions under which a move is *good* or *bad* for a particular situation according to strategy for action selection $S1$.

4.2   Mechanism for Action Selection

The interpretation of an action selection strategy gives us a characterization of the conditions under which an action is good or bad for a particular situation. Suppose the user supplies, along with an action selection strategy, a theory of action that allows the planner to determine whether these conditions hold or not for a particular situation. Then, the planner could infer what actions are good or bad for every situation, and use that information to determine what actions should be selected.

The following axiom characterizes the set of *selectable actions* for a particular situation. The predicate $Poss(a, s)$ is true provided action a can be executed at situation s (axiom 1).

$$Sel(a, s) \leftrightarrow Poss(a, s) \wedge (Good(a, s) \vee (\neg \exists b Good(b, s) \wedge \neg Bad(a, s))) \qquad (26)$$

According to the action selection mechanism described by axiom 26, an action is *selectable* at a particular situation if it is executable and *good* for that situation, or if there are no good actions for that situation and it is executable and *not bad* for that situation.

4.3   Blocks World (Continuation)

In order to interpret action selection rules, such as axioms 17 to 20, in terms of the theory of action described in section 2, we need to establish a connection between what *holds* at a situation and what is in the *database* associated with that situation. In this paper, we assume that the state associated with any situation can be described in terms of the truth values of a finite set of *frame fluents* [18] [14]. The rest of the fluents, called *defined fluents*, are described in terms of the frame fluents. The database associated with a situation determines the truth values of the frame fluents as follows: a frame fluent holds at a particular situation if and only if it is in the database associated with that situation.

$$Frame(f) \rightarrow (Holds(f, s) \leftrightarrow DB(f, s)) \qquad (27)$$

The *frame fluents* for the blocks world are those of the form $On(x, y)$ or $Clear(x)$. In addition to frame fluents, we use a number of *defined fluents*, such as *final*, *above*[7], and *tower-deadlock*.

---

[7] If we assume uniqueness of names, a complete characterization of the predicate DB for the initial and goal situations, an axiom of induction for situations, and that there is only a finite number of blocks (as we do), the definitions of $Holds(Final(x), s)$ and $Holds(Above(x, y), s)$ provided allow us to characterize the extensions of these formulas.

$$Frame(f) \leftrightarrow \exists x, y(f = On(x,y) \vee f = Clear(x)) \quad (28)$$
$$Holds(Final(x), s) \leftrightarrow (Holds(On(x,T), s) \wedge Holds(On(x,T), S_g)) \vee \quad (29)$$
$$\exists y(Holds(Final(y), s) \wedge Holds(On(x,y), s) \wedge Holds(On(x,y), S_g))$$
$$Holds(Above(x,y), s) \leftrightarrow Holds(On(x,y), s) \vee \exists z(Holds(On(x,z), s) \wedge \quad (30)$$
$$Holds(Above(z,y), s))$$
$$Holds(TD(x), s) \leftrightarrow \neg Holds(Final(x), s) \wedge \exists y(y \neq T \wedge \quad (31)$$
$$Holds(Above(x,y), s) \wedge Holds(Above(x,y), S_g))$$

$T_{BW} = T_{BW1} \bigcup \{27, \ldots 31\}$ is our extended theory of action for the blocks world. Let T_{S1} be the set of axioms $\text{INT}(S1) \bigcup \{26\} \bigcup T_{BW}$. T_{S1} is a formal model of the action selection strategy for the blocks world described at the beginning of this section. We can use this axiom set to simulate the behavior of the heuristic forward chaining planner when it is given the description of Sussman's anomaly problem T_{P1} along with the strategy for action selection T_{S1}. For example, if the constant K (maximum depth explored by the bounded depth first search strategy) is equal to 3, we can prove that the heuristic forward chaining planner only needs to explore 3 situations before finding the optimal solution (shown below).

$$T_{HFC} \bigcup T_{S1} \bigcup T_{P1} \vdash Sol(\{Move(A,C,T), Move(B,T,C), Move(A,T,B)\})$$

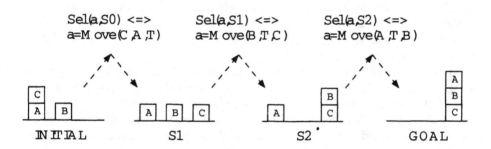

Fig. 1. Heuristic forward chaining planner using action selection strategy $S1$ for solving Sussman's anomaly problem. There is a single selectable action for every situation.

The reason for which the planner only needs to explore three situations before finding an optimal solution is the following. In the initial situation S_0, block C can be moved to final position. Action selection rule 17 implies that $Move(C, A, T)$ is a good action. Blocks A and B are not in tower-deadlock position and cannot be moved to final position, therefore there are no other good actions for the initial situation. Thus, the action selection mechanism (axiom 26) implies that $Move(C, A, T)$ is the only selectable action for S_0.

Let S_1 be $Result(Move(C, A, T), S_0)$. Block B can be moved to final position in S_1. The rest of the blocks are not in tower-deadlock position and cannot be moved to final position. Therefore, $Move(B, T, C)$ is the only good and thus selectable action for S_1.

Let S_2 be $Result(\{Move(C, A, T), Move(B, T, C)\}, S_0)$. Block A can be moved to final position in S_2. The rest of the blocks are in final position. Therefore, $Move(A, T, B)$ is the only selectable action for S_2.

$Result(\{Move(C, A, T), Move(B, T, C), Move(A, T, B)\}, S_0)$ is the first situation found by the planner that satisfies the $Goal$ predicate. Therefore, axiom 10 implies that the action sequence $\{Move(C, A, T), Move(B, T, C), Move(A, T, B)\}$ is the solution returned by the planner.

5   Related Work

Various techniques have been used to exploit domain knowledge for planning. HTN (hierarchical task network) planners [29] use domain knowledge in the form of task decomposition schemas which goes beyond the specification of preconditions and effects of actions used by classical planners. Domain knowledge has also been expressed in the form of search control knowledge. In particular, knowledge bases of forward chaining rules have been used to guide search. SOAR was the first system to use this approach [17], and a refined version of it is a prominent part of PRODIGY [28]. A similar rule-based approach to search control has also been incorporated into UCPOP [2]. The main disadvantage of the rule-based approach used by these systems is that their search control rules are specified in terms of implementation details of their planning algorithms. This is not the case for the action selection rules presented in this paper, which are expressed in terms of domain knowledge only.

In [11], a problem solver guided by *negative heuristics* (which tell a system what *not to do*) is described. The heuristics are specified in PROLOG, and relate the goal to the current state and anticipated action. They are designed to eliminate actions which clearly do not contribute to the goal. Four negative heuristics for the blocks world, which eliminate part of the search and are subsumed by axiom 18 in this paper, are proposed.

In [24], a forward chaining planner, which uses a regression based theorem prover and an iterative deepening search strategy, is proposed. The planner requires the following types of information from the user: (1) a predicate $goal(s)$, which is true if situation s satisfies the conditions of the goal for which a plan is sought; (2) a set of *action precondition and successor state axioms* for the primitive actions of the domain; and (3) a predicate $badSituation(s)$, which is true if situation s is considered to be a bad situation for the planner to consider. The planner is implemented in GOLOG [13], and it has been extended to deal with concurrent actions and incomplete initial situations [6].

The representation scheme proposed in this paper is more expressive than those used in [11] and [24], in the sense that it allows the representation of *positive heuristics* (the predicate *good* tells a system what to do), and heuristics that establish preferences among actions (the predicate *better* establishes a partial order among actions). The predicate *badSituation(s)* allows pruning the search space by characterizing those situations from which a successful plan cannot be reached, but it does not allow guiding the search in promising directions as the predicate *good* does in our formalization.

The heuristics for the blocks world used in [6] prune approximately the same set of situations as action selection rules 17 to 19. In particular, the definition of *good-tower* is equivalent to our concept of *final position*. However, the heuristics in [6] do not consider the concept of *tower-deadlock position*, and therefore they cannot be used to discriminate between actions that move arbitrary blocks to the table (which are not necessarily optimal and can be postponed) and actions that move blocks in tower deadlock position to the table (which are necessary and should be executed right away). This is the meaning of action selection rule 20. For example, the heuristics in [6] do not establish a preference between the actions $Movetotable(d)$ and $Movetotable(g)$ in the situation resulting from performing the sequence of actions $\{Movetotable(m), Movetotable(p), Movetotable(n), Movetotable(f)\}$ in the initial situation of the problem described in [6]. However, if action $Movetotable(g)$ is chosen the resulting plan contains one action more than the optimal plan. Action selection rule 20 allows characterizing $Movetotable(d)$ as a good action, because block d is in tower deadlock position, and $Movetotable(g)$ as a non bad action.

Our planner has not been designed to solve planning problems with incomplete initial situations. However, the declarative formalization of action selection strategies proposed in this paper is adequate for dealing with open world planning problems [6]. For example, if we add the definitions of $Final(x,s)$, $Above(x,y,s)$ and $TD(x,s)$ to the formalization of the blocks world presented in [6], action selection strategy $S_2 = \{32, \ldots, 37\}$ can be used for solving the open blocks world planning problem described in that paper[8].

$$\neg Final(x,s) \wedge On(x,y,S_g) \wedge Final(y,s) \to Good(Move(x,y),s) \qquad (32)$$

$$\neg Final(x,s) \wedge Ontable(x,S_g) \wedge \to Good(Movetotable(x),s) \qquad (33)$$

$$\neg Final(x,s) \wedge On(x,y,S_g) \wedge (\neg Final(y,s) \vee \exists z On(z,y,s)) \to \qquad (34)$$
$$Better(Movetotable(x), Move(x,w),s)$$

$$Final(x,s) \to Bad(Move(x,y),s) \qquad (35)$$

$$Final(x,s) \to Bad(Movetotable(x),s) \qquad (36)$$

$$TD(x,s) \to Good(Movetotable(x),s) \qquad (37)$$

In [1], a planning system called TLPlan, which uses first order linear temporal logic to represent search control knowledge, is described. This logic is interpreted over sequences of worlds. In particular, the *goal* and *temporal modalities* ($\bigcup$ until, $\square$ always, $\diamond$ eventually, and $\bigcirc$ next) are used to assert properties of world sequences. A *search control formula* describing the search control strategy to be used by the planner is specified by the user in this logic. This formula describes

---

[8] Some other changes to the formalization in [6] are required as well. For example, the definition of the predicate $Goal(s)$ should be replaced by $Goal(s) \leftrightarrow \neg \exists xy(On(x,y,S_g) \wedge \neg On(x,y,s)) \wedge \neg \exists x(Ontable(x,S_g) \wedge \neg Ontable(x,s))$. Axioms 21, 26 and a new axiom describing the state associated with the goal situation S_g should be added as well. Space limitations do not allow a more detailed explanation.

properties the sequences of worlds generated by applying successful plans to the initial situation should satisfy. The planner uses a *progression algorithm* which serves as the basis for an incremental mechanism that allows checking whether a plan prefix, generated by forward chaining, could lead to a plan that satisfies the search control formula. Interesting experiments in which TLPlan is shown to perform better than state of the art planners, such as BlackBox [10], IPP [12], SatPlan [9], GraphPlan [3], PRODIGY [28] and UCPOP [2] in various test domains using search control formulas are described.

TLPlan is an interesting example of a heuristic forward chaining planner, in which search control knowledge is expressed in terms of properties the sequences of worlds generated by *selectable plans* (rather than *actions*) must satisfy. The last search control formula used for the blocks world in [1] prunes approximately the same set of situations than the first three action selection rules of $S1$ (the action selection strategy proposed in section 4 of this paper). In particular, their definition of *good-tower* is equivalent to our concept of *final position*.

An advantage of our proposal is the availability of a formal model of the planner which allows limited forms of meta-reasoning, such as determining the correctness, redundancy, inconsistency or quality of different strategies for action selection. This is an important feature that may allow the planner to reject incorrect strategies, and to provide its users with feed back on how to improve their strategies. This is not possible in TLPLAN, because it does not have a formal description of its own mechanism for action selection which allows it to reason about the consequences of adopting a particular strategy.

6   Experiments

We have implemented a *heuristic forward chaining planner* which can use declarative representations of planning domains and strategies for action selection in Prolog. The planner has been applied to solve some blocks world problems using $S1$, the strategy for action selection described in section 4. The first problem set (shown in table 1) consists of 10 randomly generated blocks world problems of 25 blocks. The second problem set (shown in table 2) consists of 6 blocks world problems of different sizes. The sizes of the problems are specified in the first column of table 2. For each problem, we have computed the number of blocks that are initially in final and tower deadlock positions (columns *Final* and *TD*).

The numbers in the columns *Steps*, *Nodes*, and *Time* correspond to the number of steps of the plans found by our planner, the number of situations (nodes) explored, and the time in milliseconds spent on planning.

We have compared our results with those obtained from running the same problems in TLPlan. The numbers in the columns *Steps TLPlan*, *Nodes TLPlan*, and *Time TLPlan* correspond to the number of steps of the plans found by TLPlan, the number of situations (nodes) actually explored, and the time in milliseconds taken by TLPlan.

In order to make a fair comparison, we have discounted one from the number of nodes explored by TLPlan, because we do not count the initial situation.

It should be noted as well that we have used the domain definition and control strategy in *LinearBlocksWorld.tlp* (see http://www.uwaterloo.ca/~fbacchus). In this domain definition, the four actions (*pickup(x)*, *putdown(x)*, *stack(x,y)*, *unstack(x,y)*) are used to model the dynamics of the blocks world. We have used a single action *Move(x,y,z)*, which corresponds to two TLPlan actions. Therefore, the plans obtained by TLPlan should be twice as long as ours, and the number of nodes explored $2n + 1$, where n is the number of nodes explored by our planner. The formulas we have used to compute the numbers shown in the columns *Steps TLPlan* and *Nodes TLPlan* are $s/2$ and $(x - 1)/2$, respectively, where s is the number of steps of the plans found by TLPlan, and x is the number of nodes actually explored by TLPlan.

Comparing the numbers in the columns *Steps* and *Steps TLPlan*, it can be observed that TLPlan cannot find optimal plans (i.e., with a minimum number of steps) for 10 of the 16 problems posed. Our planner obtains optimal plans for the 16 problems. As far as planning time is concerned, our planner is faster than TLPlan. The only exceptions are the problems of sizes 15 and 19. However, the numbers of steps of the plans found by TLPlan are very far from optimality, 18 and 25 steps versus 14 and 18 steps for the optimal plans.

Table 1. Problems of 25 blocks.

Prob	Final	TD	Steps	Nodes	Time	Steps TLPlan	Nodes TLPlan	Time TLPlan
1	1	2	26	26	0	26	26	58
2	0	11	36	36	0	38	38	91
3	3	1	23	23	0	25	25	58
4	7	0	18	18	0	20	20	52
5	7	2	20	20	0	20	20	46
6	1	4	28	28	0	30	30	68
7	1	6	30	30	0	37	37	91
8	1	13	37	37	0	37	37	85
9	1	3	27	27	0	29	29	68
10	1	7	31	31	50	32	32	84

Table 2. Problems of different sizes.

Size	Final	TD	Steps	Nodes	Time	Steps TLPlan	Nodes TLPlan	Time TLPlan
5	2	0	4	11	0	5	5	4
13	1	3	15	15	0	15	15	19
15	2	0	14	274	320	18	18	26
19	2	0	18	3583	5610	25	25	47
25	5	1	22	29	50	22	22	51
50	24	0	26	26	0	26	26	158

The specification of the problems, the strategy for action selection, the Prolog code of the planner and the log files with the results of the experiments can be obtained from the author (jsierra@ii.uam.es).

7   Conclusions

We have studied the use of declarative representations of action selection strategies for planning. First, we have presented a representation scheme for the declarative formalization of strategies for action selection, which has a number of advantages. One of these advantages is the possibility of defining positive heuristics which can guide the search process in promising directions. The compositionality of our declarative representation of strategies for action selection is an important feature as well, since it allows refining an action selection strategy by simple additions of better heuristics.

Then, we have proposed a formal model of a heuristic forward chaining planner, which can take advantage of declarative representations of strategies for action selection. The availability of such a formal model not only shows the feasibility of our idea from a theoretical point of view, it also allows interesting forms of meta-reasoning about declarative formalizations of strategies for action selection, such as: (1) determining the correctness of a particular strategy (or a class of strategies) with respect to a given domain; (2) updating and composing strategic knowledge from different sources; or (3) determining whether a set of heuristics improve, are inconsistent or redundant with a particular strategy for action selection.

Finally, we have implemented a heuristic forward chaining planner in Prolog and run some experiments in order to determine whether this is indeed a practical idea. The experiments have shown that a heuristic forward chaining planner using declarative representations of strategies for action selection can improve the performance of state of the art planning systems, such as Blackbox, IPP or TLPlan.

Acknowledgment

The author would like to thank John McCarthy for interesting her on the declarative formalization of heuristics, and providing the intellectual and financial support necessary for the realization of this work. Vladimir Lifschitz contributed to simplify and refine my original formalization with valuable comments. Luc Steels shared with me very interesting ideas and encouraged me all the way.

References

1. Bacchus, F., and Kabanza, F. 2000 Using temporal logics to express search control knowledge for planning. Artificial Intelligence, 116:123-191.
2. Barrett, A., Golden, K., Penberthy, J.S., and Weld, D. 1993 UCPOP user's manual, (version 2.0). Technical report TR-93-09-06, University of Washington, Department of Computer Science and Engineering. ftp://cs.washington.edu/pub/ai.
3. Blum, A., and Furst, M. 1997. Fast planning through planning graph analysis. Artificial Intelligence, 90.
4. Clark, K. 1978. Negation as failure. Gallaire, H., and Minker, J. (Eds.) Advances in data base theory, Plemum Press, New York, 293-322.

5. Fikes, R., and Nilsson, N. 1971. STRIPS: A new approach to the application of theorem proving to problem solving. Artificial Intelligence, 2:189–208.
6. Finzi, A., Pirri, F., and Reiter, R. 2000. Open world planning in the situation calculus. In Proc. of AAAI-2000.
7. Genesereth, M. R., and Hsu, J. 1989. Partial programs. TR-89-20, Comp. Science, Stanford University.
8. Huang, Y., Selman, B., and Kautz, H. 1999. Control Knowledge in Planning: Benefits and Tradeoffs. In Proc. of AAAI-99.
9. Kautz, H., and Selman, B. 1996. Pushing the envelope: planning, propositional logic, and stochastic search. In Proc. of AAAI-96.
10. Kautz, H., and Selman, B. 1998 Blackbox: a new approach to the application of theorem proving to problem solving. http://www.research.att.com/kautz
11. Kibler, D., and Morris, P. 1981. Don't be stupid. In Proc. of IJCAI-81.
12. Koehler, J., Nebel, B., Hoffmann, J., Dimopoulos, Y. 1997 Extending Planning Graphs to an ADL subset. In European Conference on Planning, 1997, pp. 273-285.
13. Levesque, H., Reiter, R., Lespérance, Y., Lin, F., and Scherls, R. 1997. GOLOG: a logic programming language for dynamic domains. J. of Logic Programming, Special Issue on Actions, 31(1-3):59-83.
14. Lifschitz, 1990. Frames in the space of situations. Artificial Intelligence, 46:365-376.
15. Lifschitz, V. 1993. Circumscription. In Handbook of Logic in Artificial Intelligence and Logic Programming, D. Gabbay and C.J. Hogger, Ed., volume 3: Nonmonotonic Reasoning and Uncertain Reasoning. Oxford University Press.
16. Lifschitz, 1995. Nested abnormality theories. Artificial Intelligence, 74:1262-1277.
17. Laird, J., Newell, A., and Rosenbloom, P. 1987 SOAR: an architecture for general intelligence. Artificial Intelligence, 33(1):1-67.
18. McCarthy, J., and Hayes, P. 1969. Some philosophical problems from the standpoint of artificial intelligence. Machine Intelligence, 4:463-502.
19. McCarthy, J. 1959. Programs with common sense. In Mechanization of Thought Processes, Proc. of the Symp. of the National Physics Laboratory, 77–84.
20. McCarthy, J. 1980. Circumscription -a form of non-monotonic reasoning. Artificial Intelligence, 13:27–39.
21. McCarthy, J. 1985. Formalization of STRIPS in situation calculus. TR-85-9, Comp. Science, Stanford University.
22. McCarthy, J. 1986. Applications of circumscription to formalizing common sense knowledge. Artificial Intelligence, 28:89–116.
23. Reiter, R. 1993. Proving properties of states in situation calculus. *Artificial Intelligence* 64:337–351.
24. Reiter, R. 1999. Knowledge in action: logical foundations for describing and implementing dynamical systems. In preparation. Draft available at http://www.cs.toronto.edu/cogrobo.
25. Sierra, J. 1998. Declarative formalization of strategies for action selection. In Proc. of the Seventh International Workshop on Non-monotonic Reasoning.
26. Sierra, J. 1998. A declarative formalization of STRIPS. In Proc. of the Thirteenth European Conference on Artificial Intelligence.
27. Sierra, J. 1999. Declarative formalization of heuristics (taking advice in the blocks world). In Proc. of the International Conference on Computational Intelligence, Modeling, Control and Automation.
28. Veloso, M., Carbonell, et al. 1995 Integrating planning and learning: the PRODIGY architecture. JETAI, 7(1).
29. Wilkins, D. 1988 Practical planning: extending the classical AI planning paradigm. Morgan Kaufmann, San Mateo, California.

An Algorithmic Approach to Recover Inconsistent Knowledge-Bases

Ofer Arieli

Department of Computer Science, K.U.Leuven
Celestijnenlaan 200A, B-3001 Heverlee, Belgium
`arieli@cs.kuleuven.ac.be`

Abstract. We consider an algorithmic approach for revising inconsistent data and restoring its consistency. This approach detects the "spoiled" part of the data (i.e., the set of assertions that cause inconsistency), deletes it from the knowledge-base, and then draws classical conclusions from the "recovered" information. The essence of this approach is its coherence with the original (possibly inconsistent) data: On one hand it is possible to draw classical conclusions from any data that is not related to the contradictory information, while on the other hand, the only inferences allowed by this approach are those that do not contradict any former conclusion. This method may therefore be used by systems that restore consistent information and are obliged to their resource of information. Common examples of this case are diagnostic procedures that analyse faulty components of malfunction devices, and database management systems that amalgamate distributed knowledge-bases.

1 Motivation

In this paper we introduce an algorithmic approach to revise inconsistent information and restore its consistency. This approach (sometimes called "coherent" [5], or "conservative" [15]) considers contradictory data as useless, and uses only a consistent part of the original information for making inferences. To see the rationality behind this approach consider, for instance, the following set of propositional assertions:

$$KB = \{p, \ \neg p, \ \neg p \lor q, \ r, \ \neg r \lor s\}.$$

Since $\neg p$ is true in KB, so is $\neg p \lor q$ (even if q is false), and so a plausible inference mechanism should not apply here the Disjunctive Syllogism to p and $\neg p \lor q$. Intuitively, this is so since the information regarding p is contradictory, and so one should not rely on it for drawing inferences. On the other hand, applying the Disjunctive Syllogism to $\{r, \ \neg r \lor s\}$ may be justified by the fact that this subset of formulae should *not* be affected by the inconsistency in KB, therefore inference rules that are classically valid can be applied to it.

The two major goals of coherent approaches in general, and our formalism in particular, are therefore the following:

M. Ojeda-Aciego et al. (Eds.): JELIA 2000, LNAI 1919, pp. 148–162, 2000.
© Springer-Verlag Berlin Heidelberg 2000

a) Detect and isolate "spoiled" parts of the knowledge-base, i.e.: Remove from the knowledge-base subsets of assertions that cause inconsistency,

b) Draw classical conclusions in a non-trivial way from any data that is not related to the contradictory information. Such inferences should be semantically coherent with the original data, that is: Only inferences that do not contradict any previously drawn conclusions are allowed.

For achieving the goals above we consider an algorithmic approach that is based on a four-valued semantics [3,4]. Using a multiple-valued semantics is a common way to overcome the shortcomings of classical calculus (see, e.g., [3,6,7,12,13,14]), and as we shall see in what follows, four-valued semantics is particularly suitable for our purpose.

A similar algorithmic approach for recovering stratified knowledge-base, which is also based on a four-valued semantics, was introduced in [1,2]. Here we generalize and improve that approach in the sense that we consider a better search engine, and provide and algorithm that recovers *arbitrary* knowledge-bases rather than only stratified ones.

2 Background

2.1 Belnap Four-Valued Lattice

Our method is based on Belnap's well-known algebraic structure, introduced in [3,4]. This structure consists of four truth values: the classical ones (t, f), a truth value $(\bot)$ that intuitively represents lack of information, and a truth value $(\top)$ that may intuitively be understood as representing contradictions. These four elements are simultaneously ordered in two distributive lattices. In one of them, denoted by $L_4 = (\{t, f, \top, \bot\}, \leq_t)$, f is the $\leq_t$-minimal element, t is the $\leq_t$-maximal one, and $\bot, \top$ are two intermediate values that are incomparable. The partial order of this lattice may be intuitively understood as representing differences in the amount of *truth* of each element. In the other lattice, denoted by $A_4 = (\{t, f, \top, \bot\}, \leq_k)$, $\bot$ is the $\leq_k$-minimal element, $\top$ is the $\leq_k$-maximal one, and t, f are two intermediate values. The partial order $\leq_k$ of this lattice intuitively represents differences in the amount of *knowledge* (or information) that each element exhibits. We denote Belnap four-valued structure together with its two partial orders by $\mathcal{FOUR}$ (see Figure 1).

As usual, we shall denote the $\leq_t$-meet and the $\leq_t$-join of $\mathcal{FOUR}$ by $\wedge$ and $\vee$, respectively. In addition, we shall denote by $\neg$ the involution operation on $\leq_t$, for which $\neg\top = \top$ and $\neg\bot = \bot$.

2.2 Knowledge-Bases: Syntax and Semantics

The language we use here is the standard propositional one, based on the propositional constants $t, f, \top, \bot$, and the connectives $\vee, \wedge, \neg$ that correspond, respectively, to the join, meet, and the negation operations w.r.t. $\leq_t$. Atomic formulae

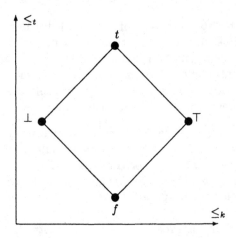

Fig. 1. Belnap lattice, $\mathcal{FOUR}$

are denoted by p, q, literals (i.e., atomic formulae or their negations) are denoted by l, and complex formulae are denoted by ψ, ϕ. Given a set S of formulae, we shall write $\mathcal{A}(S)$ to denote the set of the atomic formulae that occur in S, and $\mathcal{L}(S)$ to denote the set of the literals that occur in S ($\mathcal{A}$ and $\mathcal{L}$ denote, respectively, the set of atomic formulae and the set of literals in the language). The complement of a literal l is denoted by $\bar{l}$. An atomic formula $p \in \mathcal{A}(S)$ is called a *positive (negative) fact* of S if $p \in S$ ($\neg p \in S$). The set of all the (positive and negative) facts in S is denoted by $Facts(S)$.

The various semantic notions are defined on $\mathcal{FOUR}$ as natural generalizations of similar classical ones: A *valuation* ν is a function that assigns a truth value in $\mathcal{FOUR}$ to each atomic formula. Any valuation is extended to complex formulae in the obvious way. The set of the four-valued valuations is denoted by $\mathcal{V}$. A valuation ν *satisfies* ψ iff $\nu(\psi) \in \{t, \top\}$. t and $\top$ are called the *designated* elements of $\mathcal{FOUR}$. A valuation that satisfies every formula in a given set S of formulae is a *model* of S. A model of S will usually be denoted by M or N. The set of all the models of S is denoted by $mod(S)$.

The formulae that will be considered here are clauses, i.e.: disjunctions of literals. The following useful property of clauses is easily shown by an induction on the structure of clauses:

Lemma 1. Let ψ be a clause and ν a valuation. Then $\nu(\psi) \in \{t, \top\}$ iff there is some $l \in \mathcal{L}(\psi)$ s.t. $\nu(l) \in \{t, \top\}$.

A finite set of clauses is called a *knowledge-base*, and is denoted by KB. As the following lemma shows, representing formulae in a clause form does not reduce the generality.

Lemma 2. [1] For every formula ψ there is a finite set S of clauses such that for every valuation ν, $\nu(\psi) \in \{\top, t\}$ iff $\nu(\phi) \in \{\top, t\}$ for every $\phi \in S$.

Given a certain knowledge-base KB, we consider the $\leq_k$-minimal elements in $mod(KB)$. These models reflect the intuition that one should not assume what is not really represented in KB.

Definition 1. Let $\nu_1, \nu_2 \in \mathcal{V}$.

a) ν_1 is k-*smaller* than ν_2 iff for every atom p, $\nu_1(p) \leq_k \nu_2(p)$.
b) $\nu \in mod(KB)$ is a k-*minimal model* of KB if there is no other model of KB that is k-smaller than ν.

Example 1. Consider the following knowledge-base:

$$\mathcal{KB} = \{p, \ \neg q, \ \neg p \vee q, \ \neg p \vee h, \ q \vee r \vee s, \ q \vee \neg r \vee \neg s, \ h \vee r, \ h \vee s\}$$

The (k-minimal) models of $\mathcal{KB}$ are given in Table 1 below. We shall use $\mathcal{KB}$ for the demonstrations in the sequel.

The k-minimal models of KB will have an important role in the recovery process of KB. This may be justified by the fact that as long as one keeps the amount of information as minimal as possible, the tendency of getting into conflicts decreases.

2.3   Recovered Knowledge-Bases

Definition 2. Let $\nu \in \mathcal{V}$. Denote: $I(\nu) = \{p \in \mathcal{A} \mid \nu(p) = \top\}$. Usually we shall be interested in the assignments of ν w.r.t. a specific knowledge-base. In such cases we shall consider the following set: $I(\nu, KB) = \{p \in \mathcal{A}(KB) \mid \nu(p) = \top\}$.

As we have noted above, by "recovering a knowledge-base" we mean to turn it (in a plausible way) to a consistent one. That is:

Definition 3. A valuation ν is *consistent* if $I(\nu) = \emptyset$. A knowledge-base is consistent if it has a consistent model.

Proposition 1. [1,2] A knowledge-base is consistent iff it is classically consistent.

The recovery process is based on the following notion:

Definition 4. A *recovered knowledge-base* KB' of a knowledge-base KB is a subset of KB with a consistent model M' s.t. there is a (not necessarily consistent) model M of KB, for which $M'(p) = M(p)$ for every $p \in \mathcal{A}(KB')$.

Table 1. The (k-minimal) models of $\mathcal{KB}$

Model No.	p	q	h	r	s	k-minimal
M_1	t	$\top$	t	$\bot$	$\bot$	$+$
$M_2 - M_4$	t	$\top$	t	$\bot$	$f,t,\top$	
$M_5 - M_{16}$	t	$\top$	t	$f,t,\top$	$\bot,f,t,\top$	
$M_{17} - M_{32}$	t	$\top$	$\top$	$\bot,f,t,\top$	$\bot,f,t,\top$	
M_{33}	$\top$	f	$\bot$	t	$\top$	$+$
M_{34}	$\top$	f	$\bot$	$\top$	t	$+$
M_{35}	$\top$	f	$\bot$	$\top$	$\top$	
M_{36}	$\top$	f	f	t	$\top$	
$M_{37} - M_{38}$	$\top$	f	f	$\top$	$t,\top$	
M_{39}	$\top$	f	t	$\bot$	$\top$	$+$
M_{40}	$\top$	f	t	f	t	$+$
M_{41}	$\top$	f	t	f	$\top$	
M_{42}	$\top$	f	t	t	f	$+$
M_{43}	$\top$	f	t	t	$\top$	
M_{44}	$\top$	f	t	$\top$	$\bot$	$+$
$M_{45} - M_{47}$	$\top$	f	t	$\top$	$f,t,\top$	
M_{48}	$\top$	f	$\top$	$\bot$	$\top$	
$M_{49} - M_{50}$	$\top$	f	$\top$	f	$t,\top$	
$M_{51} - M_{52}$	$\top$	f	$\top$	t	$f,\top$	
$M_{53} - M_{56}$	$\top$	f	$\top$	$\top$	$\bot,f,t,\top$	
M_{57}	$\top$	$\top$	$\bot$	t	t	$+$
M_{58}	$\top$	$\top$	$\bot$	t	$\top$	
$M_{59} - M_{60}$	$\top$	$\top$	$\bot$	$\top$	$t,\top$	
$M_{61} - M_{64}$	$\top$	$\top$	f	$t,\top$	$t,\top$	
$M_{65} - M_{80}$	$\top$	$\top$	t	$\bot,f,t,\top$	$\bot,f,t,\top$	
$M_{81} - M_{96}$	$\top$	$\top$	$\top$	$\bot,f,t,\top$	$\bot,f,t,\top$	

Example 2. The set $\{p\}$ is a recovered knowledge-base of $KB_1 = \{p, q, \neg q\}$, but it is not a recovered knowledge-base of $KB_2 = \{p, \neg p\}$. This example demonstrates the fact that in order to recover a given inconsistent knowledge-base, it is *not sufficient* to find some of its (maximal) consistent subset(s), but it is necessary to ensure that the subset under consideration would semantically correspond to the original, inconsistent data; In our case, $\{p\}$ does not recover KB_2 even though it is a classically consistent subset of KB_2, just because of the fact that this set contradicts an information ($\neg p$) that is explicitly stated in the original knowledge-base. Therefore, the "semantical correspondence" property is not preserved in this case.[1]

---

[1] Keeping this "semantical correspondence" to the original information is one of the main differences between the present formalism and some other formalisms for restoring consistency (see, e.g., [5,6,9]).

Given an inconsistent knowledge-base KB, the idea is to choose one of its recovered knowledge-bases and to treat this set as the relevant knowledge-base for deducing classical inferences. Next we show that the set of recovered knowledge-bases of KB may be easily constructed from the set of its models:

Definition 5. Let $\nu \in \mathcal{V}$. The set that is *associated with* ν is defined as follows:

$$KB_\nu = \{\psi \in KB \mid \nu(\psi)=t \text{ and } \mathcal{A}(\psi) \cap I(\nu, KB)=\emptyset\}.$$

The set KB_ν corresponds to the (maximal) fragment of KB that can be interpreted in a consistent way by ν. Elimination of pieces of "inadequate" information in order to get a more "robust" representation of the "intended" knowledge is a common method in belief revision and argumentative reasoning (see, e.g., [5,6,9]).

Proposition 2. [1] Every set that is associated with a model of KB is a recovered knowledge-base of KB.

Proposition 2 implies that usually there will be a lot of ways to recover a given inconsistent knowledge-base. By what we have noted above, plausible candidates of being the "best" recovered knowledge-base of KB would be those sets that are associated with some k-minimal model of KB.[2]

Definition 6. A set $S \subseteq KB$ is a *preferred* recovered knowledge-base of KB if it is a maximal set that is associated with some k-minimal model of KB.

Example 3. Consider again the knowledge-base $\mathcal{KB}$ of Example 1. In the notations of Table 1, the subsets of $\mathcal{KB}$ that are associated with its k-minimal models are the following:

$$
\begin{aligned}
\mathcal{KB}_{M_1} &= \{p, \neg p \vee h, h \vee r, h \vee s\}, \\
\mathcal{KB}_{M_{33}} &= \{\neg q, h \vee r\}, \\
\mathcal{KB}_{M_{34}} &= \{\neg q, h \vee s\}, \\
\mathcal{KB}_{M_{39}} &= \{\neg q, h \vee r\}, \\
\mathcal{KB}_{M_{40}} &= \{\neg q, q \vee r \vee s, q \vee \neg r \vee \neg s, h \vee r, h \vee s\}, \\
\mathcal{KB}_{M_{42}} &= \{\neg q, q \vee r \vee s, q \vee \neg r \vee \neg s, h \vee r, h \vee s\}, \\
\mathcal{KB}_{M_{44}} &= \{\neg q, h \vee s\}, \\
\mathcal{KB}_{M_{57}} &= \{h \vee r, h \vee s\}.
\end{aligned}
$$

Thus, the preferred recovered knowledge-bases are $\mathcal{KB}_{M_1}$ and $\mathcal{KB}_{M_{40}} = \mathcal{KB}_{M_{42}}$.

3   Recovery of Inconsistent Knowledge-Bases

In this section we introduce an algorithm for recovering inconsistent knowledge-bases, and consider some of its properties.

---

[2] See [2] for some other preference criteria for choosing recovered knowledge-base.

Definition 7. Let KB be a knowledge-base, and let ν be a four-valued partial valuation defined on (a subset of) $\mathcal{A}(KB)$. The *dilution* of KB w.r.t. ν (notation: $KB \downarrow \nu$) is constructed from KB by the following transformations:

1. Deleting every $\psi \in KB$ that contains either t, $\top$, or a literal l s.t. $\nu(l) \in \{t, \top\}$,
2. Removing from every formula that remains in KB every occurrence of f, $\bot$, and every occurrence of a literal l such that $\nu(l) \in \{f, \bot\}$.

The intuition behind the dilution process resembles, in a way, that of the Gelfond–Lifschitz transformation [8]: Any data that has no effect on the rest of the process is eliminated. Thus, for instance, if a literal l in a formula ψ is assigned a designated value, then Lemma 1 assures that eventually ψ would also have a designated value, no matter what would be the values of the elements in $\mathcal{L}(\psi) \setminus \{l\}$. Hence, these elements can be disregarded in the rest of the construction, as indeed indicated by item (1) of Definition 7. The rationality behind item (2) of the same definition is similar.

Figure 2 contains a pseudo-code of the recovery algorithm. [3] [4] As we show in Theorems 1 and 2 below, given a certain knowledge-base KB as an input, the algorithm provides the valuations needed for constructing the preferred recovered knowledge-bases of KB.

It is easy to verify that the algorithm indeed halts for every knowledge-base. This is so since knowledge-bases are finite, and since for every set S of clauses and every partial valuation ν on $\mathcal{A}(S)$, we have that $\mathcal{A}(S \downarrow \nu) \subset \mathcal{A}(S)$.

Example 4. Figure 3 below demonstrates the execution of the algorithm on the knowledge-base $\mathcal{KB}$ of the canonical example (1 and 3). In this figure we denote by $p{:}x$ the fact that an atom p is assigned a value x.

In the notations of Table 1, the two leftmost paths in the tree of Figure 3 produce the k-minimal model M_1, and the other paths produce the k-minimal models M_{40} and M_{42}.[5] As noted in Example 3, these are exactly the models with whom the preferred recovered knowledge-bases of $\mathcal{KB}$ are associated. By Theorem 2, these are *all* the preferred recovered knowledge-bases of $\mathcal{KB}$.

Proposition 3. Let ν be a four-valued valuation produced by the algorithm of Figure 2 for a given knowledge-base KB. Then ν is a model of KB.

Proof: Let $\psi \in KB$. By Definition 7 and the specifications of the algorithm in Figure 2, it is obvious that at some stage of the algorithm ψ is eliminated from

---

[3] The first parameter of the first call to **Recover** is the dilution of KB w.r.t. the empty valuation. This is so in order to take care of the propositional constants that appear in KB (for instance, if $p \lor f \in KB$ then $p \in KB \downarrow \emptyset$).

[4] If the knowledge-base under consideration contains clauses that are logically equivalent to f or $\bot$ (e.g., $f \lor \bot$), then in $KB \downarrow \emptyset$ such clauses will become empty. One can easily handle such degenerated cases by adding to the algorithm a line that terminates its execution once an empty clause is detected.

[5] Later on we shall take care of the redundancy.

the set of clauses as a result of a dilution on this set. Note that a formula cannot be eliminated by successively removing every literal of it according to condition (2) of Definition 7, since the last literal that remains must be assigned a designated value. Thus there must be some $l \in \mathcal{L}(\psi)$ that is assigned a designated value. By Lemma 1, then, $\nu(\psi) \in \{t, \top\}$, and so $\nu \in mod(KB)$.   □

```
input: A knowledge-base KB.
Mods = Recover(KB↓∅, ∅);
do (∀M ∈ Mods) {
    KB_M = {ψ ∈ KB | ¬∃p ∈ A(KB) such that M(p) = ⊤};
    output(KB_M);
}

procedure Recover(S,ν)
/* S = a finite set of clauses, ν = the valuation constructed so far */
{
    if (S == ∅) then return(ν)          /* ν is a k-minimal model of KB */
    pos = {p ∈ A(S) | p ∈ S };          /* the positive facts in S */
    neg = {p ∈ A(S) | ¬p ∈ S };         /* the negative facts in S */
    if (pos ∪ neg == ∅) {
        do (∀p ∈ A(S)) {
            pick p;
            if (p ∈ L(S)) then Recover(S ∪ {p}, ν);
            if (¬p ∈ L(S)) then Recover(S ∪ {¬p}, ν);
        }
    }
    do (∀p ∈ (pos ∩ neg)) {
        pick p;
        μ(p) = ⊤;
        S' = S ↓ μ;
        do (∀q ≠ p such that q ∈ A(S) \ A(S'))
            μ(q) = ⊥;
        Recover(S', ν ∪ μ);
    }
    do (∀p ∈ (pos ∪ neg) \ (pos ∩ neg)) {
        pick p;
        if (p ∈ pos) then μ(p) = t else μ(p) = f;
        S' = S ↓ μ;
        do (∀q ≠ p such that q ∈ A(S) \ A(S'))
            μ(q) = ⊥;
        Recover(S', ν ∪ μ);
    }
}
```

Fig. 2. An algorithm for recovering knowledge-bases

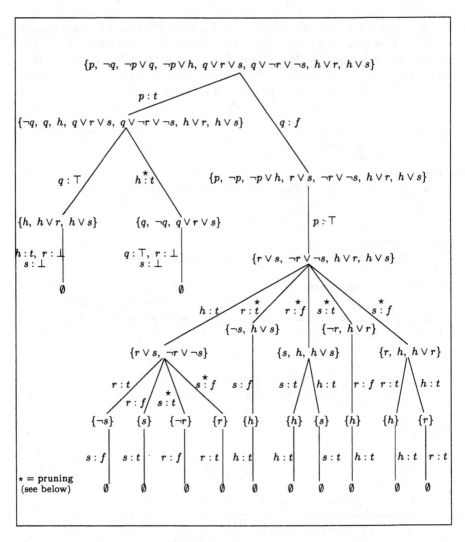

Fig. 3. Execution of the algorithm w.r.t. the canonical example

The next proposition indicates that the valuations produced by the algorithm of Figure 2 assign designated truth values only to a minimal amount of literals (no more literals than what is really necessary for providing a model for *KB*). In a sense, this means that a minimal amount of knowledge (or belief) is assumed.

Proposition 4. Let ν be a four-valued valuation produced by the algorithm of Figure 2 for a given knowledge-base *KB*. Then ν is a *choice function* on *KB*: For every $\psi \in KB$ there is exactly *one* literal $l \in \mathcal{L}(\psi)$ s.t. $\nu(l)$ is designated.

Proof: The proof is by an easy inspection on the execution of the algorithm. Consider some $\psi \in KB$. Suppose that it is eliminated at the i-th inductive call to Recover. Then all the literals $l \in \mathcal{L}(\psi)$ for which $\nu(l)$ is defined until the i-th recursive call to Recover has the property that $\nu(l) = f$ (otherwise ψ would have already been eliminated). Then there is some $l \in \mathcal{L}(\psi)$ (which is chosen during the i-th execution of Recover), for which $\nu(l) \in \{t, \top\}$, and after the next dilution ψ is eliminated, i.e.: all the rest of the literals in $\mathcal{L}(\psi)$ are assigned $\bot$. It follows, then, that every clause has a unique literal that is assigned a designated value by ν. □

Here is another evidence to the fact that only a minimal knowledge is assumed by the valuations produced by our algorithm:

Theorem 1. Let ν be a four-valued valuation produced by the algorithm of Figure 2 for a given knowledge-base KB. Then ν is a k-minimal model of KB.

Proof: First, by Proposition 3, ν is a model of KB. It remains to show, then, that ν is a k-*minimal* among the models of KB. For that consider the following set of knowledge-bases:

$$KB_0 = KB \downarrow \emptyset, \qquad KB_{i+1} = KB_i \downarrow \nu_i$$

where ν_i $(i \geq 0)$ is the partial valuation determined during the i-th recursive call to Recover.[6] Now, let us first assume that there is at least one (positive or negative) fact in KB (i.e., there is a literal $l \in \mathcal{L}(KB)$ s.t. $l \in KB$). We show that ν is a k-minimal model of KB by an induction on the number n of the recursive calls to Recover that are required for creating ν.

- $n=0$: ν_0 may assign $\top$ only to a literal l s.t. $l \in KB$ and $\bar{l} \in KB$, while all the other elements in $\mathcal{A}(KB)$ are assigned $\bot$. In this case $\top$ is the only possible value for l, and so ν is k-minimal. The same argument is true for any literal l s.t. $l \in KB$ and $\bar{l} \notin KB$ (for that l, $\nu(l) = t$). It is also obviously true for all the literals that are assigned $\bot$.
- $n \geq 1$: Let M be a model of KB. We show that $M \nless_k \nu$. Let M_1 be the reduction of M to $\mathcal{A}(KB_1)$, and suppose first that M_1 is a model of KB_1. By the induction hypothesis ν_1 is a k-minimal model of KB_1, thus there exists $p \in \mathcal{A}(KB_1)$, s.t. $M_1(p) \nless_k \nu_1(p)$, therefore $M \nless_k \nu$. The other possibility is that M_1 is not a model of KB_1. In this case there must be a clause $\psi_1 \in KB_1$ s.t. $M_1(\psi_1) \notin \{t, \top\}$. Since M is a model of KB, then by Lemma 1 there is a $\psi \in KB$ and an $l \in \mathcal{L}(\psi)$ s.t. $M(l) \in \{t, \top\}$, and $\{l\} \cup \mathcal{L}(\psi_1) \subseteq \mathcal{L}(\psi)$. But then $\nu(l) \notin \{t, \top\}$ (Otherwise, ψ is eliminated in the dilution of KB and so $\psi_1 \notin KB_1$), while $M(l) \in \{t, \top\}$. It follows that $M(l) \nless_k \nu(l)$, therefore $M \nless_k \nu$ in this case also.

To conclude, it remains to handle the case where there are no facts in KB. In this case our algorithm operates on $KB' = KB \cup \{l\}$ for some $l \in \mathcal{L}(KB)$.

---

[6] Thus, if the algorithm terminates after n recursive calls to Recover, then $\nu = \bigcup_{i=1}^{n} \nu_i$.

But now there *is* a fact in KB', and so by what we have shown above our algorithm produces a k-minimal model for KB'. Denote this model by ν'. We have to show that ν' is also a k-minimal model of KB. Indeed, ν' is clearly a model of KB. Let M be some other model of KB. If $M(l) \in \{t, \top\}$ then M is a model of KB' and so $M \not<_k \nu'$. Otherwise, $M(l) \in \{f, \bot\}$. Consider the subset of formulae of KB in which l appears as a literal: $KB(l) = \{\psi \in KB \mid l \in \mathcal{L}(\psi)\}$. Since $l \in \mathcal{L}(KB)$, it follows that $KB(l) \neq \emptyset$. Moreover, since we assume that there are no facts in KB, in particular $l \notin KB$ and $\bar{l} \notin KB$, thus $KB(l) \not\subseteq \{l, \bar{l}\}$. Now, by the definition of ν' as a valuation that is produced by our algorithm, for every $p \in \mathcal{A}(KB(l))$ s.t. $p \neq l$, we have that $\nu'(p) = \bot$. (Such p exist since $KB(l) \neq \emptyset$ and $KB(l) \not\subseteq \{l, \bar{l}\}$. These atoms are assigned $\bot$ since all the formulae in $KB(l)$ are removed after the first dilution of KB'). Now, since we assumed that $M(l) \in \{f, \bot\}$, then by Lemma 1 there must exist some $p_0 \in \mathcal{A}(KB(l))$ s.t. $M(p_0) \in \{t, \top\}$ (Otherwise $\forall \psi \in KB(l) \; M(\psi) \notin \{t, \top\}$ and so M cannot be a model of KB). Thus $M(p_0) >_k \bot = \nu'(p_0)$ and once again we have that $M \not<_k \nu'$. $\square$

Using Theorem 1 we can now show that the algorithm indeed properly recovers inconsistent knowledge-bases.

Theorem 2. For a given knowledge-base KB, the algorithm of Figure 2 produces *all* the valuations ν, for which KB_ν is a preferred recovered knowledge-base of KB.

Proof: By Theorem 1, if ν is obtained by our algorithm, then KB_ν is an element of the following set:

$$\Omega = \{KB_M \mid M \text{ is a } k\text{-minimal model of } KB\}.$$

It remains to show, therefore, that the algorithm produces valuations ν_j, for which KB_{ν_j} are the *maximal* elements of Ω. Indeed, given a k-minimal model M of KB, we show that the algorithm produces a valuation ν s.t. $I(\nu, KB) \subseteq I(M, KB)$, and therefore $KB_M \subseteq KB_\nu$.

As in Theorem 1, we denote by ν_i the partial valuation that is determined during stage i of the algorithm (thus, if the algorithm terminates after n stages, then $\nu = \cup_{i=1}^n \nu_i$), and M_i is the reduction of M to the literals on which ν_i is defined. Also, we use the following notations: $KB_0 = KB \downarrow \emptyset$, and for every $i \geq 0$, $KB_{i+1} = KB_i \downarrow \nu_i$. Now, suppose first that $Facts(KB_0) \neq \emptyset$ (i.e., there is some [positive or negative] fact in KB_0). If $\{l, \bar{l}\} \subseteq Facts(KB_0)$ for some literal l, set $\nu_0(l) = \top$ (note that in this case necessarily $M(l) = \top$ as well, since M is a model of KB and so it must assign $\top$ to all the facts of KB that are both positive and negative). Otherwise, choose some $l \in Facts(KB_0)$ s.t. $M(l) = t$ (such a literal must exist, since M is a model of KB and so it must assign designated values to the facts of KB_0), and set $\nu_0(l) = t$. If $Facts(KB_0)$ is empty, then if there is some $l \in \mathcal{L}(KB_0)$ s.t. $M(l) = t$ set $\nu_0(l) = t$ as well. Otherwise, pick some $l \in \mathcal{L}(KB_0)$ s.t. $M(l) = \bot$ and set $\nu_0(l) = t$ (there must be such a literal, since otherwise $\forall l \in \mathcal{L}(KB_0) \; M(l) \in \{\top, f\}$ and since $Facts(KB_0) = \emptyset$, this implies that M is not k-minimal, since one can easily construct a model of KB which is k-smaller than

M by changing one of the f-assignments of M to $\bot$, or one of the $\top$-assignments of M to t). Now, in order to determine ν_1 we follow a similar procedure, this time for KB_1: If $Facts(KB_1) \neq \emptyset$ then if $\{l, \bar{l}\} \subseteq Facts(KB_1)$ for some l, set $\nu_1(l) = \top$ (note that in this case necessarily $M(l) = \top$ as well, since by the construction of ν_0, we have that $KB_1 = KB \downarrow \nu_0 \subseteq KB \downarrow M_0$, and so $\{\bar{l}, l\} \subseteq KB \downarrow M_0$ as well, which means that M must assign l the value $\top$ in order to be a model of KB). Otherwise, if there is some $l \in Facts(KB_1)$ s.t. $M(l) = t$ set $\nu_1(l) = t$ as well. Otherwise, pick some $l \in Facts(KB_1)$ s.t. $M(l) \in \bot$ (again, such an l must exists. Otherwise, by the same reasons considered above, we will have a contradiction to the fact that M is a k-minimal model of KB), and set $\nu_1(l) = t$. The procedure in case that $Facts(KB_1) = \emptyset$ is the same as the one in case that $Facts(KB_0) = \emptyset$.

Now, repeat the same process until for some n, KB_n becomes empty. Let $\nu = \cup_{i=1}^n \nu_i$. The following two facts are easily verified:

1. In the process of creating ν we followed the execution of the algorithm along one path of its search tree. Hence ν is obtained by our algorithm when KB is given as its input.

2. If $\nu(l) = \top$ then $M(l) = \top$ as well (see the notes whenever $\nu_i(l) = \top$).

By (2), $I(\nu, KB) \subseteq I(M, KB)$, and so $KB_M \subseteq KB_\nu$. Thus, by (1), an output ν of the algorithm corresponds to a preferred recovered knowledge-base KB_ν of KB. □

Clearly, large knowledge-bases that contain a lot of contradictory information may be recovered in many different ways. Therefore, computing all the preferred recovered knowledge-bases in such cases might require a considerable amount of running time. It is worth noting, however, that *arbitrary recovery* of a given knowledge-base KB (i.e., producing *some* preferred recovered knowledge-base of KB) obtains quite easily. This is so since the execution time for producing the first output (valuation) is bounded by $O(|\mathcal{L}(KB)| \cdot |KB|)$; A construction of the first output requires no more than $|\mathcal{L}(KB)|$ calls to Recover (as there are no more than $|\mathcal{L}(KB)|$ picked literals), and each call takes no more than $O(|KB|)$ running time.

We conclude this section with some notes on practical ways to reduce the execution time of the algorithm.

A. Pruning of the Search Tree

Let us consider once again the search tree of Figure 3. Denote the paths in this tree from the leftmost righthand by $1, \dots, 12$. Clearly, paths 1 and 2 yield the same result. Similarly, the same valuation is produced in paths 3,6,7,11,12, and the remaining paths in the search tree also yield the same valuation. It is possible to avoid such duplications by performing a backtracking once we find out that we are constructing a valuation which is the same as another valuation that has already been produced before. Indeed, note that a path i in the search three of the algorithm corresponds to a sequence of partial valuations $\nu_0^i, \nu_1^i, \ldots, \nu_{n_i}^i$ that are constructed along its nodes. Thus, if we denote by $\mathcal{A}(KB)[\mu]$ the elements of $\mathcal{A}(KB)$ on which the partial valuation μ is defined, then it is possible to terminate the j-th flow of the algorithm (terminology: to *prune* the j-th subtree) at stage m iff there is a flow $i < j$, s.t. $\bigcup_{k=1}^{m} \mathcal{A}(KB)[\nu_k^i] = \bigcup_{k=1}^{m} \mathcal{A}(KB)[\nu_k^j]$.

Example 5. In Figure 3 the pruning locations (in paths 2, 5–12) are marked with an asterisk. Thus, only paths 1, 3, and 4 of the search tree are not pruned. They yield, respectively, the k-minimal models M_1, M_{42}, and M_{40} of $\mathcal{KB}$.[7]

Obviously, the pruning consideration might drastically improve the search mechanism of the algorithm. The tradeoff is that for checking the pruning condition we have to use much more memory space, since the algorithm has to keep tracks to valuations that correspond to previous search flows.

B. Handling Unrelated Information

There are many cases in which a new information should not affect any previous conclusion.[8] In such cases a plausible mechanism of belief revision should not retract any previous conclusion. Therefore, the general expectation is that in these cases the computational complexity of adding the new data to the knowledge-base and computing its new consequences would be relatively low. Detecting those cases and finding an appropriate methodology to handle them is sometime called "the irrelevance problem". In the next proposition we show that in cases where a totally irrelevant information arrives, it is possible to avoid executing the recovery algorithm; The new data can safely be added to any preferred recovered knowledge-base without damaging any of its properties.

Proposition 5. Let KB_1 and KB_2 be two subsets of a knowledge-base KB that satisfy the following conditions:

(a) $KB_1 \cup KB_2 = KB$,   (b) $\mathcal{A}(KB_1) \cap \mathcal{A}(KB_2) = \emptyset$,[9]   (c) KB_1 is consistent.

If S is a preferred recovered knowledge-base of KB_2, then $S \cup KB_1$ is a preferred recovered knowledge-base of KB.

Proof: For the proof we need the following result:

Lemma 5-A: [1,2] For every model M of a knowledge-base KB there is a k-minimal model M' of KB s.t. $M' \leq_k M$.[10]

---

[7] As noted in Example 3, these are exactly the models with whom the prefered recovered knowledge-bases of $\mathcal{KB}$ are associated.

[8] This is the case, for instance, where there is no evidence of any relation between the new data and the old one.

[9] In case that conditions (a) and (b) are satisfied we say that KB_1 and KB_2 are a *partition* of KB.

[10] This property is sometimes called *smoothness* [10] or *stopperdness* [11].

Suppose now that S is a preferred recovered knowledge-base of KB_2. Then it is associated with some k-minimal model ν_2 of KB_2, i.e. $S = (KB_2)_{\nu_2}$. Also, since KB_1 is classically consistent, it has a classical model, denote it ν_1. Now, consider a valuation ν that is defined for every atomic formula p as follows:

$$\nu(p) = \begin{cases} \nu_1(p) & \text{if } p \in \mathcal{A}(KB_1) \\ \nu_2(p) & \text{if } p \in \mathcal{A}(KB_2) \end{cases}$$

Since $\mathcal{A}(KB_1) \cap \mathcal{A}(KB_2) = \emptyset$, ν is well defined. It is also easy to see that ν is a model of KB, and that $KB_\nu = (KB_2)_{\nu_2} \cup (KB_1)_{\nu_1} = S \cup KB_1$. By Lemma 5-A there is a k-minimal model M of KB s.t. $M \leq_k \nu$. In particular, $I(M, KB) \subseteq I(\nu, KB)$, and so $KB_\nu \subseteq KB_M$. But $KB_\nu = S \cup KB_1$, and since S is a maximal recovered knowledge-base of KB_2, KB_ν must be a maximal recovered knowledge-base of KB. Thus $KB_M = KB_\nu = S \cup KB_1$ is a maximal recovered knowledge-base of KB and it is associated with a k-minimal model of KB. Hence $S \cup KB_1$ is indeed a preferred recovered knowledge-base of KB.   □

Note that an immediate consequence of Proposition 5 is that in case that KB is classically consistent, then KB itself is the (only) preferred recovered knowledge-base, as indeed one expects.

Example 6. Consider again our canonical example $(1, 3, 4)$. Let $\mathcal{KB}' = \mathcal{KB} \cup \{u, \neg v \lor w\}$. The prefered recovered knowledge-bases of $\mathcal{KB}'$ are simply obtained by adding $\{u, \neg v \lor w\}$ to each prefered recovered knowledge-base of $\mathcal{KB}$. I.e., the preferred recovered knowledge-bases of $\mathcal{KB}'$ are $\{p, \neg p \lor h, h \lor r, h \lor s, u, \neg v \lor w\}$ and $\{\neg q, q \lor r \lor s, q \lor \neg r \lor \neg s, h \lor r, h \lor s, u, \neg v \lor w\}$.

It follows that in many cases it is possible to drastically reduce the execution time of the algorithm: If the knowledge-base under consideration can be partitioned into two subsets such that one of them is classically consistent, then in order to recover the knowledge-base it is sufficient to activate the algorithm only on the inconsistent subset, and then to add the consistent set to every preferred recovered knowledge-base that is obtained by the algorithm.

4   Conclusion

In this work we have introduced a simple algorithmic method for restoring the consistency of inconsistent knowledge-bases. Restoration of consistent data is a key concept in many applications, such as model-base diagnostic systems, database management systems for distributed (and possibly contradicting) sources of information, and pre-processing phases of procedures for a (classical) automated deduction. In all these areas, then, the techniques discusses in this paper may be useful.

We have addressed here the propositional case in which our algorithm can easily be implemented in practice. Its computational complexity in the general case, and further practical considerations for an efficient handling of first-order languages, remain to be studied.

Acknowledgement

I would like to thank the anonymous referees for their helpful comments. This work is supported by the Visiting Postdoctoral Fellowship FWO Flanders.

References

1. O.Arieli, A.Avron. *Four-valued diagnoses for stratified knowledge-bases.* Proc. CSL'96, Selected Papers (D.Van-Dalen, M.Bezem, editors), Springer Verlag, LNCS No.1258, pages 1–17, 1997.
2. O.Arieli, A.Avron. *A model theoretic approach to recover consistent data from inconsistent knowledge-bases.* Journal of Automated Reasoning 22(3), pages 263–309, 1999.
3. N.D.Belnap. *A useful four-valued logic.* Modern Uses of Multiple-Valued Logic (G.Epstein, J.M.Dunn, editors), Reidel Publishing Company, pages 7–37, 1977.
4. N.D.Belnap. *How computer should think.* Contemporary Aspects of Philosophy (G.Ryle, editor), Oriel Press, pages 30–56, 1977.
5. S.Benferhat, D.Dubois, H.Prade. *How to infer from inconsistent beliefs without revising?* Proc. IJCAI'95, pages 1449–1455, 1995.
6. D.Dubois, J.Lang, H.Prade. *Possibilistic logic.* Handbook of Logic in Artificial Intelligence and Logic Programming (D.Gabbay, C.Hogger, J.Robinson, editors), Oxford Science Publications, pages 439–513, 1994.
7. M.Fitting. *Kleene's three-valued logics and their children.* Fundamenta Informaticae 20, pages 113–131, 1994.
8. M.Gelfond, V.Lifschitz. *The stable model semantics for logic programming.* Proc. of the 5th Logic Programming Symp. (R.Kowalski, K.Browen, editors) MIT Press, Cambridge, MA, pages 1070–1080, 1988.
9. M.Ginsberg. *Counterfactuals.* Artificial Intelligence 30(1), pages 35–79, 1986.
10. S.Kraus, D.Lehmann, M.Magidor. *Nonmonotonic reasoning, preferential models and cumulative logics.* Artificial Intelligence 44(1–2), pages 167–207, 1990.
11. D.Makinson. *General patterns in nonmonotonic reasoning.* Handbook of Logic in Artificial Intelligence and Logic Programming 3 (D.Gabbay, C.Hogger, J.Robinson, editors), Oxford Science Publishers, pages 35–110, 1994.
12. J.Pearl. *Reasoning under uncertainty.* Annual Review of Computer Science 4, pages 37–72, 1989.
13. G.Priest. *Minimally Inconsistent LP.* Studia Logica 50, pages 321–331, 1991.
14. V.S.Subrahmanian. *Mechanical proof procedures for many valued lattice-based logic programming.* Journal of Non-Classical Logic 7, pages 7–41, 1990.
15. G.Wagner. *Vivid logic.* Springer-Verlag, LNAI No.764, 1994.

Acceptance Without Minimality

Abhaya C. Nayak

Computational Reasoning Group
Department of Computing
Division of Information and Communication Sciences
Macquarie University
NSW 2109, AUSTRALIA
abhaya@mpce.mq.edu.au

Abstract. In the belief change literature, while the degree of belief (or disbelief) plays a crucial role, it is assumed that potential hypotheses that have neither been accepted nor rejected cannot be compared with each other in any meaningful manner. We start with the assumption that such hypotheses can be non-trivially compared with respect to their *plausibility* and argue that a comprehensive theory of acceptance should take into account the degree of beliefs (or disbeliefs) as well as the plausibility of such tenable hypotheses. After showing that such a comprehensive theory of acceptance based on the received principle of minimal change does not lend itself to iterated acceptance, we propose, examine and provide representation results for an alternative theory based on the principle of *rejecting the worst* that can handle repeated acceptance of evidence.

1 Introduction

The theory of belief change, originating in the classic works [AGM85, Gär88] (henceforth the *AGM Theory*) takes into account what we may term the degree or firmness of currently held beliefs. The basic idea that these theories rest on is that in assimilating new information, a rational agent should see to it that *if some currently held beliefs must be given up, then, given the option, less firmly held beliefs may be given up in favour of more firmly held beliefs*. Possibility theory [DP92], on the other hand, heavily relies on what may be termed as the degree of *disbelief*. The basic idea behind possibility theory is that in assimilating new information, a rational agent may be forced to suspend disbelief in some sentences that are currently disbelieved (i.e., their negations are believed); and in such an eventuality the agent should see to it that *given the option, the suspension of disbelief is carried out with respect to less strongly denounced propositions instead of more strongly denounced propositions*. In fact, both these approaches — belief change and possibility theory — are largely inter-translatable since the firmness of the belief in a sentence may be viewed simply as the strength of denouncement with respect to its negation.

Since each sentence is either believed or disbelieved or neither, given an agent's belief state, sentences of a language may be partitioned into three disjoint cells, namely, *beliefs* (sentences that the agents takes to be true in her model of the world), *disbeliefs* (sentences that the agent takes to be false in her model of the world) and *plausibilities*

M. Ojeda-Aciego et al. (Eds.): JELIA 2000, LNAI 1919, pp. 163–178, 2000.

(sentences that the agent is agnostic about). The measure used by the belief change camp, exemplified by, for instance, epistemic entrenchment, is primarily defined over the beliefs. The measure used by the possibility theory camp (the possibility measure), on the other hand, is primarily defined over the disbeliefs. More to the point, both these measures effectively refuse to compare different plausibilities. This is rather ironic since both these camps are rather recent entrants to the state-updating area compared to the Bayesian tradition which is primarily based on the probability of the plausibilities.[1]

It is perhaps a mistake to consider the Bayesian approach and the belief change (or, for that matter, possibility theory) as competitors: they are best viewed as complementing each other in providing us a model for the general task of accepting some new evidence. Belief change and possibility theory primarily provide a model for accepting new information that conflicts with the current knowledge. This problem has come to be known as *revision* in the literature. The account they give of accepting new information that is not in conflict with the current knowledge may be viewed as a special case that should not be taken seriously. Similarly, the Bayesian tradition may be taken as providing us a model of how to accept evidence that is consistent with the current knowledge. This problem has come to be known as *expansion*. Bayesian doctrine is more up-front about its treatment of evidence that conflicts with the current knowledge – the Bayesian doctrine is not designed to handle such evidence.

In light of the above discussion, it is apparent that a general account of acceptance should provide a non-trivial account of handling two types of evidence – disbeliefs and plausibilities – in the sense that it should be based on a measure that allows non-trivial comparison among beliefs (or disbeliefs) and among plausibilities. This purported account of acceptance may be quantitative in the Bayesian style or qualitative in the AGM style. The purpose of this paper is to provide a qualitative account of such a general theory of acceptance.

This account should satisfy certain high-level desiderata that will be explicated in more detail in the next section:

1. The theory of acceptance in question should allow the non-trivial comparison of beliefs (*mutatis mutandis* disbeliefs) on the basis of their strength or firmness,
2. The theory of acceptance in question should allow the non-trivial comparison of hypotheses that have neither been accepted nor rejected on the basis of their plausibility,
3. The construction of the purported acceptance operation should be based on rationally defensible principles
4. The properties of the purported acceptance operation should be intuitively appealing, and finally,
5. The framework used for this construction should allow for an iterated account of acceptance in a non-trivial manner.

The rest of this paper is organised as follows. In the next section, I show that when we impose comparability among plausibilities on the AGM framework, we get an operation (to be called "acceptance") that behaves like revision or abduction depending on

---

[1] In the Bayesian framework, each beliefs receive probability 1 and each disbelief gets probability 0. So there is no non-trivial comparison among beliefs (or disbeliefs). Only the comparison among the plausibilities is nontrivial.

the nature of the received evidence. In the section following it, I discuss and examine the limited nature of this operation, namely that it cannot process sequential pieces of evidence in a satisfactory manner. In the penultimate section, an alternative theory of acceptance based on the principle of rejecting the worst is presented, and its properties examined. Appropriate representation results are presented in this section. Finally I conclude with a brief discussion of how this proposed theory lends itself to an account of iterated acceptance of evidence.

2   Comparison among Plausibilities: Genesis of Abduction

In the introductory section, I argued that a theory of acceptance should take into account comparison among plausibilities, that is among sentences that are neither believed nor disbelieved by an agent. In this section, I will postulate such a comparison among plausibilities and show that this leads to an account of *abduction* or *inference to the best explanation* [Pau93] of the variety propounded by Pagnucco in [Pag96]. I will then explain how this theory of abduction can be used in a theory of acceptance and point out one of its severe limitations, namely that this account does not lend itself to an iterated account of acceptance.

The comparison among plausibilities will be modelled after the comparison among the beliefs as provided by the relation of epistemic entrenchment [GM88]. Hence I will first provide a brief introduction to the classic account of belief change [AGM85] followed by a semantic account of epistemic entrenchment [Gro88]. After that I will give an analogous account of comparison among plausibilities that will lead to Pagnucco's account of abduction [Pag96].

2.1   Belief Change

In the AGM system, a belief state is represented as a theory or belief set (i.e., a set of sentences closed under your favourite consequence operation), new information (epistemic input) is represented as a single sentence, and a state transition function, called revision, returns a new belief state given an old belief state and an epistemic input. If the input in question is not belief contravening, i.e., does not conflict with the given belief state (theory), then the new belief state is simply the consequence closure of the old state together with the epistemic input. In the other case, i.e., when the input is belief contravening, the model utilises a selection mechanism (e.g. an epistemic entrenchment relation over beliefs, a nearness relation over worlds or a preference relation over theories) in order to determine what portion of the old belief state has to be discarded before the input is incorporated into it.

From here onwards I will assume a finitary propositional object language $\mathcal{L}$.[2] Let its logic be represented by a classical logical consequence operation Cn. The yielding relation $\vdash$ is defined via Cn as: $\Gamma \vdash \alpha$ iff $\alpha \in Cn(\Gamma)$.

---

[2] A finitary language is a language generated from a finite number of atomic sentences. So the number of sentences in this language is not finite.

The AGM revision operation is required to satisfy the following rationality postulates: Let K be a belief set (a set of sentences closed under Cn), the sentence $x \in \mathcal{L}$ be the evidence, $*$ the revision operator, and K_x^* the result of revising K by x.

(1*) K_x^* is a theory
(2*) $x \in K_x^*$
(3*) $K_x^* \subseteq Cn(K \cup \{x\})$
(4*) If $K \not\vdash \neg x$ then $Cn(K \cup \{x\}) \subseteq K_x^*$
(5*) $K_x^* = K_\perp$ iff $\vdash \neg x$
(6*) If $\vdash x \leftrightarrow y$, then $K_x^* = K_y^*$
(7*) $K_{(x \wedge y)}^* \subseteq Cn(K_x^* \cup \{y\})$
(8*) If $\neg y \notin K_x^*$ then $Cn(K_x^* \cup \{y\}) \subseteq K_{(x \wedge y)}^*$

Motivation for these postulates can be found in [Gär88]. Let us call any revision operation that satisfies the above eight constraints "AGM rational". These postulates can actually be translated into constraints on a non-monotonic inference relation $\vdash$ [GM94].

The account of belief change provided here is non-constructive. A popular construction of the revision operation $*$ is obtained via the relation $\leq$ of epistemic entrenchment. This relation $\leq$ is a binary relation defined over the language $\mathcal{L}$ and the expression $x \leq y$ is meant to be read off as: *sentence y is no less firmly believed than the sentence* x. The standard conditions that $\leq$ is meant to satisfy can be found in [Gär88]. The operation $*$ can be constructed via $\leq$ in the following manner: *an arbitrary sentence y is in K_x^* just in case either y is implied by x or $(x \to \neg y) < (x \to y)$*. The principal (second) case, means that, when, relative to the evidence x, the information in $\neg y$ is less firmly held than the information in y, the sentence y should be accepted on the basis of evidence x. Instead of giving details of epistemic entrenchment, I will now provide its semantics, supplemented by visual aid, which has obvious intuitive appeal.

2.2 Semantics of Entrenchment

The semantics of epistemic entrenchment is given by what has come to be known as the "Systems of Spheres" (SOS), originally developed by Adam Grove [Gro88]. The one I will present is different in approach, but is equivalent to the construction propounded by Grove. Let $\mathcal{M}$ be the class of maximally consistent sets w of sentences in the language in question. The reader is encouraged to think of these maximal sets as worlds, models or scenarios. I will use the following expressions interchangeably: "$w \models \alpha$", "α allows w" and "$w \in [\alpha]$", where w is an element in $\mathcal{M}$ and α is either a sentence or a set of sentences.) Given the belief set K, denote by $[K]$ the worlds allowed by it, i.e., $[K] = \{w \in \mathcal{M} \mid K \subseteq w\}$. (Similarly, for any sentence x, let $[x]$ be the set of "worlds" in which x holds.)

A system of spheres is simply represented by a connected, transitive and reflexive relation (total preorder) $\sqsubseteq$ over the set $\mathcal{M}$ such that $[K]$ is exactly the set of $\sqsubseteq$-minimal worlds of $\mathcal{M}$. Intuitively, $w \sqsubseteq w'$ may be read as: w is at least as good/preferable as w' (or, w' is not strictly preferred to w).[3]

---

[3] Note the oddity: the $\sqsubseteq$-minimal world is most preferred. This is a legacy from the literature.

The relation $\sqsubseteq$ nicely captures the idea behind epistemic entrenchment: $x \leq y$ for an agent just in case, from that agent's perspective, the most preferred $\neg x$-world is no less preferred than the most preferred $\neg y$-world. More formally, $x \leq y$ iff $w_{\neg x} \sqsubseteq w_{\neg y}$ where $w_{\neg x}$ is a $\sqsubseteq$-minimal $\neg x$-world and $w_{\neg y}$ is a $\sqsubseteq$-minimal $\neg y$-world. In intuitive terms, x is less firmly believed that y just in case it is easier for one to move from one's current perspective to a $\neg x$-scenario than to a $\neg y$-scenario. It is easily verified that non-beliefs, in particular plausibilities (sentences that allow some but not all $[K]$-worlds) are all $\leq$ equivalent, and hence $\leq$ cannot discriminate among plausibilities. The reason for this is that, given two plausibilities x and y, the worlds that are $\sqsubseteq$-minimal in $[\neg x]$ and those that are $\sqsubseteq$-minimal in $[\neg y]$, being members of $[K]$, are $\sqsubseteq$-minimal worlds.

Now, we define the Grove-revision function $G*$ as: $[K_x^{G*}] = \{w \in [x]|$ for all $w' \in [x], w \sqsubseteq w'\}$, whereby $K_x^{G*} = \bigcap [K_x^{G*}]$. It turns out that the AGM revision postulates characterise the Grove revision operation $G*$.[4] A visual representation of the

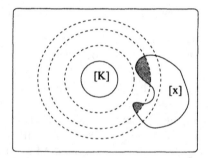

Fig. 1. Minimality Based revision – the principal case

crucial case in the Grove Construction is given in Figure 1. In this, the area marked $[x]$ represents the models allowed by the evidence x. The area $[K]$ represents the model currently entertained by the agent, and the broken circles demarcate models according to the agent's preference. The farther a model is from the centre, the less preferred it is. The shaded part of $[x]$ represents the most preferred of the models allowed by the evidence x – hence identified with $[K_x^*]$.

Viewed from this semantic angle, belief change is about preferential choice: $[K_x^{G*}]$ essentially identifies the subset to be chosen from $[x]$ as the set of worlds that are $\sqsubseteq$-best in $[x]$.

We introduce the following notation for later use.

---

[4] Readers acquainted with Grove's work will easily notice that given a system of spheres Σ, the relation $\sqsubseteq_\Sigma$ can be generated as: $w \sqsubseteq_\Sigma w'$ iff for every sphere S' that has w' as a member, there exists sphere $S \subseteq S'$ with w as a member. On the other hand, given a total preorder $\sqsubseteq$ on $\mathcal{M}$, a system of spheres $\Sigma_\sqsubseteq$ can be generated as follows: A set $S \subseteq \mathcal{M}$ is a sphere in $\Sigma_\sqsubseteq$ iff given any member w of S, if $w' \sqsubseteq w$ then w' is also a member of S. It is easily noticed that the $\sqsubseteq$-minimal worlds of $\mathcal{M}$ constitute the central sphere, and for any sentence x, the $\sqsubseteq$-minimal members of $[x]$ constitute $[K_x^{G*}]$ in the corresponding SOS.

Definition 1 *A subset T of $\mathcal{M}$ is said to be $\sqsubseteq$-flat just in case $w \sqsubseteq w'$ for all members w, w' of T. In this case, the members of T are called $\sqsubseteq$-equivalent. $w \sqsubset w'$, on the other hand, is used as an abbreviation for $(w \sqsubseteq w') \wedge (w' \not\sqsubseteq w)$*

2.3 Minimality Based Abduction

Earlier I argued that like beliefs and disbeliefs, plausibilities too can be meaningfully compared with each other. We also noticed that epistemic entrenchment does not provide a meaningful comparison among plausibilities since all the worlds validating the agent's current knowledge (namely members of $[K]$) are $\sqsubseteq$-minimal. In order to effect a non-trivial comparison among the plausibilities, therefore, it seems prudent to introduce some more structure into $[K]$. Let us accordingly give up the assumption that $[K]$ is the set of $\sqsubseteq$-minimal worlds, and instead impose the following conditions:

1. $[K] \neq \emptyset$ and
2. If $w \sqsubseteq w'$ and $w' \in [K]$ then $w \in [K]$, for every w, w' in $\mathcal{M}$.

In effect, the system of sphere represented by $\sqsubseteq$ represents an expectation ordering [GM94]. The belief state $[K]$ in this system of spheres could be any of the sphere in the system. Grove's SOS is a special case of this, namely when $[K]$ is the smallest sphere allowed by $\sqsubseteq$ – i.e., $[K]$ is the set of $\sqsubseteq$-minimal worlds. Another special case is when $[K] = \mathcal{M}$. This represents the knowledge state of an epistemically innocent agent who does not know anything about the world. But a more interesting special case is the dual of Grove's SOS: $[K] = \{w | w$ is not $\sqsubseteq$-maximal$\}$. In other words, whereas in Grove's account, $[K]$ is $\sqsubseteq$-flat, in this dual account, $\mathcal{M} \setminus [K]$ is $\sqsubseteq$-flat. If we assume a binary relation $\preceq$ over $\mathcal{L}$ defined as: $x \preceq y$ iff $w_{\neg x} \sqsubseteq w_{\neg y}$ where $w_{\neg x}$ is a $\sqsubseteq$-minimal $\neg x$-world and $w_{\neg y}$ is a $\sqsubseteq$-minimal $\neg y$-world, we get a relational measure that effectively compares plausibilities, but fails to discriminate among beliefs (and among disbeliefs). This is the mechanism that drives Pagnucco's account of abductive belief change [Pag96].

Analogous to the AGM approach to revision, expansion in Pagnucco's approach rests on minimality consideration. Given evidence x which is consistent with the current knowledge, the result of adopting x is represented by the new belief state $[K_x^+] = \{w \mid w$ is $\sqsubseteq$-minimal in $[x]\} = \{w \mid w$ is $\sqsubseteq$-minimal in $[K] \cap [x]\}$. However, since $[K]$ is not necessarily $\sqsubseteq$-flat, neither is $[K] \cap [x]$. Hence, possibly $[K_x^+] \subset [K] \cap [x]$. Thus, unlike the expansion in the AGM approach, Pagnucco's expansion operation $+$ is ampliative. In fact this operation has all the hall marks of an abductive inference. Figure 2 provides a visual representation of the abductive process suggested in [Pag96].

Pagnucco has examined the properties of this abduction operation. Let K be the current belief set, x the evidence and $+$ the abductive expansion operation. The following list fully characterises this operation.

(1$^+$) K_x^+ is a theory
(2$^+$) If $\neg x \notin K$ then $x \in K_x^+$
(3$^+$) $K \subseteq K_x^+$
(4$^+$) If $K \vdash \neg x$ then $K_x^+ = K$
(5$^+$) If $K \nvdash \neg x$ then $\neg x \notin K_x^+$

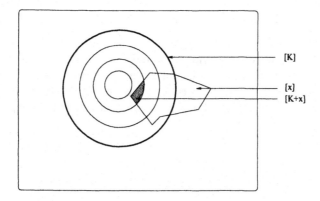

Fig. 2. Minimality based Abduction

(6^+) If $K \vdash x \leftrightarrow y$, then $K_x^+ = K_y^+$
(7^+) $K_x^+ \subseteq Cn(K_{(x \vee y)}^+ \cup \{x\})$
(8^+) If $\neg x \notin K_{x \vee y}^+$ then $K_{(x \vee y)}^+ \subseteq K_x^+$

The motivation behind these properties can be found in [Pag96].

3   Minimality Based Acceptance and Its Failure

In the last section I showed how the desire for comparision of plausibilities, combined with the minimality based belief change, leads to Pagnucco's account of abduction. In this section I will combine the AGM approach to belief change with Pagnucco's account of abduction in order to provide a comprehensive account of acceptance. Then I will show that this approach suffers from a serious setback in that it does not lend itself to an account of iterated acceptance. The next section will be devoted to an analysis of this problem of iteration, and a solution to this problem will be presented. Later on, technical exploration based on this suggestion will be performed.

3.1   Acceptance Based on Minimality

In the last section, we dispensed with the AGM idea that the belief state $[K]$ is the smallest sphere in an SOS and assumed that $[K]$ could be any sphere in the SOS. I pointed out that the AGM system (read Grove's SOS) is one special case of this, and Pagnucco's system is another special case. Now, we can combine these two accounts to offer a general account of acceptance. Roughly, what we wish the acceptance operation to do is to behave like the AGM operation when the evidence is belief contravening, and behave like the Pagnucco operator when the evidence is consistent with the current beliefs. Let us denote this minimality based acceptance operator as $\odot$ and define this operation $\odot$, given an expectation ordering $\sqsubseteq$ and an appropriate belief set K as follows:

Definition 2 (from $\sqsubseteq$ to $\odot$) *Where $\sqsubseteq$ be a total preorder on $\mathcal{M}$ and $[K]$ a sphere for $\sqsubseteq$, $[K_x^\odot]$ is defined as the set $\{w \in [x] \mid w \sqsubseteq w' \text{ for all } w' \in [x]\}$.*

It is easily verified that when the evidence x conflicts with K, the operation $\odot$ behaves like the AGM revision operator; on the other hand, if x is consistent with K, instead of behaving like the AGM revision operator, the operation $\odot$ starts behaving like Pagnucco's abductive expansion operator. This process may be visually represented as in Figure 3.

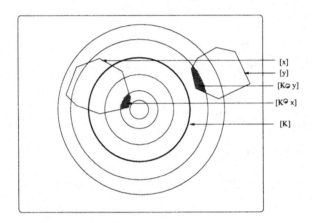

Fig. 3. Minimality based Acceptance.

3.2   Acceptance Faces the Iteration Problem

Iteration has been a well known problem in the belief change literature. Formally, a function f, in order to be iterative, simply requires that if $f(x)$ is a well defined object, then so should be $f(f(x))$. In the context of belief change, failure of the iterative property means an agent is guaranteed an initial change of mind, but not necessarily any subsequent one. Since in practice agents do not get all pieces of evidence in one go, it is highly desirable that any belief change operation, acceptance included, should have the iterative property. In the belief change lingo, it means that the belief change operation should satisfy the properties of category matching: the object that undergoes change must result in an object of the same category.

Unfortunately, however, the acceptance operation $\odot$ seriously fails on this count. There are different ways of looking at this problem. Primarily, a structured object ($[K]$, which consists of possibly many layers of $\sqsubseteq$-equivalence classes of worlds) undergoes an epistemic change in response to evidence x and results in an unstructured object ($[K_x^\odot]$, which is a single class of $\sqsubseteq$-equivalent worlds). Hence operation $\odot$ violates the principle of category matching.[5]

---

[5] Perhaps a more accurate description of the problem is the following. There are three arguments to $\odot$: an expectation ordering $\sqsubseteq$, an arbitrary sphere $[K]$ of $\sqsubseteq$ and the evidential (external) input

The practical problem is noticed very easily. Consider Figure 3. Imagine that α and β are two pieces of evidence such that $[K] \cap [\alpha] \neq \emptyset$ and $[K_\alpha^\odot] \cap [\beta] \neq \emptyset$. Assuming that $[K] \cap [\alpha]$ is not $\sqsubseteq$-flat, the first operation of $\odot$ will result in an abductive expansion. Now, in order to process evidence β we will need a system of sphere in which $[K_\alpha^\odot]$ is a sphere. But since $[K_\alpha^\odot]$ is $\sqsubseteq$-flat, given the measure $\sqsubseteq$, no matter how we permute the $\sqsubseteq$-equivalent classes, if $[K_\alpha^\odot]$ is going to be a sphere in the resultant SOS, it is going to be the central sphere. Hence we are back to a Grovian SOS, and all future expansions are going to be the non-abductive AGM expansion. Another way of looking at it is that although the desirability of a nontrivial comparison among plausibilities led to the theory of acceptance at issue here, after the first abduction, we are left only with a vacuous comparison among plausibilities.

3.3   Diagnosis and Prescription

It is clear from discussion above that iteration is desirable in the context of acceptance, and the operation $\odot$ fails on this count primarily because $[K_x^\odot]$ consists of a set of $\sqsubseteq$-equivalent worlds, in particular, the set of $\sqsubseteq$-minimal x-worlds. This has often been justified on the basis of the principle of minimality (read minimal change). Hence, in order that we may gain the ability to iterate, it is imperative to satisfy the principle of category matching. This in turn implies that we impose more structure into the set $[K_x^\odot]$, and thereby violate the principle of minimality. In this context, it is important to take into consideration a few issues:

1. What is the intuitive justification for the principle of minimality?
2. Our proposal to impose more structure into $[K_x^\odot]$ and thereby violate the principle of minimality is based on purely pragmatic ground. Can this be justified on independent grounds?
3. The discussion in the last section regarding the failure of iteration in the context of acceptance is primarily based on abduction. Is it possibly desirable to violate the principle of minimality only in the context of abduction and retain in the context of revision?

I will address these issues individually.

As to the first issue, the principle of minimality in question is essentially based on the intuitively obvious principle of choosing the best [NF98]. In order to successfully accept the evidence x, the result $[K_x^\odot]$ is required to be a subset of $[x]$. Hence, it is a matter of choosing the "right" elements of $[x]$. Since $\sqsubseteq$ reflects the agent's preference over all the worlds, members of $[x]$ included, and the $\sqsubseteq$-minimal x-worlds are deemed best among all the x-worlds, it is reasoned, the set $[K_x^\odot]$ should be identified with the set of $\sqsubseteq$-minimal x-worlds.

There are two ways of responding to the second issue. On the first count, the principle of choosing the best is a vacuous principle devoid of any prescribe content since it

---

x. In order to satisfy the principle of category matching, the output should be a pair $\sqsubseteq'$ and its arbitrary sphere $[K'] \neq [K_x^\odot]$. But since $[K_x^\odot]$ is $\sqsubseteq$-flat, there is no constructive way of generating an expectation ordering $\sqsubseteq'$ in which $[K'] = [K_x^\odot]$ is a sphere but not necessarily the the central sphere. Hence the principle of category matching is violated by $\odot$.

simply means that whatever should be chosen should be chosen. Hence no matter what one does, one cannot violate this, as it were, *analytic principle*. On the second count, there is a dual to the principle of choosing the best: the principle of *rejecting the worst* [NF98]. This principle says that *in a choice context, reject the worst available alternatives and retain the rest for further scrutiny*. This principle has no less intuitive appeal than the principle of choosing the best. Since the set $[K] \cap [x]$ (respectively, $[x]$) possibly comprises of more than two $\sqsubseteq$-equivalence classes, even after rejecting the worst members from $[K] \cap [x]$ leaves us with a set $[K_x^\ominus]$ that is *not* $\sqsubseteq$-flat, we can impose some relevant structure into $[K_x^\ominus]$ on grounds no less justifiable than the principle of minimality itself.

Finally, as to the third issue, there are at least two reasons why the principle of minimality should be violated both in the context of abduction and revision. Firstly, assuming that we employ the principle of rejecting the worst in the context of abduction, we need some special, overriding consideration to justify the principle of choosing the best (read minimality) in the context of revision. No such overriding considerations are available. This is an argument from the classic principle of insufficient reason. Secondly, and this is a pragmatic consideration, if we allow the principle of minimality to be employed in the context of revision, it is not going to solve the problem of iteration so far as acceptance is concerned. Once the agent accepts some belief contravening evidence x, the resultant $[K_x^\ominus]$ becomes $\sqsubseteq$-flat and we are back to the old problem!

I take the above discussion to justify the uniform employment of the principle of rejecting the worst in a reasoned account of acceptance.

4   Acceptance Based on Rejection

I pointed out above that the principle of minimality does not allow the theory of acceptance to extend to an iterative account. I further argued that this principle is no more justified than its dual, the principle of rejecting the worst, which, if considered, may allow an iterative account of acceptance. In this section I will develop and examine a theory of acceptance based on the principle of rejecting the worst.

4.1   The "Reject Worst Principle" and Acceptance

The principle of rejecting the worst essentially tells us that in a choice context, reject the worst among the available alternatives and retain the rest for further consideration. We must add a caveat to this in order to handle the special case when all alternatives are deemed to be equally desirable. In such a situation, all the available alternatives are worst (and also best). Since the goal is to ultimately choose some member or other from the alternatives, I will slightly weaken the principle:[6]

---

[6] There are choice contexts where an agent may want not to choose any of the available alternatives. For instance, a selection committee may want to re-advertise a position instead, if none of the interested candidates satisfy the minimum prerequisites. There are many ways of looking at it. An easy way out is to maintain that this set of candidates is *not* a set of alternatives in the first place since they do not satisfy the minimum requirement of being an alternative. There are other ways of reconciling this issue as well, but it is beyond the scope of this paper to go to the details.

— In a choice context, given that not all the available alternatives are equally desirable, reject the alternatives deemed to be worst with respect to the contextually defined selection criteria, and retain the rest for future consideration. Otherwise, reject none.

Let us denote the acceptance operation based on this principle of rejecting the worst by the symbol ∘. Figure 4 pictures how different types of evidential data (w, x, y and z) are handled by this operation. Note in particular the case of evidence y. In this case, the worst elements are rejected not from $[y]$ but from $[K] \cap [y]$. If we had rejected only the worst elements of $[y]$, the result would not have been a subset of $[K]$, and we would have lost part of the information in K, although the evidence is consistent with the current knowledge!

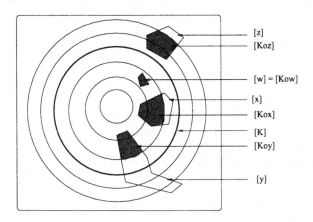

Fig. 4. Acceptance Without Minimality.

Now I will formally define how, given an appropriate total preorder $\sqsubseteq$ on $\mathcal{M}$ and an belief set K for $\sqsubseteq$, the non-minimal acceptance operation $\circ_\sqsubseteq$ (the subscript is henceforth dropped) is constructed:

Definition 3 (from $\sqsubseteq$ to $\circ$) *Where $\sqsubseteq$ be a total preorder on $\mathcal{M}$ and $[K]$ a sphere for $\sqsubseteq$*

$$[K_x^\circ] = \begin{cases} [x] & \text{if } [x] \text{ is } \sqsubseteq\text{-flat} \\ \{w \in [x] \mid w \sqsubset w' \\ \quad \text{for some } w' \in [x]\} & \text{else if } [K] \cap [x] = \emptyset \\ [K] \cap [x] & \text{else if } [K] \cap [x] \text{ is } \sqsubseteq\text{-flat} \\ \{w \in [K] \cap [x] \mid w \sqsubset w' \\ \quad \text{for some } w' \in [K] \cap [x]\} & \text{otherwise.} \end{cases}$$

This definition separates four distinct cases and treats them differently. First of all, if $[x]$ is flat, irrespective of whether it intersects $[K]$ or not, the result is simply $[x]$. This is because there is not enough structure in $[x]$ to do any more sophisticated operation. Else, if $[x]$ is "outside" $[K]$ but is not flat, then the operation $\circ$ behaves like a non-minimal belief revision [NF98]. In the third case, if $[x]$ intersects $[K]$ but the intersection itself is flat, the result simply $[K] \cap [x]$. Finally, if the intersection of $[K]$ and $[x]$ is not flat, then it behaves like a non-minimal abduction operator.

4.2 Properties of Non-minimal Acceptance

I outlined above an account of how the acceptance operator based on the principle of rejecting the worst can be constructed. Intuitive though this construction process is, it remains to be seen whether this operation has the properties required of an acceptance operator. Of the properties this operation satisfies, the following are especially interesting for reasons to be elaborated afterwards. Note the naming conventions followed: the numeric part of the name in general signifies which AGM postulate it is an analogue of and the (optional) alphabetic part signifies whether this property concerns the abductive behaviour or the revision behaviour of the operator $\circ$. For instance, the property $(7.1A\circ)$ corresponds to the AGM postulate $(7*)$ and concerns the abductive behaviour of $\circ$.

$(1\circ)$ K_x° is a theory

$(2\circ)$ $x \in K_x^\circ$

$(4\circ)$ If $K \not\vdash \neg x$ then $Cn(K \cup \{x\}) \subseteq K_x^\circ$

$(5\circ)$ $K_x^\circ = K_\perp$ iff $\vdash \neg x$

$(6R\circ)$ If $\vdash x \leftrightarrow y$, then $K_x^\circ = K_y^\circ$

$(6A\circ)$ If $K \vdash x \leftrightarrow y$, then $K_x^\circ = K_y^\circ$, given that $K \not\vdash \neg x$

$(7.1R\circ)$ If $K_x^\circ \not\subseteq Cn(x \wedge y)$ then $K_{x \wedge y}^\circ \subseteq Cn(K_x^\circ \cup \{y\})$
$\qquad$ given $K \vdash \neg x$

$(7.1A\circ)$ If $K_x^\circ \not\subseteq Cn(K \cup \{x,y\})$ then $K_{x \wedge y}^\circ \subseteq Cn(K_x^\circ \cup \{y\})$

$(7.2R\circ)$ If $K_y^\circ = Cn(y)$ then $K_{x \wedge y}^\circ \subseteq Cn(K_x^\circ \cup \{y\})$
$\qquad$ given $K \vdash \neg x$

$(7.2A\circ)$ If $K_y^\circ = Cn(K \cup \{y\})$ then $K_{x \wedge y}^\circ \subseteq Cn(K_x^\circ \cup \{y\})$
$\qquad$ given $K \not\vdash \neg x$

$(7.3R\circ)$ If $K_x^\circ \cap Cn(y) \subseteq Cn(x)$
$\qquad$ then $K_{x \wedge y}^\circ \subseteq Cn(K_x^\circ \cup \{y\})$

$(7.3A\circ)$ If $K_x^\circ \cap Cn(K \cup \{y\}) \subseteq Cn(K \cup \{x\})$
$\qquad$ then $K_{x \wedge y}^\circ \subseteq Cn(K_x^\circ \cup \{y\})$ given $K \neg \vdash \neg x$

$(8\circ)$ If $K_x^\circ \not\vdash \neg y$ then $Cn(K_x^\circ \cup \{y\}) \subseteq K_{x \wedge y}^\circ$

$(9R\circ)$ If $K \vdash \neg x$, $K_x^\circ \vdash \neg y$ but $x \not\vdash \neg y$
$\qquad$ then $K_{x \wedge y}^\circ = Cn(x \wedge y)$.

$(9A\circ)$ If $K \cup \{x\} \not\vdash \neg y$ but $K_x^\circ \vdash \neg y$
$\qquad$ then $K_{x \wedge y}^\circ \subseteq Cn(K \cup \{x,y\})$.

For an intuitive understanding of these constraints, it is helpful to view K_α° as the set of sentences that the evidence α can explain given the background knowledge K. Properties $(1\circ\text{--}6R\circ)$ are effectively basic postulates of the AGM revision operation, and justification for them can be found in [Gär88]. Postulate $(6A\circ)$ says that if two pieces of evidence contain the same information relative to, and they do not conflict with, the current knowledge, then accepting them have the same effect on the current knowledge. Note that this is a stronger postulate than $(6R\circ)$. Postulates $(7.1R\circ\text{--}7.3A\circ)$ are several variations of the AGM postulate $(7*)$. For instance, $(7.1R\circ)$ says that, when x conflicts with the current knowledge K, if x can explain certain things that cannot be classically inferred from x and y together, then everything that x and y may possibly be able to explain can be classically inferred from y together with all that x explains. On the other hand, $(7.1A\circ)$ may be paraphrased as follows: when x does not conflict with the current knowledge, if x can explain certain things that cannot be classically inferred from K, x and y together, then everything that x and y may possibly be able to explain can be classically inferred from y together with all that x explains. All these variations of $7*$ tell us under what condition a piece of evidence y loses its inferential power in presence of another piece of evidence x. Postulate $(8\circ)$ says that x and y jointly fail to explain something that follows from y in presence of what is explainable by x only if y conflicts with something that is explained by x. Finally, postulates $(9R\circ)$ and $(9A\circ)$ specify the conditions under which x and y cannot explain anything more than what can be classically inferred from them, possibly in presence of K.

4.3   Technical Results

In this section I will show that the theory of acceptance we have so far developed has the desirable features one should expect from it. I will omit the proofs due to the space limitation. Our first result is the soundness property – that $\circ$ satisfies conditions $(1\circ$– $9A\circ)$.

Theorem 1 *Let the operation $\circ$ be constructed from a given total preorder $\sqsubseteq$ on $\mathcal{M}$ and its sphere $[K]$ as specified in Definition 3. The operation $\circ$ then satisfies the basic properties $(1\circ -9A\circ)$.*

The next result (completeness result) shows that given an acceptance operation $\circ$ that satisfies $(1\circ -9A\circ)$ and a fixed belief set K, we can construct a binary relation $\sqsubseteq_{\circ,K}$ with the desired properties. (I will normally drop the subscripts for readability.) In particular, I will show that, where $\sqsubseteq$ is the relation so constructed: (1) $\sqsubseteq$ is a total preorder over $\mathcal{M}$, (2) the SOS (System of Spheres) corresponding to $\sqsubseteq$ has $[K]$ as one of its spheres.

Definition 4 (from $\circ$ to $\sqsubseteq$) *Given an acceptance operation $\circ$ and a belief set K, $w \sqsubseteq_{\circ,K} w'$ iff either (1) both $w \in [K]$ and $w' \notin [K]$ or (2) $w \in [K_x^\circ]$ whenever $w' \in [K_x^\circ]$, for every sentence x such that either (a) $K \vdash \neg x$ and both $w, w' \in [x]$ or (b) $K \nvdash \neg x$ and both $w, w' \in [K] \cap [x]$.*

Theorem 2 *Let $\circ$ be an acceptance operation satisfying $(1\circ) - (9A\circ)$ and K a belief set. Let $\sqsubseteq$ be generated from $\circ$ and K as prescribed by Definition 4. Then $\sqsubseteq$ is a total preorder on $\mathcal{M}$ such that $[K]$ is one of the spheres of $\sqsubseteq$.*

Theorems 1 and 2 jointly provide the representation result.

Furthermore, the total preorder $\sqsubseteq_{\circ,K}$ constructed from a given non minimal revision operation $\circ$ and belief set K is the desired $\sqsubseteq$ in the sense the non minimal acceptance operation constructed from it, in turn, behaves like the original operation $\circ$ with respect to the belief set K.

Theorem 3 *Let* ∘ *be a non minimal belief revision operator satisfying postulates (1* ∘
−9A∘) and K be an arbitrary belief set. Let ⊑ *be defined from* ∘ *and K in accordance
with Definition 4. Let* ∘′ = ∘⊑ *be defined from* ⊑, *in turn, via Definition 3. Then for any
sentence x (and the originally fixed belief set K) it holds that* $K_x^\circ = K_x^{\circ'}$.

Conversely, one can start with a total preorder ⊑, construct an acceptance operation
∘ from it via Definition 3 and then construct a a total preorder ⊑ from that ∘ in turn via
Definition 4, then one gets back the original relation ⊑.

Theorem 4 *Let* ⊑ *be a total preorder on* $\mathcal{M}$ *and* [K] *one of its spheres. Let* ∘ *be defined
(for K) from* ⊑ *via Definition 3. Let* ⊑′=⊑∘ *be defined from* ∘, *in turn, via Definition 4.
Then* $w \sqsubseteq w'$ *iff* $w \sqsubseteq' w'$ *for any two worlds* $w, w' \in \mathcal{M}$

5 Discussion

In this paper, first we argued that although in the literature on belief change, it is taken
for granted that there can be no meaningful comparison among tenable hypotheses that
have neither been accepted nor rejected, a case can be made for nontrivial comparison
among them on the basis of their plausibility. Equipped with a measure that can compare
among such hypotheses as well as among the beliefs (or disbeliefs, as the case may
be), we modelled a comprehensive account of acceptance pretty much in the AGM-
Grove tradition. We then showed that this operation fails to take in to account repeated
mind change on part of the agent. Accordingly, we developed an alternative theory of
acceptance based on the principle of rejecting the worst. We motivated it on the ground
that it can handle the problem of iterated acceptance.

One of the things pointed out to be crucial in order to handle the problem of iteration
is satisfaction of the principle of category matching. It is only natural that in order
to provide an iterated account of acceptance, we identify an expectation ordering that
succeeds the current expectation ordering after a piece of evidence is accepted. The
acceptance operation ∘ as described so far fails to do that. Given an expectation ordering
⊑, a belief set K and a piece of evidence x, we know what the new belief set K_x° would
be; but we do not know what expectation ordering it is a sphere of. What we precisely
need is a more general acceptance operation • that accepts as parameters an expectation
ordering ⊑, a belief set K associated with ⊑ and a piece of evidence x and returns a
new expectation ordering $\langle \sqsubseteq, K \rangle_x^\bullet$ one of whose spheres is K_x°.

In general, there are many ways of satisfying these constraints. However what we
need is a rational way of satisfying these constraints. In the literature on iterated belief change, there has been two basic approaches to solve the analogous problem, both
grounded in Spohn's seminal work [Spo88]. One, based on what has come to be known
as *conditionalisation* has been adopted in many works [Nay94, Wil94]. This approach
maintains the relative ordering of worlds that are consistent with the evidence as well
as the worlds that falsify the evidence, but gives more priority to the former class of
worlds. The other, which has come to be known as *adjustment* has been adopted by
[Wil94]. This approach on the other hand maintains the original ordering of all worlds
that are inconsistent with the new belief set, giving priority only to the worlds that

are consistent with the new belief set. In the account that follows, I adopt the former strategy.

Definition 5 *Let $\sqsubseteq$ be an expectation ordering and K be a theory such that $[K]$ is a sphere of $\sqsubseteq$. Let x be a sentence. Then $\langle \sqsubseteq, K \rangle_x^\bullet = \langle \sqsubseteq', K' \rangle$ where*

1. $K' = K_x^\circ$
2. $w \sqsubseteq' w'$ *for all worlds* w, w' *iff both*
   (a) *Either* $w \in [x]$ *or* $w' \notin [x]$, *and*
   (b) *if* $w \not\sqsubseteq w'$ *then both* $w \in [x]$ *and* $w' \notin [x]$.

The first condition, $K' = K_x^\circ$ ensures that the revised K matches with the one mandated by the acceptance operation $\circ$. The first clause of the second condition, namely *Either $w \in [x]$ or $w' \notin [x]$*, ensures that in the revised expectation ordering, worlds consistent with the evidence x are not accorded less priority than the worlds that falsify such evidence. The second clause of the second condition, namely if $w \not\sqsubseteq w'$ then both $w \in [x]$ and $w' \notin [x]$ ensures that the original priority among worlds is reversed only if it conflicts with the principle that worlds consistent with the evidence should be accorded more priority than the worlds falsifying the evidence.

I conclude this section with a quick proof that $[K_x^\circ]$ is indeed a sphere in the expectation ordering $\sqsubseteq'$ thus defined. Suppose that $w \in [K_x^\circ]$ and $w' \sqsubseteq' w$ but $w' \notin [K_x^\circ]$. Since $w \in [K_x^\circ]$, surely $w \in [x]$. Since $w' \sqsubseteq' w$, it follows that either $w' \in [x]$ or $w \notin [x]$. Hence it follows that $w' \in [x]$. However $w' \notin [K_x^\circ]$ where from it follows that $w' \not\sqsubseteq w$. It follows from the second clause of the second condition that $w \notin [x]$ contradicting the earlier result that $x \in [x]$. ∎

References

[AGM85]   Carlos E. Alchourrón, Peter Gärdenfors, and David Makinson. On the logic of theory change: Partial meet contraction and revision functions. *Journal of Symbolic Logic*, 50:510–530, 1985.

[DP92]   Didier Dubois and Henri Prade. Belief change and possibility theory. In Peter Gärdenfors, editor, *Belief Revision*, pages 142–182. Cambridge University Press, 1992.

[Gär88]   Peter Gärdenfors. *Knowledge in Flux: Modeling the Dynamics of Epistemic States*. Bradford Books, MIT Press, Cambridge Massachusetts, 1988.

[GM88]   Peter Gärdenfors and David Makinson. Revisions of knowledge systems using epistemic entrenchment. In *Proceedings of the Second Conference on Theoretical Aspect of Reasoning About Knowledge*, pages 83–96, 1988.

[GM94]   Peter Gärdenfors and David Makinson. Nonmonotonic inference based on expectations. *Artificial Intelligence*, 65:197–245, 1994.

[Gro88]   Adam Grove. Two modellings for theory change. *Journal of Philosophical Logic*, 17:157–170, 1988.

[Nay94]   Abhaya C. Nayak. Iterated belief change based on epistemic entrenchment. *Erkenntnis*, 41:353–390, 1994.

[NF98]   Abhaya C. Nayak and Norman Y. Foo. Reasoning without minimality. In Hing-Yan Lee and Hiroshi Motoda, editors, *Proceedings of the Fifth Pacific Rim International Conference on Artificial Intelligence (PRICAI-98)*, pages 122–133. Springer Verlag, 1998.

[Pag96]   Maurice Pagnucco.   *The Role of Abductive Reasoning within the Process of Belief revision.*   PhD thesis, University of Sydney, 1996. (www.comp.mq.edu.au/~ morri/Papers/morri.Phd.ps.gz).

[Pau93]   Gabrielle Paul. Approaches to abductive reasoning: An overview. *Artificial Intelligence Review,* 7:109–152, 1993.

[Spo88]   Wolfgang Spohn. Ordinal conditional functions: A dynamic theory of epistemic states. In William L. Harper and Brian Skryms, editors, *Causation in Decision, Belief Change, and Statistics, II,* pages 105–134. Kluwer Academic Publishers, 1988.

[Wil94]   Mary-Anne Williams. Transmutations of knowledge systems. In Jon Doyle, Erik Sandewall, and Pietro Torasso, editors, *Proceedings of the Fourth International Conference on Principles of Knowledge Representation and Reasoning,* pages 619–629. Morgan Kaufmann, 1994.

Reduction Theorems for
Boolean Formulas Using Δ-Trees

Gloria Gutiérrez, Inma P. de Guzmán, Javier Martínez, Manuel Ojeda-Aciego, and Agustín Valverde

Dept. Matemática Aplicada. Universidad de Málaga.
P.O. Box 4114. E-29080 Málaga, Spain.
Phone: +34 95 213 2871, Fax: +34 95 213 2746
aciego@ctima.uma.es

Abstract. A new tree-based representation for propositional formulas, named Δ-tree, is introduced. Δ-trees allow a compact representation for negation normal forms as well as for a number of reduction strategies in order to consider only those occurrences of literals which are relevant for the satisfiability of the input formula. These reduction strategies are divided into two subsets (meaning- and satisfiability-preserving transformations) and can be used to decrease the size of a negation normal form A at (at most) quadratic cost. The reduction strategies are aimed at decreasing the number of required branchings and, therefore, these strategies allow to limit the size of the search space for the SAT problem.

1 Introduction

Efficient representations for formulas in negation normal form (nnfs) are necessary in order to describe and implement efficient algorithms on this kind of formulas. The ability to reason on specifications written in a language as close as possible to natural language is important for information sciences; thus, reasoning efficiently on nnfs is interesting because these formulas are easier to obtain from specifications given in natural language.

Formulas in conjunctive normal form (cnf or in clause form) are usually interpreted as lists of clauses, and formulas in disjunctive normal form (dnf) are interpreted as lists of cubes; these interpretations allow efficient descriptions and implementations of algorithms to study satisfiability (e.g. linear ordered resolution). In this work we use the generalization of these interpretations to nnfs given by the Δ-trees, that is, we use *trees* of clauses and cubes. Specifically, nnfs are represented as trees of clauses and cubes such that each clause-node in the tree is an implicant of the formula represented by its scope and, similarly, each cube-node is an implicate of the formula represented by its scope. The new representation is named Δ-tree because its nodes are built up from Δ-lists [2]. After defining the notion of Δ-tree, the operators Norm and Δ-**Tree** are introduced which, respectively, associate a nnf to each Δ-tree and vice versa. In addition, it can be shown that this correspondence preserves equivalence and,

M. Ojeda-Aciego et al. (Eds.): JELIA 2000, LNAI 1919, pp. 179–192, 2000.

therefore, we can easily extend the concepts of validity and satisfiability to Δ-trees.

We introduce the concept of restricted Δ-tree (generalizing the well-known concept of restricted cnf in which clauses with repeated or contradictory literals are not allowed and subsumed clauses are omitted), which involves only restricted clauses and cubes in the representation and, in addition, prohibits that a single literal is both an implicant and an implicate of the same subformula.

Later, we introduce meaning-preserving transformations, with at most quadratic complexity, which eliminate the conclusive or simple nodes and usually reduces the size of the input Δ-tree. Roughly speaking, a conclusive node in a Δ-tree is one which can be substituted by a logical constant preserving the meaning of the whole tree, and a simple node in a Δ-tree satisfies that the subformula it represents is equivalent to a literal; thus, we introduce the so-called restricted Δ-tree, which generalized the concept of restricted cnf. In addition, several satisfiability-preserving transformations are presented with generalize the one literal rule and the pure literal rule from the clausal framework. Some of these transformations were introduced in [2], and described using the so-called $\widehat{\Delta}$-sets. The fact that $\widehat{\Delta}$-sets are no longer necessary when working with Δ-trees is extremely interesting when implementing the method, since the simple data structure of Δ-tree stores both the information about the structure of the formula and its associated $\widehat{\Delta}$-sets.

Finally, the last section includes some experimental results from an implementation of the method described in [2] based on Δ-trees.

2   Preliminary Concepts and Definitions

Throughout the rest of the paper, we will work with a classical propositional language, $\mathcal{L}$, over a denumerable set of propositional variables, $\mathcal{V}$, and connectives $\{\neg, \wedge, \vee\}$, the semantics for this language being the standard one. We will write $A \equiv B$ to denote that A and B are logically equivalent, and $\Omega \models A$ to denote that A is a logical consequence of Ω, that is, any model of Ω is a model of A. We will use the usual notions of literal (propositional variable or the negation of a propositional variable), clause (disjunction of literals), cube (conjunction of literals), and negation normal form (a formula in which the negations are only in the literals):

In this paper, we will always use cubes and clauses ordered by the lexicographic order in the set of literals, denoted $\mathcal{V}^{\pm}$.

- A literal ℓ is an *implicant* of a formula A if $\ell \models A$.
- A literal ℓ is an *implicate* of a formula A if $A \models \ell$.

We will use the standard notion of tree and address of a node in a tree [6]. An address η in the syntactic tree T_A of a formula A will also mean, when no confusion arises, the subformula of A corresponding to the node of address η in T_A; ε will denote the address of the root node.

We will also use finite lists written in juxtaposition, with the standard notation, nil, for the empty list. If λ and λ' are lists, $\ell \in \lambda$ denotes that ℓ is an element of λ; and $\lambda \subseteq \lambda'$ means that all elements of λ are elements of λ'. The conjugate of a literal ℓ is denoted as $\bar{\ell}$, with the standard meaning, that is, $\bar{p} = \neg p$ and $\bar{\neg}p = p$. If $\lambda = \ell_1 \ell_2 \ldots \ell_n$ is a list of literals, then $\bar{\lambda} = \bar{\ell_1} \bar{\ell_2} \ldots \bar{\ell_n}$

3   The Δ-Trees

In this section we introduce the concept of Δ-tree as an alternative representation of nnfs:

Definition 1 (Δ-tree). *A Δ-tree T in $\mathcal{L}$ is a labeled tree in the set*

$$\mathcal{H} = \big\{[\alpha]\lambda \mid \lambda \in \texttt{List}(\mathcal{V}^{\pm}) \cup \{\bot\}\big\} \cup \big\{[\beta]\lambda \mid \lambda \in \texttt{List}(\mathcal{V}^{\pm}) \cup \{\top\}\big\}$$

inductively defined by the three properties below:

1. *The leaves in a Δ-tree are elements in $\mathcal{H}$.*
2. *Let T_1, ..., T_m be Δ-trees whose roots are $[\beta]\lambda_1, \cdots, [\beta]\lambda_m$ and $[\alpha]\lambda \in \mathcal{H}$, then the tree*

$$[\alpha]\lambda$$
$$T_1 \quad \ldots \quad T_m$$

   is a (conjunctive) Δ-tree.
3. *Let T_1, ..., T_m be Δ-trees whose roots are $[\alpha]\lambda_1, \cdots, [\alpha]\lambda_m$ and $[\beta]\lambda \in \mathcal{H}$, then the tree*

$$[\beta]\lambda$$
$$T_1 \quad \ldots \quad T_m$$

   is a (disjunctive) Δ-tree.

Every Δ-tree T can be interpreted as a propositional formula A in nnf. This interpretation also allows to identify the subtrees of T with subformulas of A. The idea is just to consider each α-node (resp. β-node) as a conjunction (resp. disjunction) with the literals in λ as immediate successors in addition to the subformulas represented by its immediate successors, T_i, in the Δ-tree; the nnf so obtained from a Δ-tree T will be denoted by $\texttt{Norm}(T)$. In the case of an empty clause or an empty cube we have $[\alpha]\text{nil} \equiv \top$ and $[\beta]\text{nil} \equiv \bot$, that is why the definition does not include the cases $[\alpha]\top$ and $[\beta]\bot$.

We can go the other way round as well, and generate a Δ-tree representative for each nnf. But, in order to be able to generalize the reductions to the Δ-trees, we want to have more information than this in the lists λ, we want to have the Δ-lists. In the next section we present a short summary of Δ-lists. These were firstly introduced in [1], and have been recently used in the development of a large set of reduction strategies for studying the satisfiability of non-clausal propositional formulas [2].

3.1 A Short Review of Δ-Lists

We associate to each nnf A a pair of lists of literals denoted $\Delta_0(A)$ and $\Delta_1(A)$, the so-called associated Δ-lists of A.

In a nutshell, $\Delta_0(A)$ and $\Delta_1(A)$ are, respectively, lists of implicates and implicants of A.

Definition 2 (Δ-lists). *Given a nnf A, $\Delta_0(A)$ and $\Delta_1(A)$ are elements of $\mathrm{List}(\mathcal{V}^{\pm})\cup\{\top,\bot\}$ called Δ-lists associated with A, recursively defined as follows:*

$$\Delta_0(\ell) = \ell \qquad\qquad\qquad \Delta_1(\ell) = \ell$$
$$\Delta_0(\bot) = \bot \qquad\qquad\qquad \Delta_1(\bot) = \mathrm{nil}$$
$$\Delta_0(\top) = \mathrm{nil} \qquad\qquad\qquad \Delta_1(\top) = \top$$
$$\Delta_0\left(\bigwedge_{i=1}^{n} A_i\right) = \biguplus_{i=1}^{n}\Delta_0(A_i) \qquad \Delta_1\left(\bigwedge_{i=1}^{n} A_i\right) = \bigcap_{i=1}^{n}\Delta_1(A_i)$$
$$\Delta_0\left(\bigvee_{i=1}^{n} A_i\right) = \bigcap_{i=1}^{n}\Delta_0(A_i) \qquad \Delta_1\left(\bigvee_{i=1}^{n} A_i\right) = \biguplus_{i=1}^{n}\Delta_0(A_i)$$

In the definition above there are two versions of the union operator, and this can be explained because of the intended interpretation of these sets and Theorem 1 below:

1. Elements in Δ_0 are considered to be conjunctively connected. Namely, if ℓ and $\overline{\ell} \in \Delta_0(A)$, then $\Delta_0(A)$ simplifies to $\bot$. This way, we obtain a set of implicates which can be thought of as a cube.
2. Elements in Δ_1 are considered to be disjunctively connected. Namely, if ℓ and $\overline{\ell} \in \Delta_1(A)$, then $\Delta_1(A)$ simplifies to $\top$. This way, we obtain a set of implicants which can be thought of as a clause.

The next theorem states that elements of $\Delta_0(A)$ are implicates of A, and that elements of $\Delta_1(A)$ are implicants of A. It follows easily by structural induction from the definition of Δ-lists.

Theorem 1 ([2]). *Let A be a nnf and ℓ be a literal in A then:*

1. *If $\ell \in \Delta_0(A)$, then $A \models \ell$ and, equivalently, $A \equiv \ell \wedge A$.*
2. *If $\ell \in \Delta_1(A)$, then $\ell \models A$ and, equivalently, $A \equiv \ell \vee A$.*

As an easy consequence of the previous theorem we get the following corollary, defining a meaning-preserving substitution for a formula A whose result contains only one occurrence of any literal in the Δ-lists of A.

Corollary 1. *Let A a nnf and ℓ a literal in A. Then:*

1. *If $\ell \in \Delta_0(A)$, then $A \equiv A[\ell/\top, \overline{\ell}/\bot] \wedge \ell$.*
2. *If $\ell \in \Delta_1(A)$, then $A \equiv A[\ell/\bot, \overline{\ell}/\top] \vee \ell$.*

Remark 1. The substitution defined in the corollary above never increases the size of A; actually, the size is always decreased but in the following cases:

1. If A is a conjunctive formula such that $\ell \in \Delta_0(A)$, and there is only one occurrence of ℓ.
2. If A is a disjunctive formula such that $\ell \in \Delta_1(A)$, and there is only one occurrence of ℓ.

3.2   Back to the Δ-Trees

Given a nnf A, the operator Δ-**Tree** generates a Δ-tree whose nodes are the Δ-lists associated to A.

Definition 3 (Operator Δ-Tree). *Let A be a nnf, we generate a Δ-tree by using the operator Δ-**Tree***, *recursively defined as follows:*

1. *Let A be a clause, $A \neq \bot$, then Δ-**Tree**$(A) = [\beta]\Delta_1(A)$.*
2. *Let A be a non-literal cube such that $A \neq \top$ and A is not a literal, then Δ-**Tree**$(A) = [\alpha]\Delta_0(A)$.*
3. *Let A be a disjunctive nnf, and let $A_1, \ldots, A_n$, with $n \geq 1$, be the non-literal disjuncts of A, then*

$$\Delta\text{-}\mathbf{Tree}(A) = \frac{[\beta]\Delta_1(A)}{\Delta\text{-}\mathbf{Tree}(A_1) \quad \ldots \quad \Delta\text{-}\mathbf{Tree}(A_n)}$$

4. *Let A be a conjunctive nnf, and let $A_1, \ldots, A_n$, with $n \geq 1$, be the non-literal conjuncts of A, then*

$$\Delta\text{-}\mathbf{Tree}(A) = \frac{[\alpha]\Delta_0(A)}{\Delta\text{-}\mathbf{Tree}(A_1) \quad \ldots \quad \Delta\text{-}\mathbf{Tree}(A_n)}$$

Example 1. Consider $A = ((p \wedge (\bar{p} \vee (q \wedge \bar{r}))) \vee q \vee r) \wedge ((\bar{p} \wedge q) \vee (p \wedge q)) \wedge ((\bar{q} \wedge p) \vee r)$, where every node η has associated the pair $(\Delta_0(\eta), \Delta_1(\eta))$

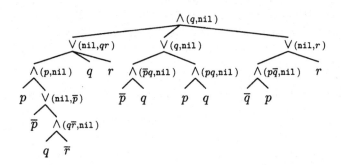

For the formula A above we have that Δ-Tree(A) is:

$$[\alpha]q$$

$[\beta]qr$	$[\beta]$nil	$[\beta]r$
$[\alpha]p$	$[\alpha]\overline{p}q$ $[\alpha]pq$	$[\alpha]p\overline{q}$
$[\beta]\overline{p}$		
$[\alpha]q\overline{r}$		

Note that for the previous example $\text{Norm}(\Delta$-Tree$(A))$ is *not* equal to A, for a new literal q is attached as an immediate successor of the root node, making explicit that q is an implicate of the formula. Anyway, operators Norm and Δ-Tree are inverse, up to equivalence, as stated in the following result.

Theorem 2. *Let A be a nnf. Then $A \equiv \text{Norm}(\Delta$-Tree$(A))$.*

It is remarkable the idea that, in some sense, the structure of Δ-tree allows to substitute reasoning with literals by reasoning on clauses and cubes.

4   Restricted Δ-Trees

In this section, meaning-preserving transformations are introduced which allow to reduce the size of a Δ-tree and get a normal form for it. These transformations extend to Δ-trees the definitions of Δ_0-*conclusive*, Δ_1-*conclusive* and ℓ-*simple* given for nnfs in [2].

4.1   Subformulas Which Can Be Substituted by Constants

The result of Corollary 1 is extended to Δ-trees, in that not only literals, but also subformulas can be substituted by the constants $\top$ or $\bot$. The operators $\Phi_\bot$ and $\Phi_\top$ on Δ-trees reduce a Δ-tree by deleting its *redundant* nodes, that is, those nodes which can be substituted by logical constants in a meaning-preserving way.

Definition 4 (0-conclusive node). *Let η be a node of a Δ-tree T is said to be 0-conclusive if it satisfies any of the following conditions:*

- *It is labeled with $[\alpha]\bot$.*
- *It is a monary node labeled with $[\beta]$nil.*
- *It is labeled with $[\alpha]\lambda$, it has an immediate successor $[\beta]\lambda'$ which is a leaf and $\overline{\lambda'} \subseteq \lambda$.*
- *It is labeled with $[\alpha]\lambda$, its predecessor is labeled with $[\beta]\lambda'$ and $\lambda \cap \lambda' \neq \emptyset$.*

The operator $\Phi_\bot$ searches for and deletes the 0-conclusive nodes by applying the following steps:

- If η is labeled with $[\alpha]\bot$ and $\eta \neq \varepsilon$, then $\Phi_\bot$ deletes η.
- If η is a monary node labeled with $[\beta]$nil, then $\Phi_\bot$ deletes η and collapses its ancestors with its (only) succesor.
- If η is labeled with $[\alpha]\lambda$, it has an immediate successor $[\beta]\lambda'$ which is a leaf and $\overline{\lambda'} \subseteq \lambda$, then $\Phi_\bot$ substitutes η by $\alpha[\bot]$.
- If η is labeled with $[\alpha]\lambda$, its predecessor is labeled with $[\beta]\lambda'$ and $\lambda \cap \lambda' \neq \emptyset$, then $\Phi_\bot$ deletes η.

Intuitively, the previous definition detects those nodes in the Δ-tree which, in some sense, can be substituted by $\bot$ without affecting the meaning. The effective deletion of those nodes is made by an operator, $\Phi_\bot$.

Theorem 3. *Let T be a Δ-tree, the operator $\Phi_\bot$ has quadratic complexity in the worst case, and $\Phi_\bot(T)$ has no 0-conclusive nodes and, in addition, $T \equiv \Phi_\bot(T)$.*

The 1-conclusive nodes and the operator $\Phi_\top$ are defined by duality, interchanging α and β, and replacing $\bot$ by $\top$.

4.2 Simple Leaves

In order to get to a *restricted* Δ-tree it is also necessary to detect which leaves are redundant, in the sense that do not represent proper clauses or cubes, but literals.

Definition 5 (Simple node). *Let T be a non-leaf Δ-tree, and let η be a leaf in T. We say that η is simple if it is labeled with either $[\alpha]\ell$ or $[\beta]\ell$, where $\ell \in \mathcal{V}^\pm$.*

Theorem 4. *Let T be a Δ-tree, then there exists an operator Φ_ℓ, with linear complexity in the worst case, such that $\Phi_\ell(T)$ is a Δ-tree without simple leaves and, in addition, $T \equiv \Phi_\ell(T)$.*

4.3 Updated Δ-Trees

A useful property of the operator Δ-Tree is that, given a nnf A, in Δ-Tree(A) the label of each $[\alpha]$ (resp. $[\beta]$) node is the Δ_0- (resp. Δ_1-)list associated to the subformula that it represents. However, this property need not hold when some transformation has already been applied on T.

Definition 6 (Updated node, updated tree). *Let T be a Δ-tree, and let η be a node of T that is neither a leaf nor the root. Let $[\Theta]\lambda$ be the label of the predecessor of η, and let $[\Theta]\lambda_1, \ldots, [\Theta]\lambda_n$ be the labels of its immediate successors. We say that η can be updated if it satisfies some of the next conditions:*

1. *It is labeled with $[\overline{\Theta}]$nil and $\bigcap_{i=1}^n \{\lambda_1, \ldots, \lambda_n\} \not\subset \lambda$.*
2. *It is labeled with $[\overline{\Theta}]\ell$ for some $\ell \in \mathcal{V}^\pm$ and satisfies both $\ell \notin \lambda$ and $\ell \in \bigcap_{i=1}^n \{\lambda_1, \ldots, \lambda_n\}$.*

We say that a tree T is updated if it has no nodes that can be updated.

In order to obtain an updated Δ-tree, we have to drive upwards all those literals that can be generated by intersections; this operation is done by the operator **Update**.

Theorem 5. *If T is a Δ-tree, there exists an operator* **Update**, *with quadratic complexity in the worst case, such that* **Update**(T) *is updated and, in addition,* **Update**$(T) \equiv T$

4.4   Restricted Δ-Trees

Definition 7 (Restricted tree). *Let T be a Δ-tree. If T is updated and it has neither 0-conclusive nodes nor 1-conclusive nodes nor simple leaves, then it is said to be restricted.*

The operators defined in the previous sections allow us to transform every Δ-tree in another equivalent and restricted one.

Definition 8 (Operator Restrict). *If T is a Δ-tree,* **Restrict** *traverses T and in every node it tests whether the node is 0-conclusive, or 1-conclusive, or a simple leaf, or a node that can be updated, and in this case applies the corresponding operator in* $\{\Phi_\perp, \Phi_\top, \Phi_\ell, \text{Update}\}$.

From Theorems 3–5 we immediately obtain the following result.

Theorem 6. *Let T be a Δ-tree, then* **Restrict**(T) *is restricted and, in addition,* $T \equiv$ **Restrict**(T).

Example 2. Given the formula $A = (p \vee q) \wedge (r \vee s) \wedge ((p \wedge q) \vee p)$, whose associated Δ-tree is

$$[\alpha]p$$
$$[\beta]pq \qquad [\beta]rs \qquad [\beta]p$$
$$[\alpha]pq$$

An application of the operator $\Phi_\perp$ (node 3 can be reduced) leads to

$$[\alpha]p$$
$$[\beta]pq \qquad [\beta]rs \qquad [\beta]p$$

Now, operator Φ_ℓ is applied to node 3, and we obtain

$$[\alpha]p$$
$$[\beta]pq \qquad [\beta]rs$$

Finally, operator $\Phi_\top$ can be applied again, for the occurrence of p in the root allows to reduce that in node 1, giving the restricted Δ-tree

$$[\alpha]p$$
$$[\beta]rs$$

which, using the operator **Norm**, leads to the formula $p \wedge (r \vee s)$.

5   Equisatisfiability of Δ-Trees

In this section several satisfiability-preserving transformations are introduced which allow to reduce the size of a Δ-tree T. These transformations are called *complete reduction*, *subreduction* (reduction of Δ-subtrees) and a *purity rule*.

Recall that the substitution of a logical constant for a literal ℓ, denoted $A[\ell/\top, \bar{\ell}/\bot]$, represents the formula obtained from A substituting all occurrences of ℓ by $\top$, and all occurrences of $\bar{\ell}$ by $\bot$. We extend this notion to Δ-trees using the definition below:

Definition 9 (Substitutions on Δ-trees). *Let T a Δ-tree, then $T[\ell/\top, \bar{\ell}/\bot]$ denotes the Δ-tree obtained traversing T and applying the following transformations:*

- *If $\ell \in \lambda$ and $[\beta]\lambda$ is the label of $\eta \neq \varepsilon$, then the subtree rooted at η in T is deleted.*
- *If $\bar{\ell} \in \lambda$ and $[\alpha]\lambda$ is the label of $\eta \neq \varepsilon$, then the subtree rooted at η in T is deleted.*
- *If $\ell \in \lambda$ and $[\alpha]\lambda$ is the label of η in T, then ℓ is deleted from λ.*
- *If $\bar{\ell} \in \lambda$ and $[\beta]\lambda$ is the label of η in T, then $\bar{\ell}$ is deleted from λ.*
- *If $\ell \in \lambda$ and $[\beta]\lambda$ is the label of ε, then $T[\ell/\top, \bar{\ell}/\bot] = \top$.*
- *If $\bar{\ell} \in \lambda$ and $[\alpha]\lambda$ is the label of ε, then $T[\ell/\top, \bar{\ell}/\bot] = \bot$.*

The following easy-to-prove lemma states that the definition we have just given coincides with the usual meaning of substitution in formulas.

Lemma 1. *Let T be a Δ-tree. Then $\mathrm{Norm}(T[\ell/\top, \bar{\ell}/\bot]) \equiv \mathrm{Norm}(T)[\ell/\top, \bar{\ell}/\bot]$.*

Given a Δ-tree T and a set of literals Γ, we will denote by $T[\Gamma/\top, \overline{\Gamma}/\bot]$ the Δ-tree obtained by substituting all the literals of Γ by $\top$, and their opposite by $\bot$.

5.1   Complete Reduction

The first satisfiability-preserving transformation we are introducing is called *complete reduction*, and can be seen as a generalization of the one literal rule in the Davis-Putnam algorithm for satisfiability. We first define what a completely reducible Δ-tree is and, then, the corresponding theorem about complete reduction is stated.

Definition 10 (Completely reducible Δ-tree). *If T is a Δ-tree and its root is $[\alpha]\lambda$ with $\lambda \neq \mathrm{nil}$, we say that T is completely reducible.*

Theorem 7. *Let T be a completely reducible Δ-tree with root $[\alpha]\lambda$ and let Γ be the set $\{\ell_i \mid \ell_i \in \lambda\}$. Then T is satisfiable iff $T[\Gamma/\top, \overline{\Gamma}/\bot]$ is satisfiable. Furthermore, if I is a model of $T[\Gamma/\top, \overline{\Gamma}/\bot]$, then any extension I' of I satisfying $I'(\ell) = I(\ell)$ if $\ell \notin \Gamma$, and $I'(\ell) = 1$ if $\ell \in \Gamma$, is a model of T.*

5.2   Subreduction

All the transformations performed by the operator `Restrict` only use the information of a node and its immediate succesors. The next transformation uses the information in a node to simplify all its descendants.

Theorem 8. *Let T be a Δ-tree and η a node of T. If $[\Theta]\lambda$ is the label of η, $\ell \in \lambda$ and there is an ancestor η' of η verifying one of the following conditions*

  1. $[\overline{\Theta}]\lambda'$ is the label of η', and $\ell \in \lambda'$
  2. $[\Theta]\lambda'$ is the label of η', and $\overline{\ell} \in \lambda'$

Then the Δ-tree T' obtained by deleting the subtree rooted at η in T is equivalent to T.

It is important to notice that $\text{Norm}(\eta)$ need not be equivalent to $\bot$ or $\top$ (depending on Θ), but the Δ-trees obtained after the substitution are equivalent.

The next theorem states how a Δ-tree can be reduced when Theorem 8 cannot be applied.

Theorem 9. *Let T be a Δ-tree and η a node of T. If $[\Theta]\lambda$ is the label of η, $\ell \in \lambda$ and there is an ancestor η' of η verifying one of the following conditions*

  1. $[\Theta]\lambda'$ is the label of η', and $\ell \in \lambda'$, or
  2. $[\overline{\Theta}]\lambda'$ is the label of η', and $\overline{\ell} \in \lambda'$

Then the Δ-tree T' obtained by erasing the literal ℓ in λ is equivalent to T.

By using Theorems 8 and 9 we can define the operator `SubReduce` as follows:

Definition 11 (Operator `Subreduce`). *Let T be a Δ-tree, then $\text{SubReduce}(T)$ is the Δ-tree obtained traversing T in a reverse depth-first order (from leaves to the root, and from right to left) and performing the transformations given by Theorems 8 and 9.*

The following theorem, a simple consequence of Theorems 8 and 9, states that `SubReduce` implements a meaning-preserving substitution.

Theorem 10. *Let T be a Δ-tree. Then $\text{SubReduce}(T) \equiv T$.*

Note that for all literal ℓ in $\text{SubReduce}(T)$, no occurrence of ℓ and $\overline{\ell}$ appear in the scope of ℓ. Therefore, only the relevant occurrences of literals are maintained after applying subreduction to a formula.

5.3   Pure Literal

The concept of pure literal for nnfs in [9] can be immediately extended for Δ-trees, by using Theorem 2.

If A is a nnf and ℓ is a pure literal, then A is satisfiable iff $A[\ell/\top]$ is satisfiable. This result can also be extended for Δ-trees.

A more general concept, that includes the previous one, is the concept of Δ-pure literal.

Definition 12. *Let T a Δ-tree. We say that ℓ is a Δ-pure literal in T if, when traversing the Δ-tree in depth-first order, the first occurrence of either ℓ or $\bar{\ell}$ is ℓ in all the branches.*

Theorem 11. *Let T a Δ-tree and ℓ a Δ-pure literal in T. Then T is satisfiable iff $T[\ell/\top, \bar{\ell}/\bot]$ is satisfiable.*

Example 3. Given the nnf $A = (r \vee \bar{s}) \wedge (((\bar{p} \vee q) \wedge (\bar{p} \vee \bar{s})) \vee ((\bar{r} \vee ((\bar{q} \vee p) \wedge (\bar{s} \vee \bar{q}))) \wedge (\bar{q} \vee s \vee r))) \wedge (((p \wedge r) \vee (p \wedge \bar{s})) \wedge q) \vee s)$, the associated Δ-tree is

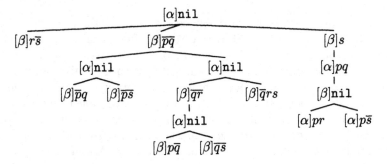

The operator SubReduce gives the Δ-tree

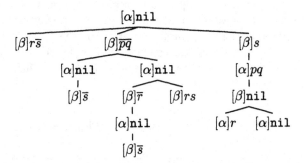

Now, the operator $\Phi_\top$ deletes the subtree rooted at node 311 and the nodes 2211 and 21 to obtain the Δ-tree:

Using the operator SubReduce we obtain the Δ-tree on the left and finally, $\Phi_\top$ applied once again on node 21 gives the Δ-tree on the right:

$$[\alpha]\texttt{nil}$$

$[\beta]\overline{rs}$	$[\beta]\overline{pqs}$	$[\beta]s$

$[\alpha]\texttt{nil}$ $[\alpha]pq$

$[\beta]\overline{r}$

$$[\alpha]\texttt{nil}$$

$[\beta]\overline{rs}$	$[\beta]\overline{pqrs}$	$[\beta]s$

$[\alpha]pq$

Finally, Φ_T applied once again on node 21 gives the Δ-tree

Using the operator Norm we obtain the formula $(r\vee\overline{s})\wedge(\overline{p}\vee\overline{q}\vee\overline{r}\vee\overline{s})\wedge((p\wedge q)\vee s)$.

6   Experimental Results

We have written a straightforward implementation for the Macintosh port of the interpreter of Objective CAML (an ML-like functional language) in order to obtain a rapid prototype of a theorem prover. Δ-trees have been used to implement the reductions just described, together with a naive branching rule based on the Davis-Putnam procedure; namely, a formula A is splitted into two subformulas $A[p/\mathsf{T}]$ and $A[p/\bot]$, where p is the first variable occurring in A.

As our method is specially focused on non-cnf formulas we have run the prover, named TAS, on the IFIP benchmarks for hardware verification [3]. The results obtained, using a Power Macintosh G3 with 64 Mb of memory and 233 Mhz, are compared with those obtained in [7], for he also uses there a reduction-like strategy (which he calls *simplification*), in his experiments he used a Sun SuperSPARK. In Table 1, we compare our implementation with the results obtained by Isabelle [8] (a well-known interactive prover, written in Standard ML) and Beatrix (a sicstus Prolog implementation in the spirit of lean tableau theorem proving). As several strategies were used in the cited work, in fairness to Isabelle and Beatrix, we compare our running time with their best absolute results no matter the strategy used.

Table 1. TAS vs Beatrix and Isabelle.

Problem	Isabelle	Beatrix	TAS	Problem	Isabelle	Beatrix	TAS
ex2	1.3	0.0	0.00	mul	130.9	0.2	0.07
transp	0.2	0.0	0.00	rip02	1.6	0.0	0.03
risc	9.8	0.6	0.05	rip04	994.5	0.5	0.38
counter	68.8	0.1	0.13	rip06	-	3.0	2.75
hostint1	96.5	0.2	0.10	rip08	-	18.2	17.18

It is important to remark that the results obtained are by far much better than those of Isabelle, showing that not only the scaling factor in problems such as ripOn can be reduced but also that absolute run time values are comparable to those obtained by Beatrix, which shortens the gap between lean theorem proving in Prolog and standard theorem proving in ML-like languages. In Table 2 some more results are compared with the run time of Beatrix, where an important speed-up when using TAS can be noticed.

Table 2. Run time (seconds) on other IFIP benchmarks.

Problem	Beatrix	TAS	Problem	Beatrix	TAS	Problem	Beatrix	TAS
d3 (satisf.)	0.1	0.17	dk17	3.0	0.38	sqn	11.2	0.43
misg	0.7	0.35	z5xpl	4.1	0.38	add1	12.2	1.20
ztwaalf1	0.8	0.80	f51m	5.7	0.48	dc2	12.5	0.40
mp2d	1.1	1.03	pitch	5.7	2.55	mul03	20.1	1.03
dk27	2.2	0.07	vg2	7.0	2.82	rd73	30.4	1.27
z4	2.3	1.53	alu	7.1	3.98	root	33.7	0.67
rom2	2.5	3.03	x1dn	7.2	3.37	alupla20	618.1	31.72
table	2.8	2.72	z9sym	9.8	4.07			

To make the comparison more interesting we also chose to run TAS on the Random 3-Sat benchmark, although TAS has not been neither designed nor optimised for cnf formulas. Table 3 shows the results for the standard random distribution of 3-SAT, where 3_sat(V,C) means that samples had C clauses, with 3 literals selected uniformly among V variables and each literal negated with probability 0.5.

We show our results together with the results of two different flavours of Beatrix, the 'standard' one (in which the usual β-rule is used) and the 'lemmaizing' version (an asymmetric rule for a limited form of cut).

$$\frac{S, \beta_1 \quad S, \beta_2}{S, \beta} \; \text{Std} \qquad\qquad \frac{S, \beta_1 \quad S, \overline{\beta_1}, \beta_2}{S, \beta} \; \text{Lem}$$

One can easily see that, although our implementation has been run on a interpreter (as far as we know no compiler for CAML is still available for Macs) the performance of TAS is in between the two flavours of Beatrix. The speedup factor of TAS w.r.t. the standard version of Beatrix is about 2 for formulas with 32 variables and about 3.5 for formulas with 64 variables, whereas the better performance of the lemmaizing version of Beatrix averages 1.63 for 32 variables and 2.72 for 64 variables.

These results are neither surprising, for the standard version of Beatrix is just a tableau system improved with a particular case of our reductions, nor discouraging, for the branching rule we have implemented is just a raw DPLL-like procedure.

It is worth to note that, although the computational pay-off of the reductions implemented in TAS results in poor runtimes for the formulas in the first row of the table, the negative effect disappears as the size of the formulas is increased.

Table 3. TAS vs Beatrix on Random 3-SAT.

C/V	Problem	Beatrix	Bea-Lem	TAS	Problem	Beatrix	Bea-Lem	TAS
3	3_sat(32,96)	0.3	0.2	0.80	3_sat(64,192)	1.4	1.0	7.55
4	3_sat(32,128)	3.9	1.2	2.07	3_sat(64,256)	334.6	38.4	98.31
4.25	3_sat(32,136)	6.1	1.8	3.03	3_sat(64,272)	554.3	56.4	188.81
4.5	3_sat(32,144)	6.9	2.1	3.53	3_sat(64,288)	1,050.9	72.0	216.64
5	3_sat(32,160)	8.2	2.4	3.90	3_sat(64,320)	568.6	60.0	141.72
6	3_sat(32,192)	7.7	2.6	3.71	3_sat(64,384)	240.3	39.4	90.88

7   Conclusions

We have introduced Δ-trees for propositional formulas. This representation allows a compact representation for well-formed formulas as well as for a number of reduction strategies in order to consider only those occurrences of literals which are relevant for the satisfiability of the input formula. It is important to notice that this structure can be also extended to other non-classical logics where the TAS methodology works. Finally, the reduction strategies have been implemented and tests are reported which show the relative good performance of our implementation of the techniques introduced.

References

1. G. Aguilera, I. P. de Guzmán, and M. Ojeda-Aciego. Increasing the efficiency of automated theorem proving. *Journal of Applied Non-Classical Logics*, 5(1):9–29, 1995.
2. G. Aguilera, I. P. de Guzmán, M. Ojeda-Aciego, and A. Valverde. Reductions for non-clausal theorem proving. *Theoretical Computer Science*, 2000. To appear. Available at http://www.ctima.uma.es/aciego/TR/TAS-tcs.pdf.
3. L.J. Claesen, editor. *Formal VLSI correctness verification—VLSI design methods*, volume 2. Elsevier, 1990.
4. I. P. de Guzmán, M. Ojeda-Aciego, and A. Valverde. Implicates and reduction techniques for temporal logics. *Ann. Math. Artificial Intelligence* 27:3–23, 1999.
5. I. P. de Guzmán, M. Ojeda-Aciego, and A. Valverde. Multiple-valued tableaux with Δ-reductions. In *Proc. of the Intl. Conf. on Artificial Intelligence, ICAI'99*, pages 177–183. C.S.R.E.A., 1999.
6. J.H. Gallier. *Logic for Computer Science: Foundations for Automatic Theorem Proving*. Wiley & Sons, 1987.
7. F. Massacci. Simplification: a general constraint propagation technique for propositional and modal tableaux. In *Proceedings of Tableaux'98*, pages 217–231. Lect. Notes in Artificial Intelligence 1397, 1998.
8. L.C. Paulson. *Isabelle: A Generic Theorem Prover*. Springer, 1994. LNCS 828.
9. P. W. Purdom, Jr. Average time for the full pure literal rule. *Information Sciences*, 78:269–291, 1994.

Simultaneous Rigid Sorted Unification*

Pedro J. Martín and Antonio Gavilanes

Dep. de Sistemas Informáticos y Programación. Universidad Complutense de Madrid
{pjmartin, agav}@sip.ucm.es

Abstract. In this paper we integrate a sorted unification calculus into free variable tableau methods for logics with term declarations. The calculus we define is used to close a tableau at once, unifying a set of equations derived from pairs of potentially complementary literals occurring in its branches. Apart from making the deduction system sound and complete, the calculus is terminating and so, it can be used as a decision procedure. In this sense we have separated the complexity of sorts from the undecidability of first order logic.

1 Introduction

In the context of logical systems, sorts are widely accepted as a means of increasing efficiency, reducing the search space, and allowing more natural representations. Two main approaches have been followed in the incorporation of sorts to logics. Usually, sorts behave *statically* when sorts properties -sort hierarchies and sort declarations for operations- are fixed in the signature [1,14,13].

On the other hand, for the purpose of natural language understanding it results interesting to design inference systems which are capable of deducing taxonomic information, that is, the reasoning process may actually alter the sorts properties such as hierarchies [8]. In this sense, sorts behave *dynamically* when the information about sorts and individuals co-exists within the same formal framework [5,6]. The greatest expressivity is achieved when the sort declarations of operations are expressed by means of a new formula constructor. Thus the so called *logics with term declarations* [15] arise as logical systems including, in a single formalism, a classical many sorted logic together with all the information it entails (relations between sorts and sort declarations for function symbols).

This paper follows a research line involved in the construction of tableau methods for logics with term declarations [7,10,11]. Instead of defining new inference rules, we separate sorts from first order logic using a sorted unification calculus. The calculus is required to unify a set of equations derived from pairs of potentially complementary literals occurring in the branches of a tableau. Free variables present two difficulties to be considered when designing the sorted calculus. Firstly, variables are attached to sorts restricting their domain [15,5,6], so we can only apply substitutions that are *well-sorted*. This means that the (static) sort of every substituted variable and the (dynamic) sort of the respective substituting term must be the same. Second, free variables behave rigidly

---

* Research supported by the Spanish Project TIC98-00445-C03-02 "TREND".

M. Ojeda-Aciego et al. (Eds.): JELIA 2000, LNAI 1919, pp. 193–208, 2000.

and so they can only be instanced once [3]. Then we have to consider the sort information occurring in the whole tableau even when closing a single branch.

In this paper, we improve our previous results with a calculus that fulfills the following properties:

1. It is *simultaneous*, so a tableau can be (globally) closed *at once*. As we will see, the search space can be more efficiently pruned when we consider all the branches at the same time.

2. It is quite *simple*. It suitably combines a standard unification procedure and just four sorted rules (two symmetric non-failure rules and their failure versions). Moreover, the applicability conditions of the rules are quite simple.

3. It is *terminating*, then we have separated the complexity of sorts away from the undecidability of first order logic. Then, the calculus can be used as a decision procedure because it is enough to traverse a finite search space. Moreover termination allows more elegant soundness and completeness proofs.

The paper is organized as follows. Section 2 presents the Logic with Term Declarations and some results about its ground tableau methods. In Section 3 we introduce free variable tableaux and the notion of rigid sorted unification (RSU) problem. Section 4 presents a calculus for solving these RSU-problems and its main properties; it is extended to a global version for solving simultaneous rigid sorted unification ($SRSU$)-problems in Section 5. Section 6 integrates this last calculus into a new free variable tableau system. We finish with a discussion of the achieved results. Due to lack of space most of the proofs have been omitted. They can be found in [9]

2   The Logic with Term Declarations LTD

LTD extends the ordinary first-order predicate logic by introducing a new formula constructor $t \in s$ (called *term declaration*) which expresses that the term t has sort s. In LTD operations have no static sort, then, a LTD-signature Σ consists of a finite set S of sorts s, and *unsorted* sets $\mathcal{C}$, $\mathcal{F}$ and $\mathcal{P}$ of constant, function and predicate symbols respectively, the last ones of elements with arity. Only variables are attached to a fixed sort; they belong to one of the countable sets of the sorted family $X = (X^s)_{s \in S}$.

The sets of Σ-terms $T(\Sigma)$ and Σ-formulas $F(\Sigma)$ are defined as in first-order logic, but including term declarations. For example, $\forall x^s(x^s \in s')$ is a formula expressing that the sort s is a subsort of s', while $\forall x^s(f(x^s) \in s')$ expresses that the range of the function f in the s-domain is a set of s'-elements. A set of formulas $\mathcal{L}$ is called a $\in$-theory, or simply a *theory*, if it is composed of term declarations. Substitutions are finite replacements of variables for terms, written in the form $[t_1/x_1^{s_1}, \ldots, t_n/x_n^{s_n}]$.

A Σ-structure $\mathcal{D}$ in LTD is a total domain D containing a family of domains $\{D^s \mid s \in S\}$, and sets of interpretations $\{c^{\mathcal{D}} \in D \mid c \in \mathcal{C}\}$, $\{f^{\mathcal{D}} : D^n \to D \mid f^n \in \mathcal{F}\}$, $\{P^{\mathcal{D}} : D^n \to \{\underline{t}, \underline{f}\} \mid P^n \in \mathcal{P}\}$, for symbols of Σ. Considering that we do not have sort declarations in the signature, domains can possibly be empty; it is only known that $\bigcup D^s \subseteq D$.

A valuation for $\mathcal{D}$ is a sorted family $\rho = (\rho^s)_{s \in S}$ of finite mappings ρ^s : $X^s \to D^s$ of the form $[\rho^s(x_1^s)/x_1^s, \ldots, \rho^s(x_n^s)/x_n^s]$; $dom(\rho^s) = \{x_1^s, \ldots, x_n^s\}$ is the domain of ρ^s, and $dom(\rho) = \bigcup_{s \in S} dom(\rho^s)$ is the domain of ρ. Note that $dom(\rho^s) = \emptyset$ if $D^s = \emptyset$. As usual, $\rho[d/x^s]$ will denote the valuation that assigns d to x^s and behaves as ρ elsewhere.

The semantic value $[\![t]\!]_\rho^{\mathcal{D}}$ of a term t in a Σ-interpretation $\langle \mathcal{D}, \rho \rangle$ is defined as usual and it exists whenever $var(t) \subseteq dom(\rho)$. The boolean value $[\![\varphi]\!]_\rho^{\mathcal{D}}$ of a formula φ in $\langle \mathcal{D}, \rho \rangle$ exists if $free(\varphi) \subseteq dom(\rho)$ and it is defined as usual for first-order formulas, except for:

$$- \ [\![\forall x^s \varphi]\!]_\rho^{\mathcal{D}} = \begin{cases} \underline{t} & \text{if } [\![\varphi]\!]_{\rho[d/x^s]}^{\mathcal{D}} = \underline{t}, \text{ for all } d \in D^s \\ \underline{f} & \text{otherwise.} \end{cases}$$

$$- \ [\![\exists x^s \varphi]\!]_\rho^{\mathcal{D}} = \begin{cases} \underline{t} & \text{if there exists } d \in D^s \text{ such that } [\![\varphi]\!]_{\rho[d/x^s]}^{\mathcal{D}} = \underline{t} \\ \underline{f} & \text{otherwise.} \end{cases}$$

$$- \ [\![t \in s]\!]_\rho^{\mathcal{D}} = \begin{cases} \underline{t} & \text{if } [\![t]\!]_\rho^{\mathcal{D}} \in D^s \\ \underline{f} & \text{otherwise.} \end{cases}$$

In the sequel when we write $[\![t]\!]_\rho^{\mathcal{D}}$ (resp. $[\![\varphi]\!]_\rho^{\mathcal{D}}$), we assume $var(t) \subseteq dom(\rho)$ ($free(\varphi) \subseteq dom(\rho)$), which trivially holds for ground terms (sentences).

Next we outline a ground tableau method for LTD. The completeness proof of the free variable tableau versions we present will be based on lifting the completeness of the ground method. Suppose that Σ has been extended to a signature $\overline{\Sigma}$, with a countable set of new constants. The rules α and β are defined as in classical first-order tableaux [4]. For γ and δ rules we define:

$$\gamma) \quad \begin{array}{c} \forall x^s \varphi \\ t \in s \\ \hline \varphi[t/x^s] \end{array} \qquad\qquad \delta) \quad \begin{array}{c} \exists x^s \varphi \\ \hline \varphi[c/x^s] \\ c \in s \end{array}$$

In γ, t is a ground term; in δ, c is a new constant not occurring in the branch. Note how the sort information is managed dynamically in LTD, and term declarations are used ($t \in s$) or introduced ($c \in s$) in the branch expansion.

Definition 1 *A branch B of a tableau is closed if an atomic contradiction φ and $\neg \varphi$ (φ atomic) appears in B. A tableau is closed if all its branches are closed.*

Theorem 2 (Soundness and Completeness) *[7] Given a set of Σ-sentences Φ, Φ has a closed tableau if and only if Φ is not satisfiable.*

Example 3 *Let Σ be a signature composed of the sorts s, s', the constant a, the unary function symbol f and the binary predicate symbol P. In order to have a more pleasant and direct understanding of the following sentences, we would like to refer to sort s as representing human beings, s' as kind people, $f(\square)$ as giving the father of $\square$, and $P(\square, \diamond)$ as expressing that $\square$ gets along with $\diamond$. Suppose that 1: a is a human being ($a \in s$), 2: which does not get along with* its father $(\neg P(a, f(a)))$, *3: every kind human being gets along with everybody* $(\forall x^s (x^s \in s' \rightarrow \forall y^s P(x^s, y^s)))$ *and 4: the father of every human being is a human being* $(\forall x^s f(x^s) \in s)$. *Then it is obvious, as the following closed ground tableau shows, that* ¬5: *some human beings are not kind* $(\exists x^s (\neg x^s \in s'))$.

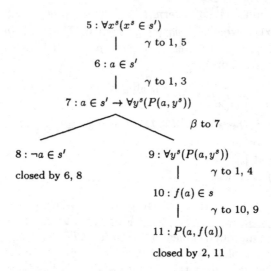

$$5 : \forall x^s (x^s \in s')$$

γ to 1, 5

$$6 : a \in s'$$

γ to 1, 3

$$7 : a \in s' \rightarrow \forall y^s (P(a, y^s))$$

β to 7

$$8 : \neg a \in s'$$

closed by 6, 8

$$9 : \forall y^s (P(a, y^s))$$

γ to 1, 4

$$10 : f(a) \in s$$

γ to 10, 9

$$11 : P(a, f(a))$$

closed by 2, 11

LTD is not more expressive than first order logic (sorts can be expressed as unary predicates [16]), but it allows more pleasant representations and deductions. In the example above, the formalization and the tableau can be expressed in first order logic, but at the cost of: (1) using more complex formulas (e.g. formula 3 would be transformed into $\forall x(S(x) \rightarrow (S'(x) \rightarrow \forall y(S(y) \rightarrow P(x, y))))$ that produces more branches to be closed) and (2) decreasing the efficiency because we loose the sort information in the γ-applications (e.g. x in the previous formula could be instanced to the term $f(f(f(a))))$.

Even if we used static ordered sorts, the formalization of $x^s \in s'$ would need the sort $s \cap s'$, making the signature dependent on the problem. Furthermore we can consider a different sort hierarchy in each branch of the tableau. In this sense, term declarations improve static ordered sorts as well.

3   Free Variable Tableaux

Now we will assume that the extended signature $\overline{\Sigma}$ also contains a countable set of new function symbols. The free variable tableau method defines the following new rules for quantifications:

$$\gamma')\quad \frac{\forall x^s \varphi}{\varphi[y^s/x^s]} \qquad\qquad \delta')\quad \frac{\exists x^s \varphi}{\varphi[f(x_1^{s_1}, \ldots, x_n^{s_n})/x^s]}{f(x_1^{s_1}, \ldots, x_n^{s_n}) \in s}$$

In γ', y^s is a new variable in the tableau; in δ', f is a new function symbol applied to the free variables occurring in the branch.

Obviously, free variables of a tableau may be substituted. As variables are sorted, the application of a substitution is sound in those contexts which ensure that the sort of every substituted variable is preserved. In *LTD*, theories play the role of these syntactic contexts.

Definition 4 (Well-Sorted Substitution) *A substitution* $[t_1/x_1^{s_1}, \ldots, t_n/x_n^{s_n}]$ *is well-sorted w.r.t. a theory* $\mathcal{L}$*, if* $(t_i \in s_i) \in \mathcal{L}$*,* $1 \leq i \leq n$*. A substitution* τ *is well-sorted w.r.t. a tableau* T *with branches* $B_1, \ldots, B_n$*, if the restriction of* τ *to the free variables of* B_i*, that is* $\tau|_{free(B_i)}$*, is well-sorted w.r.t. the theory included in* B_i*,* $1 \leq i \leq n$*.*

Well-sorted substitutions can be safely applied to free variable tableaux. Denote by $\mathcal{S}1$ the tableau system composed of α, β, γ', δ' and the substitutivity rule *sub* defined by:

sub) If T *is a free variable tableau and* τ *is an idempotent substitution well-sorted w.r.t.* T *then* $T\tau$ *is a free variable tableau*

The concepts of closed branch and closed tableau are defined as in Definition 1. Then we can prove the soundness and completeness of $\mathcal{S}1$; these proofs are very similar to those presented in [10] (see this paper for more explanations about the importance of idempotency in the rule *sub* and how to overcome empty domains -due to empty domains, soundness and completeness of $\mathcal{S}1$ are not stated as symmetric results; other approaches about how to overcome the problems of empty domains can be found in [2,16]).

Theorem 5 (Soundness of $\mathcal{S}1$) *Given a set of* Σ*-sentences* Φ*, if* Φ *has a closed free variable tableau then* Φ *is not satisfiable in structures with non-empty domains, for every sort.*

Theorem 6 (Completeness of $\mathcal{S}1$) *Given a set of* Σ*-sentences* Φ*, if* Φ *is not satisfiable then* Φ *has a closed free variable tableau.*

As in classical first-order tableaux [4], improving ground tableaux involves to restrict the application of the rule *sub* and use it only for closing branches. This results in the integration of a unification calculus which finds well-sorted unifiers for potentially complementary literals occurring in a branch. However, in order to perform a complete deduction system, unifiers must be structured in a particular form, as the following example shows.

Example 7 *Let* T *be the closed ground sketch of tableau presented below on the left and* T' *be the free variable tableau built as* T *on the right.*

$$a \in s$$

$$\neg P(a)$$

$$\forall x^s \ (x^s \in s')$$

$$\forall u^{s'} \ P(u^{s'})$$

$$|$$

$$a \in s'$$

$$|$$

$$P(a)$$

$$a \in s$$

$$\neg P(a)$$

$$\forall x^s \ (x^s \in s')$$

$$\forall u^{s'} \ P(u^{s'})$$

$$|$$

$$x^s \in s'$$

$$|$$

$$P(u^{s'})$$

T' should be closed by solving the unification problem $u^{s'} \simeq a$ corresponding to the single branch of T'; however this problem cannot be solved by any well-sorted substitution w.r.t. the theory presented in the branch $\{a \in s, x^s \in s'\}$. Nevertheless there is a sequence of unitary idempotent substitutions $\sigma = [x^s/u^{s'}][a/x^s]$, relating both tableaux, which is gradually well-sorted, in the sense that each unitary component is well-sorted after the application of the preceding ones in the sequence. So σ can be applied to T' using the rule sub twice. The sequence σ emphasizes the idea of an existing order in the application of the rule sub to T', corresponding to the order of γ-applications to T.

Therefore we will define a unification calculus lifting any closed ground tableau to a closed free variable one, by deriving a *sequence* of well-sorted unitary substitutions. Previously we define a concept of triangularity which captures the order of γ-applications to ground tableaux; then we adapt the notion of well-sortedness to sequences.

Definition 8 *A sequence of unitary substitutions $[t_1/x_1^{s_1}] \ldots [t_n/x_n^{s_n}]$ is triangular if it satisfies:*

1. $var(t_i) \cap \{x_1^{s_1}, \ldots, x_i^{s_i}\} = \emptyset, 1 \le i \le n$
2. $x_i \ne x_j, 1 \le i < j \le n$.

Definition 9 *Let $\sigma = \sigma_1 \ldots \sigma_n$, $\mathcal{L}$ and T be a triangular sequence of unitary substitutions, a theory and a free variable tableau, respectively. We say that σ is well-sorted w.r.t. $\mathcal{L}$ (resp. T), if σ_i is well-sorted w.r.t. $\mathcal{L}\sigma_1 \ldots \sigma_{i-1}$ (resp. $T\sigma_1 \ldots \sigma_{i-1}$), $1 \le i \le n$.*

Note that well-sorted sequences w.r.t. tableaux can be soundly applied using the rule *sub*, by gradually applying each of its unitary components. So, in Example 7, $[x^s/u^{s'}][a/x^s]$ is well-sorted w.r.t. T' and can be used to close it. Consequently we must design a calculus that obtains well-sorted sequences instead of a unique idempotent well-sorted substitution.

4   Rigid Sorted Unification

In this section we present how to solve unification problems arising when closing a single branch. Specifically, a *Rigid Sorted Unification* (shortly *RSU-)problem* has the following structure:

Given a finite theory $\mathcal{L}$ and a finite set of equations Γ, is there a well-sorted sequence of unitary substitutions w.r.t. $\mathcal{L}$ that unifies Γ?

For solving *RSU*-problems, we define the unification calculus $\mathcal{C}$. The non-failure rules of $\mathcal{C}$ have the form

$$\frac{\Gamma \quad \sigma_1 \ldots \sigma_n}{\Gamma' \quad \sigma_1 \ldots \sigma_n \sigma'}$$

where Γ, Γ' are sets of (oriented) equations and $\sigma_1 \ldots \sigma_n$, $\sigma_1 \ldots \sigma_n \sigma'$ are sequences of unitary substitutions. $\mathcal{C}$ is composed of ten rules: six standard rules for syntactic unification (tautology, decomposition, orientation, application, clash and cycle [16]) plus the following four ones:

The Sorted Rules of $\mathcal{C}$

(LW) Left Weakening

$$\frac{x^s \simeq t', \Gamma \quad \sigma_1 \ldots \sigma_n}{t \simeq t', \Gamma \quad \sigma_1 \ldots \sigma_n[t/x^s]}$$

if $(t \in s) \in \mathcal{L}\sigma_1 \ldots \sigma_n$ and $x^s \notin var(t)$

(RW) Right Weakening

$$\frac{y^{s'} \simeq x^s, \Gamma \quad \sigma_1 \ldots \sigma_n}{y^{s'} \simeq t, \Gamma[t/x^s] \sigma_1 \ldots \sigma_n[t/x^s]}$$

if $(t \in s) \in \mathcal{L}\sigma_1 \ldots \sigma_n$ and $x^s \notin var(t)$

(FWF) Functional Weakening Failure

$$\frac{x^s \simeq f(t_1, \ldots, t_n), \Gamma \, \sigma_1 \ldots \sigma_n}{\text{Fail}}$$

if there is no formula $t \in s$ in $\mathcal{L}\sigma_1 \ldots \sigma_n$ such that $x^s \notin var(t)$

(VWF) Variable Weakening Failure

$$\frac{y^{s'} \simeq x^s, \Gamma \quad \sigma_1 \ldots \sigma_n}{\text{Fail}}$$

if there is no formula $t \in s$ in $\mathcal{L}\sigma_1 \ldots \sigma_n$ such that $x^s \notin var(t)$, nor $t \in s'$ such that $y^{s'} \notin var(t)$

When solving *RSU*-problems, the application of standard rules has always preference. Furthermore we assume that there exists a terminating algorithm $\mathcal{A}$ for syntactic unification, transforming a set of equations Γ into *Fail* or a solved set of equations, by the non-deterministic application of the six standard rules. In this sense, the algorithm $\mathcal{A}$ behaves as a black box and we do not take care of the non-determinism its rules entail. Incorporating auxiliary calculi for solving some well-stated problems has been used in many other areas [3,12].

Definition 10 *Let Γ be a set of equations and $\sigma = \sigma_1 \ldots \sigma_n$ a sequence of unitary substitutions. One $\mathcal{C}$-standard step is the application of the algorithm $\mathcal{A}$ to the pair $\langle \Gamma, \sigma \rangle$ until Fail or a solved set of equations Γ' is reached. One $\mathcal{C}$-sorted step is the application of a sorted rule to the pair $\langle \Gamma, \sigma \rangle$ using a theory. One $\mathcal{C}$-step* *is one C-standard or C-sorted step. We write $\langle \Gamma, \sigma_1 \ldots \sigma_n \rangle \vdash_C \langle \Gamma', \sigma_1 \ldots \sigma_n \sigma_{n'} \rangle$ ($n' \in \{n, n+1\}$) (resp. $\langle \Gamma, \sigma_1 \ldots \sigma_n \rangle \vdash_C$ Fail) to express one non-failure (resp. failure) C-step.*

We say that the calculus C unifies a set of equations Γ w.r.t. a theory $\mathcal{L}$ by the sequence of unitary substitutions $\sigma_1 \ldots \sigma_n$, or $\sigma_1 \ldots \sigma_n$ is a C-unifier for Γ w.r.t. $\mathcal{L}$, if there exists a chain of C-steps, alternating C-standard and C-sorted steps, starting with $\langle \Gamma, \emptyset \rangle$ and finishing with $\langle \emptyset, \sigma_1 \ldots \sigma_n \rangle$.

Note that C-standard steps do not append elements to the sequence of unitary substitutions, and they can possibly be empty if the set of equations is still in solved form after one C-sorted step. Note also that C-sorted steps are always applied to sets of equations in solved form.

The computation of a solution to a RSU-problem can be viewed as the search for C-unifiers in a C-derivation tree: nodes are either pairs $\langle \Gamma, \sigma \rangle$ or failure leaves *Fail*, and branches alternate C-standard and C-sorted steps. Branching in a node only occurs due to (explicit) non-determinism in C-sorted steps; the non-determinism derived from syntactic unification is implicit in the algorithm $\mathcal{A}$. Leaves are either successful pairs $\langle \emptyset, \sigma \rangle$ or failure leaves *Fail*. As we will see, a failure node after one C-standard step allows to cut the branch expansion of that node, while after one C-sorted step, allows to cut the branch expansion of its parent.

Example 11 *Suppose $\mathcal{L} = \{a \in s, y^{s'} \in s, z^{s''} \in s, b \in s'\}$ and $\Gamma = \{f(x^s) \simeq f(b)\}$. The C-derivation tree for this RSU-problem is:*

$$\langle f(x^s) \simeq f(b), \ \emptyset \rangle$$

$$| \ C\text{-standard step}$$

$$\langle x^s \simeq b, \ \emptyset \rangle$$

LW   LW   LW

$\langle a \simeq b, \ [a/x^s] \rangle$ $\langle z^{s''} \simeq b, \ [z^{s''}/x^s] \rangle$ $\langle y^{s'} \simeq b, \ [y^{s'}/x^s] \rangle$

C-standard step $|$ C-standard step $|$ $|$ C-standard step

Fail $\langle z^{s''} \simeq b, \ [z^{s''}/x^s] \rangle$ $\langle y^{s'} \simeq b, \ [y^{s'}/x^s] \rangle$

FWF $|$ $|$ LW

Fail $\langle b \simeq b, \ [y^{s'}/x^s][b/y^{s'}] \rangle$

$|$ C-standard step

$$\langle \emptyset, \ [y^{s'}/x^s][b/y^{s'}] \rangle$$

The first branch finishes in a failure node after one C-standard step, and the second one, after one C-sorted step. The third branch obtains the unique C-unifier $[y^{s'}/x^s][b/y^{s'}]$.

4.1   Properties of the Calculus $\mathcal{C}$

First we show that the unification calculus $\mathcal{C}$ is terminating for every *RSU*-problem.

Theorem 12 (Termination) *The $\mathcal{C}$-derivation tree of every RSU-problem is finite.*

The calculus $\mathcal{C}$ is sound in the sense that given a set of equations Γ and a theory $\mathcal{L}$, every $\mathcal{C}$-unifier is a solution to the corresponding *RSU*-problem.

Theorem 13 (Soundness) *Let $\Gamma, \mathcal{L}$ and σ be a set of equations, a theory and a sequence of unitary substitutions, respectively. If $\mathcal{C}$ unifies Γ w.r.t. $\mathcal{L}$ by σ then:*
(i) σ is well-sorted w.r.t. $\mathcal{L}$
(ii) σ unifies Γ.

The completeness of $\mathcal{C}$ should read as follows: if there is a well-sorted sequence of unitary substitutions σ w.r.t. $\mathcal{L}$ unifying Γ then $\mathcal{C}$ unifies Γ w.r.t. $\mathcal{L}$ by a sequence τ which is more general than σ^{*}. But we are only interested in lifting a particular class of sequences of unitary substitutions, those sequences σ derived from a closed ground tableau $\mathcal{T}$ in the following way. Let $\mathcal{T}'$ be a free variable tableau built as $\mathcal{T}$, then σ is obtained by appending unitary substitutions to the sequence which correspond to the γ-applications to $\mathcal{T}$; that is, if $\forall x^{s}\varphi$ and $t \in s$ is used in $\mathcal{T}$ then we add $[t'/x^{s}]$ to the beginning of the current σ, where $t' \in s$ is the term declaration associated to $t \in s$ occurring in $\mathcal{T}'$. In Example 7, we would obtain $[x^{s}/u^{s'}][a/x^{s}]$. These sequences are ground and can be captured by the concept of *hyperwell-sortedness*. Only hyperwell-sorted sequences will be considered in the completeness of $\mathcal{C}$.

Definition 14 *A triangular sequence of unitary substitutions $[t_{1}/x_{1}{}^{s_{1}}]\ldots[t_{n}/x_{n}{}^{s_{n}}]$ is hyperwell-sorted w.r.t. a theory $\mathcal{L}$, if $(t_{i} \in s_{i}) \in \mathcal{L}, 1 \leq i \leq n$.*

In a hyperwell-sorted sequence, the order of the substitutions is not relevant because the declaration of the replaced term explicitly appears in the theory. It is immediate that every hyperwell-sorted sequence is also well-sorted; the inverse is not true, for example $[a/x^{s}][a/u^{s'}]$ is well-sorted but not hyperwell-sorted w.r.t. the theory $\{a \in s, x^{s} \in s'\}$.

For proving completeness, we examine the standard and the sorted case. For the former, we suppose that the algorithm $\mathcal{A}$ for syntactic unification is complete, so it fails whenever the given set of equations is not syntactically unifiable, and it succeeds giving a solved set of equations, otherwise. For the latter, we prove the following results. First the next technical lemma states that extracting and moving a unitary component through a sequence, from its place to the beginning, preserves hyperwell-sortedness and does not change the substitution.

---

[*] Sequences of unitary substitutions are compared through the respective substitutions resulting from composing their unitary components.

Lemma 15 *Let $\sigma_1 \ldots \sigma_n$ be a sequence such that $\sigma_i = [t_i/x_i^{s_i}]$, $1 \le i \le n$. For a fixed $m \in \{1, \ldots, n\}$ we define $\sigma_i' = [t_i[t_m/x_m^{s_m}]/x_i^{s_i}]$, $1 \le i \le m-1$. If $\sigma_1 \ldots \sigma_n$ is hyperwell-sorted w.r.t. a theory $\mathcal{L}$ then:*

1. *$\sigma_1' \ldots \sigma_{m-1}'\sigma_{m+1} \ldots \sigma_n$ is hyperwell-sorted w.r.t. $\mathcal{L}\sigma_m$*
2. *$\sigma_m\sigma_1' \ldots \sigma_{m-1}'\sigma_{m+1} \ldots \sigma_n = \sigma_1 \ldots \sigma_n$.*

The next two lemmas prove completeness of the sorted case. If a set of equations is unifiable by a hyperwell-sorted sequence then one non-failure $\mathcal{C}$-sorted step can be taken because we can extract a unitary component from the sequence, as in Lemma 15. This step is always feasible since in a hyperwell-sorted sequence the declaration of every replaced term explicitly appears in the theory, wherever it occurs in the sequence. Conversely, if one failure $\mathcal{C}$-sorted step proceeds then the set of equations is not unifiable by any hyperwell-sorted sequence.

Lemma 16 (Sorted Completeness) *Let Γ and $\mathcal{L}$ be a solved non-empty set of equations and a theory, respectively. Let $\tau = \tau_1 \ldots \tau_n$ be a hyperwell-sorted sequence w.r.t. $\mathcal{L}$ that unifies Γ. Then there exists a set of equations Γ' and a unitary substitution σ such that $\langle \Gamma, \emptyset \rangle \vdash_C \langle \Gamma', \sigma \rangle$. Moreover there exists another hyperwell-sorted sequence $\theta_1 \ldots \theta_k$ w.r.t. $\mathcal{L}\sigma$ unifying Γ'.*

Lemma 17 (Sorted Failure) *Let Γ and $\mathcal{L}$ be a solved set of equations and a theory, respectively. If $\langle \Gamma, \emptyset \rangle \vdash_C$ Fail after one $\mathcal{C}$-sorted step then there is not a well-sorted, therefore neither hyperwell-sorted, sequence w.r.t. $\mathcal{L}$ unifying Γ.*

Theorem 18 (Completeness). *Let Γ and $\mathcal{L}$ be a set of equations and a theory, respectively. Let $\sigma_1 \ldots \sigma_n$ be a hyperwell-sorted sequence w.r.t. $\mathcal{L}$ unifying Γ. Then there exists a $\mathcal{C}$-unifier for Γ w.r.t. $\mathcal{L}$.*

Then we can solve a given RSU-problem by examining its $\mathcal{C}$-derivation tree.

Corollary 19 *The RSU-problem is decidable.*

Proof. Given a RSU-problem and its associated $\mathcal{C}$-derivation tree:

(i) answer *yes* whenever there is a successful leaf. This answer is correct by Theorem 13,

(ii) answer *no* whenever every branch ends in a failure node. In this case there is no hyperwell-sorted sequence w.r.t. the theory, by Theorem 18. Although the notions of hyperwell-sortedness and well-sortedness are not equal, this answer is correct because their mutual existence is equivalent, as the following result proves. ∎

Theorem 20 *There exists a hyperwell-sorted sequence w.r.t. $\mathcal{L}$ unifying Γ if and only if there exists a well-sorted sequence w.r.t. $\mathcal{L}$ unifying Γ.*

5   Simultaneous Rigid Sorted Unification

In the sequel, $\mathcal{T}$ is a free variable tableau with branches $B_1, \ldots, B_m$.

Rigid sorted unification can be introduced in a tableau system in two different ways. In a first approach, we can use the calculus $\mathcal{C}$ to close only a single branch each time; this approach, followed in [10] but using a non-terminating variant of the calculus $\mathcal{C}$, presents a clear disadvantage. The point is that well-sortedness w.r.t a branch is not equivalent to well-sortedness w.r.t. the whole tableau, because free variables can occur repeated in different branches. In fact, not every local well-sorted unifier (w.r.t. the theory included in the branch to be closed) is well-sorted w.r.t. $\mathcal{T}$, so an extra test is needed to check that the obtained local $\mathcal{C}$-unifier is applicable to (well-sorted w.r.t.) $\mathcal{T}$. Observe that this test can only fail or succeed after the local $\mathcal{C}$-unifier has been totally built.

In a second approach, we can try to close the whole tableau in a single step, looking for a simultaneous well-sorted unifier. In this setting, we try to unify a set of equations Γ composed of one pair of potentially complementary literals from each branch of $\mathcal{T}$. A simultaneous calculus avoids the disadvantage of the local calculus because it considers all the branches at once; so it implicitly incorporates the previous extra test every time the sequence is extended. In this sense, a simultaneous calculus prunes the search space more than a local calculus, because it does not extend wrong sequences that are not going to become well-sorted w.r.t. the whole tableau.

Following this approach, the *Simultaneous Rigid Sorted Unification* (shortly *SRSU*)-*problem* arises:

Given a free variable tableau $\mathcal{T}$ and a finite set of equations Γ, is there a well-sorted sequence w.r.t. $\mathcal{T}$ that unifies Γ?

For solving *SRSU*-problems, we define the calculus $\mathcal{D}$. It is a natural extension of $\mathcal{C}$, in the sense that it takes care of all the branches of $\mathcal{T}$ when a new unitary substitution is added to the sequence. The calculus $\mathcal{D}$ is composed of the six standard rules for syntactic unification and the natural extension of the previous $\mathcal{C}$-sorted rules. For example:

(LW) Left Weakening
$$\frac{x^s \simeq t', \ \Gamma \quad \sigma_1 \ldots \sigma_n}{t \simeq t', \ \Gamma \quad \sigma_1 \ldots \sigma_n[t/x^s]}$$

if $x^s \notin var(t)$ and for each B_j $(x^s \in free(B_j \sigma_1 ... \sigma_n) \Rightarrow (t \in s) \in B_j \sigma_1 ... \sigma_n)$

$\mathcal{D}$ is used similarly to $\mathcal{C}$, that is alternating standard and sorted steps until the set of equations to be unified is empty. Then the notions of $\mathcal{D}$-step (standard or sorted), $\mathcal{D}$-unifier and $\mathcal{D}$-derivation tree can be defined as we did in the previous section, but using a free variable tableau instead of a single theory. Moreover we can prove that the calculus $\mathcal{D}$ satisfies the same properties.

Theorem 21 (Termination) *The $\mathcal{D}$-derivation tree of every SRSU-problem is finite.*

The calculus $\mathcal{D}$ only builds well-sorted sequences w.r.t. a free variable tableau $\mathcal{T}$ that unify the initial set of equations Γ. Hence, we can answer *yes* to the corresponding *SRSU*-problem, whenever a $\mathcal{D}$-unifier exists.

Theorem 22 (Soundness) *Let $\Gamma, \mathcal{T}$ and σ be a set of equations, a free variable tableau and a sequence of unitary substitutions, respectively. If $\mathcal{D}$ unifies Γ w.r.t. $\mathcal{T}$ by σ then:*

(i) σ is well-sorted w.r.t. $\mathcal{T}$
(ii) σ unifies Γ.

As in the previous section, in a tableau system we are not interested in any sequence that can be inferred from a closed ground tableau. To this end hyperwell-sortedness is extended to tableaux and the completeness theorem is stated.

Definition 23 *A triangular sequence of substitutions $[t_1/x_1^{s_1}] \ldots [t_n/x_n^{s_n}]$ is hyperwell-sorted w.r.t. a free variable tableau $\mathcal{T}$, if $x_i^{s_i} \in free(B) \implies (t_i \in s_i) \in B$, $1 \leq i \leq n$, for every branch B.*

Theorem 24 (Completeness) *Let Γ and $\mathcal{T}$ be a set of equations and a free variable tableau, respectively. Let $\sigma_1 \ldots \sigma_n$ be a hyperwell-sorted sequence w.r.t. $\mathcal{T}$ unifying Γ. Then there exists a $\mathcal{D}$-unifier for Γ w.r.t. $\mathcal{T}$.*

It is important to note that we can not solve a given *SRSU*-problem by examining the associated finite $\mathcal{D}$-derivation tree (cfr. Corollary 19) because a similar result to Theorem 20 does not always hold for the simultaneous case, as the next example shows.

Example 25 *Let $\mathcal{T}$ be the sketch of a free variable tableau below. The sequence $[a/z^{s'}][a/x^s]$ is well-sorted w.r.t. $\mathcal{T}$ and unifies $\{a \simeq x^s\}$, so $[a/z^{s'}][a/x^s]$ is a solution to the related SRSU-problem and $\mathcal{T}$ could be closed. However it does not correspond to a closed ground tableau; in fact, there is not a hyperwell-sorted sequence, nor a $\mathcal{D}$-unifier neither, because, in the first branch, x^s had to be bound to the constant a while, in the second one, to $z^{s'}$.*

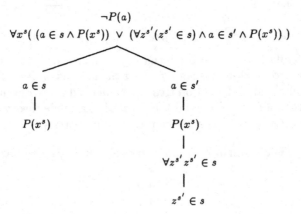

This example has two consequences. On one hand, the calculus $\mathcal{D}$ does not completely solve the *SRSU*-problem, although the completeness of the tableau system will not be affected. On the other hand, the decidability of the *SRSU*-problem remains open.

6 Free Variable Tableaux with Simultaneous Rigid Sorted Unification

Now we use the calculus $\mathcal{D}$ for defining the tableau system $\mathcal{S}2$ which is composed of the rules $\alpha, \beta, \gamma', \delta'$ and the new closure rule:

(SRSU-Closure Rule) A free variable tableau $\mathcal{T}$ with branches $B_1, \ldots, B_m$ is closed if there exist a set of equations $\Gamma = \{L_1 \simeq L'_1, \ldots, L_m \simeq L'_m\}$, where $L_i \simeq L'_i$ corresponds to a pair of potentially complementary literals occurring in B_i, and a $\mathcal{D}$-unifier w.r.t. $\mathcal{T}$ unifying Γ

We use the system $\mathcal{S}2$ for building closed tableaux as follows:

1. Expand non-deterministically the tableau, using the rules $\alpha, \beta, \gamma', \delta'$.

2. Define a set of equations Γ by selecting one pair of potentially complementary literals from every branch of the current tableau. Build the finite $\mathcal{D}$-derivation tree for Γ w.r.t. the current tableau. If a $\mathcal{D}$-unifier exists then the tableau is closed, using the *SRSU-closure rule*; otherwise, try with another set of equations, if there exists another choice, or go back to 1.

Observe that the unique step taking sorts into account (step 2) always finishes -it can be seen as a decision procedure. Therefore we have separated the complexity of sorts away from the undecidability of first order logic.

Theorem 26 (Soundness of $\mathcal{S}2$) *For every set of Σ-sentences Φ, if Φ has a closed free variable tableau then Φ is not satisfiable in structures with non-empty domains, for every sort.*

Theorem 27 (Completeness of $\mathcal{S}2$) *For every set of Σ-sentences Φ, if Φ is not satisfiable then Φ has a closed free variable tableau.*

Example 28 *We use the system $\mathcal{S}2$ to solve the problem of Example 3. First we apply rules γ and β to build the free variable tableau $\mathcal{T}$:*

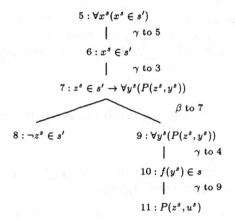

Second we use the calculus $\mathcal{D}$ to unify the set of equations $\Gamma = \{z^s \simeq x^s, z^s \simeq a, u^s \simeq f(a)\}$ w.r.t. $\mathcal{T}$. Observe that $\mathcal{D}$ has to succeed because Γ is unified by the hyperwell-sorted sequence $[f(y^s)/u^s][a/y^s][a/z^s][a/x^s]$ (this is the sequence that relates $\mathcal{T}$ to the ground tableau of Example 3). Next we show a successful $\mathcal{D}$-derivation for Γ w.r.t. $\mathcal{T}$:

$$
\text{LW} \quad \frac{\{z^s \simeq x^s, z^s \simeq a, u^s \simeq f(a)\}}{\{a \simeq x^s, a \simeq a, u^s \simeq f(a)\}} \quad [a/z^s]
$$

$$
\text{LW} \quad \frac{\{x^s \simeq a, u^s \simeq f(a)\}}{\{a \simeq a, u^s \simeq f(a)\}} \quad [a/z^s][a/x^s]
$$

$$
\text{LW} \quad \frac{\{u^s \simeq f(a)\}}{\{f(y^s) \simeq f(a)\}} \quad [a/z^s][a/x^s][f(y^s)/u^s]
$$

$$
\text{LW} \quad \frac{\{y^s \simeq a\}}{\{a \simeq a\}} \quad [a/z^s][a/x^s][f(y^s)/u^s][a/y^s]
$$

$$
\emptyset
$$

Let us compare the simultaneous calculus $\mathcal{D}$ w.r.t. a local approach (cfr. beginning of Section 5) consisting of a) the local calculus $\mathcal{C}$ applied to each branch independently and b) a test for checking whether a $\mathcal{C}$-unifier is a well-sorted sequence w.r.t. the whole tableau $\mathcal{T}$. Then we must solve the following two problems:

1) $\{z^s \simeq x^s\}$ w.r.t. the theory $\{a \in s, x^s \in s'\}$

2) $\{z^s \simeq a, u^s \simeq f(a)\}$ w.r.t. the theory $\{a \in s, x^s \in s', f(y^s) \in s\}$

In the second problem, we can apply the $\mathcal{C}$-rule LW, using the declaration $f(y^s) \in s$, to obtain the unitary substitution $\sigma = [f(y^s)/z^s]$. However, any sequence extending σ will not be well-sorted w.r.t. $\mathcal{T}$ (the test will fail, but only once the $\mathcal{C}$-unifier has been totally built!) and so the $\mathcal{C}$-derivation subtree following this step is useless. In this sense the calculus $\mathcal{D}$ is more efficient because it prevents the extension of wrong sequences that are not going to become well-sorted w.r.t. the whole tableau.

7   Conclusions and Related Work

We have presented the logic with term declarations *LTD*. This is an order-sorted logic which extends the classical first-order logic by introducing a new formula constructor $t \in s$, allowing the dynamic declaration of the term t as an element of sort s. Logics with terms declarations already appeared in [5,15,16]. There variables can be restricted to non unitary sorts; for example, $x^{s \cap s'}$ denotes an individual of the intersection sort $s \cap s'$. In *LTD*, this sorted variable can be expressed including the term declaration $x^s \in s'$ where needed.

Apart from our previous papers, tableau methods only concern [16]. [5] and [15] consider resolution based methods, the former in a more general framework. In these two papers, sorted variables behave as universal in the involved unification processes, in contrast to the rigid approach used in tableaux.

When dealing with free-variable tableau versions for *LTD*, the first question to be solved is how to define sound substitutions of variables in tableaux. This concept is the key to perform a proper integration of any sorted unification calculus into a tableau system. In [10] we proved that some possible attempts to define a substitutivity rule (cfr. [16]) fall into error. In this sense, the (decidability) results about rigid sorted unification presented in [16] seem to be useless for tableaux because its calculus is sound and complete w.r.t. an unsafe well-sortedness definition; that is, the application of its involved unifiers in its calculus produces unsound tableau systems. For this reason, decidability results for a sorted unification method useful for tableaux remained open till now.

Regarding our previous paper [10], there are two main differences. First, [10] presented a local unification calculus that required an extra test to check well-sortedness w.r.t. the whole tableau; second such calculus was not terminating. Now we have defined the simultaneous unification calculus $\mathcal{D}$ which implicitly incorporates the extra test every time a sequence is extended. In this sense, we have also shown that the calculus $\mathcal{D}$ prunes more efficiently the search space. Moreover $\mathcal{D}$ is terminating, so it can be successfully integrated in a tableau system unlike the calculus presented in [10]. Observe that non-terminating unification calculi are useless within a tableau system because they can never end when trying a non-unifiable problem.

The calculus $\mathcal{D}$ also improves [10] in other minor points. It has less rules with simpler applicability conditions. Due to termination, the technique used for proving the completeness of $\mathcal{D}$ is different and it strongly simplifies the tedious proof for the calculus presented in [10]. Now we easily state completeness proving that the existence of hyperwell-sorted solutions can be preserved in the $\mathcal{D}$-unification process.

At present, we are working on a prototype of the tableau system $S2$. As in this paper, we proceed by steps: first implementing the previous sorted calculus $\mathcal{C}$, then the calculus $\mathcal{D}$, and finally, incorporating $\mathcal{D}$ to free variable tableaux. As future work, it would be useful to design efficient strategies to transform the non-deterministic calculus $\mathcal{D}$ into a real decision procedure.

References

1. A. G. Cohn. *A more expressive formulation of many sorted logic.* Journal of Automated Reasoning 3, 113–200, 1987.
2. A. G. Cohn. *A many sorted logic with possibly empty sorts.* CADE'11. LNCS 607, 633–647, 1992.
3. A. Degtyarev, A. Voronkov. *What you always wanted to know about rigid E-unification.* Journal of Automated Reasoning 20(1), 47–80, 1998.
4. M. Fitting. *First-Order Logic and Automated Theorem Proving* (2 edition). Springer, 1996.
5. A. M. Frisch. *The substitutional framework for sorted deduction: fundamental results on hybrid reasoning.* Artificial Intelligence 49, 161–198, 1991.
6. A. Gavilanes, J. Leach, P. J. Martín, S. Nieva. *Reasoning with preorders and dynamic sorts using free variable tableaux.* AISMC-3. LNCS 1138, 365–379, 1996.
7. A. Gavilanes, J. Leach, P. J. Martín, S. Nieva. *Semantic tableaux for a logic with preorders and dynamic declarations.* TABLEAUX'97 (Position paper), CRIN 97-R-030, 7–12, 1997.
8. O. Herzog et al. *LILOG-Linguistic and logic methods for the computational understanding of german.* LILOG-Report 1b, IBM Germany, 1986.
9. P. J. Martín, A. Gavilanes. *Simultaneous sorted unification for free variable tableaux: an elegant calculus.* TR-SIP 86/98. 1998.
10. P. J. Martín, A. Gavilanes, J. Leach. *Free variable tableaux for a logic with term declarations.* TABLEAUX'98. LNAI 1397, 202–216. 1998.
11. P. J. Martín, A. Gavilanes, J. Leach. *Tableau methods for a logic with term declarations.* Journal of Symbolic Computation 29, 343–372, 2000.
12. R. Nieuwenhuis, A. Rubio. *Theorem proving with ordering and equality constrained clauses.* Journal of Symbolic Computation 19, 321–351, 1995.
13. M. Schmidt-Schauss. *Computational Aspects of an Order Sorted Logic with Term Declarations.* LNAI 395, Springer, 1989.
14. C. Walther. *A Many-sorted Calculus based on Resolution and Paramodulation.* Research Notes in Artificial Intelligence. Pitman, 1987.
15. C. Weidenbach. *A sorted logic using dynamic sorts.* MPI-I-91-218, 1991.
16. C. Weidenbach. *First-order tableaux with sorts.* Journal of the Interest Group in Pure and Applied Logics 3(6), 887–907, 1995.

Partially Adaptive Code Trees

Alexandre Riazanov and Andrei Voronkov

University of Manchester
{riazanov,voronkov}@cs.man.ac.uk

Abstract. Code trees [8] is an indexing technique used for implementing several indexed operations on terms in the theorem prover Vampire [5]. Code trees offer greater flexibility than discrimination trees. In this paper we review a new, considerably faster, version of code trees based on a different representation of the query term. We also introduce a partially adaptive version of code trees.

Keywords: automated theorem proving, subsumption, matching, term indexing, code trees

1 Introduction

In [8] code trees, a new indexing technique for forward subsumption, was presented. In order to implement efficiently forward subsumption on a large set of clauses a general subsumption algorithm is specialised at run time for each particular clause in the set. The specialised version of the algorithm is represented as a sequence of instructions of some abstract machine. Such codes are integrated into an indexing structure — a code tree, which allows one to perform subsumption check by the whole set of clauses at once. Although code trees can be considered as a differently presented version of discrimination trees, the compilation-based approach gives some serious advantages. Code sequences for indexed terms are rather flexible objects as they allow various equivalence-preserving transformations to be performed on the index. This flexibility enables invention and formulation of new optimisations. Exploiting the notion of abstract machine makes description of the indexing technique more machine-oriented and its efficient implementation feasible.

Although experiments with the original version of code trees have shown high effectiveness of the compilation-based approach, a case study revealed that the original formulation of this technique leaves space for significant improvements. In this paper we discuss several improvements implemented in version 0.0 of *Vampire* [5] that has won CASC-16 [7] in the MIX division and CASC-17 in the FOF division.

The main improvement was achieved by changing the representation of query clauses. The original version [8] deals with query terms represented as tree-like structures. It has been discovered that the *flatterm* [1] representation of query clauses eliminates the need for some operations in code trees and also makes the expensive operation of term comparison faster. We will describe the new version of code trees in Section 3.

M. Ojeda-Aciego et al. (Eds.): JELIA 2000, LNAI 1919, pp. 209–223, 2000.

Apart from the representation of queries, another shortcoming in the original version of code trees is worth special attention. There are two factors that can increase the efficiency of indexing techniques: *early detection of failure* and *better sharing* of structure (or in our case sharing of code). In the case of code trees early detection of failure can be achieved by applying term comparison instructions as early as possible. At the same time this can deteriorate sharing to a very high extent, so that the size of a code tree grows as much as 10 times on some benchmarks. In Section 4 we describe a partially adaptive version of code trees in which both early detection of failure and better code sharing is achieved by moving term comparison instructions up and down the tree during the compilation of indexed terms. Moreover, we can change the comparison instructions to achive better sharing. We call the resulting version of code trees *partially adaptive* because the tree can adapt to insertion of new instructions by changing itself. The ability to partially adapt code trees with small overhead shows their advantage over the more standard data structures used for forward subsumption and similar clause retrieval operations, for example discrimination trees [3]. Finally, in Section 5 we describe experiments with partially adaptive code trees.

2   Preliminaries

We assume acquaintance with the basic notions of terms, substitutions and clauses. A clause C_1 *subsumes* a clause C_2 if there exists a substitution θ such that $C_1\theta$ is a subset of C_2. In [8] indexing for multiliteral clauses was done by composition of indexes for their literals. Since our current approach to dealing with multiliteral clauses does not differ from the one of [8], it is sufficient to consider only the unit clause case in order to illustrate our main optimizations. In the case of unit clauses, subsumption can be reformulated as the matching problem on terms. We say that a term t_2 *matches* a term t_1 if there exists a substitution θ such that $t_1\theta = t_2$. In this case we will also say that t_1 *subsumes* t_2.

We will follow the general framework of term indexing presented in [4]. In general, the term indexing problem can be formulated as follows. Given a set of terms I, called the set of *indexed terms* and a single term t, called the *query term*, we have to retrieve quickly each term $s \in I$ such that a *retrieval condition* R holds between s and t, i.e. we have $R(s,t)$. For the purpose of this paper the retrieval condition R is *forward subsumption*: $R(s,t)$ holds if s subsumes t. The *term indexing problem* consists of finding a datastructure, called the *index* which allows one to perform efficiently the following operations: *term retrieval*, i.e. finding all (or some) $s \in I$ that are in relation R with the query term t, and *index maintenance*: changing the index when terms are inserted into or deleted from, the set of indexed term.

A *code tree* is a datastructure for term indexing. The main idea of code trees is as follows. Let F be a procedure for performing forward subsumption, so $F(s,t)$ returns true is s subsumes t. For each indexed term $s \in I$ we specialize

F by fixing its first argument to s. This specialized procedure is denoted by F_s, thus we have $F_s(t) = F(s,t)$ for all terms s and t. The procedure F_s for each indexed term $s \in I$ is represented as a sequence of instructions of an *abstract subsumption machine*. There is a small number of instructions, some of them have parameters. Then the procedures $\{F_s \mid s \in I\}$ are combined into a larger set of instructions F_I, called the *code tree for I*. The set of instructions F_I is better viewed as a tree rather than a sequence, hence the name *code tree*. The set of instructions F_I is a procedure that can be executed on any query term t such that $F_I(t) \leftrightarrow (\exists s \in I) F_s(t)$.

3   Code Trees for the Flatterm-Based Representation of Query Terms

In this section we describe a version of code trees obtained by adapting the original one of [8] to the new representation of queries. Following [8], we start from considering compilation of terms for the case of forward subsumption by one clause. To represent our algorithms formally, we will need quite a few definitions.

3.1   Positions in Term

If t is a term, $top(t)$ denotes the *top symbol* of t defined as follows:

$$top(t) = \begin{cases} t, & \text{if } t \text{ is a variable or constant;} \\ f, & \text{if } t = f(t_1, \ldots, t_n). \end{cases}$$

We call a *position* any finite sequences of natural numbers, including the empty sequence, denoted by λ. The notion of *position in a term t* and the *subterm of t at a position p*, denoted t/p, are given by the following definition.

1. the empty position λ is a position in t and $t/\lambda = t$.
2. if $t/p = f(t_1, \ldots, t_n), n > 0$, then $p.1, \ldots, p.n$ are positions in t and $t/(p.i) = t_i$ for all $i \in \{1, \ldots, n\}$.

$Pos(t)$ will denote the set of all positions in t. For technical purposes we we extend $Pos(t)$ by a special object ε called the *end position* in t. The set $Pos(t) \cup \{\varepsilon\}$ will be denoted by $Pos^+(t)$. When it is necessary to tell the end position from other positions, we call the positions from $Pos(t)$ *proper positions*. *Size* of a term t, denoted $|t|$, is defined as the number of proper positions in t. We denote by $<$ the lexicografic ordering on positions extended in the following way: $p < \varepsilon$ for any proper position p. To perform traversal of a term t we will need two operations on proper term positions: $next_t$ and $after_t$, which can be informally explained as follows. Represent the term t as a tree and imagine a term traversal in the left-to-right, depth-first direction. Suppose $t/p = s$. Then $t/next_t(p)$ is the subterm of t visited immediately after s, and $t/after_t(p)$ is the subterm visited immediately after traversal of all subterms of s. Formally, let $\lambda = p_1 < \ldots < p_n < p_{n+1} = \varepsilon$ be all positions in t. Then $next_t(p_i) = p_{i+1}$ for all $i \leq n$. The definition of $after_t$ is as follows: $after_t(\lambda) = \varepsilon$ and for $1 < i < n$ $after_t(p_i) = p_j$, and j is the smallest number such that $j > i$ and for all $i < k < j$ the position p_i is a prefix of p_k.

As it was mentioned, our new code trees are interpreted on queries represented as *flatterms*. In Vampire we use an *array-based* version of flatterms. A term t is represented by an array of the size $|t|$. Let $p_1 < \ldots < p_n$ be all positions in t. Then the i-th element of the array is a pair $\langle s, j \rangle$, where $s = top(t/p_i)$ and $p_j = after_t(p_i)$.

In can be seen that computation of our major operations on positions, $next_q$ and $after_q$, can be done very efficiently on such a representation. $next_q$ is computed by a simple incrementation of the corresponding subscript, so $next_q(p_i) = p_{i+1}$, and the subscript of $after_q(p_i)$ is given in the ith element explicitly. Another serious advantage of this representation in comparison with tree-like terms is that equality of two subterms q/p_i and q/p_j can be checked efficiently, without using stack operations.

For technical purposes we introduce a new set of variables $*_1, *_2, \ldots$, called the *technical variables*. A term containing no technical variables will be called an *ordinary term*. Let $\lambda = p_0 < p_1 < \ldots < p_n$ be all proper positions in t. Then for $i \in \{0, \ldots, n\}$, $pos_i(t)$ will denote p_i.

Let $p_{k_1} < \ldots < p_{k_m}$ be all such proper positions in t that $top(t/p_{k_i})$ is a variable. The i-th *variable position* in t, denoted by $vp_i(t)$, is defined as $vp_i(t) = p_{k_i}$. For $i > m$ $vp_i(t)$ is undefined. The *technical skeleton* of a term t, denoted by $tsk(t)$, is the term obtained from t be replacing the subterm of t at the ith variable position by the technical variable $*_i$, for all i. For example, the technical skeleton of $f(x_1, a, g(x_1, x_2))$ is $f(*_1, a, g(*_2, *_3))$.

The *variable equivalence relation* for a term t, denoted $\mathcal{E}_t$, is the equivalence relation on $\{1, \ldots, m\}$ such that: $\langle i, j \rangle \in \mathcal{E}_t$ if and only if $top(t/vp_i(t)) = top(t/vp_j(t))$. For example, the variable equivalence relation for $f(x_1, a, g(x_1, x_2))$ consists of two equivalence classes: $\{1, 2\}$ and $\{3\}$. The pair $\langle tsk(t), \mathcal{E}_t \rangle$ will be called the *technical abstraction* of t. Note the two terms have the same technical abstraction if and only if they are variants of each other. If $\mathcal{B}$ is a binary relation, $\mathcal{B}^{\approx}$ denotes the transitive, reflexive and symmetric closure of $\mathcal{B}$. If $\mathcal{E}$ is an equivalence relation and $\mathcal{B}$ is such a binary relation that $\mathcal{B}^{\approx} = \mathcal{E}$, then $\mathcal{B}$ is called a *frame* of $\mathcal{E}$. A frame is called *minimal* if no proper subset of it is a frame. Throughout the rest of the paper we consider only equivalence relations over finite sets of the form $\{1, \ldots, m\}$. A finite sequence $\langle u_1, v_1 \rangle, \ldots, \langle u_k, v_k \rangle$ of pairs of integers is called a *computation sequence* for $\mathcal{E}$ if the relation $\{\langle u_1, v_1 \rangle, \ldots, \langle u_k, v_k \rangle\}$ is a minimal frame of $\mathcal{E}$ and $u_i < v_i$ for all $i \in \{1, \ldots, k\}$. Such a computation sequence is called *canonical* if each u_i is the minimal element of its equivalence class in $\mathcal{E}$ and for $i < j$ $v_i < v_j$. Note that the canonical computation sequence is uniquely defined.

3.2 Compilation for Forward Subsumption by One Clause

We are going to solve the following problem: given a term t and a query term q we have to check if t subsumes q. Figure 1 shows a deterministic algorithm that does the job.

procedure $Subsume(t, q)$
begin
　/* First phase: term traversal */
　let $subst$ be an array for storing positions in q;
　$pos_t := \lambda$;
　$pos_q := \lambda$;
　while $pos_t \neq \varepsilon$
　　if $tsk(t)/pos_t = *_i$
　　　then
　　　　$subst[i] := pos_q$;
　　　　$pos_q := after_q(pos_q)$;
　　　　$pos_t := after_t(pos_t)$;
　　　else /* t/pos_t is not a variable */
　　　　if $top(t/pos_t) = top(q/pos_q)$
　　　　　then
　　　　　　$pos_q := next_q(pos_q)$;
　　　　　　$pos_t := next_t(pos_t)$;
　　　　　else return failure;
　　　fi;
　　fi;
　end while;
　/* Second phase: comparison of terms */
　let $\langle u_1, v_1 \rangle, \ldots, \langle u_n, v_n \rangle$ be the canonical computation sequence for $\mathcal{E}_t$.
　$i := 1$;
　while $i \leq n$
　　if $q/subst[u_i] \neq q/subst[v_i]$
　　　then return failure;
　　　else $i := i + 1$;
　end while
　return success;
end

Fig. 1. A one-to-one subsumption algorithm

Following [8] we specialise this general subsumption algorithm $Subsume$ for each indexed term t, obtaining its specialized version $Subsume_t$. The specialized version has the property $Subsume_t(q) = Subsume(t, q)$, for each query term q. The specialized algorithm is represented as a sequence of instructions of an abstract machine. In other words, we *compile* the term into code of the abstract machine. Then this code is submitted, together with the query term q, to the

procedure $Subsume_t(q)$
begin

$p := \lambda;$	$initl : Initialize(l_1)$
if $top(q/p) \neq f$ **return** failure;	$l_1 :$ $Check(f, l_1, faill)$
$p := next_q(p);$	
if $top(q/p) \neq g$ **return** failure;	$l_2 :$ $Check(g, l_3, faill)$
$p := next_q(p);$	
$subst[1] := p;$	$l_3 :$ $Put(1, l_4, faill)$
$p := after_q(p);$	
$subst[2] := p;$	$l_4 :$ $Put(2, l_5, faill)$
$p := after_q(p);$	
if $top(q/p) \neq h$ **return** failure;	$l_5 :$ $Check(h, l_6, faill)$
$p := next_q(p);$	
$subst[3] := p;$	$l_6 :$ $Put(3, l_7, faill)$
$p := after_q(p);$	
$subst[4] := p;$	$l_7 :$ $Put(4, l_8, faill)$
$p := after_q(p);$	
if $q/subst[1] \neq q/subst[3]$ **return** failure;	$l_8 :$ $Compare(1, 3, l_9, faill)$
if $q/subst[1] \neq q/subst[4]$ **return** failure;	$l_9 :$ $Compare(1, 4, l_{10}, faill)$
return success;	$l_{10} :$ $Success$
end	$faill : Failure$

Fig. 2. The algorithm *Subsume* specialized for the term $t = f(g(x_1, x_2), h(x_1, x_1))$

Fig. 3. The corresponding sequence of instructions

interpreting procedure. Before presenting technical details let us consider one simple example.

Example 1. Let $t = f(g(x_1, x_2), h(x_1, x_1))$ be the compiled term. The specialised version of the matching algorithm for this term is shown in Figure 2.

This specialized version can be rewritten in a more formal way using special instructions *Initialize*, *Check*, *Put*, *Compare*, *Success* and *Failure* as shown in Figure 3. The semantics of these instructions should be clear from the example, but will also be formally explained later.

3.3   Abstract Subsumption Machine

Now we are ready to describe the abstract machine, its instructions, compilation process, and interpretation formally. Memory of the abstract machine is divided into the following "registers":

1. substitution register *subst* which is an array of positions in the query term;
2. register p for storing the current position in the query term;
3. a register *instr* for storing the label of the current instruction.

To identify instructions in code we will use special objects — labels. We distinguish two special labels: *initl*, and *faill*. A *labeled instruction* will be written as a pair of the form $l : I$, where l is a label and I is the instruction itself. The instruction set of our abstract machine consists of *Initialize, Check, Put, Compare, Success* and *Failure*. *Success* and *Failure* have no arguments. Other instruction have the following form:

- *Initialize*(l_1), where l_1 is a label;
- *Check*(f, l_1, l_2), where f is a function symbol and l_1, l_2 are labels;
- *Put*(n, l_1, l_2), where n is a positive integer and l_1, l_2 are labels;
- *Compare*(m, n, l_1, l_2), where m, n are positive integers and l_1, l_2 are labels.

For convenience, we define two functions on instructions, *cont* and *back*. On all the above instructions *cont* returns l_1 and *back* returns l_2. Intuitively, *cont* is the label of the instructions that should be executed after the current instruction (if this instruction succeeds), and *back* is the label of the instruction that is executed if the current instruction fails.

The semantics of the instructions is shown in Figure 4. At the moment the last argument of *Put* is dummy. It will be used when we discuss the case of many indexed terms.

$Initialize(l_1)$	$p := \lambda;$ **goto** l_1	$Check(s, l_1, l_2)$	**if** $top(q/p) = s$ **then** $\quad p := next_q(p);$ $\quad$ **goto** l_1 **else goto** l_2
$Put(n, l_1, l_2)$	$subst[n] := p;$ $p := after_q(p);$ **goto** l_1	$Compare(m, n, l_1, l_2)$	**if** $q/subst[m] = q/subst[n]$ **then goto** $l_1;$ **else goto** l_2
$Success$	**return** success	$Failure$	**return** failure

Fig. 4. Semantics of instructions in code sequences

For a given indexed term t, compilation of instructions for $Subsume_t$ results in a set of labeled instructions, called the *code for* t. It consists of two parts: *traversal code* and *compare code* plus three standard instructions: $initl : Initialize(l_1)$, $succl : Success$ and $faill : Failure$.

Suppose $p_1 < p_2 < \ldots < p_m$ are all positions in t. The *traversal code* for t is the set of instructions $\{l_1 : I_1, \ldots, l_m : I_m\}$, where l_i's are labels and I_i's are defined as follows:

$$I_i = \begin{cases} Check(top(t/p_i), l_{i+1}, faill), & \text{if } t/p_i \text{ is not a variable} \\ Put(k, l_{i+1}, faill), & \text{if } tsk(t)/p_i = *_k \end{cases}$$

Let $\langle u_1, v_1 \rangle, \ldots, \langle u_n, v_n \rangle$ be the canonical computation sequence for $\mathcal{E}_t$. Then the *compare code* for t is the set of instructions $l_{m+i} : Compare(u_i, v_i, l_{m+i+1}, faill)$ for $i \in \{1, \ldots, n\}$, where $l_{m+n+1} = succl$. In Figure 3 from example 1 instructions $l_1 - l_7$ and l_8, l_9 form the traversal and compare code correspondingly.

The code for t is executed on the query term according to the semantics of instructions shown in Figure 4, beginning with the instruction *Initialize*. It is unlikely that the following statement will surprise anybody: execution of the code for t on any query term q terminates and returns *success* if and only if t subsumes q. Observe that code for t has a linear structure: instructions can be executed sequentially. In view of this observation we will call code for t also the *code sequence* for t.

3.4 Code Trees for Many-to-One Subsumption

Recall that our main problem is to find if any term t in a large set T of indexed terms subsumes a given query term q. Using compilation described in the previous subsection, one can solve the problem by the execution of code for all terms in T. This solution is inappropriate for large sets of terms. However, code sequences for terms can still be useful as we can share many instructions from code for different terms. We rely on the following observation: in most instances in automated theorem proving the set T contains many terms having similar structure. Code sequences for similar terms often have long coinciding prefixes. It is natural to combine the code sequences into one indexing structure, where the equal prefixes of code sequences are shared. Due to the tree-like form of such structures we call them *code trees*. Nodes of code trees are instructions of the abstract subsumption machine. Linking of different code sequences is done by setting appropriate values to the *cont* and *back* arguments of the instructions. A branch of such tree is a code sequence for some indexed term interleaved by some instructions of code sequences for other indexed terms. Apart from reducing memory consumption, combining code sequences in one index results in tremendous improvements in time-efficiency since during a subsumption check shared instructions are executed once for several terms in the indexed set. To illustrate this idea let us compare the code sequences for the terms $t_1 = f(f(x_1, x_2), f(x_1, x_1))$ and $t_2 = f(f(x_1, x_2), f(x_2, x_2))$.

$initl : Initialize(l_1)$	$initl : Initialize(l_1)$
$l_1 : Check(f, l_2, faill)$	$l_1 : Check(f, l_2, faill)$
$l_2 : Check(f, l_3, faill)$	$l_2 : Check(f, l_3, faill)$
$l_3 : Put(1, l_4, faill)$	$l_3 : Put(1, l_4, faill)$
$l_4 : Put(2, l_5, faill)$	$l_4 : Put(2, l_5, faill)$
$l_5 : Check(f, l_6, faill)$	$l_5 : Check(f, l_6, faill)$
$l_6 : Put(3, l_7, faill)$	$l_6 : Put(3, l_7, faill)$
$l_7 : Put(4, l_8, faill)$	$l_7 : Put(4, l_8, faill)$
$l_8 : Compare(1, 3, l_9, faill)$	$l_8 : Compare(2, 3, l_9, faill)$
$l_9 : Compare(1, 4, l_{10}, faill)$	$l_9 : Compare(2, 4, l_{10}, faill)$
$l_{10} : Success$	$l_{10} : Success$

Sharing the first eight instructions of this results in the following code:

C :

$initl : Initialize(l_1)$
$l_1 : Check(f, l_2, faill)$
$l_2 : Check(f, l_3, faill)$
$l_3 : Put(1, l_4, faill)$
$l_4 : Put(2, l_5, faill)$
$l_5 : Check(f, l_6, faill)$
$l_6 : Put(3, l_7, faill)$
$l_7 : Put(4, l_8, faill)$
$l_8 : Compare(1, 3, l_9, l_{11})$ $l_{11} : Compare(2, 3, l_9, faill)$
$l_9 : Compare(1, 4, l_{10}, faill)$ $l_{12} : Compare(2, 4, l_{10}, faill)$
$l_{10} : Success$

We can execute this code as follows. First, the eight shared instructions are executed. If none of them results in failure, we continue by executing instructions l_8, l_9, l_{10}. If the *Success* instruction l_{10} is reached the whole process terminates with success. Otherwise, if any of the equality checks l_8, l_9, failed, we have to backtrack and resume the execution from the instruction l_{11}.

In general, to maintain a code tree for a dynamicaly changing set T, one has to implement two operations: integration of new code sequences into the tree, when a term is added to T, and removal of sequences when a term is deleted from T. The integration of a code sequence CS into a code tree CT can be done as follows. We move simultaniously along the sequence CS and a branch of CT beginning from the *Initialize* instructions. If the current instruction I_T in CT coincides with the current instruction I_S in CS up to the label arguments, we skip the instructions following labels in their *cont* arguments. If I_T differs from I_S we have to consider two cases:

1. If $back(I_T)$ is not the *Failure* instruction, in the code tree we move to this instruction and continue integration.
2. If $back(I_T)$ is *Failure*, we set the *back* argument of I_T to the label of I_S. Thus, the rest of the code sequence CS together with the passed instructions in CT forms a new branch in the tree.

Removal of obsolete branches is also very simple: we remove from the code all unshared instructions corresponding to the removed term and link the remaining instructions in appropriate manner. Due to postponing *Compare* instructions, code trees maitained in this manner have an important property: traversal codes for any terms having the same technical skeleton are shared completely.

Code trees are executed nearly the same way as code sequences, but with one difference due to possible backtrack points. As soon as an instruction with a backtrack argument is found, we store its backtrack argument and the current position in the query term in special stacks *backtrPos* and *backtrInstr*. Semantics of instructions in code trees is shown in Figure 5

It is worth noting that all operations in the semantics of instructions can be executed very efficiently on flatterms.

$Initialize(l_1)$ $p := \lambda$;
            $backtrPos := empty\ stack$;
            $backtrInstr := empty\ stack$;
            goto l_1

$Check(s, l_1, l_2)$ **if** $top(q/p) = s$
            then
                $push(l_2, backtrInst)$;
                $push(p, backtrPos)$;
                $p := next_q(p)$;
                goto l_1
            else **goto** l_2

$Put(n, l_1, l_2)$ $push(l_2, backtrInst)$;
            $push(p, backtrPos)$;
            $subst[n] := p$;
            $p := after_q(p)$;
            goto l_1

$Compare$ **if** $q/subst[m] = q/subst[n]$
(m, n, l_1, l_2) **then**
                $push(l_2, backtrInst)$;
                $push(p, backtrPos)$;
                goto l_1
            else **goto** l_2

$Success$ **return** success

$Failure$ **if** $backtrPos$ is empty
            then **return** failure
            else
                $p = pop(backtrPos)$;
                goto $pop(backtrInst)$

Fig. 5. Semantics of instructions in code trees

To conclude the section we descibe here the differences between this version of code trees and that of [8]. These differences make the execution of code trees significantly faster:

1. The original version of code trees contained 6 more instructions:
   (a) The flatterm representation of queries made it possible to get rid of the stack instructions *Push* and *Pop* heavily used in the original version to encode term-traversal related operations.
   (b) Effect of the *Right* and *Down* instructions is now part of the semantics for *Check* and *Put*. This saves space and time: instead of fetching two instructions we only need to fetch one (instructions are interpreted, so there is an overhead in fetching the next instruction).
   (c) Due to better organization of backtracking, the instructions *Fork* and *Restore* used for the maintanence of backtracking are not needed any more.
2. The execution of any instruction except *Compare* requires constant time. The most expensive *Compare* instruction requires comparison of two subterms of the query term. Due to the flatterm representation of the query term, *Compare* instructions are now executed more efficiently.

4   Partially Adaptive Code Trees

From the discussion in the previous section the reader could get a feeling that code trees are slightly optimized discrimination trees. In this section we discuss an optimization which is essentially impossible on discrimination trees. This optimization, *partially adaptive code trees*, shows greater flexibility of code trees as compared to discrimination trees or substitution trees [2,6].

It is believed that a greater amount of sharing, and hence efficiency, can be gained by using *adaptive* indexing structures (see [4]). An example of such structure is *substitution trees* [2] or *adaptive automata* (see [4]). The idea of adaptive structures is that the order of the query term traversal is not fixed in advance, so indexing structures can adapt themselves to new orders of traversal when indexed terms are added or deleted. The price paid for adaptiveness is quite high, so it is not clear that adaptive structures can be more efficient than the standard ones. The index maintainance becomes more complex, choosing a wrong order can actually slow down execution, and it is difficult to ensure that the order is good: usually, the problem of optimality of a given structure is coNP-complete (see [4]). In the case of code trees for forward subsumption, the use of adaptive structures requires tree-like representation of query terms and, consequently, a larger set of instructions.

However, the flexibility of code trees allows one to make them *partially adaptive*, without changing the order of traversal of query terms. The main idea is to use the fact that *Compare* instructions commute with many other instructions, and thus can be moved up and down the tree (with essentially no overhead in the index maintainence). To illustrate this idea, consider the term $t_1 = f(x_1, x_1, x_2, x_2)$ and the following code sequences C_1, C'_1:

C_1 :

$initl : Initialize(l_1)$
$l_1 : Check(f, l_2, faill)$
$l_2 : Put(1, l_3, faill)$
$l_3 : Put(2, l_4, faill)$
$l_4 : Put(3, l_5, faill)$
$l_5 : Put(4, l_6, faill)$
$l_6 : Compare(1, 2, l_7, faill)$
$l_7 : Compare(3, 4, l_8, faill)$
$l_8 : Success$

C'_1 :

$initl : Initialize(l_1)$
$l_1 : Check(f, l_2, faill)$
$l_2 : Put(1, l_3, faill)$
$l_3 : Put(2, l_4, faill)$
$l_4 : Compare(1, 2, l_5, faill)$
$l_5 : Put(3, l_6, faill)$
$l_6 : Put(4, l_7, faill)$
$l_7 : Compare(3, 4, l_8, faill)$
$l_8 : Success$

The code sequence C_1 is computed by our compilation algorithm. The code sequence C'_1 is obtained from C_1 by moving the instruction $Compare(1, 2, \ldots)$ up the sequence. Such a lifting of some *Compare* instructions serves two purposes. The first one is earlier detection of failure. For example, execution of the code C_1 on the query term $q = f(a, b, a, a)$ determines failure after 7 instructions, while C'_1 fails after 5 instructions.

The second purpose of moving instructions up the tree is that it can increase sharing of code when new code sequences are integrated into code trees. Moreover, since *Compare* are potentially expensive instructions, sharing of them is especially desirable. For example, consider the term $t_2 = f(x_1, x_1, a, x_2)$ and two equivalent code sequences for t_2:

$$C_2:$$
$initl : Initialize(l_1)$
$l_1 : Check(f, l_2, faill)$
$l_2 : Put(1, l_3, faill)$
$l_3 : Put(2, l_4, faill)$
$l_4 : Check(a, l_5, faill)$
$l_5 : Put(3, l_6, faill)$
$l_6 : Compare(1, 2, l_7, faill)$
$l_7 : Success$

$$C_2':$$
$initl : Initialize(l_1)$
$l_1 : Check(f, l_2, faill)$
$l_2 : Put(1, l_3, faill)$
$l_3 : Put(2, l_4, faill)$
$l_4 : Compare(1, 2, l_5, faill)$
$l_5 : Check(a, l_6, faill)$
$l_6 : Put(3, l_7, faill)$
$l_7 : Success$

Combining C_1 with C_2 gives us the following code tree:

$$T:$$
$initl : Initialize(l_1)$
$l_1 : Check(f, l_2, faill)$
$l_2 : Put(1, l_3, faill)$
$l_3 : Put(2, l_4, faill)$
$l_4 : Put(3, l_5, l_9)$ $l_9 : Check(a, l_{10}, faill)$
$l_5 : Put(4, l_6, faill)$ $l_{10} : Put(2, l_8, faill)$
$l_6 : Compare(1, 2, l_7, faill)$ $l_{11} : Compare(1, 2, l_8, faill)$
$l_7 : Compare(3, 4, l_8, faill)$
$l_8 : Success$

Combining C_1' with C_2' gives us a code tree with less instructions:

$$T':$$
$initl : Initialize(l_1)$
$l_1 : Check(f, l_2, faill)$
$l_2 : Put(1, l_3, faill)$
$l_3 : Put(2, l_4, faill)$
$l_4 : Compare(1, 2, l_5, faill)$
$l_5 : Put(3, l_6, l_9)$ $l_9 : Check(a, l_{10}, faill)$
$l_6 : Put(4, l_7, faill)$ $l_{10} : Put(2, l_8, faill)$
$l_7 : Compare(3, 4, l_8, faill)$
$l_8 : Success$

Execution of the code tree T on the query term $f(a, b, a, a)$ fails after 10 instructions, while execution of T' fails only after 5.

Under some circumstances, *Compare* instructions can also be moved down the tree, for the same purpose of increasing sharing. We will illustrate this later, when we discuss the algorithm of insertion into code trees. Thus, the new code trees can adapt themselves to the insertion of new code sequences by moving some instructions up and down the tree (but without changing the order of traversal of the query term). This is why we call them *partially adaptive*.

Apart from moving *Compare* instructions, other equivalence-preserving transformations of code sequences can be used to improve sharing. This optimization is based on the observation that different computation sequences can be used for computing an equivalence relation. When encoding the technical equivalence $\mathcal{E}_t$ by a sequence of *Compare* instructions we can use any computation sequence for $\mathcal{E}_t$ instead of the canonical one.

Example 2. Let us illustrate this idea by an example. Consider the terms $t_1 = f(x_1, x_2, x_2, x_2)$ and $t_2 = f(x_1, x_1, x_1, x_2)$. The canonical computation sequences for $\mathcal{E}_{t_1}$ and $\mathcal{E}_{t_2}$ are $\langle 2, 3 \rangle, \langle 2, 4 \rangle$ and $\langle 1, 2 \rangle, \langle 1, 3 \rangle$. The correponding *Compare* instructions in the code sequences for t_1 and t_2 cannot be shared. However, the equivalence relation $\mathcal{E}_{t_2}$ can be computed by the sequence $\langle 2, 3 \rangle, \langle 1, 3 \rangle$, so that the instructions *Compare*$(2, 3, \ldots)$ can be shared resulting in the following code tree for $\{t_1, t_2\}$:

$\quad$ *initl* : *Initialize*(l_1)
$\quad l_1$: *Check*$(f, l_2, faill)$
$\quad l_2$: *Put*$(1, l_3, faill)$
$\quad l_3$: *Put*$(2, l_4, faill)$
$\quad l_4$: *Put*$(3, l_5, faill)$
$\quad l_5$: *Compare*$(2, 3, l_6, faill)$
$\quad l_6$: *Put*$(4, l_7, l_9)$
$\quad l_7$: *Compare*$(3, 4, l_8, faill)$ $\qquad l_9$: *Compare*$(1, 3, l_8, faill)$
$\quad l_8$: *Success*

Note that the semantics of instructions in partially adaptive code trees is the same as in the standard code trees. The only difference between the two versions of code trees is in their maintenance: the compilation of code sequences and their insertion into a code tree.

Now specialising the algorithm on a given term may produce several different codes. We have to fix a strategy of chosing an appropriate code sequence for a given term in presence of a code tree. The choice of the strategy must reflect our two main goals: better degree of sharing and earlier detection of failure. Moreover, we often have to modify the tree itself significantly since some code sequences in the tree are to be adapted to the new code sequences being integrated. Thus, the situation is more complex than with the basic version, compilation should be done simulataneously with modifying the tree. Our third goal is efficiency of maintainence: the insertion into and deletion from code trees should be fast.

In view of the third goal, the deletion algorithm we use in Vampire is very simple. After having deleted a code sequence from a tree we do not try to modify the trees by shifting *Compare* instructions. This means that the code tree for a set of indexed terms T can change when we insert a code sequence for a new indexed term t, and then immediately delete this code sequence.

We will now focus on the algorithm for insertion into code trees. We do not define the algorithm here, but only describe it informally and give an illustrating example. The algorithm is similar to the standard insertion algorithm into code trees (or discrimination trees), but with the following difference. First, we make insertion by ignoring *Compare* instructions at all. Second, we shift some *Compare* instructions down the tree. Third, we insert remaining *Compare* instructions from the new code.

Example 3. Consider a code tree for the set $\{f(x_1, x_1, x_2, x_2), f(x_1, x_1, a, b)\}$:

$initl : Initialize(l_1)$
$l_1 : Check(f, l_2, faill)$
$l_2 : Put(1, l_3, faill)$
$l_3 : Put(2, l_4, faill)$
$l_4 : Compare(1, 2, l_5, faill)$
$l_5 : Put(3, l_6, l_9)$ $l_9 : Check(a, l_{10}, faill)$
$l_6 : Put(4, l_7, faill)$ $l_{10} : Check(b, l_8, faill)$
$l_7 : Compare(3, 4, l_8, faill)$
$l_8 : Success$

Suppose that we insert into the set the new term $t = f(x_1, x_2, x_1, x_1)$. The code sequence for this term consists of the traversal code

$initl : Initialize(m_1)$
$m_1 : Check(f, l_2, faill)$
$m_2 : Put(1, l_3, faill)$
$m_3 : Put(2, l_4, faill)$
$m_4 : Put(3, l_5, faill)$
$m_5 : Put(4, l_6, faill)$

followed by a sequence of *Compare* corresponding to a computation sequence for the equivalence relation $\mathcal{E}_t$ consising of two classes $\{1, 3, 4\}$ and $\{2\}$.

If we ignore the *Compare* instructions in the code tree, then the nodes m_1, m_2, m_3, m_4, m_5 would be merged into the nodes l_1, l_2, l_3, l_5, l_6, respectively. But between l_3 and l_4 the tree contains the instruction $l_4 : Compare(1, 2, l_5, faill)$, and $\langle 1, 2 \rangle$ does not belong to $\mathcal{E}_t$. So, we have to move $Compare(1, 2, l_5, faill)$ down the tree. The instruction $l_7 : Compare(3, 4, l_8, faill)$ can be shared, since $\langle 3, 4 \rangle$ belongs to $\mathcal{E}_t$. To compute the equivalence relation $\mathcal{E}_t$, we should add either $\langle 1, 3 \rangle$ or $\langle 1, 4 \rangle$ to the computation sequence $\langle 3, 4 \rangle$. So, we obtain the following code tree:

$initl : Initialize(l_1)$
$l_1 : Check(f, l_2, faill)$
$l_2 : Put(1, l_3, faill)$
$l_3 : Put(2, l_5, faill)$
$l_5 : Put(3, l_6, l_{11})$ $l_{11} : Compare(1, 2, l_9, faill)$
$l_6 : Put(4, l_7, faill)$ $l_9 : Check(a, l_{10}, faill)$
$l_7 : Compare(3, 4, l_4, faill)$ $l_{10} : Check(b, l_8, faill)$
$l_4 : Compare(1, 2, l_8, l_{12})$ $l_{12} : Compare(1, 3, l_8, faill)$
$l_8 : Success$

5   Experiments

Our experiments have shown that in many cases making code trees partially adaptive gives significant reduction of the total number of executed instructions, though it may give an increase in the number of executed expensive *Compare* instructions. We compared overall performance of the system with the partialy adaptive version of code trees and the basic version on 75 problems from the MIX division of CASC-16 [7]. Note that both compared versions are based on the flatterms, the old version dealing with tree-like queries is unfortunately not available for comparison. The problems were run with the time limit of 10 minutes on a PC with a Pentium III 500MHz processor. We restricted memory usage by 300Mb and the number of kept clauses by 100000. In the table below we give times consumed by the optimised version and the basic one ($time_o$ and $time_b$ correspondingly). To make comparison, we calculate percentage of difference ($diff$) between times consumed by the versions w.r.t. the best time. Negative value of $diff$ indicates cases when the optimised version showed worse results. From the whole benchmark suit 34 problems were selected by the following criteria: (1) one of the versions works at least 30 seconds on the selected problems with the given limits, (2) absolute value of $diff$ must exceed 1%

problem	diff	$time_o$	$time_b$	problem	diff	$time_o$	$time_b$
alg003-1	3.33%	41.13	42.5	lcl005-1	-1.4%	92.45	91.17
alg004-1	9.86%	37.9	41.64	lcl015-1	-1.26%	72.09	71.19
boo020-1	-3.05%	42.2	40.95	lcl016-1	-1.27%	71.27	70.37
cid003-1	9.92%	35.67	39.21	lcl017-1	2.47%	72.05	73.83
cid003-2	6.36%	46.97	49.96	lcl020-1	10.74%	77.41	85.73
civ002-1	-2.54%	86.59	84.44	lcl021-1	5.28%	76.89	80.95
col077-1	-1.38%	30.79	30.37	lcl099-1	4.62%	42.81	44.79
grp054-1	1.48%	51.24	52	lcl105-1	4.31%	39.15	40.85
grp073-1	-1.39%	34.22	33.75	lcl122-1	1.14%	80.34	81.26
grp106-1	-1.19%	47.61	47.05	lcl125-1	-1.01%	36	35.64
grp107-1	-1.11%	70.55	69.77	lcl127-1	3.99%	57.02	59.3
grp108-1	-1.15%	56.93	56.28	lcl129-1	3.7%	32.42	33.61
grp110-1	-1.54%	40.2	39.59	lcl166-1	5.54%	77.36	81.65
grp111-1	-1.1%	53.86	53.27	lcl167-1	10.48%	77.26	85.36
lat002-1	-2.53%	61.13	59.62	prv008-1	9.65%	166.47	182.55
lat005-3	-1.75%	61.31	60.25	rng025-1	49.18%	31.98	47.71
lat005-4	-4.15%	60.43	58.02	rng034-1	11.46%	50.69	56.5

References

1. J. Christian. Flatterms, discrimination nets, and fast term rewriting. *Journal of Automated Reasoning*, 10(1):95–113, February 1993.
2. P. Graf. Substitution tree indexing. In J. Hsiang, editor, *Rewriting Techniques and Applications*, volume 914 of *Lecture Notes in Computer Science*, pages 117–131, 1995.
3. William W. McCune. Experiments with discrimination-tree indexing and path indexing for term retrieval. *Journal of Automated Reasoning*, 9(2):147–167, 1992.
4. I.V. Ramakrishnan, R. Sekar, and A. Voronkov. Term indexing. In A. Robinson and A. Voronkov, editors, *Handbook of Automated Reasoning*. Elsevier Science and MIT Press, 2000. To appear.
5. A. Riazanov and A. Voronkov. Vampire. In H. Ganzinger, editor, *Automated Deduction—CADE-16. 16th International Conference on Automated Deduction*, Lecture Notes in Artificial Intelligence, pages 292–296, Trento, Italy, July 1999.
6. J.M.A. Rivero. *Data Structures and Algorithms for Automated Deduction with Equality*. Phd thesis, Universitat Politècnica de Catalunya, Barcelona, May 2000.
7. G. Sutcliffe. The CADE-16 ATP system competition. *Journal of Automated Reasoning*, 2000. to appear.
8. A. Voronkov. The anatomy of Vampire: Implementing bottom-up procedures with code trees. *Journal of Automated Reasoning*, 15(2):237–265, 1995.

On Dialogue Systems with Speech Acts, Arguments, and Counterarguments

Henry Prakken

Institute of Information and Computing Sciences
Utrecht University, The Netherlands
http://www.cs.uu.nl/staff/henry.html

Abstract. This paper proposes a formal framework for argumentative dialogue systems with the possibility of counterargument. The framework allows for claiming, challenging, retracting and conceding propositions. It also allows for exchanging arguments and counterarguments for propositions, by incorporating argument games for nonmonotonic logics. A key element of the framework is a precise definition of the notion of relevance of a move, which enables flexible yet well-behaved protocols.

1   Introduction

In recent years, dialogue systems for argumentation have received interest in several fields of artificial intelligence, such as explanation [2], AI and law [4, 6], discourse generation [5], multi-agent systems [10, 1], and intelligent tutoring [9]. These developments justify a formal study of such dialogue systems; this paper contributes to this study by an attempt to integrate two relevant developments in the fields of argumentation theory and artificial intelligence.

In argumentation theory, formal dialogue systems have been developed for so-called 'persuasion' or 'critical discussion'; see e.g. [8, 14]. In persuasion, the initial situation is a conflict of opinion, and the goal is to resolve this conflict by verbal means. The dialogue systems regulate the use of speech acts for such things as making, challenging, accepting, withdrawing, and arguing for a claim. The proponent of a claim aims at making the opponent concede his claim; the opponent instead aims at making the proponent withdraw his claim. A persuasion dialogue ends when one of the players has fullfilled their aim. Logic governs the dialogue in various ways. For instance, if a participant is asked to give grounds for a claim, these grounds have to logically imply the claim. Or if a proponent's claim is logically implied by the opponent's concessions, the opponent is forced to accept the claim, or else withdraw some of her concessions.

Although such dialogue systems make an interesting link between the (static) logical and (dynamic) dialogical aspects of argumentation, they have one important limitation. The underlying logic is deductive, so that players cannot reply to an argument with a counterargument, since such a move presupposes a nonmonotonic, or defeasible logic. Yet in actual debates it is very common to attack one's opponent's arguments with a counterargument. This is where a recent development in AI becomes relevant, viz. the modelling of nonmonotonic, or defeasible

M. Ojeda-Aciego et al. (Eds.): JELIA 2000, LNAI 1919, pp. 224–238, 2000.

reasoning in the form of dialectical argument games; e.g. [7, 13, 11]. Such games model defeasible reasoning as a dispute between a proponent and opponent of a proposition. The proponent starts with an argument for it, after which each player must attack the other player's previous argument with a counterargument of sufficient strength. The initial proposition is provable if the proponent has a winning strategy, i.e., if he can make the opponent run out of moves in whatever way she attacks. Clearly, this dialectical setup fits well with the above-mentioned dialogue system applications. The main aim of this paper is to incorporate these argument games in protocols for persuasion dialogue. This results in a subtype of persuasion dialogues that in [11] were called 'disputes'.

The following example illustrates these observations.

Paul: My car is safer than your car. (*persuasion: making a claim*)
Olga: Why is your car safer? (*persuasion: asking grounds for a claim*)
Paul: Since it has an airbag. (*persuasion: offering grounds for a claim; dispute: stating an initial argument*)
Olga: That is true, (*persuasion: conceding a claim*) but I disagree that this makes your car safe: the newspapers recently reported on airbags expanding without cause. (*dispute: stating a counterargument*)
Paul: I also read that report (*persuasion: conceding a claim*) but a recent scientific study showed that cars with airbags are safer than cars without airbags, and scientific studies are more reliable than sporadic newspaper reports. (*dispute: rebutting a counterargument, and arguing about strength of conflicting arguments*)
Olga: OK, I admit that your argument is stronger than mine. (*persuasion: conceding a claim*) However, your car is still not safer, since its maximum speed is much higher. (*dispute: alternative counterargument*)

A second aim of this paper is to study the design of argumentative dialogue systems. Although most current systems are carefully designed, their underlying principles are often hard to see. Therefore, I shall in Section 2 propose a general framework for disputational protocols, based on intuitive principles. In Section 3 I shall instantiate it with a particular protocol (illustrated in Section 4), after which I conclude with a discusison in Section 5.

2   A Framework for Disputational Protocols

2.1   Elements and Variations

In the present framework, the initial situation of a persuasion dialogue is a conflict of opinion between two rational agents about whether a certain claim is tenable, possibly on the basis of shared background knowledge. The goal of a persuasion dialogue is to resolve this conflict by rational verbal means. The dialogue systems should be designed such that they are likely to promote this goal. Differences between the various protocols might be caused by different opinions on how this goal can be promoted, but also by, for example, different contexts in which dialogues take place (e.g. legal, educational, or scientific dispute), or by limitations of such resources as time or reasoning capacity.

The present framework fixes the set of participants; two players are assumed, a *proponent* and an *opponent* of an initial claim. According to [14], dialogue systems regulate four aspects of dialogues:

- *Locution rules* (what moves are possible)
- *Structural rules* (when moves are legal)
- *Commitment rules* (The effects of moves on the players' commitments);
- *Termination rules* (when dialogues terminates and with what outcome).

For present purposes a fifth element must be distinguished, viz. the *underlying logic for defeasible argumentation*. On all five points the framework must allow for variations. In particular, the framework should leave room for:

- allowing one or allowing several moves per turn (*unique-move* vs. *multi-move* protocols);
- different choices on whether players can move alternatives to their earlier moves (*unique-response* vs. *multi-response* protocols);
- different underlying argument games (but all for justification);
- various sets of speech acts (but always including claims and arguments);
- different rules for legality of dialogue moves. In particular,
  - different views on inconsistent commitments
  - automatic vs forced commitment to implied commitments
- different rules for the effects of moves on the commitments of the players;
- different termination and winning criteria.

On the other hand, some conditions are hardwired in the framework. Most importantly, every move must somehow have a bearing to the main claim. This is realised by two other principles: every move must be a *reply* to some other move, being either an *attack* or a *surrender*, and every move should be *relevant*.

2.2   The Framework

The framework defines the notion of a protocol for dispute (*PPD*).

Definition 1. *[Protocols for persuasion with dispute]. A protocol for persuasion with dispute (PPD) consists of the following elements. (L, Players, Acts, Replies, Moves, PlayerToMove, Comms, Legal, Disputes, Winner), as defined below.*

I now define and comment on each of the elements of a protocol for dispute.

- L is a notion of [11], viz. a protocol for disputes based on a logic for defeasible argumentation. $wff(L)$ is the set of all well-formed formulas of L's language and $Args(L)$ the set of all its well-formed arguments. For any set $T \subseteq wff(L)$, $Args_L(T) \subseteq Args(L)$ are all L-arguments constructible on the basis of the input information T. Below, L will often be left implicit.

Logics for defeasible argumentation (cf. [12]) formalise nonmonotonic reasoning as the construction and comparison of possibly conflicting arguments. They define the notions of an argument and of conflict between arguments, and assume or define standards for comparing arguments. The output is a classification of arguments as, for instance, 'justified', 'defensible' or 'overruled'. One way to define argumentation logics is, as noted above, in the form of argument games. In [11] I showed how these games can be 'dynamified' in that the information base is not given in advance but constructed during the dispute. For present purposes this is very important, since in persuasion dialogues this typically happens.

The format of both arguments games and protocols for dispute is very similar to that of *PPD*'s. The main elements missing are the set *Act* and the functions *Replies* and *Comms*, since these formalisms have no room for speech acts.

- *Players* = $\{P, O\}$. $\overline{Player} = O$ iff *Player* = *P*, and *P* iff *Player* = *O*.
- *Acts* is the set of speech acts. $\{claim\ \varphi,\ argue(\Phi,\ so\ \varphi)\} \subseteq Acts$ (here, $\Phi \subseteq wff(L)$, $\varphi \in wff(L)$ and $(\Phi,\ so\ \varphi) \in Args(L)$). Acts have a performative and a content part. Note that each protocol has a *claim* and an *argue* act.
- *Replies* : $Acts \longrightarrow Pow(Acts)$

  is a function that assigns to each act its possible replies. It is defined in terms of two other functions of the same type, *Attacks* and *Surrenders*. These functions jointly satisfy the following conditions. For any $A, B \in Acts$:
  1. $B \in Replies(A)$ iff $B \in Attacks(A)$ or $B \in Surrenders(A)$;
  2. $Attacks(A) \cap Surrenders(B) = \emptyset$;
  3. If $B \in Surrenders(A)$, then $Replies(B) = \emptyset$;
  4. If $B \in Attacks(A)$, then $Replies(B) \neq \emptyset$.

Intuitively, an attacking reply is a challenge to the replied-to act, while a surrendering reply gives up the possibility of attack. For instance, challenging a claim, responding to a challenge with an argument for the claim, and stating a counterargument are attacking replies, while retracting a proposition in reply to a challenge and conceding a proposition in reply to a claim are surrenders.

- *Moves* is the set of all well-formed moves. All moves are initial or replying moves. An *initial move* is of the form $M_1 = (Player, Act)$, and a *replying move* is of the form $M_i = (Player, Act, Move)$ $(i > 1)$. $Player(M_i)$ denotes the first element of a move M_i, $Act(M_i)$ its second element and $Move(M_i)$ its third element. If $Move(M_i) = M_j$, we say that M_i is a *reply to*, or *replies to* M_j, and that M_j is the *target* of M_i.

  Now the set *Moves* is recursively defined as the smallest set such that if $Player \in Players$, $Act \in Acts$ and $M_i \in Moves$, then $(Player, Act) \in Moves$ and $(Player, Act, M_i) \in Moves$.
- *PlayerToMove* determines the player to move at each stage of a dialogue. Let $Pow^*(Moves)$ be the set of all finite sequences of subsets of *Moves*. Then

  $PlayerToMove$: $Pow^*(Moves) \longrightarrow Players$

such that $PlayerToMove(D) = P$ if $D = \emptyset$; else

1. $PlayerToMove(D) = P$ iff the dialogical status of M_1 is 'out';
2. $PlayerToMove(D) = O$ iff the dialogical status of M_1 is 'in'.

The *PlayerToMove* function is completely defined by the framework: proponent begins a dispute and then a player keeps moving until s/he has changed the 'dialogical status' of the initial claim (to be defined below) his or her way. This function is hardwired in the framework since the *Legal* function of the framework requires moves to be relevant, and a move will (roughly) be defined to be relevant iff it can change the dialogical status of the initial move. Clearly, this does not leave room for other *PlayerToMove* functions than the above one.

– *Comms* is a function that assigns to each player at each stage of a dialogue a set of propositions to which the player is committed at that stage.

   $Comms$: $Pow^*(Moves) \times Players \longrightarrow Pow(wff(L))$.

   such that $Comms_\emptyset(P) = Comms_\emptyset(O)$.

Note that $Comms_\emptyset(p)$ can be nonempty (although it must have the same content for P and for O). This allows for an initially agreed or assumed basis for discussion. Note also that the framework does not require consistency of a player's commitments. This is since some protocols allow inconsistency, after which the other player can demand retraction of one of the sources of inconsistency.

– *Legal* is a function that for any dialogue specifies the legal moves at that point, given the dialogue so far and the players' commitments. Let C_p ($p \in$ *Players*) stand for $Pow(wff(L)) \times p$. Then

   $Legal$: $Pow^*(Moves) \times C_P \times C_O \longrightarrow Pow(Moves)$

   (Below I will usually leave the commitments implicit).
   This function is constrained as follows. For all $M \in Moves$ and all $D \in Pow^*(Moves)$, if $M_i \in Legal(D)$, then:
   1. If $D = \emptyset$, then M_i is an initial move and $Act(M_i)$ is of the form $claim(\varphi)$;
   2. $Move(M_i) \in D$;
   3. $Act(M_i)$ is a reply to $Act(Move(M_i))$;
   4. If M_i and M_j ($j < i$) are both replies to $M_k \in D$ and $M_j \in D$, then $Act(M_{i+1}) \neq Act(M_j)$;
   5. If $Act(M_i)$ is of the form $Argue(A)$ then M_i's counterpart in the L-dispute L_i associated with D_i is legal in L_i;
   6. M_i is relevant in D.

Condition 1 says that a dispute always starts with a claim. Condition 2 says the obvious thing that a replied-to move must have been moved in the dialogue. Condition 3 says that an act can only be moved if it is a reply to the act moved in the replied-to move. Condition 4 states the obvious condition that if a player backtracks, the new move must be different from the first move.

The last two conditions are crucial. Condition 5 incorporates the underlying disputational protocol L, by requiring *argue* moves to conform to the legality rules of this protocol. L_i is the proof-theoretical 'subdispute' of D in which the *argue* move occurs. Note that thus the framework assumes that with each sequence of PPD-moves an L-dispute can be associated. Particular protocols must specify the details.

Finally, Condition 6 every move to be relevant. Relevance, to be defined below, is the framework's key element in allowing maximal freedom (including backtracking and postponing replies) while yet ensuring focus of a dispute.

- *Disputes* is the set of all sequences $M_1, \ldots, M_n$ of moves such that for all i:
    1. $Player(M_i) = PlayerToMove(M_1, \ldots, M_{i-1})$,
    2. $M_i \in Legal(M_1, \ldots, M_{i-1})$.
- *Winner* is a function that determines the winner of a dialogue, if any:

    $Winner$: $Disputes \longrightarrow Players$

    The winning function is constrained by the following condition.
    - If $Winner(D) = p$, then $PlayerToMove(D) = \bar{p}$ and $Legal(D) = \emptyset$;

Thus, to win it must hold that the other player has run out of moves. The rationale for this is the relevance condition (to be defined next); as long as a player can make relevant moves, s/he should not be losing. Note that termination is defined implicitly, as the situation where a player-to-move has no legal moves.

I now turn to relevance. This notion is defined in terms of the dialogical status of a move (either 'in' or 'out'), which captures whether its mover has been able to 'defend' the move against attacks. A move can be in in two ways: the other player can have conceded it, or all attacks of the other player have been successfully replied to (where success is determined recursively). As for conceding a move, the general framework only states two necessary conditions:

- If a move M is conceded in D, then it has a surrendering reply in D.
- If M is conceded in D, it is conceded in all continuations of D.

The reason why these conditions are not sufficient lies in the most natural treatment of replies to arguing moves. In Section 3 we shall see that an arguing move has several elements (premises, conclusion, inference rule), some of which can be surrendered but others attacked at the same time. Therefore the notion of conceding a move must be fully defined in particular dialogue systems.

Definition 2. *[Dialogical status of moves] A move M of a dialogue D is either in or out in D. It is in in D iff*

1. *M is conceded in D; or else*
2. *all attacking moves in D that reply to it are out in D.*

Now a move is relevant iff any attacking alternative would change the status of the initial move of the dialogue. This can be captured as follows.

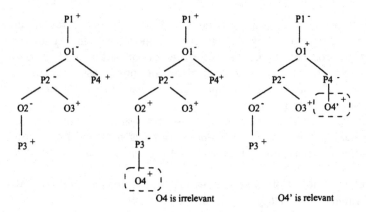

Fig. 1. Dialogical status of moves.

Definition 3. *[Relevance.] A move in a dialogue D is a* relevant target *iff any attacking reply to it changes the dialogical status of D's initial move. A move is* relevant *in D iff it replies to a relevant target in D.*

Note that a reply to a conceded move is never relevant.

To illustrate these definitions, consider figure 1. The first dispute tree shows the situation after P_4. The next tree shows the dialogical status of the moves when O has continued with replying to P_3: this move does not affect the status of P_1, so O_4 is irrelevant. The final tree shows the situation where O has instead replied to P_4: then the status of P_1 has changed, so O_4' is relevant.

3   An Instantiation of the Framework

To illustrate the general framework, I now instantiate it with a specific protocol.

The Underlying Disputational Protocol The disputational protocol L is that of liberal disputes as defined in [11], instantiated with proof-theoretical rules for sceptical argumentation. Liberal disputes allow an argument as long as it is relevant. In [11] it is shown that this protocol satisfies certain 'soundness' and 'fairness' properties with respect to the underlying argumentation logic.

Besides a set *Args* of constructible arguments, L also assumes a binary relation of *defeat* among arguments. An argument *strictly defeats* another if the first defeats the second but not the other way around. Now Dung's argument game says that proponent begins with an argument and then players take turns as follows: proponent's arguments strictly defeat their targets, while opponent's arguments defeat their targets. In addition, proponent is not allowed to repeat his moves in one 'dialogue line' (a dispute without backtracking moves). The precise definition of the notions of an argument, conflict and comparison of arguments are not essential, and therefore I keep these elements semiformal, using obvious logical symbols in the examples, with both material ($\supset$) and defeasible ($\Rightarrow$) implication. But the protocol assumes that arguments can be represented as a premises-conclusion pair Φ, so φ, where $\Phi \subseteq wff(L)$ are the premises and $\varphi \in wff(L)$ is the conclusion of the argument.

Speech Acts The set of speech acts is defined as follows.

Acts	Attacks	Surrenders
claim φ	*why* φ	*concede* φ
why φ	*argue* Φ, *so* φ	*retract* φ
concede φ		
retract φ		
argue(Φ, *so* φ)	*why* φ_i ($\varphi_i \in \Phi$)	*concede* φ_i ($\varphi_i \in \Phi$)
	argue(Φ', *so* φ')	*concede*(Φ *implies* φ)
concede(Φ *implies* φ)		

Here $\Phi, \Phi' \subseteq wff(L)$, $\varphi, \varphi' \in wff(L)$, and ($\Phi$, *so* φ) and (Φ', *so* φ') $\in Args(L)$.

The *claim*, *why*, *retract* and *concede* φ moves are familiar from MacKenzie-style dialogue systems. The *argue* move is present in e.g. [4] and [14]. The conceding an inference move is adapted from [4]. Its effect is to give up the possibility of counterargument. Note that an argument can be replied to by replying to one of its premises or to its inference rule, or by a counterargument.

Commitment Rules   The commitment rules are as follows. Let $D_i = M_1, \ldots, M_i$ be any sequence of moves, and let $Player(M_i) = p$.

- If $Act(M_i) = claim\ \varphi$ or *concede* φ, then $Comms_{D_i}(p) = Comms_{D_{i-1}}(p) \cup \{\varphi\}$.
- If $Act(M_i) = argue(\Phi,\ so\ \varphi)$, then $Comms_{D_i}(p) = Comms_{D_{i-1}}(p) \cup \Phi \cup \{\varphi\}$.
- If $Act(M_i) = retract\ \varphi$ then $Comms_{D_i}(p) = Comms_{D_{i-1}}(p)/\{\varphi\}$.
- In all other cases the commitments remain unchanged.

The effects of claims, concessions and retractions are obvious. As for the effects of moving arguments, note that their conclusion is not also added to the mover's commitments. This is since some dialectical proof theories, including the present-used one, sometimes allow a player to attack himself. In [14] the material implication is also added to the commitments of the argument's mover. Although this works fine if the underlying is monotonic, in the present approach, which allows defeasible arguments, this is different.

Legality of Moves The definition of the *Legal* function is completed as follows. For all $M \in Moves$ and all $D \in Pow^*(Moves)$, $M_i \in Legal(D)$ iff the above conditions and the following conditions are satisfied.

7. Each move must leave the mover's commitments classically consistent;
8. If $Act(M_i) = concede\ \varphi$, then
   (a) $Comms_{D_{i-1}}(Player_i) \not\vdash \varphi$;
   (b) $Comms_{D_{i-1}}(Player_i)$ do not justify $\neg\varphi$;

9. If $Act(M_i) = retract\ \varphi$, then
   (a) $\varphi \in Comms_{D_{i-1}}(Player_i)$; and
   (b) φ was explicitly added to $Comms_{D_{i-1}}(Player_i)$.
10. If $Act(M_i) = why\ \varphi$, then $Comms_{D_{i-1}}(Player_i)$ do not justify φ.
11. If $Act(M_i) = argue(\Phi,\ so\ \varphi)$, then
    (a) all preceding moves $M_j \in D$ with $Act(M_j) = why\ \varphi_i$ ($\varphi_i \in \Phi$) are out;
    (b) If M_i replies to an argue move M_j, then M_j has no child $concede(\Phi,\ so\ \varphi)$.

As for Condition 7, note that a commitment set which supports two conflicting defeasible arguments does not have to be classically inconsistent. Whether it is, depends on the underlying logic for constructing arguments. Many logics allow the consistent expression of examples like 'Tweety is a bird, birds generally fly, but Tweety does not fly'. This enables such moves as "I concede your argument as the general case, but in this case I have a counterargument .."

Condition 8a says that a proposition may only be conceded if the mover is not committed to it. (This allows conceding a proposition that is defeasibly implied by the player's own commitments.) Condition 8b forbids conceding a proposition if the opposite is justified by the player's own commitments.

Condition 9 is obvious. Condition 10 allows retractions of 'explicit' commitments only. This forces a player to explicitly indicate how an implied commitment is retracted. Condition 11a forbids moving arguments of which the premises are under challenge. This is [8]'s way to avoid arguments that "beg the question". Finally, Condition 11b says that if an argument was already conceded, no counterargument can be stated any more.

Conceding a Move Next I complete the definition of conceding a move.

Definition 4 (Conceding a move). *A move M in a dialogue D has been conceded iff*

- $Act(M) \neq argue(A)$ *and M has a surrendering child; or*
- $Act(M) = argue(A)$ *and both all premises and the inference rule of A have been conceded.*

Associated *L*-Disputes Next the notion of an *L*-dispute associated with a *PPD*-dispute must be defined. This notion is used in determining legality of counterarguments, but it can also serve to study logical properties of winning criteria. The idea is that during a *PPD*-dispute an *L*-dispute of arguments and counterarguments is constructed. A technical problem is that *argue* replies to *why* moves extend an argument backwards, by replacing one of its premises with an argument for this premise. To account for this, we must first define the notions of a combination of two arguments and of a modification of an argument.

Definition 5. *[Combinations of arguments.] Let $(A = S,\ so\ \varphi)$ and $(B = S',\ so\ \psi)$ be two arguments such that $\psi \in S$. Then $A \otimes B = (S/\{\psi\}) \cup S',\ so\ \varphi$.*

Definition 6. *[Modification of arguments.] For any arguments A, B and C, A is a modification of A; if B is a modification of A and $B \otimes C$ is defined, then $B \otimes C$ is a modification of A; nothing else is a modification of A. We also say that A modifies A, and B modifies A in $A \otimes B$. And an argue move modifies another argue move if the argument moved by the first modifies the argument moves by the second move.*

For any move $M = (p, \alpha, m)$ and arguments a and b, $M[a/b] = (p, argue(b), m)$ if $\alpha = argue(a)$; otherwise $M[a/b] = M$. Likewise for initial moves.

Now the notion of the L-part of a dispute can be defined.

Definition 7. *[L-disputes of a PPD-dispute.] For any PPD-dispute D, the associated L-dispute $L(D)$ is a sequence of argue moves defined as follows.*

1. $T(\emptyset) = \emptyset$;
2. If $Act(M_{i+1}) \neq argue(A)$ for any A, then $T(D_{i+1}) = T(D_i)$;
3. If $Act(M_{i+1}) = argue(A)$ for some A, then
   (a) If M_{i+1} replies to an argue move M_j, then $T(D_{i+1}) = T(D_i), M'_{i+1}$, where M'_{i+1} is M_{i+1} except that it replies to the move in $T(D_{i+1})$ modified by M_j;
   (b) If $M_{i+1} = (p, \alpha, m)$ replies to a why move replying to a claim, then $T(D_{i+1}) = T(D_i), (p, \alpha)$;
   (c) If M_{i+1} replies to a why φ move replying to an $argue(B)$ move M_j, then
      i. If T_i contains any argue moves M_k resulting from modifications of M_j such that their arguments C still have a premise φ, then $T(D_{i+1}) = T^*(D_i)$, where $T^*(D_i)$ is obtained from $T(D_i)$ by replacing C in all such M_k with $C \otimes A$, and then adjusting the targets of moves when these targets have been changed.
      ii. Else $T(D_{i+1}) = T(D_i), M'_j$, where M'_j is obtained from M_j by replacing B with $B \otimes A$.

So the construction of an L-dispute starts with the empty set, and each *PPD*-move other than an *argue* move leaves its content unchanged. As for *argue PPD*-moves, two cases must be distinguished, whether it replies to another *argue* move or to a *why* move. In the first case the argue move can simply be added to the L-dispute, but the second case is more complex. Again two cases must be considered. If the replied-to *why* move itself replied to the initial claim, then the *argue* is the root of a new dialectical tree, so the move to which it replies must be omitted, to turn it into an initial move. Finally, if the replied-to *why* φ move challenged the premise of an argument B, then again two cases must be considered. If the L-dispute contains modifications of A that still contain premise φ, then these modifications (if not equal to B itself) were triggered by a *why* attack on another premise of B. In that case φ must in all these modifications be replaced with the premises of A. Note that if no such other *why* attacks were made, this boils down to modifying B itself. (Note also that if in T, M_i replies to M_j, and M_j is then modified by M_k, from then on M_i replies to the modified move.) If, however, no modification of B in the L-dispute contains a premise φ, then moving A was an alternative to an earlier reply to the *why* φ move. Then we must add to the L-dispute an alternative modification of the original $argue(B)$ move, with $B \otimes A$ (note that the original move was not in T_i any more).

In general an L-dispute is a collection of trees, since a *why* reply to the initial move can be answered with alternative arguments. So the condition of the general framework that an *argue* move M is legal in the associated L-dispute means that M is legal in the tree contained in this dispute that itself contains $Move(M)$.

Winning As for winning, several definitions are conceivable. Part of the aim of the present framework is to provide a setting in which the alternatives can be compared. In the present protocol I simply turn the necessary conditions of the general framework into a necessary-and-sufficient condition.

Definition 8. *For any dispute D, $Winner(D) = p$ iff $PlayerToMove(D) = \overline{p}$ and $Legal(D) = \emptyset$.*

It is immediate that if p wins D, then M_1 in D is labelled p's way. However, the same does not always hold for the associated L-dispute. Consider the following dispute D (In the examples below I leave the replied-to move implicit if it is the preceding move, and P_i and O_i stand for turns of a player.)

P_1: *claim* p O_1: *why* p
P_2: *argue*$(q, q \Rightarrow p, so\ p)$ O_2: *argue*$(q, r, q \wedge r \Rightarrow \neg p, so\ \neg p)$
P_3: *why* r O_3: *concede* p (to P_1)

Now P has won, but $T(D) = P_2, O_2$, in which P_2 is *out* and O_2 is *in*. So a player can lose by unforced surrenders.

It also holds that if O has won, P is not committed to his main claim any more. This is since if all other moves have become illegal for P, he can still surrender to O's initial *why* attack. However, it does not hold that if P has won, O is always committed to P's main claim φ. This is since O might have moved an argument with premise $\neg\varphi$ and in the course of the dispute retracting φ may have become irrelevant and thus illegal, so that conceding M_1 has also become illegal. Future research should reveal whether this is a problematic property of the protocol.

4   Examples of Dialogues

Example 1. Most argumentation logics do not allow counterarguments to deductively valid arguments. If such a logic underlies our protocol, then conceding the premises of such an argument can cause a loss. Consider

P_1: *claim* p O_1: *why* p
P_2: *argue*$(q, q \supset p, so\ p)$ O_2: *concede* q, *concede* $q \supset p$

Now O is still to move, and her only legal moves are *concede*$(\{q, q \supset p\}$ *implies* $q)$ and *concede* p, after which moves P_1 is still *in* so O cannot move.

Example 2. The next example (on the Nixon diamond) shows that a player can lose with a poor move even if the player's own commitments support a valid counterargument. Suppose $Comms_0(p) = \{Qx \Rightarrow Px, Qn\}$ and consider

P_1: *claim* $\neg Pn$ O_1: *why* $\neg Pn$

P_2: *argue*$(Rn, Rx \Rightarrow \neg Px,$ *so* $\neg Pn)$ O_2: *concede* Rn, *concede* $Rx \Rightarrow \neg Px$,

              concede $\neg Pn$ (to P_1).

Now P wins while O could instead of conceding P_1 have attacked it with *argue*$(Qn, Qx \Rightarrow Px,$ *so* $Pn)$. Note also that if O had not conceded P_2's premises, then conceding $\neg Pn$ would have violated condition 8b on move legality.

Example 3. The next dispute shows that a player can sometimes use the other player's commitments against that player (the commitments are shown each time when they have changed).

Move	$Comms_D(P)$	$Comms_D(O)$
	$\{s \Rightarrow \neg q, r \wedge t \Rightarrow p\}$	$\{s \Rightarrow \neg q, r \wedge t \Rightarrow p\}$
P_1: *claim* p	$\{s \Rightarrow \neg q, r \wedge t \Rightarrow p, p\}$	
O_1: *why* p		
P_2: *argue*$(r, s, r \wedge s \Rightarrow p,$ *so* $p)$	$\{s \Rightarrow \neg q, r \wedge t \Rightarrow p, p,$ $r, s, r \wedge s \Rightarrow p\}$	
O_2: *concede* r, *argue*$(q, q \Rightarrow t, t \Rightarrow \neg s,$ *so* $\neg s)$		$\{s \Rightarrow \neg q, r \wedge t \Rightarrow p, q, r$ $q \Rightarrow t, t \Rightarrow \neg s, \neg s\}$

At this point, O's commitments justify p, since they contain an implicit argument for p. Suppose P next moves this argument. Then O can in turn use a counterargument supported by P's commitments.

P_3: *argue*$(r, q, q \Rightarrow t,$ $r \wedge t \Rightarrow p,$ *so* $p)$	$\{s \Rightarrow \neg q, r \wedge t \Rightarrow p, p,$ $r, s, r \wedge s \Rightarrow p,$ $r, t, q \Rightarrow t\}$	
O_3: *argue*$(s, s \Rightarrow \neg q,$ *so* $\neg q)$		$\{s \Rightarrow \neg q, r \wedge t \Rightarrow p, q, r$ $q \Rightarrow t, t \Rightarrow \neg s, \neg s\}$

And the dispute continues.

Example 4. Next I illustrate the construction of an L-dispute. I first list a *PPD*-dispute and then the construction of the associated L-dispute.

P_1: *claim* p O_1: *why* p

P_2: *argue*$(q, q \Rightarrow p,$ *so* $p)$ O_2: *argue*$(r, r \Rightarrow \neg p,$ *so* $\neg p)$

P_3: *argue*$(s, t, s \wedge t \Rightarrow p,$ *so* $p)$ (P_3 jumps back to O_1)

            O_3: *why* s

P_4: *argue*$(u, u \Rightarrow s,$ *so* $s)$ O_4: *argue*$(v, v \Rightarrow \neg u,$ *so* $\neg u)$

P_5: $argue(x, x \Rightarrow s, \text{ so } s)$ (P_5 jumps back to O_3)

O_5: $argue(y, y \Rightarrow \neg x, \text{ so } \neg x)$

P_6: $argue(z, z \Rightarrow \neg r, \text{ so } \neg r)$ (P_6 jumps back to O_2)

(O_6 jumps back to P_3) O_6: $why \ t$

P_7: $argue(k, k \Rightarrow t, \text{ so } t)$

The associated L-dispute is constructed as follows. The first two arguments are added with P_2 and O_2, so (denoting disputes with their last move and listing the replied-to moves between square brackets):

$$T(P_2) = P_2$$
$$T(O_2) = P_2, O_2[P_2]$$

So far, T contains just one dialectical tree. A second tree is created by P_3, which is an alternative *argue* reply to O's *why* attack on P's main claim. Hence

$$T(P_3) = P_2, O_2[P2], P_3$$

With P_4 the first modification of an argument in T takes place. P_3's argument is combined with P_4's argument for s (displayed with overloaded $\otimes$).

$$T(P_4) = P_2, O_2[P2], P_3 \otimes P_4$$

O_4 simply adds a new argument, which replies to P_3 as modified by P_4.

$$T(O_4) = P_2, O_2[P2], P_3 \otimes P_4, O_4[P_3 \otimes P_4]$$

P_5 splits the second tree in T into two alternative trees, by giving an alternative backwards extension of its root. Then O_5 simply extends the newly created tree, after which P_6 extends the first tree in T.

$$T(P_5) = P_2, O_2[P2], P_3 \otimes P_4, O_4[P_3 \otimes P_4], P_3 \otimes P_5$$
$$T(O_5) = P_2, O_2[P2], P_3 \otimes P_4, O_4[P_3 \otimes P_4], P_3 \otimes P_5, O_5[P_3 \otimes P_5]$$
$$T(P_6) = P_2, O_2[P2], P_6[O_2], P_3 \otimes P_4, O_4[P_3 \otimes P_4], P_3 \otimes P_5, O_5[P_3 \otimes P_5]$$

Finally, P_7 illustrates an interesting phenomenon. It replaces the second premise of O_3 with an argument; however, O_3 was already modified twice in two alternative ways with respect to its first premise, so P_7 actually modifies both of these modifications of O_3. This results in the following final L-dispute. (Note also that the targets of O_4 and O_5 have been replaced with their extended versions.)

P_2:	$q, q \Rightarrow p, \text{ so } p$
O_2:	$r, r \Rightarrow \neg p, \text{ so } \neg p \ [P_2]$
P_6:	$z, z \Rightarrow \neg r, \text{ so } \neg r \ [O_2]$
$P_3 \otimes P_4 \otimes P_7$:	$u, u \Rightarrow s, k, k \Rightarrow t, s \wedge t \Rightarrow p, \text{ so } p$
$P_3 \otimes P_5 \otimes P_7$:	$x, x \Rightarrow s, k, k \Rightarrow t, s \wedge t \Rightarrow p, \text{ so } p$
O_4:	$v, v \Rightarrow \neg u, \text{ so } \neg u \ [P_3 \otimes P_4 \otimes O_7]$
O_5:	$y, y \Rightarrow \neg x, \text{ so } \neg x \ [P_3 \otimes P_5 \otimes O_7]$

5   Discussion

Alternative Instantiations To discuss some alternative instantiations of the framework, note first that alternative definitions of winning may be possible, for instance, in terms of what is implied by the players' commitments. Secondly, as for maintaining consistency of one's commitments, some protocols allow inconsistency but give the other party the option to demand resolution of the conflict; a similar *resolve* move is possible if a commitment is explicitly retracted but still implied by the remaining commitments [8, 14]. Thus the burden of proving inconsistency or implicit commitment is placed upon the other party. Finally, as for replies to *why* moves, the obligation to reply to it with an argument for the challenged claim could be made dependent on questions of the burden of proof.

Features and Restrictions of the Framework The framework of this paper is flexible in some respects but restricted in some other respects. It is flexible, firstly, since it allows for different sets of speech acts, and different commitment rules, underlying logics and winning criteria. It is also 'structurally' flexible, in that it allows for backtracking, including jumping to earlier branches, and for postponing replies to move (even indefinitely if the move has become irrelevant). This flexibility is induced by the notion of relevance.

However, the framework also has some restrictions. For instance, the condition of relevance prevents the moving in one turn of alternative ways to change the status of the main claim. Further, the requirement that each move replies to a preceding move excludes some useful moves, such as lines of questioning in cross-examination of witnesses, with the goal of revealing an inconsistency in the witness testimony. Typically, such lines of questioning do not want to reveal what they are aiming at. The same requirement also excludes invitations to retract or concede [8, 14]. Finally, the framework only allows two-player disputes, leaving no room for, for example, arbiters or judges.

Related Research There have been some earlier proposals to combine formal dialogue systems with argumentation logics. Important early work was done by Loui [7], although he focussed less on speech act aspects. A major source of inspiration for the present research was Tom Gordon's model of civil pleading in anglo-american law [4] (cf. also [6]). Gordon presents a particular protocol rather than a framework. The same holds for a recent proposal in the context of multi-agent negotiation systems [1]. Finally, [3] shows how protocols for multi-party disputes can be formalised in situation calculus. Brewka focuses less on dialectical and relevance aspects but more on describing the 'current state' of a dispute and how it changes. His approach paves the way for, for instance, formal verification of consistency of protocols.

Conclusion This paper has presented a formal framework for persuasion dialogues with counterargument, and has given one detailed instantiation. Unlike earlier work, the framework is based on some general design principles, notably the distinction of attacking and surrendering replies to a move, and the notions of dialogical status and relevance of moves. The framework's instantiation also provided a still generic notion of an argument-counterargument dispute associated with a persuasion dialogue; I expect that this notion will provide a basis for investigating logical properties of the protocol, especially of its winning conditions.

Being a first attempt to provide a general framework, the focus of this paper has been more on definition than on technical exploration. Much work needs to be done on investigating its properties. In fact, one aim of this paper was to make this further work possible.

References

[1] L. Amgoud, N. Maudet, and S. Parsons. Modelling dialogues using argumentation. In *Proceedings of the Fourth International Conference on MultiAgent Systems*, Boston, MA, 2000.

[2] T.J.M. Bench-Capon, D. Lowes, and A.M. McEnery. Using Toulmin's argument schema to explain logic programs. *Knowledge Based Systems*, pages 177–183, 1991.

[3] G. Brewka. Dynamic argument systems: a formal model of argumentation processes based on situation calculus. *Journal of Logic and Computation*, 2000. To appear.

[4] T.F. Gordon. *The Pleadings Game. An Artificial Intelligence Model of Procedural Justice*. Kluwer Academic Publishers, Dordrecht/Boston/London, 1995.

[5] F. Grasso, A. Cawsey, and R. Jones. Dialectical argumentation to solve conflicts in advice giving: a case study in the promotion of healthy nutrition. *International Journal of Human-Computer Studies*, 2000. To appear.

[6] J.C. Hage, R.E. Leenes, and A.R. Lodder. Hard cases: a procedural approach. *Artificial Intelligence and Law*, 2:113–166, 1994.

[7] R.P. Loui. Process and policy: resource-bounded non-demonstrative reasoning. *Computational Intelligence*, 14:1–38, 1998.

[8] J.D. MacKenzie. Question-begging in non-cumulative systems. *Journal of Philosophical Logic*, 8:117–133, 1979.

[9] N. Maudet and D. Moore. Dialogue games for computer-supported collaborative argumentation. In *Proceedings of the Workshop on Computer-Supported Collaborative Argumentation for Learning Communities*, Stanford, 1999.

[10] S. Parsons, C. Sierra, and N.R. Jennings. Agents that reason and negotiate by arguing. *Journal of Logic and Computation*, 8:261–292, 1998.

[11] H. Prakken. Relating protocols for dynamic dispute with logics for defeasible argumentation. *Synthese*, 2000. To appear in special issue on New Perspectives in Dialogical Logics.

[12] H. Prakken and G.A.W. Vreeswijk. Logical systems for defeasible argumentation. In D. Gabbay, editor, *Handbook of Philosophical Logic*. Kluwer Academic Publishers, Dordrecht/Boston/London, 2000. Second edition, to appear.

[13] G.A.W. Vreeswijk. The computational value of debate in defeasible reasoning. *Argumentation*, 9:305–341, 1995.

[14] D.N. Walton and E.C.W. Krabbe. *Commitment in Dialogue. Basic Concepts of Interpersonal Reasoning*. State University of New York Press, Albany, NY, 1995.

Credulous and Sceptical Argument Games for Preferred Semantics

Gerard A.W. Vreeswijk and Henry Prakken

Institute of Information and Computing Sciences
Utrecht University, The Netherlands
{gv,henry}@cs.uu.nl

Abstract. This paper presents dialectical proof theories for Dung's preferred semantics of defeasible argumentation. The proof theories have the form of argument games for testing membership of *some* (credulous reasoning) or *all* preferred extensions (sceptical reasoning). The credulous proof theory is for the general case, while the sceptical version is for the case where preferred semantics coincides with stable semantics. The development of these argument games is especially motivated by applications of argumentation in automated negotiation, mediation of collective discussion and decision making, and intelligent tutoring.

1   Introduction

An important approach to the study of nonmonotonic reasoning is that of logics for defeasible argumentation (for an overview see [25]). Within this approach, a unifying perspective is provided by the work of [9] and [4] (below called the 'BDKT framework'). It takes as input a set of arguments ordered by a binary relation of 'attack', and it produces as output one or more 'argument extensions', which are maximal (in some sense) sets of arguments that survive the competition between all input arguments. A definition of argument extensions can be regarded as an *argument-based semantics* for defeasible reasoning. BDKT have developed various alternative such semantics, and investigated their properties and interrelations. They have also shown how many nonmonotonic logics can be recast in their framework. Thus their framework serves as a unifying framework not only for defeasible argumentation but also for nonmonotonic reasoning in general.

The BDKT framework exists in two versions. The version of [9] completely abstracts from the internal structure of arguments and the nature of the attack relation, while the version of [4] is more concrete. It regards arguments as sets of assumptions that can be added to a theory formulated in a monotonic logic in order to derive defeasible conclusions, and it defines attack in terms of a notion of contrariness of assumptions.

Besides a definition of argument-extensions, it is also important to have a test for extension membership of individual arguments, i.e., to have a *proof theory* for the semantics. A natural (though not the only) form of such proof theories

M. Ojeda-Aciego et al. (Eds.): JELIA 2000, LNAI 1919, pp. 239–253, 2000.

is the dialectical form of an *argument game* between a defender and challenger of an argument [18, 29, 8, 5, 26, 24, 14]. The defender starts with an argument to be tested, after which each player must attack the other player's arguments with a counterargument of sufficient strength. The initial argument is provable if its defender has a winning strategy, i.e., if he can make the challenger run out of moves in whatever way she attacks. The precise rules of the argument game depend on the semantics which the proof theory is meant to capture.

For [4]'s assumption-based version dialectical proof theories have been studied by [15]. However, for [9]'s abstract version only the so-called 'grounded (sceptical) semantics' has been recast in dialectical style, viz. by [8]. Grounded semantics is sceptical in the sense that it always induces a unique extension of admissible arguments: in case of an irresolvable conflict between two arguments, it leaves both arguments out of the extension. For the other semantics of [9], which in case of irresolvable conflicts all induce multiple extensions, dialectical forms must still be developed. This paper contributes to this development: it presents a dialectical argument game for perhaps the most important multiple-extension semantics of [9], so-called preferred semantics. In fact, we shall present two results: a proof theory for membership of *some* preferred extension (credulous reasoning) and the same for membership of *all* preferred extensions (*sceptical* reasoning, although only for the case where preferred semantics coincides with stable semantics).

It should be motivated why proof theories for the most abstract version of the BDKT framework are important besides their counterparts for the assumption-based version. Kakas & Toni's work is very relevant when arguments can be cast in assumption-based form. In many applications this is possible, but in other applications this is different. For instance, argumentation has been used as a component of negotiation protocols, where arguments for an offer should persuade the other party to accept the offer [16, 20]. Argumentation is also part of some recent formal models and computer systems for dispute mediation [10, 11, 6], and it has been used in computer programs for intelligent tutoring: for instance, in a system (Belvedere) that teaches scientific reasoning [27] and in systems that teach argumentation skills to law students, e.g. [1]'s CATO system and [28]'s ARGUE system. Now in many applications of these types, arguments have a structure that cannot be naturally cast in assumption-based form. For instance, they can be linked pieces of unstructured natural-language text (cf. Belvedere or Gordon's ZENO system), or they consist of analogical uses of precedents, such as CATO's arguments. It is especially for such applications that proof theories for [9]'s abstract framework are relevant.

It should also be motivated why a proof-theory for preferred semantics is important despite the pessimistic results on computational complexity recorded by [7]. To start with, these pessimistic results concern worst-case scenarios, and cases might be identified where computation of preferred semantics is still feasible. Moreover, as demonstrated by e.g. [21, 18], logics for defeasible argumentation provide a suitable basis for resource-bounded reasoning: dialogues corresponding to such logics can be interrupted at any time such that the intermediate outcome is still meaningful. Finally, there is a possible use of argument-based proof theories which does not suffer from the computational complexity, viz. in automated mediation and tutoring. In, for instance, mediation systems for negotiation or collective decision making, and also in systems for intelligent tutoring, the search for arguments and counterarguments is not performed by the computer, but by the users of the system, who input their arguments into the system during a discussion. In such applications the argument-based proof theory can be used as a protocol for dispute: it checks whether the users' moves are legal, and it determines given only the arguments constructed by the users, which of the participants in a dispute is winning. (See e.g. [23] for a logical study of this use of dialectical proof theories).

Finally, we must motivate why argument-game versions are important besides other argument-based proof theories, such as [21]'s proof theory for his system, which is based on preferred semantics. This has to do with applications in fields like mediation and tutoring. In these fields, argumentation has been used as a component of several computational dialogue systems based on speech acts, such as models of legal procedure, [10, 13, 3, 17], discourse generation systems [12], multi-agent negotiation systems [20, 2], and intelligent tutoring [19]. In our opinion, the dialectical form of an argument game is ideally suited for embedding in such dialogue systems (see [22] for a formal study of such embeddings).

The structure of this paper is as follows. In Section 2 we provide an overview of the basics of the BDKT framework. In Section 3 we discuss with the help of examples which features our argument games should have. Then we define the credulous argument game in Section 4 and the sceptical game in Section 5, after which we discuss some limitations in Section 6.

2   Definitions and Known Results

In this section we review the basics of the BDKT framework, as far as needed for present purposes. The input of the system is a set of arguments ordered by an attack relation.

Definition 1. *(Argument system [9]). An argument system $\mathcal{A}$ is a pair*

$$\mathcal{A} = \langle X, \leftarrow \rangle, \tag{1}$$

where X is a set of arguments, and $\leftarrow$ is a relation between pairs of arguments in X. The expression $a \leftarrow b$ is pronounced "a is attacked by b," "b is an attacker of a," or "b is a counterargument of a".

Example 1. The pair $\mathcal{A} = \langle X, \leftarrow \rangle$ with arguments

$$X = \{a, b, c, d, e, f, g, h, i, j, k, l, m, n, p, q\}$$

and $\leftarrow$ as indicated in Figure 1 is an (abstract) example of an argument system. It accommodates a number of interesting cases and anomalies, and will therefore be used as a running example throughout this paper.

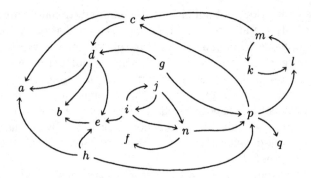

Fig. 1. Attack relations in the running example.

In practical applications it is necessary to further specify the internal structure of the arguments and the relation ←. See e.g. [30]. However, for the purpose of this paper it not necessary to do so; at present it suffices to know that there are arguments, and that some arguments attack other arguments.

The output of the system is one or more argument extensions, which are sets of arguments that represent a maximally defendable point of view. The different semantics of the BDKT framework define different senses of 'maximally defendable'. We list the definitions of two of them, stable and preferred semantics.

1. An argument a is *attacked* by a set of arguments B if B contains an attacker of a. (Not all members of B need attack a.)
2. An argument a is *acceptable* with respect to a set of arguments C, if every attacker of a is attacked by a member of C: for example, if $a \leftarrow b$ then $b \leftarrow c$ for some $c \in C$. In that case we say that c defends a, and also that C defends a.
3. A set S of arguments is *conflict-free* if no argument in S attacks an argument in S.
4. A conflict-free set S of arguments is *admissible* if each argument in S is acceptable with respect to S.
5. A set of arguments is a *preferred extension* if it is a $\subseteq$-maximal admissible set.
6. A conflict-free set of arguments is a *stable extension* if it attacks every argument outside it.

The following results of [9] will be used in the present paper.

Known results. *(from [9])*

1. *Each admissible set is contained in a $\subseteq$-maximally admissible set*
2. *Every stable extension is preferred.*
3. *Not every preferred extension is stable.*
4. *Stable extensions do not always exist; preferred extensions always exist.*
5. *Stable and preferred extensions are generally not unique.*

3   The Basic Ideas Illustrated

In this section we discuss with the help of examples which features our argument games should have.

Our game for testing membership of *some* extension is based on the following idea. By definition, a preferred extension is a $\subseteq$-maximal admissible set. It is known that each admissible set is contained in a maximal admissible set, so the procedure comes down to trying to construct an admissible set 'around' the argument in question. If this succeeds we know that the admissible set, and hence the argument in question, is contained in a preferred extension.

Suppose now we wish to investigate whether a is preferred, i.e., belongs to a preferred extension. We know that it suffices to show that the argument in question is admissible. The idea is to start with $S = \{a\}$, which most likely is not admissible. (Because S is small, and small sets are usually conflict-free but not admissible.) So other arguments must be found (or constructed) in order to complete S into an admissible set.

Procedure. *(Constructing an admissible set). Let a be an argument for which we try to construct an admissible set. This task can best be divided in two sub-tasks:*

Task 1: *Let us suppose this task is performed by person **PRO**, who assumes*
construc- *a constructive role by trying to show that a is contained in an ad-*
tion. *missible set. To this end, **PRO** examines if there are arguments that attack his arguments constructed thus far. If there is such an argument, **PRO** tries to attack it by trying to construct an argument that attacks the original attacker (acceptability). If **PRO** has found such an argument, it must be consistent with his previous arguments (conflict-freeness).*

***PRO**'s role is purely defensive: his goal is to incorporate defenders against attacks constructed thus far — not to extend his collection of arguments per sé. To the contrary, in fact: **PRO**'s goal is to keep his collection of arguments as small as possible, because **PRO** is more vulnerable if he (or she)[1] has more arguments to defend.*

Task 2: *This task is performed by person **CON**, who assumes a critical role*
criti- *by trying to find counterarguments to arguments advanced by **PRO**.*
cism. *In a way, **CON**'s aim is to 'make **PRO** talk' in the sense that **PRO** is more vulnerable if he has more arguments to defend.*

*The procedure formulated here is not necessarily adversarial: one way to look at it is to say that **CON** helps **PRO** by attending him to arguments that might invalidate **PRO**'s collection of admissible arguments.*

---

[1] From here on we will use the generic masculine form, intending no bias.

Example 2. (Straight failure). Consider the argument system that was presented at the beginning of this paper. Suppose **PRO**'s task is to show that a is preferred. Since preferred extensions are maximally admissible sets it suffices for **PRO** to show that a is admissible, i.e., that a is contained in an admissible set.

The first action of **PRO** is simply putting forward a:

If a can't be criticized, i.e., if there are no attackers, then $S = \{a\}$ is admissible, and **PRO** succeeds. However, since $a \leftarrow h$,

CON forwards h:

Now it is up to **PRO** to defend a by finding arguments against h. There are no such arguments, so that **PRO** fails to construct an admissible set 'around' a. So a is not admissible, hence not preferred.

Example 3. (Straight success). Suppose that **PRO** wants to show that b is admissible.

The first action of **PRO** is putting forward b:

CON attacks b with d:

PRO defends this attack with g:

Since **CON**'s attack on b
with d has failed, **CON**
returns to b and attacks it
again, this time with e:

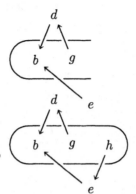

PRO defends b again, this
time with h. Since **CON** is
unable to find other
argument against b, g or h,
PRO may now close S:

Example 4. (Even loop success). Suppose that **PRO** wants to show that f is
admissible.

The first action of **PRO** is
putting forward f:

CON attacks f with n:

PRO defends this attack
with i:

CON attacks i with j:

PRO defends i with i itself
(so that i is self-defending).
CON is unable to put
forward other arguments
that attack f or i so that
PRO closes S:

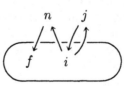

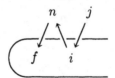

This example shows that **PRO** must be allowed to repeat his arguments, while
CON must be forbidden to repeat **CON**'s arguments (at least in the same 'line
of dispute'; see further below)

Example 5. (Odd loop failure). Suppose that **PRO** wants to show that m is admissible.

The first action of **PRO** is putting forward m:

CON attacks m with l:

PRO defends this attack with p:

CON attacks p with h:

PRO backtracks and removes p from S. He then tries to defend l with k instead:

CON attacks k with m (and, as a bonus, introduces an inconsistency in S):

PRO has no other arguments in response to l and m, so that he is unable to close S into an admissible set. So m is not contained in an admissible set. Note that we cannot allow **PRO** to reply to m with l, since otherwise the set that **PRO** is constructing 'around' m is not conflict-free, hence not admissible. So we must forbid **PRO** to repeat **CON**'s moves. On the other hand, this example also shows that **CON** should be allowed to repeat **PRO**'s moves, since such a repetition reveals a conflict in **PRO**'s position.

Example 6. (The need for backtracking). Consider next an argument system with five arguments a, b, c, d and e and attack relations as shown in the graph.

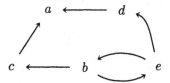

This example shows that we must allow **CON** to backtrack. Suppose **PRO** starts with a, **CON** attacks a with d, and **PRO** defends a with e. If **CON** now attacks e with b, **PRO** can defend e by repeating e itself. However, **CON** can backtrack to a, this time attacking it with c, after which **PRO**'s only move is defending a with b. Then **CON** can repeat **PRO**'s move e, revealing that **PRO**'s position is not conflict-free.

Repetition Let us summarise our observations about repetition of moves. If **PRO** can defend an argument by using one of his previous arguments that is not backtracked, then should **PRO** do that? Further, does it make sense for **PRO** to repeat arguments advanced by **CON**? The same questions can be asked for repetitions by **CON**.

   i. It makes sense for **PRO** to repeat itself (if possible), because **CON** might fail to find or produce a new attacker against **PRO**'s repeated argument. If so, then **PRO**'s repetition closes a cycle of even length, of which **PRO**'s arguments are admissible.
  ii. **CON** should repeat **PRO** (if possible), because it would show that **PRO**'s collection of arguments is not conflict-free.
 iii. **PRO** should not repeat **CON**, because it would introduce a conflict into **PRO**'s own collection of arguments.
  iv. It does not make sense if **CON** repeats itself, because **PRO** has already shown to have adequate defense for **CON**'s previous arguments.

Finally, we show that **CON** should be allowed to repeat **CON**'s arguments when they are from different 'lines' of a dispute. A *dispute line* is a dispute where each move replies to the immediately preceding move; i.e., in a dispute line no backtracking is allowed.

Example 7. (repetition from different lines)

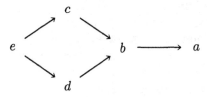

Suppose **PRO** starts a dispute for a and **CON** attacks a with b. Then **PRO** has two alternative ways to defend a, viz. with c and with d, but **CON** must be allowed to reply to each of them with e.

4   The Credulous Argument Game Defined

We now turn to the formal definition of our argument games, starting with the credulous game. During a dispute a tree of dispute lines is constructed. This can be illustrated with the following format of disputes, taken from [29].

Example 2:
1. | **PRO** : a
2. || **CON** : $h\ddagger$

Example 3:
1. | **PRO** : b
2. || **CON** : d
3. ||| **PRO** : $g\dagger$
4. || **CON** : e
5. ||| **PRO** : $h\ddagger$

Example 4:
1. | **PRO** : f
2. || **CON** : n
3. ||| **PRO** : i
4. |||| **CON** : j
5. ||||| **PRO** : i (iv)

Example 5:
1. | **PRO** : m
2. || **CON** : l
3. ||| **PRO** : p
4. |||| **CON** : $h\dagger$
5. ||| **PRO** : k
6. |||| **CON** : m (iii)

The vertical bars "|||" indicate the *level* of the dispute, i.e., the depth of the tree. E.g., in Ex. 3, **PRO** responded to a response of **CON** (level 3), after which **CON** backtracks (level 2) to try a new argument against b.

The "$\dagger$"-symbol means that the player cannot respond to the last argument of the other player, while the "$\ddagger$"-symbol means that the player is unable to respond to all arguments of the other player presented thus far. A number in the range (i-iv) means that a next move of the player would make no sense on the basis of the corresponding repetition guideline.

Rules and Correspondence   To establish a precise correspondence between disputes and preferred extensions, it is necessary to make the terminology more precise and to define the rules under which a dispute is conducted.

- A *move* is simply an argument (if the first move) or else an argument attacking one of the previous arguments of the other player.
- Both parties can *backtrack*.
- An *eo ipso* (meaning: "you said it yourself") is a move that uses a previous non-backtracked argument of the other player.
- A *block* is a move that places the other player in a position in which he cannot move.
- A *two-party immediate response dispute* (TPI-dispute) is a dispute in which both parties are allowed to repeat **PRO**, in which **PRO** is not allowed to repeat **CON**, and in which **CON** is allowed to repeat **CON** iff the second use is in a different line of the dispute. **CON** wins if he does an *eo ipso* or blocks **PRO**. Otherwise, **PRO** wins.

A main argument of a TPI-dispute is *defended* if the dispute is won by **PRO**.

Proposition 1. (Soundness and completeness of the credulous game). *An argument is in some preferred extension iff it can be defended in every TPI-dispute.*

Proof. By definition of preferred extensions it suffices to show that an argument is admissible iff it can be defended in every dispute.

First suppose that a can be defended in every dispute. This includes disputes in which **CON** has opposed optimally. Let us consider such a dispute. Let A be the arguments that **PRO** used to defend a. (in particular $a \in A$.) If A is not conflict-free then $a_i \leftarrow a_j$ for some $a_i, a_j \in A$, and **CON** would have done an *eo ipso*, which is not the case. If A is not admissible, then $a_i \leftarrow b$ for some $a_i \in A$ while $b \not\leftarrow A$. In that case, **CON** would have used b as a winning argument, which is also not the case. Hence A is admissible.

Conversely, suppose that $a \in A$ with A admissible. Now **PRO** can win every dispute by starting with a, and replying with arguments from A only. (**PRO** can do this, because all arguments in A are acceptable wrt A.) As long as **PRO** picks his arguments from A, **CON** cannot win by *eo ipso*, because A is conflict-free. So a can be defended in dispute.

5   The Sceptical Argument Game Defined

Above, **PRO** tries to show that the main argument is contained in a preferred set. This is known as *credulous* reasoning. If **PRO** wishes to verify whether the main argument is contained in *all* preferred sets, then **PRO** does *sceptical* reasoning. Before defining an argument game for this kind of reasoning, we must first explain why for sceptical reasoning it is relevant to study preferred semantics besides [9]'s grounded semantics, which is also meant for sceptical reasoning. The reason is that grounded semantics is too weak to capture certain types of sceptical conclusions.

Example 8. (Floating arguments.) Consider the arguments a, b, c and d with the attack relations as shown in the picture.

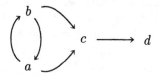

Since no argument is unattacked, the grounded extension is empty. However, this example has two preferred extensions, $\{a, d\}$ and $\{b, d\}$, and both of them contain d.

Next we illustrate that there are cases where an argument system has a unique preferred extension but not all of its elements are contained in the grounded extension.

Example 9. Consider four arguments a, b, c and d with the attack relations as shown in the picture.

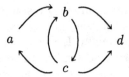

The unique preferred extension is $\{c\}$, so c is sceptically preferred, but the grounded extension is empty, since none of the arguments are unattacked.

We now define the sceptical argument game. A result for sceptical reasoning can be obtained by observing that a dispute is symmetric, since **CON** also may be given the task to construct an admissible set, viz. for the attackers he uses. If **CON** succeeds, he has shown that there exists at least one admissible set not including the main argument.

Proposition 2. (Soundness and completeness of the sceptical game). *In argument systems where each preferred extension is also stable, an argument is in all preferred extensions iff it can be defended in every TPI-dispute, and none of its attackers can be defended in every TPI-dispute.*

Proof. This result can be proven on the basis of the previous proposition, and by the fact that a stable extension attacks every argument outside it.

Consider any argument system where all preferred extensions are stable. For the only-if-part of the equivalence, consider any argument a that is in all preferred extensions. Then (by assumption that these extensions are also stable) all attackers of a are attacked by all such extensions, so by conflict-freeness of preferred extensions, none of these attackers is in any such extension. But then none of a's attackers is credulously provable.

For the if part, Let a be any argument that is credulously provable and such that none of a's attackers are credulously provable. Then none of these attackers is in any preferred extension, so (by assumption that these extensions are also stable) they are attacked by all such extensions. But then a is defended by all these extensions, so they all contain a.

The following example shows that this result does not hold in general.

Example 10. Consider:

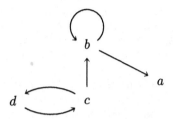

a is contained in one preferred extension, viz. $E_1 = \{a, c\}$, but not in the other preferred extension, which is $E_2 = \{d\}$. Note that the self-attacking argument b prevents a from being a member of E_2 although b is not itself a member. The problem is that E_2 does not attack b so that a is not acceptable with respect to E_2. This situation cannot arise when all preferred extensions are stable, since then they attack all arguments outside them.

6   Discussion

The present paper has provided simple and intuitive argument games for both credulous and sceptical reasoning in preferred semantics. However, there are still some limitations and drawbacks.

A limitation is, of course, that the sceptical game is not sound and complete in general. A first drawback is the fact that the sceptical game actually consists of two parallel games, which is less elegant in applications in mediation and tutoring systems. In future research we hope to improve the games in both respects.

Another drawback is that in some cases proofs are infinite. This is obvious when an argument has an infinite number of attackers, but even otherwise some proofs are infinite, as in the following example.

Example 11. (Infinite attack chain.) Consider an infinite chain of arguments $a_1, \ldots, a_n, \ldots$ such that a_1 is attacked by a_2, a_2 is attacked by a_3, and so on.

$$a_1 \longleftarrow a_2 \longleftarrow a_3 \longleftarrow a_4 \longleftarrow a_5 \longleftarrow \cdots$$

PRO can win a game for a_1 (or for any other argument) since **CON** is never able to move a block, but **PRO** neither has a blocking move available.

Nevertheless, it is easy to verify that with a finite set of arguments all proofs are finite.

References

[1] V. Aleven and K.D. Ashley. Evaluating a learning environment for case-based argumentation skills. In *Proceedings of the Sixth International Conference on Artificial Intelligence and Law*, pages 170–179, New York, 1997. ACM Press.

[2] L. Amgoud, N. Maudet, and S. Parsons. Modelling dialogues using argumentation. In *Proceedings of the Fourth International Conference on MultiAgent Systems*, Boston, MA, 2000.

[3] T.J.M. Bench-Capon. Specification and implementation of Toulmin dialogue game. In *Legal Knowledge-Based Systems. JURIX: The Eleventh Conference*, pages 5–19, Nijmegen, 1998. Gerard Noodt Instituut.

[4] A. Bondarenko, P.M. Dung, R.A. Kowalski, and F. Toni. An abstract, argumentation-theoretic approach to default reasoning. *Artificial Intelligence*, 93(1-2):63–101, 1997.

[5] G. Brewka. A reconstruction of Rescher's theory of formal disputation based on default logic. In A. Cohn, editor, *Proceedings of the 11th European Conference on Artificial Intelligence*, pages 366–370. John Wiley & Sons, Ltd., 1994.

[6] G. Brewka. Dynamic argument systems: a formal model of argumentation processes based on situation calculus. *Journal of Logic and Computation*, 2000. To appear.

[7] Y. Dimopoulos, B. Nebel, and F. Toni. Preferred arguments are harder to compute than stable extensions. In T. Dean, editor, *Proc. of the 16th Int. Joint Conf. on Artificial Intelligence (IJCAI99)*, pages 36–41. Morgan Kaufmann, 1999.

[8] P.M. Dung. Logic programming as dialog game. Unpublished paper, Division of Computer Science, Asian Institute of Technology, Bangkok, 1994.

[9] P.M. Dung. On the acceptability of arguments and its fundamental role in nonmonotonic reasoning, logic programming, and n-person games. *Artificial Intelligence*, 77(2):321–357, 1995.

[10] T.F. Gordon. *The Pleadings Game: An Artificial Intelligence Model of Procedural Justice*. Kluwer, Dordrecht, 1995. Revised edition of the author's Ph.D. thesis (same title), Technische Hochschule, Darmstadt, 1993.

[11] T.F. Gordon, N. Karaçapilidis, H. Voss, and A. Zauke. Computer-mediated cooperative spatial planning. In H. Timmermans, editor, *Decision Support Systems in Urban Planning*, pages 299–309. E & FN SPON Publishers, London, 1997.

[12] F. Grasso, A. Cawsey, and R. Jones. Dialectical argumentation to solve conflicts in advice giving: a case study in the promotion of healthy nutrition. *International Journal of Human-Computer Studies*, 2000. To appear.

[13] J.C. Hage, R.E. Leenes, and A.R. Lodder. Hard cases: a procedural approach. *Artificial Intelligence and Law*, 2:113–166, 1994.

[14] H. Jakobovits and D. Vermeir. Dialectic semantics for argumentation frameworks. In *Proceedings of the Seventh International Conference on Artificial Intelligence and Law*, pages 53–62, New York, 1999. ACM Press.

[15] A.C. Kakas and F. Toni. Computing argumentation in logic programming. *Journal of Logic and Computation*, 9:515–562, 1999.

[16] S. Kraus, K. Sycara, and A. Evenchik. Reaching agreements through argumentation: a logical model and implementation. *Artificial Intelligence*, 104:1–69, 1998.

[17] A.R. Lodder. *DiaLaw: On Legal Justification and Dialogical Models of Argumentation*. Kluwer, Dordrecht, 1999.

[18] R.P. Loui. Process and policy: Resource-bounded nondemonstrative reasoning. *Computational Intelligence*, 14(1):1–38, 1998.

[19] N. Maudet and D. Moore. Dialogue games for computer-supported collaborative argumentation. In *Proceedings of the Workshop on Computer-Supported Collaborative Argumentation for Learning Communities,* Stanford, 1999.

[20] S. Parsons, C. Sierra, and N.R. Jennings. Agents that reason and negotiate by arguing. *Journal of Logic and Computation,* 8:261–292, 1998.

[21] J.L. Pollock. *Cognitive Carpentry. A Blueprint for How to Build a Person.* MIT Press, Cambridge, MA, 1995.

[22] H. Prakken. On dialogue systems with speech acts, arguments, and counterarguments. 2000. These proceedings.

[23] H. Prakken. Relating protocols for dynamic dispute with logics for defeasible argumentation. *Synthese,* 2000. To appear in special issue on New Perspectives in Dialogical Logics.

[24] H. Prakken and G. Sartor. Argument-based extended logic programming with defeasible priorities. *Journal of Applied Non-classical Logics,* 7:25–75, 1997.

[25] H. Prakken and G.A.W. Vreeswijk. Logical systems for defeasible argumentation. To appear in D. Gabbay (ed.), Handbook of Philosophical Logic, 2nd edition. Kluwer Academic Publishers, Dordrecht etc., 2000.

[26] G.R. Simari, C.I. Chesñevar, and A.J. Garcia. The role of dialectics in defeasible argumentation. In *Proceedings of the XIV International Conference of the Chilean Computer Science Society,* 1994.

[27] D. Suthers, A. Weiner, J. Connelly, and M. Paolucci. Belvedere: engaging students in critical discussion of science and public policy issues. In *Proceedings of the Seventh World Conference on Artificial Intelligence in Education,* pages 266–273, 1995.

[28] H.B. Verheij. Automated argument assistance for lawyers. In *Proc. of the Seventh Int. Conf. on Artificial Intelligence and Law,* pages 43–52, New York, 1999. ACM.

[29] G.A.W. Vreeswijk. Defeasible dialectics: A controversy-oriented approach towards defeasible argumentation. *The Journal of Logic and Computation,* 3(3):3–27, 1993.

[30] G.A.W. Vreeswijk. Abstract argumentation systems. *Artificial Intelligence,* 90:225–279, 1997.

A General Approach to Multi-agent Minimal Knowledge

Wiebe van der Hoek[1], Jan Jaspars[2], and Elias Thijsse[*3]

[1] Computer Science, Utrecht University
wiebe@cs.uu.nl
[2] Free Lance Logician, Amsterdam
jaspars@wins.uva.nl
[3] Computational Linguistics & AI, Tilburg University
thysse@kub.nl

Abstract. We extend our general approach to characterizing information to multi-agent systems. In particular, we provide a formal description of an agent's knowledge containing exactly the information conveyed by some (honest) formula φ.

Only knowing is important for dynamic agent systems in two ways. First of all, one wants to compare different states of knowledge of an agent and, secondly, for agent a's decisions, it may be relevant that (he knows that) agent b does not know more than φ.

There are three ways to study the question whether a formula φ can be interpreted as minimal information. The first method is semantic and inspects 'minimal' models for φ (with respect to some order $\leq$ on states). The second one is syntactic and searches for stable expansions, minimal with respect to some language $\mathcal{L}^*$. The third method is a deductive test, known as the disjunction property. We present a condition under which the three methods are equivalent.

Then, we show how to construct the order $\leq$ by collecting 'layered orders'. We then focus on the multi-agent case and identify languages $\mathcal{L}^*$ for several orders $\leq$, and show how they yield different notions of honesty for different multi-modal systems. Finally, some consequences of the different notions are discussed.

Classification. Knowledge representation, Non-classical logics.

1 Introduction

What is a knowledge state? To answer this question, we give a general approach to characterizing information in a modal context. In particular, we want to obtain a formal description of an agent's knowledge containing *exactly*, that is, at least

---

* This author's research partly took place during his sabbatical at CSLI, Stanford USA, which is gratefully acknowledged. He also thanks Michael Dunn (Indiana University Logic Group) and Arnis Vilks (Handelshochschule Leipzig) for kind invitations and useful comments.

M. Ojeda-Aciego et al. (Eds.): JELIA 2000, LNAI 1919, pp. 254–268, 2000.

but not more than the information conveyed by some formula φ, in other words, the case in which φ is the agent's *only* knowledge. Characterizing an agent's exact knowledge state is important in dynamic agent systems in several ways. First of all, when the system evolves, one might wish to compare the different states of one agent: which actions (or, more specifically moves) optimally extend his knowledge? Secondly, in multi-agent systems, agent a may wish to be sure that *all that* b knows is φ, and exploit the fact that b does *not know more than that*. Finally, when such agents start to *exchange information*, they must be aware of principles governing their communication: Usually utterances are intended to convey minimal knowledge with respect to some domain (Grice's maxim of quantity).

Formulas φ representing all that the agent knows, are called *honest*. For the one-agent case, some observations about only knowing and honesty are well-accepted. For instance, where purely objective formulas are rendered honest, a typical example of a dishonest formula is $\varphi = (\Box p \vee \Box q)$: if an agent claims to only know φ, he would know something that is stronger than φ (i.e., either $\Box p$ or $\Box q$). A more sophisticated analysis of honesty generally depends on the epistemic background logic. What is especially important here, is which intro-spective capacities we are ready to attribute to the agent. For example, if the background logic contains the axiom of *positive* introspection $\Box\psi \rightarrow \Box\Box\psi$ we can infer $\Box\Box p$ if only p is known. This seems innocent since the inferred knowledge is still related to the initial description p. On the other hand, if we accept the axiom of *negative* introspection $\neg\Box\psi \rightarrow \Box\neg\Box\psi$, then we can infer knowledge concerning q, for example $\Box\neg\Box q$, from only knowing p. This knowledge cannot be derived from only knowing $p \wedge q$, which intuitively represents *more* knowledge than only knowing p. As we stressed in [6], this kind of inferences effects the treatment of honesty for different modal systems.

For the multi-agent case, intuition seems to be much less clear. Of course, where objective formulas are all honest in the one agent case, this property is easily convertible to formulas with no operator $\Box_a$, when considering honesty for agent a. Hence, a can honestly claim to only know $\Box_b p \vee \Box_b q$, for $b \neq a$. But if $\Box_a$ re-occurs in the scope of $\Box_b$, the resulting formula $\Box_b p \vee \Box_b \Box_a q$ becomes *dishonest* again if $\Box_b$ represents knowledge. With mixed operators, in particular in the presence of negation, matters soon get fuzzy.

Studies of 'only knowing' ([3,11]) and 'all I know' ([8]) have largely been re-stricted to particular modal systems, such as **S5**, **S4** and **K45**. Recently Halpern [2] has also taken other modal systems such as **K**, **T** and **KD45** into account. Although his approach suggests similar results for e.g. **KD4**, in [6] we adopted a more general perspective: *given any modal system, how to characterize the min-imal informational content of modal formulas*. For *multi-agent only knowing*, we only know of a (more or less) general approach by Halpern ([2]), putting a no-tion of 'possibility' to work on tree models and, for the **S5**$_m$ case, enriching the language with modal operators Q_i^ζ, for any formula ζ and agent i.

In this paper, besides *arbitrary* normal multi-modal systems we prefer to use standard Kripke models, instead of Fagin and Vardi's knowledge structures, and Halpern's tree models. We try to obtain this general view by putting our framework of [6] to work for the multi-agent case. In order to appreciate this fully, the reader has to realize that, in general, there are three ways to study the question whether a formula φ allows for a minimal interpretation (if $\Box\psi$ allows for such a minimal interpretation, ψ is called honest), and, if so, what can be said about the consequences of φ under this interpretation.

The first approach is a semantic one: Given a formula φ, try to identify models for φ that carry the least information. This approach requires a suitable order $\leq$ between states (i.e., model-world pairs) in order to identify minimal (or, rather, *least*) elements. For the simple (universal) S5-models the order coincides with the superset-relation between sets of worlds. Our challenge here is to give a general definition of such an order, which suits any multi-modal system. The second approach is mainly syntactic in nature and presupposes a sublanguage $\mathcal{L}^*$ of 'special' formulas. Given a consistent formula φ, we then try to find a maximally consistent set containing φ with a smallest $\mathcal{L}^*$-part. This approach can be identified as the search for so-called stable expansions, which are related to maximally consistent sets in a straightforward way. The last approach is purely deductive, and is also known as the disjunction property (DP): φ allows for a minimal interpretation if for any disjunction in $\mathcal{L}^*$ that can be derived from φ, one disjunct is derivable from φ.

In [6], we were able to formulate a condition under which the three approaches mentioned above are equivalent. This paves the way to focus on defining 'suitable' orders on information states in a general way, rather than trying to establish the equivalences of the characterizations for specific orders, again and again. The information orders on states that we consider are induced by *layered orders* $\leq_n$ between states, where n settles the depth of the equivalence.

For the one-agent case, we obtained minimality results with respect to the following languages (by considering appropriate orders on states):

- $\mathcal{L}^* = \Box\mathcal{L}$, where $\mathcal{L}$ is the full modal language (*general honesty*). However, it appears that under this choice, almost every formula is honest in the systems **K**, **K4**, **KD** and **KD4**. On the other hand, for many other systems there are no honest formulas. So for most systems, the notion of general honesty is trivial: all or no formulas are honest.
- $\mathcal{L}^* = \Box\mathcal{L}^+$, where $\mathcal{L}^+$ is the modal language where no $\Box$ occurs in the scope of a negation (*positive honesty*). For **S5**, the corresponding notion of honesty coincides with the approach in [3]. Moreover, for all systems except **K**, **KD**, **K4** and **KD4**, this notion of honesty is not trivial.

For the multi-agent case, there are many more options. Generalizing the first language above gives rise to a notion of honesty which encounters, *mutatis mutandis*, the same problem of trivialization as states for the one agent case. The second, so-called positive language can be generalized in different ways, which for most systems lead to nontrivial notions of honesty. The anomalous cases are still the *weak doxastic logics* (generalizing an observation of Halpern in [2]): in $\mathbf{KD_m}$ and $\mathbf{KD4_m}$ all formulas φ for which $\Box_a\varphi$ is consistent, are honest; in $\mathbf{K_m}$ and $\mathbf{K4_m}$, even all formulas are honest. This means that in $\mathbf{KD4_m}$, for example, agent a can honestly claim that he only knows whether he knows p, which we believe to be counter-intuitive.

In this paper, we present three generalizations of positive honesty: one corresponds to the 'a-objective' language (the formulas do not contain $\Box_a$), one to the 'a-positive' language (no $\Box_a$ in the scope of negation), and one that combines these two.

2 Modal Logical Preliminaries

Let us agree on some technicalities. Our multi-modal language $\mathcal{L}$ or $\mathcal{L}^A$ has finitely many modal operators $\Box_1, \Box_2, \ldots, \Box_m$, over a finite set of atoms $\mathcal{P} = \{p, q, r \ldots\}$, using the classical connectives $\neg$, $\wedge$ and $\vee$. Here $\{1, 2, \ldots, m\}$ encodes the set of agents $\mathcal{A}$. We use a for an arbitrary agent in $\mathcal{A}$ on which we focus. The operator $\Box_a$ denotes "Agent a has the information that ...", which may involve knowledge, belief or any other propositional attitudes. The dual modal operators $\Diamond_1, \Diamond_2, \ldots, \Diamond_m$ are introduced by definition: $\Diamond_a \varphi = \neg \Box_a \neg \varphi$. Given a set of formulas Γ, we define a's knowledge about Γ by $\Box_a \Gamma = \{\Box_a \varphi \mid \varphi \in \Gamma\}$ and a's knowledge in Γ by $\Box_a^- \Gamma = \{\varphi \mid \Box_a \varphi \in \Gamma\}$.

A measure of modal complexity of formulas, called *modal depth*, has the usual recursive definition: $d(p) = 0$ (for $p \in \mathcal{P}$), $d(\neg \varphi) = d(\varphi)$, $d(\varphi \wedge \psi) = d(\varphi \vee \psi) = \max\{d(\varphi), d(\psi)\}$ and $d(\Box_i \varphi) = d(\varphi) + 1$. We often consider the sublanguage of formulas of limited modal depth: $\mathcal{L}_{(n)} = \{\varphi \in \mathcal{L} \mid d(\varphi) \leq n\}$. So, $\mathcal{L}_{(0)}$ is the purely propositional subset of $\mathcal{L}$ (void of modal operators). Other sublanguages of interest will be defined in the sequel.

We use multi-modal Kripke models $\langle W, R_1, \ldots, R_m, V \rangle$ or $\langle W, \boldsymbol{R}, V \rangle$ to interpret $\mathcal{L}$; here $w R_a v$ or $v \in R_a[w]$ means that given world w, world v is an epistemic alternative to a. Truth is relative to a model-world pair ('state', for short). The connectives $\neg$, $\wedge$ and $\vee$ are interpreted as usual; the modal operators also get the classical interpretation: $M, w \models \Box_a \varphi$ iff for all $v \in R_a[w] : M, v \models \varphi$. The theory of a state $\langle M, w \rangle$ is $\mathrm{Th}(M, w) = \{\varphi \mid M, w \models \varphi\}$. If the model is obvious from the context, we will omit it and simply write $w \models \varphi$. Consequence is defined relative to a given set of models $\mathcal{S}$: $\Gamma \models_\mathcal{S} \varphi$ iff $M, w \models \varphi$ for all $M \in \mathcal{S}$ s.t. $M, w \models \Gamma$. States are assumed to be related by what we call an *information order* $\leq^a$ for any agent a; for the time being $\leq^a$ is only required to be a pre-order (i.e. reflexive and transitive). A major question is which formulas are preserved moving from w to w' if $w \leq^a w'$. It will prove important to single out so-called *persistent* sublanguages of such formulas, in particular those that are rich enough to reversely characterize the information order.

The inference relation $\vdash$ is obtained relative to a modal system $\mathbf{S}$, which at least contains classical propositional logic and the rule defining the minimal system $\mathbf{K}$: $\Gamma \vdash \varphi \Rightarrow \Box_a \Gamma \vdash \Box_a \varphi$. Formulas φ and ψ are *equivalent in* $\mathbf{S}$, if both $\varphi \vdash_\mathbf{S} \psi$ and $\psi \vdash_\mathbf{S} \varphi$. The logics $\mathbf{S}$ that we consider have the nice property that $\mathcal{L}_{(n)}$ is *finitary*: since $\mathcal{P}$ is finite, $\mathbf{S}$ induces only finitely many equivalence classes. A set Γ is $\mathbf{S}$-consistent if for some φ: $\Gamma \nvdash_\mathbf{S} \varphi$; Γ is maximal $\mathbf{S}$-consistent ($\mathbf{S}$-m.c.)

if **S**-consistent, though it cannot properly be extended to a larger **S**-consistent set. A formula φ is a theorem of **S** if $\emptyset \vdash_{\mathbf{S}} \varphi$, also written as $\vdash_{\mathbf{S}} \varphi$.

Here are some familiar axioms and their corresponding condition on the accessibility relation: **T** $\Box_a\varphi \to \varphi$ (*reflexivity*); **D** $\Box_a\varphi \to \Diamond_a\varphi$ (*seriality*); **4** $\Box_a\varphi \to \Box_a\Box_a\varphi$ (*transitivity*); **5** $\Diamond_a\varphi \to \Box_a\Diamond_a\varphi$ (*Euclidicity*); **B** $\varphi \to \Box_a\Diamond_a\varphi$ (*symmetry*); **G** $\Diamond_a\Box_a\varphi \to \Box_a\Diamond_a\varphi$ (*confluence*).

Axiom **4** is also known as "positive introspection"; axiom **5** in its equivalent form $\neg\Box_a\varphi \to \Box_a\neg\Box_a\varphi$ is known as "negative introspection". Unimodal systems (involving only one agent's modality) are characterized by their constituting rules and axioms: **K**, **KD**, **KD4**, **KD45**, etc. If **S** is a standard modal system, then **S**$_m$ is its m-agent counterpart. A state verifies a logic **S** if it verifies all the theorems of **S**.

3 Minimal Information in Multi-modal Logic

Suppose we have an information order $\leq^a$ on states. When do we consider the information φ to be minimal for agent a? We suggest that φ constitutes *minimal information for a*, or that φ is *a-honest*, if $\Box_a\varphi$ is true in a least state $\langle M, w \rangle$.

Definition 1. *A formula φ is a-honest with respect to* **S** *and $\leq^a$ iff there is an* **S**-*state $\langle M, w \rangle$ such that $M, w \models \Box_a\varphi$ and*

$$M', w' \models \Box_a\varphi \Rightarrow M, w \leq^a M', w' \text{ for all } \mathbf{S}\text{-states } \langle M', w' \rangle.$$

This characterization of minimal information may however not always be convenient. In some cases one would prefer a syntactic characterization, a deductive test, or a combination of these. This can be achieved by relating the information order $\leq^a$ to a proper sublanguage $\mathcal{L}^a$ through *persistence* [for all $\varphi \in \mathcal{L}^a$: $M, w \leq^a M', w' \Rightarrow (M, w \models \varphi \Rightarrow M', w' \models \varphi)$] and a converse of this, called *characterization* [for all $\varphi \in \mathcal{L}^a (M, w \models \varphi \Rightarrow M', w' \models \varphi) \Rightarrow M, w \leq^a M', w'$]. We are now able to propose alternative approaches to minimality:

(1) Formula φ has a $\leq^a$-least verifying **S**-state (i.e. there exists a state $\langle M, w \rangle$ verifying **S** such that $M, w \models \Box_a\varphi$ and for all states $\langle M', w' \rangle$ verifying **S**: $M', w' \models \Box_a\varphi \Rightarrow M, w \leq^a M', w'$).

(2) Formula φ has an $\mathcal{L}^a$-smallest **S**-m.c. expansion (i.e. there exists a maximal **S**-consistent Γ such that $\varphi \in \Gamma$ and for all **S**-m.c. Δ: $\varphi \in \Delta \Rightarrow \Gamma \cap \mathcal{L}^a \subseteq \Delta$).

(3) Formula φ has **S**-DP with respect to $\mathcal{L}^a$, i.e. φ is **S**-consistent and for every $\psi_1, \psi_2, \ldots \psi_k \in \mathcal{L}^a$: $\varphi \vdash_{\mathbf{S}} (\psi_1 \vee \cdots \vee \psi_k) \Rightarrow$ for some $i \leq k$: $\varphi \vdash_{\mathbf{S}} \psi_i$.

Theorem 1. *Let $\mathcal{L}^a$ be a characteristic persistent sublanguage of $\mathcal{L}$ with respect to $\leq^a$. Then the minimal information equivalences hold for $\mathcal{L}^a$ and $\leq^a$, i.e., the conditions (1), (2) and (3) above are equivalent. More specifically, given the condition, φ is a-honest with respect to $\leq^a$ and* **S** *iff (all statements are equivalent):*

- $\Box_a\varphi$ *has a $\leq^a$-least* **S**-*state*
- $\Box_a\varphi$ *has an $\mathcal{L}^a$-smallest* **S**-*m.c. expansion*

- φ has a $\Box_a^- \mathcal{L}^a$-smallest a-stable expansion
- $\Box_a \varphi$ has **S**-DP over $\mathcal{L}^a$.

Here, Σ is a-stable if $\Sigma = \Box_a^- \Gamma$ for some **S**-m.c. Γ, and Σ is a stable expansion of ψ if Σ is stable and $\psi \in \Sigma$. Linking some of these criteria, we can say that φ is (a-)honest if true in some world (w) in an appropriate model (M) where *part of the knowledge in that world is minimal* (since then $\Box_a^- \mathrm{Th}(M, w)$ is its $\Box_a^- \mathcal{L}^a$-smallest stable expansion), which seems a fairly intuitive notion. When the equivalent notions of Theorem 1 hold for a particular order $\leq^a$, system **S** and language $\mathcal{L}^a$, we say that $\leq^a$ and $\mathcal{L}^a$ *determine* a notion of honesty in **S**. We can also link up the semantic definition of honesty with deduction, providing a perhaps even more intuitive characterization:[1]

Corollary 1. *Let $\mathcal{L}^a$ be persistent and characterizing for $\leq^a$. Then φ is a-honest with respect to $\leq^a$ and **S** iff there is an **S**-state $\langle M, w \rangle$ such that:*

- $M, w \models \Box_a \varphi$ and $\forall \psi \in \mathcal{L}^a : M, w \models \psi \Rightarrow \Box_a \varphi \vdash_{\mathbf{S}} \psi$
- *or, equivalently,* $\forall \psi \in \mathcal{L}^a : M, w \models \psi \Leftrightarrow \Box_a \varphi \vdash_{\mathbf{S}} \psi$

All this makes clear that we 'just' have to specify which part of the knowledge is involved. More formally speaking, we have to pinpoint the right information order $\leq^a$, or, equivalently, its characterizing persistent sublanguage $\mathcal{L}^a$. This, however, is a non-trivial problem, since surely not every information order *has* such a characterizing persistent sublanguage. For example, if $\leq^a$ is mere identity or even isomorphism of models, not even the entire language suffices to characterize the model (up to isomorphism). Also, pursuing our results for the single-agent approach, we know that unlimited bisimulation is too strong a requirement, *vide* [6]. As we showed in our earlier paper, a layered, limited kind of bisimulation is preferable. Two technically correct orders in the single agent case will be generalized in the next subsections. Although the initial, so-called general information order is not intuitively sound, it serves as a first step to more profound information orders. But we start by generalizing an umbrella result for such layered pre-orders.

3.1 Layered Information Orders

An information order and its characterizing persistent language can be obtained along fairly general patterns from the underlying layered orders and their characterizing persistent languages. This is a very convenient tool for many orders to follow, since we can restrict attention to one simple layer at the time.

Suppose $\leq_n^a$ is a pre-order on the set of model-world pairs for each natural number n ('layer n'). From now on, assuming $M = \langle W, \boldsymbol{R}, V \rangle$ and $M' = \langle W', \boldsymbol{R}', V' \rangle$, the base case will be defined as $M, w \leq_0^a M', w' \Leftrightarrow V(w) = V'(w')$. Then we define $\leq^a$ for any layered order $\leq_n^a$ by:

$$M, w \leq^a M', w' \Leftrightarrow \forall n \in \mathbb{N} \; \forall v' \in R_a'[w'] \; \exists v \in R_a[w] : M, v \leq_n^a M', v'.$$

---

[1] This characterization was triggered by a question of Arnis Vilks.

We say that $\leq^a$ is *induced by* $\leq_n^a$ if the above equivalence holds. Finally, let $\mathcal{L}^a$ be a sublanguage and $\mathcal{L}_{(n)}^a = \mathcal{L}^a \cap \mathcal{L}_{(n)}$ be its subset of formulas of modal depth up to n. The following Lemma explains how a persistence and characterization result for languages with finite depth and layered orders can be lifted to the full language and the induced order. Lemma 1 will be implicitly used throughout the paper.

Lemma 1 (Collecting). *If $\mathcal{L}_{(n)}^a$ is persistent and characterizing for $\leq_n^a$, and $\mathcal{L}^a$ is closed under $\vee$, then $\Box_a \mathcal{L}^a$ is persistent and characterizing for $\leq^a$.*

We will now inspect orders inspired by Ehrenfeucht-Fraïssé games (see, for example, [1,4]).

3.2   The Multi-modal General Information Order

In the first Ehrenfeucht-Fraïssé order the underlying, layered order is in fact an equivalence relation ("EF-equivalence"). Define $\simeq_n^{\mathcal{A}}$ recursively (recall that $M, w \simeq_0 M', w' \Leftrightarrow V(w) = V'(w'))$ [2] by: $M, w \simeq_{n+1} M', w'$ iff

- $M, w \simeq_n M', w'$ &
- $\forall i \in \mathcal{A} \, \forall v' \in R_i'[w'] \, \exists v \in R_i[w] : M, v \simeq_n M', v'$ (*back*) &
- $\forall i \in \mathcal{A} \, \forall v \in R_i[w] \, \exists v' \in R_i'[w'] : M, v \simeq_n M', v'$ (*forth*)

Then the *general information order* $\sqsubseteq^a$ is induced by $\simeq_n$. By a rather straightforward induction, one shows that $\mathcal{L}_{(n)}^{\mathcal{A}}$ is characteristic and persistent for $\simeq_n^{\mathcal{A}}$, and hence the collecting lemma gives that $\Box_a \mathcal{L}$ is persistent and characterizing with respect to $\sqsubseteq^a$. So, the information equivalences hold for $\sqsubseteq^a$ and $\Box_a \mathcal{L}$. We say that φ is *generally a-honest* if $\Box_a \varphi$ has a $\sqsubseteq^a$-least model. This implies the usual equivalences, i.e., $\sqsubseteq^a$ and $\Box_a \mathcal{L}$ determine a notion of a-honesty in **S**, for any modal system **S**.

However, as we noticed in [6], this notion of honesty is, though technically correct, intuitively a rather poor one. It also leads to excessive trivialization. In *weak doxastic logics* such as $\mathbf{KD_m}$ and $\mathbf{KD4_m}$, *all* formulae φ such that $\Box_a \varphi$ is consistent, are generally a-honest. For $\mathbf{K_m}$ and $\mathbf{K4_m}$, we can go a step further: all formulas are honest, as Halpern [2] notices for the first system.

In (relatively) *strong logics*, however, i.e. systems with some form of negative introspection (such as **5**, **B** and **G**), there are virtually *no* honest formulas. Because then there surely are non-theorems $\Box_a \varphi_i$ such that $\vdash \Box_a \varphi_1 \vee \Box_a \varphi_2$ which leads to an easy violation of the Disjunction Property. For example, p is generally a-dishonest in $\mathbf{S5_m}$: note that the formula $\Box_a \Diamond_b \neg p \vee \Box_a \Diamond_a \Box_b p$ is an instantiation of **4**, so this formula is derivable from $\Box_a p$ in $\mathbf{S5_m}$, whereas neither of its disjuncts is.

So, as in the unimodal case, not much is left. Among the epistemic logics only a few systems such as $\mathbf{T_m}$ and $\mathbf{S4_m}$ survive. But even then $\sqsubseteq^a$ has counterintuitive effects: growth of information does not lead to less uncertainty, as it should be.

---

[2] The superscript $\mathcal{A}$ is omitted whenever clear from context.

4   Generalizations of the Positive Order

From the single agent case we know that only the positive information order is intuitively sound with respect to honesty. In a positive information order we merely want to preserve positive knowledge of one or more agents. In other words, we disregard negative knowledge, i.e. knowledge of not knowing. We, typically, encounter back simulation on the underlying layers, but usually no forth simulation, or only a restricted form of forth simulation.

It is not *a priori* clear which notion of positive information order is involved. We will discuss several options in what follows, of which the last is the more general one, using both underlying layers from the general and from the so-called 'objective' information order. We start by discussing a rather straightforward generalization from the single agent case.

4.1   Positive Honesty

The positive information order only preserves positive knowledge of agent of a. It is the most obvious generalization of one-agent positive honesty. The formulas of the characterizing language do not have negative occurrences of $\Box_a$, and so, by definition, no $\Diamond_a$ as well. Formally, let $\mathcal{L}^{+a}$ consist of those $\varphi \in \mathcal{L}$ for which φ does not contain $\Box_a$ in the scope of $\neg$. Formulas in $\mathcal{L}^{+a}$ are called a-*positive*.[3] So, $\Box_a p \vee \neg \Box_b q$, $\Box_a \neg p$ and $\Box_a p \wedge \neg q$ are members of $\mathcal{L}^{+a}$, but $\neg \Box_a p$ and $\Diamond_a p \vee \Box_b q$ are not.

Now consider $\mathcal{L}^a = \Box_a \mathcal{L}^{+a}$. This is a correct generalization of the single agent positive language, which by itself is a generalization of the so-called objective one-agent formulas which suit **S5**. We will call the elements of $\Box_a \mathcal{L}^{+a}$ a-*positive knowledge formulas*. What is the corresponding $\leq^a$? Essentially, the underlying order displays the back direction of the EF-equivalence for all agents, operating on a-positive formulas until subformulas are reached that are $\Box_a$-free, where full EF-equivalence for all agents except a takes over. Then, $M, w \leq^{+a}_{n+1} M', w'$ iff:

- $M, w \sim^{\mathcal{A} \backslash \{a\}}_{n+1} M', w'$ &
- $\forall i \in \mathcal{A} \, \forall v' \in R'_i[w'] \, \exists v \in R_i[w] : M, v \leq^{+a}_n M', v'$ (*back*)

Let the *positive information order* $\leq^{+a}$ be induced by $\leq^{+a}_n$. Then $\mathcal{L}^{+a}_{(n)}$ is characteristic and persistent for $\leq^{+a}_n$, so the collecting lemma guarantees that $\Box_a \mathcal{L}^{+a}$ is persistent and characterizing with respect to $\leq^{+a}$. Thus, we obtain the following.

Theorem 2. *The minimal information equivalences hold for $\leq^{+a}$ and $\Box_a \mathcal{L}^{+a}$.*

Now, φ is called *positively a-honest* if $\Box_a \varphi$ has a $\leq^{+a}$-least model.

Thus, we have that $\leq^{+a}$ and $\Box_a \mathcal{L}^{+a}$ determine a notion of a-honesty in **S**, for any system **S**. So, the notion of positive honesty is technically sound, that is, there is a persistent language that characterizes the positive information order, and it seems a proper extension of the unimodal case. It avoids problems

---

[3] BNF-definitions of the languages considered are given at the end of this paper.

objective honesty encounters, such as one noticed by Halpern [2] for $\mathbf{S5_m}$ (or, more precisely, extensions of $\mathbf{KB4_m}$): suppose p is some fact totally unrelated to a formula φ (for example, p may not occur in φ), then $\Box_a\varphi \vdash \Box_a p \lor \Box_a \Box_b \neg \Box_b \Box_a p$. It is clear, however, that each of the disjuncts itself does not follow from $\Box_a\varphi$. Yet φ may constitute innocent knowledge, e.g. $\Box_b q$. But for our notion of positive honesty, this counter-example to the DP test is avoided by the restriction that the disjuncts should be in the a-positive knowledge language; here, obviously, $\Box_b \neg \Box_b \Box_a p \notin \mathcal{L}^{+a}$.

Yet we do not want to exclude other possible notions of honesty *a priori*, and therefore now turn to one studied earlier.

4.2   Objective Honesty

To make a different start in formalizing multi-agent positive honesty, we return to Halpern's [2] definition of a-objective formulas and the notion of honesty connected to it. Halpern reserves the notion objective honesty for the two strong doxastic systems $\mathbf{K45_m}$ and $\mathbf{KD45_m}$. This seems harmless for these two systems. Our main concern is that developing a whole apparatus for just two modal systems, and again different ones for others, leads to an approach which lacks generality and in fact conceals much of the general pattern. In fact, in Halpern's approach it is not clear *why* a-objective formulas might be suitable for the two systems mentioned. We think that we can in fact explain much of the reasons for its feasibility.

The idea of a-objective knowledge is that agent a only has knowledge of information 'outside' of a, i.e. knowledge of facts and other agents' knowledge. Such other agents' knowledge may again involve a's knowledge, but still counts as external for a. This is easily formalized when we start with the *a-objective* (that is, wide scope a-operator-free) formulas: let $\mathcal{L}^{-a}$ consist of those $\varphi \in \mathcal{L}$ for which φ does not contain wide scope $\Box_a$. In other words, in an a-objective formula, every $\Box_a$ and $\Diamond_a$ has to be in the scope of a $\Box_b$ or $\Diamond_b$ ($b \neq a$). Examples: $\Box_a p \lor \Box_b q$, $\Box_a \neg p$ are not in $\mathcal{L}^{-a}$, but $\neg \Box_b p$ and $\Diamond_b(p \lor \neg \Box_a q)$ are.

So where does the agent a's knowledge enter the story? Here she is: consider $\mathcal{L}^a = \Box_a \mathcal{L}^{-a}$. A formula is then called an *a-objective knowledge formula* if it is of the form $\Box_a \varphi$ with $\varphi \in \mathcal{L}^{-a}$.

The corresponding Ehrenfeucht-Fraïssé order $\leq^a$ can be obtained from the underlying layered order, which is again an equivalence relation. The recursive clause for $\simeq_n^{-a}$ is the following: $M, w \simeq_{n+1}^{-a} M', w'$ iff

- $M, w \simeq_n^{-a} M', w'$ &
- $\forall i \neq a \forall v' \in R_i'[w'] \exists v \in R_i[w] : M, v \simeq_n^{A} M', v'$ (*back*) &
- $\forall i \neq a \forall v \in R_i[w] \exists v' \in R_i'[w'] : M, v \simeq_n^{A} M', v'$ (*forth*)

So, $\simeq_{n+1}^{-a}$ not only uses the general EF-equivalence relation on layer n, its overall formulation is close to that of the general information order, be it that it shares the exclusion of agent a with the positive information order.

One can now prove that $\mathcal{L}_{(n)}^{-a}$ is characteristic and persistent for $\simeq_n^{-a}$. Thus, if we define *objective information order* $\leq^{-a}$ to be induced by $\simeq_n^{-a}$, the collecting lemma guarantees that $\Box_a \mathcal{L}^{-a}$ is persistent and characterizing with respect to $\leq^{-a}$.

Theorem 3. *The information equivalences hold for $\leq^{-a}$ and $\Box_a \mathcal{L}^{-a}$.*

This again implies that $\leq^{-a}$ and $\Box_a \mathcal{L}^{-a}$ determine a notion of (objective) a-honesty in **S**, for any system **S**.

As can been seen from the format of the a-objective formulas, agent a's knowledge is not taken into account. For fully introspective this is unproblematic in the one-agent case: there one can show that positive knowledge formulas can be reduced to disjunctions of objective knowledge formulas, which implies that for each system containing **K45**, objective honesty amounts to positive honesty.

It should be emphasized that this equivalence only holds for the one agent case and full introspective knowledge. For more agents there is no such reduction, since an objective knowledge formula need not be (equivalent to) a positive one, e.g. $\Box_a \Box_b \neg \Box_a p$ is an a-objective knowledge formula which is not related to any a-positive knowledge formula whatsoever. If we want to generalize this equivalence to fully introspective multi-agent systems, we have to relax the notion of positive formula somewhat, as will be done in the next subsection.

4.3   Positive-Objective Honesty

We want to generalize objective knowledge to what we consider to be a more adequate notion of multi-modal honesty. The *a-positive-objective* formulas can, roughly, be characterized as having no wide scope negative occurrence of $\Box_a$ operators. Again assume for simplicity's sake that we only consider formulas where every $\Diamond_i$ is replaced by $\neg \Box_i \neg$. Let $\mathcal{L}^{\pm a}$ consist of those $\varphi \in \mathcal{L}$ for which every $\Box_a$ in φ in the scope of $\neg$ is also in the scope of a $\Box_i$ with $i \neq a$. Thus, $\mathcal{L}^{\pm a}$ can also be regarded as the closure of $\mathcal{L}^{-a}$ under the operations $\wedge$, $\vee$ and $\Box_a$. Examples: $\Box_a p \vee \Box_b q$, $\Box_a p \wedge \neg \Box_b q$ and $\Box_a \Box_b \neg \Box_a p$ are members of $\mathcal{L}^{\pm a}$, but $\neg \Box_a p$ and $\Box_a \neg \Box_a \neg p \vee \Box_b q$ are not.

Once again, what is the corresponding $\leq^a$? For evaluating formulas, we essentially want to have recursive back moves for agent a in the EF-order, until a-objective formulas are reached, and then proceed with the a-objective equivalence. So, more formally, the recursive step in $\leq_n^{\pm a}$ is defined by $M, w \leq_{n+1}^{\pm a} M', w'$ iff:

- $M, w \sim_{n+1}^{-a} M', w'$ &
- $\forall v' \in R_a'[w'] \exists v \in R_a[w] : M, v \leq_n^{\pm a} M', v'$ (*back*)

Then the *a-positive-objective information order* $\leq^{\pm a}$ is induced by $\leq_n^{\pm a}$.

Now consider $\mathcal{L}^a = \Box_a \mathcal{L}^{\pm a}$. Notice that $\mathcal{L}^{\pm a}$ extends both $\mathcal{L}^{+a}$ and $\mathcal{L}^{-a}$, thus generalizes both the positive and the objective approach. Since $\mathcal{L}_{(n)}^{\pm a}$ is characteristic and persistent for $\leq_n^{\pm a}$, the collecting lemma shows that $\Box_a \mathcal{L}^{\pm a}$ is persistent and characterizing for $\leq^{\pm a}$. Now φ is called *positive-objectively* a-*honest* when $\Box_a \varphi$ has a $\leq^{\pm a}$-least model.

Theorem 4. *The minimal information equivalences hold for $\leq^{\pm a}$ and $\Box_a \mathcal{L}^{\pm a}$.*

This implies that $\leq^{\pm a}$ and $\Box_a \mathcal{L}^{\pm a}$ determine a notion of a-honesty in **S**. We can show that for fully introspective systems the extension from objective to positive-objective formulas is immaterial, since then again a's positive-objective knowledge can be reduced to a-objective knowledge. So, objective honesty and positive-objective honesty coincide for $\mathbf{K45_m}$ and $\mathbf{KD45_m}$, and $\mathbf{S5_m}$. Although positive, objective, and positive-objective honesty agree on (one agent) **S5**, they, surprisingly, do not on $\mathbf{S5_m}$ ($m > 1$). Since $\Box_a p \vee \Box_a \Box_b \neg \Box_b \Box_a p$ is derivable in $\mathbf{S5_m}$, there are virtually no (positive-)objectively honest formulas in this system. However, we have already seen that for $\mathbf{S5_m}$, the positive information order seems correct.

5   Relating and Evaluating Types of Honesty

In the previous section we noticed that for fully introspective systems the different types of honesty may actually coincide, depending on the number of agents m. But before checking examples and assessing the intuitive correctness of these notions, some more general observations can be made.

The types of honesty distinguished in this paper are ordered as indicated in Figure 1. This hierarchy easily follows from DP, using the fact that $\Box_a \mathcal{L}^{+a} \cup \Box_a \mathcal{L}^{-a} \subseteq \Box_a \mathcal{L}^{\pm a} \subseteq \Box_a \mathcal{L}$.

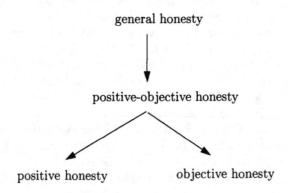

Fig. 1. Relating notions of honesty

This reduces the number of checks to be made for specific examples. In general, *dishonesty* can be shown fairly easily by using the relevant DP, but it may be harder to show *honesty* more or less directly. It is not *prima facie* clear how to prove honesty, since DP then has to be checked for an infinite set of formulas. Also, minimality of stable expansions encounters similar problems and finding the least model may be non-trivial, which is related to the complexity of the information orders. Presumably, for many relevant multi-modal systems these intricate orders have simple counterparts.

To assess the adequacy of the notions of honesty, we picked a number of examples and checked their (dis)honesty for the four types proposed, and for the modal systems $S4_m$, $K45_m$, $S5_m$ ($m > 1$). In addition to hierarchy constraints and the observation about collapse, we already noticed that there are hardly any generally honest formulas for $K45_m$ and $S5_m$.

For the former system also inconsistent formulas are vacuously generally a-honest. Neither full information nor inconsistent information is of much interest here. Moreover, for $S5_m$ we also noticed large (positive-)objective dishonesty. Therefore, the in some sense maximally honest formulas (characterizing innocent partial knowledge) display the left-hand pattern in Table 1 ('pob' denotes positive objective honesty, etc.). This pattern manifests itself in many formulas that are also intuitively honest for agent a: p, $\Box_i p$, $\dots$. The most challenging cases are disjunctions of (negated) knowledge formulas. As we will see, whether or not they are intuitively honest largely depends on the agency of the knowing subject. So, also the following formulas are indeed maximally a-honest: $\Box_b p \vee \Box_b q$, $\Box_a p \vee \Diamond_a q$, $\Box_b p \vee \Diamond_a q$, $\Box_b p \vee \Diamond_b q$, and $p \vee q$. The other extreme are the totally dishonest formulas displaying the pattern on the right, exemplified by the paradigm $\Box_a p \vee \Box_a q$.

Table 1. Patterns of maximal (left) and minimal (right) honesty

	$S4_m$	$K45_m$	$S5_m$		$S4_m$	$K45_m$	$S5_m$
gen	+	-	-	*gen*	-	-	-
pob	+	+	-	*pob*	-	-	-
obj	+	+	-	*obj*	-	-	-
pos	+	+	+	*pos*	-	-	-

There are many (34) intermediate cases. A very common pattern here is the one in which honesty only depends on the amount of introspection attributed to the agents, witnessed by the pattern on the left below. Examples of formulas with this honesty pattern (displayed in Table 2, left) are $\Box_a p \vee \Box_b q$, $\Box_a p \vee \Diamond_b q$, and $\Box_a p \vee q$. Also, honesty may depend on the type and not on the modal systems under inspection, as with the formula $\Box_a p \vee \Box_a \Box_b \Diamond_a q$, showing the pattern on the right in Table 2.

Finally, two more complicated patterns can be obtained by the formulas $\Box_a p \vee \Box_a \Diamond_a q$ (on the left) and $\Box_a p \vee \Box_a \Diamond_a \Box_a q$ (right) in Table 3.

The tentative conclusion from inspecting these examples is that positive honesty seems to be the intuitively correct notion for multi-modal systems.

Table 2. Introspection (left) and type (right) dependent honesty patterns

	$S4_m$	$K45_m$	$S5_m$		$S4_m$	$K45_m$	$S5_m$
gen	+	-	-	*gen*	-	-	-
pob	+	-	-	*pob*	-	-	-
obj	+	-	-	*obj*	-	-	-
pos	+	-	-	*pos*	+	+	+

Table 3. Some other patterns of honesty

	$S4_m$	$K45_m$	$S5_m$		$S4_m$	$K45_m$	$S5_m$
gen	-	-	-	*gen*	-	-	-
pob	+	+	-	*pob*	+	-	-
obj	+	+	-	*obj*	+	-	-
pos	+	+	+	*pos*	+	-	-

6   Conclusion

We have given generalizations of information orders for multi-agent only know-ing, which apply to arbitrary modal systems and ordinary Kripke models. Using a general theorem relating information orders and their corresponding (sub-) languages, we were able to identify several equivalent characterizations of hon-esty. In particular, we have explored the general information order and some positive and objective information orders. So-called positive honesty seems the intuitively correct notion here.

An interesting question for future research concerns the transfer of techniques developed for the single agent case to multi-modal systems. For example, one might try to adapt the amalgamation techniques as used in [6] to prove, by means of the disjunction property, honesty in **S4** and weaker systems. It is also interesting to generalize the test procedure as proposed and proved correct in [2] for objective honesty to other types of honesty.

There are many ways to extend the multi-agent perspective on only knowing. For instance, one might give up the assumption that all agents use the same logic and move to heterogeneous systems. Also, a notion of *group honesty* is as yet unexplored. Finally, we like to investigate multi-agent honesty from a more constructive perspective: can we give a procedure to generate a minimal model for a given formula? And, can we extend the partial approach of [5] to the multi-agent case?

References

1. H.-D. Ebbinghaus & J. Flum, *Finite model theory*, Springer–Verlag, Berlin, 1995.
2. J.Y. Halpern, 'Theory of Knowledge and Ignorance for Many Agents', in *Journal of Logic and Computation*, **7** No. 1, pp. 79–108, 1997.

3. J.Y. Halpern & Y. Moses, 'Towards a theory of knowledge and ignorance', in Kr. Apt (ed.) *Logics and Models of Concurrent Systems*, Springer–Verlag, Berlin, 1985.
4. M. Hennessy & R.Milner, 'Algebraic laws for Nondeterminism and Concurrency', *Journal of the* ACM 32, pp.137-161, 1985.
5. W. van der Hoek, J.O.M. Jaspars, & E.G.C. Thijsse, 'Honesty in Partial Logic'. *Studia Logica*, **56** (3), 323–360, 1996. Extended abstract in proceedings of KR'94.
6. W. van der Hoek, J.O.M. Jaspars, & E.G.C. Thijsse, 'Persistence and Minimality in Epistemic Logic', *Annals of Mathematics and Artificial Intelligence*, **27** (1999), pp. 25–47, 2000. Extended abstract in J. Dix, U. Furbach, L. Fariñas del Cerro (eds.), *Logics in Artificial Intelligence. Proceedings JELIA'98*, Springer Verlag, LNAI 1489.
7. G.Lakemeyer, 'All they know: a study in multi-agent auto epistemic reasoning', IJCAI'93, pp. 376–381, 1993
8. H.J. Levesque, 'All I know: a study in auto-epistemic logic', in *Artificial Intelligence*, **42**(3), pp. 263–309, 1990.
9. G. Schwarz & M. Truszczyński, 'Minimal knowledge problem: a new approach', *Artificial Intelligence* **67**, pp. 113–141, 1994.
10. R. Parikh, 'Monotonic and nonmonotonic logics of knowledge', *Fundamenta Informaticae* 15, pp. 255-274, 1991
11. M. Vardi, 'A model-theoretic analysis of monotonic knowledge', IJCAI85, pp. 509–512, 1985.

Appendix: Defining Languages

Here we give explicit BNF definitions of the (sub)languages considered in this paper. Before doing that we summarize the languages in Table 4:

Table 4. Symbols, names and informal descriptions of languages

Name	Language	Condition
$\mathcal{L} = \mathcal{L}^A$	full, general	no restriction
$\mathcal{L}^{-a}$	a-objective	only $\square_a$ in scope $\square_{i \neq a}$
$\mathcal{L}^{+a}$	a-positive	no $\square_a$ in scope of $\neg$
$\mathcal{L}^{\pm a}$	a-positive-objective	$\square_a$ only in scope $\neg$ if in scope $\square_{i \neq a}$

The languages are now defined by the following BNF expressions:

Table 5. Languages and their BNFs

Name	BNF definition
$\mathcal{L}^A$	$\varphi ::= p \ (p \in \mathcal{P}) \mid \neg\varphi \mid \varphi \wedge \varphi \mid \square_i\varphi \ (i \in \mathcal{A})$
$\mathcal{L}^{-a}$	$\varphi_0 ::= p \ (p \in \mathcal{P}) \mid \neg\varphi_0 \mid \varphi_0 \wedge \varphi_0 \mid \square_i\varphi \ (i \in \mathcal{A} - \{a\})$
$\mathcal{L}^{+a}$	$\varphi_1 ::= \varphi \ (\varphi \in \mathcal{L}^{A \setminus \{a\}}) \mid \varphi_1 \wedge \varphi_1 \mid \varphi_1 \vee \varphi_1 \mid \square_i\varphi_1 \ (i \in \mathcal{A})$
$\mathcal{L}^{\pm a}$	$\varphi_2 ::= \varphi_0 \mid \varphi_2 \wedge \varphi_2 \mid \varphi_2 \vee \varphi_2 \mid \square_a\varphi_2$

A Modal Logic for Network Topologies

Rogier M. van Eijk, Frank S. de Boer, Wiebe van der Hoek, and John-Jules Ch. Meyer

Institute of Information and Computing Sciences, Utrecht University,
P.O. Box 80.089, 3508 TB Utrecht, The Netherlands
{rogier, frankb, wiebe, jj}@cs.uu.nl

Abstract. In this paper, we present a logical framework that combines modality with a first-order quantification mechanism. The logic differs from standard first-order modal logics in that quantification is not performed inside the states of a model, but the states in the model themselves constitute the domain of quantification. The locality principle of modal logic is preserved via the requirement that in each state, the domain of quantification is restricted to a subset of the entire set of states in the model. We show that the language is semantically characterised by a generalisation of classical bisimulation, called history-based bisimulation, consider its decidability and study the application of the logic to describe and reason about the topologies of multi-agent systems.

1 Introduction

Over the last years an increasing interest can be observed in large-scale distributed computing systems that consist of heterogeneous populations of interacting entities. Examples from practice include for instance the electronic market places in which buyers and sellers come together to trade goods. This trend can also be observed in the fields of computer science and artificial intelligence with the current focus on *multi-agent systems* [14]. In these systems, an *agent* constitutes an autonomous entity that is capable of perceiving and acting in its environment and additionally has a social ability to communicate with other agents in the system. In *heterogeneous* multi-agent systems, the agents are assumed to be of different plumage, each having their individual expertise and capabilities. Moreover, in *open* multi-agent systems, new agents can be dynamically integrated [5].

One of the issues in open heterogeneous multi-agent systems is the *agent location problem*, which denotes the difficulty of finding agents in large populations [13]. For instance, given an agent that needs to accomplish a particular task that it is incapable of performing all by itself, the problem amounts to finding an agent that has the expertise and capabilities to join in this task. In these systems, it is typically impossible for the individual agents to maintain a complete list of the agents that are present. That is, each of the agents has a list of other agents that it knows of, but due to the dynamics of the system this list is normally not exhaustive. Hence, the agent needs to communicate with the other agents in the system in order to come to know about new agents that it is currently not aware of to exist. This enables the agent to extend its individual circle of acquaintances.

M. Ojeda-Aciego et al. (Eds.): JELIA 2000, LNAI 1919, pp. 269–283, 2000.

The purpose of this paper is to develop a formal logic to describe and reason about network topologies, like for instance the topology of multi-agent systems. Formally, such a network topology can be represented by a directed graph where the nodes in the graph denote the entities in the system and the edges make up the acquaintance relation, describing what entitites know each other. Seen from a logical point of view, these graphs constitute Kripke frames, which are employed in the semantics of modal logic [11]. This observation naturally leads to an approach of describing network topologies by means of modal logic, which is corroborated by the fact that we want to describe and reason about network topologies with respect to a local perspective, that is, from the viewpoint of a particular entity in the topology. Basic modal logic however is not fit to describe and reason about network topologies, as it does not have the expressive power to distinguish between *bisimilar* structures, like for instance loops and their unfoldings, which clearly induce different network topologies.

In this paper, we present an extension of the basic modal logic with variables and a first-order quantification that complies with the locality principle of modal logic. It differs from the standard first-order modal logics [6] in that there is no quantification *inside* the states of a model. Instead, the *states* in the model themselves constitute the domain of quantification; i.e., the logic covers a mechanism of binding variables to states in a model. Such variable binding mechanisms are also gaining attention in the field of *hybrid languages*, which are languages originally developed with the objective to increase the expressiveness of tense logics [4]. Our framework can be viewed upon as a formalisation of hybrid languages in terms of an equational theory in which we can reason about the equalities (and inequalities) of states of a model. We preserve the locality principle of modal logic via the requirement that in each state the domain of quantification is restricted to a subset of the entire set of states in the model.

Moreover, we define a semantic characterisation of the logic, which is based on a generalisation of the classical notion of bisimulation equivalence. Instead of relating states, this generalised type of bisimulation relates tuples that are comprised of a state together with a sequence of states. In the semantic characterisation, these additional sequences are employed to represent variable bindings that are generated during the evaluation of formulae.

The remainder of this paper is organised as follows. In Section 2, we start with considering basic modal logic and graded modal logic, and argue that these are not well-fit as logics for network topologies. In Section 3, we develop the syntax and semantics of a general modal logic with an implicit bounded quantification mechanism. Subsequently, in Section 4, we establish a semantic characterisation of the logic, while the decidability of the logic is discussed in Section 5. Additionally, in Section 6 we consider the application of the logic to describe and reason about the topologies of multi-agent systems. Finally, we wrap up in Section 7 where we provide some directions for future research.

2   Towards a Logic for Network Topologies

The most straightforward logic to describe and reason about network topologies is standard first-order logic. However, rather than taking the bird's-eye perspective, our aim is to reason about network topologies from a *local point of view*. The following example explains the difference between these two perspectives.

Example 1 (*Local point of view*)
Consider an agent w that knows two agents v_1 and v_2, which each in turn, know one other agent. The structures $\mathcal{M}$ and $\mathcal{N}$ in Figure 1 constitute two of the possible situations. In situation $\mathcal{M}$, the two acquaintances of the acquaintances of w are distinct,

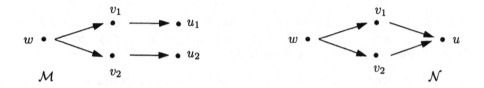

Fig. 1. Network topologies with indirect acquaintances

while in $\mathcal{N}$ these two acquaintances are one and the same agent. From an external point of view these two structures are clearly distinct. However, what if we consider them from the local perspective of w? The crucial observation here is that whereas v_1 and v_2 are among the agents that are known by w, the agents u_1 and u_2 are not. Consequently, as w does not know the identity of either u_1 and u_2, it cannot decide whether they are the same or distinct. In other words, as far as w is concerned, the actual situation could be the one depicted by $\mathcal{M}$ as well as the one depicted by $\mathcal{N}$. However, standard first-order logic can obviously distinguish between these two situations.

Our purpose is to develop a (fragment of first-order) logic that is fit to reason about network topologies from a local perspective.

2.1 Basic Modal Logic

Languages that are designed to describe and reason about relational structures from a local perspective, are the *languages of modal logic*. The basic modal language can be defined as follows.

Definition 2 (*Basic modal language $\mathcal{L}_0$*)
Formulae φ in the language $\mathcal{L}_0$ are generated using the following BNF-grammar:

$$\varphi ::= \top \mid \varphi_1 \wedge \varphi_2 \mid \neg\varphi \mid \Diamond\varphi.$$

A modal formula is either equal to $\top$, the conjunction of two modal formulae, the negation of a modal formula, or the operator $\Diamond$ followed by a modal formula. It is the operator $\Diamond$ that gives the language the modal flavour; it has various readings like for instance the interpretation of expressing *possibility*. The dual $\Box$ of this operator, which is defined as $\neg\Diamond\neg$, can be thought of denoting *necessity*. Finally, we assume the usual abbreviations $\perp$ for $\neg\top$, $\varphi_1 \vee \varphi_2$ for $\neg(\neg\varphi_1 \wedge \neg\varphi_2)$ and $\varphi_1 \rightarrow \varphi_2$ for $\neg\varphi_1 \vee \varphi_2$. Note that we do not consider propositional variables here.

The basic modal logic is used to reason about relational structures, especially about relational structures that are referred to as Kripke structures.

Definition 3 (*Kripke structures*)
A structure for the language $\mathcal{L}_0$, which is also called a *Kripke structure*, is a tuple of the form:

$$\mathcal{M} = \langle W, r \rangle,$$

where W constitutes the domain of the structure, which elements are referred to as *states, nodes, worlds* or *agents*, and $r \subseteq W \times W$ denotes an *accessibility* relation on W. For each state $w \in W$ we use the notation $r(w)$ to denote the set $\{u \in W \mid r(w, u)\}$.

The interpretation of modal formulae is given in the following truth definition.

Definition 4 (*Truth definition for $\mathcal{L}_0$*)
Given a structure $\mathcal{M} = \langle W, r \rangle$, a state $w \in W$ and a formula $\varphi \in \mathcal{L}_0$, the truth definition $\mathcal{M}, w \models \varphi$ is given by:

$$\mathcal{M}, w \models \top$$
$$\mathcal{M}, w \models \varphi_1 \wedge \varphi_2 \Leftrightarrow \mathcal{M}, w \models \varphi_1 \text{ and } \mathcal{M}, w \models \varphi_2$$
$$\mathcal{M}, w \models \neg\varphi \qquad \Leftrightarrow \mathcal{M}, w \not\models \varphi$$
$$\mathcal{M}, w \models \Diamond\varphi \qquad \Leftrightarrow \exists v \in r(w) : \mathcal{M}, v \models \varphi$$

Additionally, we have $\mathcal{M} \models \varphi$ if for all $w \in W$ it holds that $\mathcal{M}, w \models \varphi$.

Kripke structures can be viewed upon as representing network topologies: the elements of W constitute the nodes in the network and the relation r defines the accessibility relation; e.g., $r(w, u)$ denotes that w has *access* to u, or that u is an *acquaintance* of w, or that w *knows* u, or that w can *communicate* to u, and so on. The modal logic $\mathcal{L}_0$ can then be used to describe these topologies. For instance, the formula $\Diamond\Diamond\top$ expresses that there exists an acquaintance of an acquaintance. That is, $\mathcal{M}, w \models \Diamond\Diamond\top$ holds in case there exist v and u such that $r(w, v)$ and $r(v, u)$. The basic language $\mathcal{L}_0$ is however not rich enough for adequate descriptions of network topologies. Consider for instance the two structures $\mathcal{M}$ and $\mathcal{N}$ in Figure 2. In the structure $\mathcal{M}$, there is an agent that

Fig. 2. Different number of direct acquaintances

knows two different agents, while in the structure $\mathcal{N}$ only one agent is known. From the perspectives of w and v these two structures clearly denote distinct situations, as we assume that agents know the identities of their acquaintances and hence, can distinguish between the situation that their circle of acquaintances is comprised of two agents from the situation that this circle consists of only one agent. However, the basic language $\mathcal{L}_0$ lacks the expressive power to distinguish between both networks; i.e., there does not exist a formula $\varphi \in \mathcal{L}_0$ with $\mathcal{M}, w \models \varphi$ and $\mathcal{N}, v \not\models \varphi$. Formally, this follows from the fact that these structures are bisimilar.

2.2   Graded Modal Logic

Graded modal logic is an extension of the basic modal language that deals with numbers of successors [10]. Rather than one modal operator $\Diamond$ the graded language contains a set $\{\Diamond_n \mid n \geq 0\}$ of operators. A formula of the form $\Diamond_n \varphi$ expresses that there exist more than n accessible worlds in which φ holds. Hence, graded modal logic can distinguish between the above models $\mathcal{M}$ and $\mathcal{N}$. For instance, we have $\mathcal{M}, w \models \Diamond_1 \top$ but $\mathcal{N}, v \not\models \Diamond_1 \top$.

Graded modal languages are still not suitable to describe network topologies. For instance, consider the two structures $\mathcal{M}$ and $\mathcal{N}$ in Figure 3, which denote a loop and its unfolding, respectively. In $\mathcal{M}$, there is an agent that knows only itself, whereas in

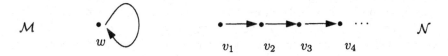

Fig. 3. Loop and its unfolding

$\mathcal{N}$ there is an agent that knows another agent that knows another agent that knows yet another agent ... and so on. However, whereas we believe that an adequate logic for network topology should be able to distinguish between these two structures, it can be shown that graded modal logic does not possess the expressive power.

3   Modal Logic with Bounded Quantification

Our analysis of the reason why basic modal logic and its extension with graded modal-ities are not adequate to describe network topologies, is that they lack a mechanism of dealing with *identities*. For instance, if we reconsider the structure $\mathcal{M}$ from Figure 2, then although v_1 and v_2 have no distinguishable property that is expressible in the lan-guage $\mathcal{L}_0$, there is one significant intrinsic difference between them and that is their identity; i.e., they are two distinct states in the topology.

Our approach in developing a logic for network topologies therefore consists in extending the basic modal logic with a mechanism of dealing with *state identity*. That is, the language $\mathcal{L}_0$ is expanded with a collection *Var* of variables that are used as state identifiers. In order to be able to instantiate these variables we additionally introduce a form of *implicit bounded quantification*. We refer to this language as $\mathcal{L}_1$.

Definition 5 (*The extended modal language $\mathcal{L}_1$*)
Given a set *Var* of variables, terms t and formulae φ are generated using the following BNF-grammar:

$$t ::= self \mid x$$

$$\varphi ::= (t_1 = t_2) \mid \varphi_1 \wedge \varphi_2 \mid \neg\varphi \mid \Diamond\varphi \mid \exists x\varphi,$$

where x ranges over the variables of *Var*.

We assume the usual abbreviation $\forall x\varphi$ for $\neg\exists x\neg\varphi$. The formula $\top$ can be modelled as the formula $self = self$. A formula φ is called a *sentence* if it contains no free variables, i.e., all variables x in φ occur in the scope of a quantifier $\exists x$.

The language $\mathcal{L}_1$ extends the language $\mathcal{L}_0$ with variables to denote the identities of states; there is additionally a special constant *self* that always denotes the *current* state. An atomic formula is of the form $t_1 = t_2$, expressing that two terms denote the same state. Additionally, a formula of the form $\exists x\varphi$ expresses that there exists a state (which is denoted by x) for which φ holds.

Although the syntax of the language $\mathcal{L}_1$ closely resembles the syntax of first-order modal logic [6], there is a fundamental difference in the semantics of both languages. In first-order modal logic, quantification is performed *inside* the states of a model. That is, each state constitutes a model in itself as it contains a domain over which the existential quantifier $\exists$ can quantify. However, in the present logic, the states of a model *themselves* constitute the domain of quantification. Moreover, there is a second fundamental difference, namely in the range of quantification. Whereas in first-order modal logic, the existential quantifier ranges over the entire domain, in our logic it is restricted to range over a *subdomain*, namely over the states that are directly reachable via the accessible relation. The ratio behind this is that for instance in the setting of multi-agent topologies, the accessible agents are precisely the agent whose identities are known. Moreover, it gives rise to a form of *implicit bounded quantification* that complies with the local character of modal logic: like one is not allowed to go from one state to an arbitrary state, only to an *accessible* state, one cannot instantiate variables with arbitrary states but only with states that are accessible.

To obtain a framework that is as general as possible (and that perhaps can be applied to other areas besides network topologies), we explicitly distinguish between the accessibility relation and the domains of quantification. That is, we introduce the notion of a *neighbourhood relation* which defines for each state the collection of states over which can be quantified in this state.

Definition 6 (*Structures for the language $\mathcal{L}_1$*)
A structure for $\mathcal{L}_1$ is a tuple that is of the form:

$$\mathcal{M} = \langle W, r, n \rangle,$$

where W constitutes the domain of the structure, $r \subseteq W \times W$ denotes an *accessibility* relation on W and $n \subseteq W \times W$ denotes a *neighbourhood* relation on W. For each state $w \in W$, we use $n(w)$ to denote the set $\{u \in W \mid n(w, u)\}$ of states in the neighbourhood.

A *network topology* is then a special type of structure, namely a structure $\langle W, r, n \rangle$ that satisfies $n = r$ and additionally $w \in n(w)$, for all $w \in W$. Thus, network topologies are structures in which the neighbourhood relation coincides with the accessibility relation and each state is part of its own neighbourhood (and thus also accessible from itself). The rationale behind the latter requirement is that we assume that each agent in a network knows itself (cf. [8]).

In order to interpret formulae from the language $\mathcal{L}_1$, we need to extend the truth definition of $\mathcal{L}_1$ with a mechanism of interpreting variables. We achieve this via the standard notion of an assignment function.

Definition 7 (*Assignment function*)
Given a structure $\mathcal{M} = \langle W, r, n \rangle$ an *assignment function* f is a partial function of type $Var \rightarrow W$ with finite domain, which maps variables to states in the structure. The set $\langle W \rangle$ consists of all assignment functions over W. The empty assignment function, which is undefined for all inputs, is denoted by $\langle \rangle$. Moreover, given an assignment f, a state $w \in W$ and a variable $x \in Var$, we define the *variant* $f[x \mapsto w]$ *of* f to be the function defined by:

$$f[x \mapsto w](y) = \begin{cases} w & \text{if } y \equiv x \\ f(y) & \text{otherwise} \end{cases}$$

where $\equiv$ stands for syntactic equality.

The interpretation of terms and formulae in the language $\mathcal{L}_1$ are given via the following truth definition.

Definition 8 (*Truth definition for $\mathcal{L}_1$*)
Given a structure $\mathcal{M} = \langle W, r, n \rangle$, a state $w \in W$, and an assignment $f : Var \rightarrow W$, we define the interpretation of terms t in $\mathcal{L}_1$ as follows:

$$I_{w,f}(t) = \begin{cases} w & \text{if } t \equiv self \\ f(t) & \text{otherwise} \end{cases}$$

The truth definition $\mathcal{M}, w, f \models \varphi$ is given by:

$$\mathcal{M}, w, f \models (t_1 = t_2) \Leftrightarrow I_{w,f}(t_1) = I_{w,f}(t_2)$$
$$\mathcal{M}, w, f \models \varphi_1 \wedge \varphi_2 \Leftrightarrow \mathcal{M}, w, f \models \varphi_1 \text{ and } \mathcal{M}, w, f \models \varphi_2$$
$$\mathcal{M}, w, f \models \neg\varphi \Leftrightarrow \mathcal{M}, w, f \not\models \varphi$$
$$\mathcal{M}, w, f \models \Diamond\varphi \Leftrightarrow \exists v \in r(w) : \mathcal{M}, v, f \models \varphi$$
$$\mathcal{M}, w, f \models \exists x\varphi \Leftrightarrow \exists v \in n(w) : \mathcal{M}, w, f[x \mapsto v] \models \varphi$$

Additionally, we have $\mathcal{M}, w \models \varphi$ if for all assignments f it holds that $\mathcal{M}, w, f \models \varphi$. Finally, we have $\mathcal{M} \models \varphi$ if for all $w \in W$ it holds that $\mathcal{M}, w \models \varphi$.

Note the difference in the truth definition between the operators $\Diamond$ and $\exists$ with respect to the point of evaluation: in the truth definition of the former operator there is a shift in perspective, viz. from w to v, whereas in the latter, the point of view w remains fixed. In other words, $\exists$ quantifies over the current neighbourhood while the operator $\Diamond$ is used to change the current scope of quantification. Additionally, note that the constant *self* constitutes a *non-rigid designator* [6] in the sense that its denotation differs among the states in a structure; in particular, in each state the denotation of this designator is the state itself.

We could say that the logic $\mathcal{L}_1$ exhibits a separation between the mechanisms of *structure traversal* and *variable instantiation*; that is, the operator $\Diamond$ is used to make shifts of perspective along the accessibility relation, while the operator $\exists$ is employed to instantiate variables with states in the neighbourhood. The general set up of the semantic framework enables us to consider modality and quantification in isolation as well as to explore their interplay. For instance, we are in the position to examine to what extent the language $\mathcal{L}_1$ can express connections between the accessibility and the neighbourhood relation: e.g., it can express the property that the neighbourhood relation is a subrelation of the accessibility relation. That is, for all structures $\mathcal{M} = \langle W, r, n \rangle$ and states $w \in W$ the following holds: $\mathcal{M}, w \models \forall x \Diamond(x = self) \Leftrightarrow n(w) \subseteq r(w)$. Secondly, this does not hold the other way around; in Corollary 16, we state that there does not exist a formula that expresses $r(w) \subseteq n(w)$, for all w. However, a straightforward refinement of the language would be an extension with the inverse operator of $\Diamond$, which has a natural interpretation in the context of network topologies, as it denotes the *is-known-by* relation. The interpretation of this operator, which we denote by $\Diamond^{-1}$, is as follows:

$$\mathcal{M}, w, f \models \Diamond^{-1}\varphi \Leftrightarrow \exists v : w \in n(v) \text{ and } \mathcal{M}, v, f \models \varphi.$$

Given a structure $\mathcal{M} = \langle W, r, n \rangle$, for which we assume $w \in n(w)$, for all $w \in W$, the following holds. For all states $w \in W$:

$$\mathcal{M}, w \models \exists x(x = self \land \Box(\exists y(y = self \land \Diamond^{-1}(x = self \land \exists z(z = y)))))$$
$$\Leftrightarrow$$
$$r(w) \subseteq n(w).$$

To obtain some further familiarity with the language $\mathcal{L}_1$, let us consider several properties of network topologies that we can express with it.

Example 9

- First of all, the formula $\exists x(x = self)$, which can be thought of expressing "knowing yourself", is valid in any network topology.
- Secondly, the formula $\exists x(x = self \land \Box \Diamond x = self)$ is true in a state in case all accessible states have in turn access to this state. In other words, it expresses "everyone that I know, knows me".
- Additionally, the formula $\exists xy(\neg(x = y) \land \Diamond(x = self \land \neg \Diamond y = self) \land \Diamond(y = self \land \neg \Diamond x = self))$ is true in a particular state, in case there are two distinct accessible states that are not accessible to one another. Informally, it can be thought of as expressing "I know two agents that do not know each other".
- Finally, we illustrate that quantification does not commute with modality. Consider the formula $\exists x \Box(x = self)$, which is true in a state in case there is exactly one accessible state, and as in network topologies the accessibility relation is reflexive, can be thought of expressing "I know of only myself". On the other hand, the formula $\Box \exists x(x = self)$, which can be thought of expressing "everyone that I know, knows itself", is valid in any network topology.

4   Semantic Characterisation

In this section, we study the expressiveness of the extended modal language $\mathcal{L}_1$. In particular, we address the issue what properties the language can express and what properties are beyond its expressive power. The central restult of this study is a *semantic characterisation* of the language, which amounts to the identification of the conditions under which two structures satisfy precisely the same formulae of $\mathcal{L}_1$.

For the basic modal language $\mathcal{L}_0$ the semantic characterisation is given by the the notion of a *bisimulation* [3,9]. That is, two structures satisfy the same modal formulae from $\mathcal{L}_0$ if and only if they are related by a bisimulation.[1] The language $\mathcal{L}_1$ combines the standard modal logic $\mathcal{L}_0$ with a bounded quantification mechanism. In order to deal with variable instantiation we employ the notion of an *injective sequence*.

Definition 10 (*Sequences*)

- Given a set W of states, a sequence $\mathbf{w} = [w_1 \cdots w_n]$ over W is called *injective* if $w_i = w_j$ implies $i = j$, for all $1 \leq i, j \leq n$. We employ the notation $[W]$ to denote the set of all injective sequences over W. Additionally, for all $U \subseteq W$, we say $w \in U \setminus \mathbf{w}$ in case w is an element of U but does not occur in $\mathbf{w}$. We use the notation $\mathbf{w}_i$ to denote the i-th element of $\mathbf{w}$. Finally, $[]$ denotes the empty sequence.
- The operator $\bullet : [W] \times W \rightarrow [W]$ appends states to sequences of states; i.e., $[w_1 \cdots w_n] \bullet w = [w_1 \cdots w_n w]$, provided that w does not occur in $[w_1 \cdots w_n]$.

Injective sequences can be thought of as *abstractions* of assignment functions, which just contain that information that is needed in the semantic characterisation. That is, each assignment function $f : Var \rightarrow W$, which we assume to be of finite range, can be represented by an injective sequence consisting of the elements in the range of f in some particular order. This representation thus abstracts from the particular domain of the function f.

We are now in the position to define the notion of a *history-based bisimulation*, which extends the notion of a bisimulation with a mechanism that handles bounded quantifications. For technical convenience only, we assume that the variables in formulae are bound only once.[2] That is, we do not consider formulae of the form $\exists x(\varphi \wedge \exists x \psi)$. This is not a real restriction as we can always take an alphabetic variant of these formulae: $\exists x(\varphi \wedge \exists y(\psi[y/x]))$ where y is a fresh variable, which is logically equivalent.

Definition 11 (*History-based bisimulation*)
Given the models $\mathcal{M} = \langle W, r^{\mathcal{M}}, n^{\mathcal{M}} \rangle$ and $\mathcal{N} = \langle U, r^{\mathcal{N}}, n^{\mathcal{N}} \rangle$, a relation

$$Z \subseteq (W \times [W]) \times (U \times [U])$$

is called a *history-based bisimulation*, if $(w, \mathbf{w})Z(u, \mathbf{u})$ implies the following:

---

[1] Properly, this is not true; one has to assume the image finiteness property or to consider ultra-filter extensions.

[2] This simplifies the condition (n-bisim) in Definition 11, as it allows us to restrict to extensions of sequences rather having to account for removals of states as well.

(self) $w = \mathbf{w_i}$ iff $u = \mathbf{u_i}$
(var) $\mathbf{w_i} \in n^{\mathcal{M}}(w)$ iff $\mathbf{u_i} \in n^{\mathcal{N}}(u)$
(r-bisim) if $w' \in r^{\mathcal{M}}(w)$ then $\exists u' \in r^{\mathcal{N}}(u)$ with $(w', \mathbf{w})Z(u', \mathbf{u})$
(n-bisim) if $w' \in n^{\mathcal{M}}(w) \setminus \mathbf{w}$ then $\exists u' \in n^{\mathcal{N}}(u) \setminus \mathbf{u}$ with $(w, \mathbf{w} \bullet w')Z(u, \mathbf{u} \bullet u')$

and vice versa for (r-bisim) and (n-bisim), where the roles of $(w, \mathbf{w})$ and $(u, \mathbf{u})$ are interchanged. Additionally, we define wZu to hold in case $(w, [])Z(u, [])$.

Example 12 (*History-based bisimulation*)
To illustrate the notion of a history-based bisimulation, let us return to the structures $\mathcal{M}$ and $\mathcal{N}$ depicted in figure 3, where we assume that the neighbourhood relation coincides with the accessibility relation. The language $\mathcal{L}_1$ distinguishes between these two structures, consider for instance the formula $\exists x(x = self)$.

We argue that there does not exist a history-based bisimulation Z with wZv_1. For suppose that such a relation exists then $(w, [])Z(v_1, [])$ and condition **n-bisim** requires $(w, [w])Z(v_1, [v_2])$ and subsequently by **r-bisim** we obtain $(w, [w])Z(v_2, [v_2])$. However, this is in contradiction with condition **var** as $w \in n(w)$ while $v_2 \notin n(v_2)$. Hence, we conclude that such a relation Z does not exist.

This simple case shows why the bisimulation is called *history-based*: the sequences $[w]$ and $[v_2]$ represent *histories* of states that have been encountered in neighbourhoods while traversing the structures $\mathcal{M}$ and $\mathcal{N}$ along their accessibility relation. If the elements of these sequences are encountered again, that is, are in the neighbourhood of the present state w in $\mathcal{M}$, this should be mimicked in $\mathcal{N}$, that is, are in the neighbourhood of the present state v_2.

If we restrict ourselves to finite structures, the notion of a history-based bisimulation is decidable. Note that it is crucial here that injective sequences do not contain repetitions of states.

Observation 13 (*Decidability of history-based bisimulation*)
Given structures $\mathcal{M}$ and $\mathcal{N}$ with *finite* domains, for all states $w \in \mathcal{M}$ and $u \in \mathcal{N}$, it is *decidable* whether there exists a history-based bisimulation Z with wZu.

It is worth remarking here that the notion of a history-based bisimulation is quite different from the notion of a *history-preserving bisimulation* [7]. The latter is a very strong notion saying that two states are history-preserving bisimilar in case they are related by a bisimulation and additionally, the respective substructures consisting of the states that can reach the state via the accessibility relation, are isomorphic.

Before we phrase the semantic characterisation of the language $\mathcal{L}_1$ in theorem 15, we define the notion of an image finite state.

Definition 14 (*Image-finiteness*)
Given a structure $\mathcal{M} = \langle S, r, n \rangle$ we let r^* denote the reflexive, transitive closure of r. A state $w \in S$ is called *r-image finite* if $r(v)$ is finite for all v with $(w, v) \in r^*$, and is called *n-image finite* if $n(v)$ is finite for all v with $(w, v) \in r^*$. Moreover, w is called *image finite* if it is both r-image finite and n-image finite.

Properly, we do not need the assumption of image-finiteness, as analogous to the proof of the semantic characterisation of standard modal logic, we could use ultrafilter extensions [3]. However, for the sake of simplicity we adopt this property here.

Theorem 15 (*Semantic characterisation*)
Given two structures $\mathcal{M}$ and $\mathcal{N}$, for all states w from $\mathcal{M}$ and u from $\mathcal{N}$ the following holds:

(i) if wZu for some history-based bisimulation Z then for all sentences $\varphi \in \mathcal{L}_1$ we have $\mathcal{M}, w \models \varphi \Leftrightarrow \mathcal{N}, u \models \varphi$

(ii) if w, u are image finite and $\mathcal{M}, w \models \varphi \Leftrightarrow \mathcal{N}, u \models \varphi$ for all sentences $\varphi \in \mathcal{L}_1$, then wZu for some history-based bisimulation Z.

Because of space limitations, we do not give a proof of this non-trivial result. Instead, we consider some applications of the result. First of all, consider the models $\mathcal{M}$ and $\mathcal{N}$ from Figure 1, where we assume that the accessibility relation and the neighbourhood relation coincide. The language $\mathcal{L}_1$ cannot distinguish between these models. This follows from the fact that there exists a history-based bisimulation between $\mathcal{M}, w$ and $\mathcal{N}, w$. Secondly, the language $\mathcal{L}_1$ cannot express the property that the accessibility relation is contained in the neighbourhood relation, as stated in the following result.

Corollary 16 There does not exist a formula $\varphi \in \mathcal{L}_1$ such that for all structures $\mathcal{M} = \langle W, r, n \rangle$ and states $w \in W$ we have: $\mathcal{M}, w \models \varphi \Leftrightarrow r(w) \subseteq n(w)$.

5   Decidability

In this section, we discuss the decidability of the language $\mathcal{L}_1$.

5.1   The Guarded Fragment

In this section, we examine the connection of our logic with the guarded fragment of first-order logic [1]. This logic, which satisfies the property of being decidable, consists of first-order formulae that are build from arbitrary atoms, boolean operators and finally, quantifications of the following format:

$$\exists \mathbf{y}(R\mathbf{y}\mathbf{x} \wedge \varphi(\mathbf{x}, \mathbf{y})),$$

where R is a particular predicate and $\mathbf{y}$ and $\mathbf{x}$ are sequences of variables. The semantic characterisation of the guarded fragment is defined in terms of a guarded bisimulation. That is, any formula ψ is equivalent to a formula in the guarded fragment if and only if ψ is invariant for guarded bisimulations. This notion is defined below.

Definition 17 A *guarded bisimulation* between two models $\mathcal{M}$ and $\mathcal{N}$ is a non-empty set F of finite partial isomorphisms that satisfies the following conditions. For all $f : X \rightarrow Y$ in F, we have

- for all guarded sets Z in $\mathcal{M}$ there exists g in F with domain Z such that g and f agree on $X \cap Z$

– for all guarded sets W in $\mathcal{N}$ there exists g in F with range W such that g^{-1} and f^{-1} agree on $Y \cap W$

where a set V is called *guarded* in a model in case there exist $a_1, \ldots, a_n$ (with repetitions, possibly) such that $V = \{a_1, \ldots, a_n\}$ and for some relation R we have that $R(a_1, \ldots, a_n)$ is true in the model.

We argue that $\mathcal{L}_1$ does not fall inside the guarded fragment. Consider the two structures $\mathcal{M}$ and $\mathcal{N}$ in Figure 2. The set F consisting of the partial isomorphisms $\{w \mapsto v, w_1 \mapsto v_1\}$ and $\{w \mapsto v, w_2 \mapsto v_1\}$ constitutes a guarded bisimulation between $\mathcal{M}$ and $\mathcal{N}$. Hence, there is no formula in the guarded fragment that distinguishes between these two structures. However, in our language $\mathcal{L}_1$ there is for instance the formula $\psi = (\exists x \exists y (\neg x = y))$ with $\mathcal{M}, w \models \psi$ and $\mathcal{N}, u \not\models \psi$. So, $\psi \in \mathcal{L}_1$ is not invariant for guarded bisimulations and therefore is not equivalent to a formula in the guarded fragment. So, we establish the following result.

Observation 18 (*Relation with guarded fragment*)
The language $\mathcal{L}_1$ is not contained in the guarded fragment of first-order logic.

5.2   Hybrid Languages

Our framework has connections with the work on what are called *hybrid languages*, which are languages that like $\mathcal{L}_1$ also combine modality with first-order quantification mechanisms [4,2]. In particular, hybrid languages extend the basic modal language $\mathcal{L}_0$, with a collection of *nominals* that are used to label states in models. These nominals are propositional formulae that are true at exactly one state in a model, and so to speak are employed as global *unique* names for states. Further extensions additionally incorporate operators of the form $@_i$ to *jump* to the state that is denoted by the nominal i, as well as operators to *bind* nominals. Here we consider the two fundamental ones of these binding operators; viz. the hybrid operator $\downarrow x$ and the hybrid existential quantifier, which we denote as $\bar{\exists} x$ to distinguish it from the quantifier $\exists x$ from $\mathcal{L}_1$.

First of all, the quantifier $\downarrow x$ binds the variable x to the current state of evaluation. It can be defined in the language $\mathcal{L}_1$ as follows:

$$\downarrow x \varphi \ = \ \exists x (x = \textit{self} \wedge \varphi).$$

Moreover, it corresponds to existential quantification in the class of structures in which the neighbourhood of states is given by the state itself; that is, in the class:

$$\{\mathcal{M} \mid \mathcal{M} \models \exists x (x = \textit{self} \wedge \forall y (y = x))\}.$$

Additionally, the hybrid quantifier $\bar{\exists} x$ ranges over the *entire* set of states in a structure. If we consider this operator in our framework, it corresponds to existential quantification in the class of structures in which the neighbourhood relation is universal, meaning that each state is in the neighbourhood of any other state. This class can be defined as follows:

$$\{\langle W, r, n \rangle \mid n = W \times W\}.$$

The language $\mathcal{L}_1$ is not expressive enough to characterise the above class of models, as this type of existential quantification assumes an external view on models rather than the local view that has been taken in our framework.

Finally, we mention the hybrid operator $@_x$ that is used to jump to the state denoted by the variable x. The truth definition of this operator can be given as follows:

$$\mathcal{M}, w, f \models @_x\varphi \Leftrightarrow \mathcal{M}, f(x), f \models \varphi.$$

This operator has no counterpart in our framework due to the fact that in each state, it allows going to states that are not necessary reachable via the accessibility relation. This is in contrast with one of our underlying assumptions, saying that in a state one cannot go to arbitrary states but only to an accessible one.

5.3 Finite Model Property and Decidability

The language $\mathcal{L}_1$ does not satisfy the finite model property, which is due to the fact that it can compel infinite neighbourhoods. Let $R(x, y)$ stand for the formula:

$$\Diamond(x = self \wedge \Diamond y = self),$$

which expresses that from the accessible state x the state y is accessible. Subsequently, let φ denote the conjunction of the following formulae $\exists x(x = x)$, which expresses that a neighbourhood is nonempty, $\forall x(\neg R(x, x))$ expressing the irreflexivity of the relation R, $\forall x \forall y \forall z((R(x, y) \wedge R(y, z)) \rightarrow R(x, z))$ denoting transitivity and $\forall x \exists y(R(x, y))$ expressing seriality. If this formula is true in a particular state w then the neighbourhood of this state is infinite. The construction of this neighbourhood $\{v_1, v_2, v_3, \ldots\}$ is sketched in figure 4.

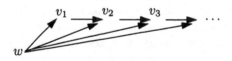

Fig. 4. An infinite neighbourhood

Moreover, it follows that the validity problem of the the language $\mathcal{L}_1$ is undecidable. In fact, this is a direct consequence of the result claimed in [2], which says that the hybrid language consisting of the basic modal language $\mathcal{L}_0$ extended with variables and the operator $\downarrow x$, is undecidable. The claim then follows from the fact that this hybrid language is a sublanguage of $\mathcal{L}_1$; i.e., hybrid formulae of the form $\downarrow x\varphi$ can be modelled in $\mathcal{L}_1$ as $\exists x(x = self \wedge \varphi)$.

The interesting question now arises of the role of the constant *self* in this result. Currently, we are investigating the expressivity and complexity of the language $\mathcal{L}_1$ without this constant. Here, we only mention that this sublanguage does not satisfy the finite model property either, which can be shown in a similar manner as above, using the following definition:

$$R(x, y) = \Diamond(\exists u(u = x) \wedge \Diamond(\exists u(u = y))).$$

Thus, $R(x, y)$ expresses that y is known in a state that can be accessed from some accessible state (with respect to the current state) in which x is known.

6   Topologies of Multi-agent Systems

In this section, we consider an application of our logic to the description of multi-agent topologies. More specifically, we show how our basic notion of quantification can be used in reasoning about the ambiguities of *names*; that is, situations in which one agent is known by other agents under different names.

Formally, we extend our language $\mathcal{L}_1$ with a countable set C of names, with typical element c. A term t in the extended language, which is called $\mathcal{L}_2$, is thus either a variable x, the constant *self*, or a name $c \in C$. Formulae are defined as in Definition 5 and they are interpreted over the following structures.

Definition 19 A *multi-agent topology* over the set of names C is a structure:

$$\langle W, r, I \rangle,$$

where W is a set of states, or agents, $r \subseteq W \times W$ denotes the accessibility relation, and I is a total function which assigns to each $w \in W$ an interpretation $I(w)$ of each name $c \in C$, that is, $I(w) \in C \to W$.

The definition of the truth of a formula φ in the extended language $\mathcal{L}_2$ involves a straightforward adaptation of the truth definition of the language $\mathcal{L}_1$ and is therefore omitted. Instead, we explain here the use of quantification in the description of the ambiguities to which names may give rise. First, we observe that without quantification we cannot describe phenomena like that one agent is known by different agents under different names. For example, given an agent w, we cannot describe the situation that $I(w)(c) = I(w')(c)$, for some $(w, w') \in r$, simply because the modal operators induce a "context switch", that is, a different interpretation of the names. However this situation can be described using quantifiers simply by the formula:

$$\exists x(x = c \wedge \Diamond(x = c)).$$

So, we bind the value of the constant c to the variable x, and use the fact that the interpretation of the variables is fixed, that is, does not change when "moving" from one agent to another.

In practice, we may assume without loss of generality that the set C of names is finite. Under this assumption we can, without loss of expressive power, restrict to *bounded* quantification of the form:

$$\exists x(x = c \wedge \varphi).$$

For this language the validity problem is decidable. We are currently working on a decision procedure that is based on a semantic tableau construction.

7 Future Research

Many issues remain to be studied such as expressivity and complexity results and the development of a complete axiomatization of (sublanguages of) the language $\mathcal{L}_1$. Other topics of interest include the introduction of predicates to describe properties of agents, for example properties expressing security aspects. Additionally, we want to investigate the introduction of the inverse $\Diamond^{-1}$ and the reflexive, transitive closure $\Diamond^*$ of the operator $\Diamond$ for describing properties of network topologies. A final issue is the study of the connection with epistemic logic [12].

References

1. H. Andréka, J. van Benthem, and I. Németi. Modal languages and bounded fragments of predicate logic. ILLC Research Report ML-1996-03, Institute for Logic, Language and Computation, 1996.
2. C. Areces, P. Blackburn, and M. Marx. A road-map on complexity for hybrid logics. In *Proceedings of Annual Conference of the European Association for Computer Science Logic (CSL'99)*, volume 1683 of *Lecture Notes in Computer Science*. Springer-Verlag, 1999.
3. J.F.A.K van Benthem. *Modal Logic and Classical Logic*. Bibliopolis, Napels, 1983.
4. P. Blackburn and J. Seligman. Hybrid languages. *Journal of Logic, Language and Information*, 4:251–272, 1995.
5. R.M. van Eijk, F.S. de Boer, W. van der Hoek, and J.-J.Ch. Meyer. Open multi-agent systems: Agent communication and integration. In N.R. Jennings and Y. Lespèrance, editors, *Intelligent Agents VI, Proceedings of 6th International Workshop on Agent Theories, Architectures, and Languages (ATAL'99)*, volume 1757 of *Lecture Notes in Artificial Intelligence*, pages 218–232. Springer-Verlag, Heidelberg, 2000.
6. M. Fitting and R.L. Mendelsohn. *First-Order Modal Logic*. Kluwer Academic Publishers, Dordrecht, The Netherlands, 1998.
7. U. Goltz, R. Kuiper, and W. Penczek. Propositional temporal logics and equivalences. In *Proceedings of Concur'92*, volume 630 of *Lecture Notes in Computer Science*, pages 222–236, Berlin, 1992. Springer-Verlag.
8. A.J. Grove and J.Y. Halpern. Naming and identity in epistemic logics. part I: The propositional case. *Journal of Logic and Computation*, 3(4):345–378, 1993.
9. M. Hennessy and R. Milner. Algebraic laws for nondeterminism and concurrency. *Journal of Association of Computer Machinery*, 32:137–162, 1985.
10. W. van der Hoek. On the semantics of graded modalities. *Journal of Applied Non Classical Logics*, 2(1):81–123, 1992.
11. G.E. Hughes and M.J. Cresswell. *An Introduction to Modal Logic*. Methuen and Co. Ltd, London, 1968.
12. J.-J.Ch. Meyer and W. van der Hoek. *Epistemic Logic for Computer Science and Artificial Intelligence*, volume 41 of *Cambridge Tracts in Theoretical Computer Science*. Cambridge University Press, 1995.
13. O. Shehory. A scalable agent location mechanism. In *Intelligent Agents VI - Proceedings of 6th International Workshop on Agent Theories, Architectures, and Languages (ATAL'99)*, volume 1757 of *Lecture Notes in Artificial Intelligence*, pages 162–172. Springer-Verlag, Heidelberg, 2000.
14. M. Wooldridge and N. Jennings. Intelligent agents: theory and practice. *The Knowledge Engineering Review*, 10(2):115–152, 1995.

Avoiding Logical Omniscience by Using Subjective Situations

Antonio Moreno[1], Ulises Cortés[2], and Ton Sales[2]

[1] Departament d'Enginyeria Informàtica i Matemàtiques - Universitat Rovira i Virgili (URV)
Carretera de Salou, s/n. 43006-Tarragona, Spain
amoreno@etse.urv.es
[2] Dep. de Llenguatges i Sistemes Informàtics - Universitat Politècnica de Catalunya (UPC)
C/Jordi Girona, 1-3. 08034-Barcelona, Spain
{ia,sales}@lsi.upc.es

Abstract. The beliefs of the agents in a multi-agent system have been formally modelled in the last decades using *doxastic logics*. The *possible worlds model* and its associated *Kripke semantics* provide an intuitive semantics for these logics, but they commit us to model agents that are *logically omniscient*. We propose a way of avoiding this problem, using a new kind of entities called *subjective situations*. We define a new doxastic logic based on these entities and we show how the belief operators have some desirable properties, while avoiding logical omniscience. A comparison with two well-known proposals (Levesque's *logic of explicit and implicit beliefs* and Thijsse's *hybrid sieve systems*) is also provided.

1 Introduction

In the last decade *doxastic modal logics* have been considered the most appropriate formal tool for modelling the beliefs of the agents composing a multi-agent system ([1]). The standard way of providing a meaning to the modal formulas of these logics is to use the *possible worlds model* ([2]) and its associated *Kripke semantics* ([3]). This semantics is quite natural and intuitive, but it is well known that the agents modelled in this framework are *logically omniscient* ([4]). Therefore, this semantics is unsuitable to model the beliefs of realistic, non-ideal agents. The aim of our work is to provide a plausible way of modelling the beliefs of non-logically omniscient agents, while keeping the essence and the beauty of the possible worlds model and the Kripke semantics.

This article[1] is structured as follows. In section 2 we give an intuitive explanation of our approach to the logical omniscience problem, which is based in a new kind of entities called *subjective situations*. In a nutshell, a subjective situation is the perception that an agent has of a certain state of affairs. These situations, as will be explained below, will take the role of possible worlds. In section 3, a formalization of subjective situations in the framework of doxastic propositional logic is made. Section 4 is devoted to a study of the behaviour of the modal belief operators, that extends and generalizes our previous results ([5]). It is shown how their properties do indeed correspond with

---

[1] This research has been supported by the CICYT project *SMASH: Multi-agent systems and its application to hospital services (TIC96-1038-C04-04)*.

M. Ojeda-Aciego et al. (Eds.): JELIA 2000, LNAI 1919, pp. 284–299, 2000.
© Springer-Verlag Berlin Heidelberg 2000

our intuitions about what should be an adequate formalization of the doxastic attitude of a non-ideal, non-logically omniscient agent. In section 5, a comparison of our proposal with two well-known approaches (Levesque's *logic of explicit and implicit beliefs* ([6]) and Thijsse's *hybrid sieve systems* ([7])) is performed. The paper finishes with a brief summary and the bibliographical references.

2   Motivation of Subjective Situations

The most popular way of dealing with the logical omniscience issue is to change the concept of what a *possible world* is (see [8] for a detailed review of the most interesting approaches to the problem of logical omniscience). Regardless of the way in which the concept of possible world is modified, there is a kernel that never changes: the formal representation of a possible world is not related in any way with the notion of *agent*. Thus, it may be said that all the approaches in the literature present an *objective* view of what a possible world is (*i.e.* a world is the same for all the agents, is independent of them). In a standard *Kripke structure*, the only item that depends on each agent is its *accessibility relation* between possible worlds.

The traditional meaning assigned to the accessibility relation R_i of an $Agent_i$ is that it represents the uncertainty that $Agent_i$ has about the situation in which it is located (*e.g.* $(w_0 R_5 w_1)$ means that $Agent_5$ cannot distinguish between worlds w_0 and w_1). This situation is quite peculiar, because the formulae that are true in two worlds that are linked by an accessibility relation are, in principle, totally unrelated (*i.e.* given a Kripke structure, there is no relationship between the accessibility relation between states and the function that assigns truth values to the basic propositions in each of them).

Our proposal may be motivated by the following scenario. Imagine two people (α and β) that are watching a football match together. In a certain play of the game, a fault is made and the referee awards a penalty kick. α thinks that the referee is right, because it has noticed that the fault was made inside the penalty area (let us represent this fact with proposition P); at the same time, β is thinking that the referee was wrong because, in its perception of the situation, the fault was made just an inch outside the penalty area. How can this situation (and the beliefs of the two agents) be formally represented?

Following the standard approach, we could model the fact that α believes P and β believes $\neg P$ by assuming that in all the (objectively described) worlds considered as possible in the current state by α the proposition P holds, whereas in all the worlds considered as possible by β (β's doxastic alternatives) P is false. This account of each agent's doxastic state does not seem very satisfactory to us, at least for two reasons:

- It does not tell us how each agent's perception of the situation influences in its own beliefs. An agent is supposed to eliminate instantly from its set of doxastic alternatives all those (completely specified) possible worlds in which a basic proposition has a truth value that does not match the agent's current beliefs. It would be more plausible to have a framework in which the agent kept a partial description of the situation in which it is located, and in which it could use the facts that it keeps perceiving from the environment in order to keep increasing and refining its beliefs ([9], [10]).

- Assuming that the fault was indeed made inside the penalty area, most philosophers would argue that α not only *believes* P but also *knows* it (being P true in the real world), whereas β believes $\neg P$ but can not possibly know $\neg P$, being it actually false[2]. Thus, in a somehow *magical* way, one agent would have some knowledge (that would coincide with reality) whereas the other wouldn't.

In our opinion, this state of affairs (the actual situation, comprising both the football match and the agents, along with their beliefs) may not be adequately described with a simple assignment of truth values to the basic propositions. Even if we had an accurate description of the real world, does it really matter very much whether the fault was made inside the penalty area in order to model the beliefs of the two agents involved in the scene?

The situation (s) is obviously the same for the two agents α and β (they are watching the same match together). From α's point of view, the description of s should make true proposition P; however, from β's perspective, in the present situation P should be considered false. Obviously, there would be many aspects of s in which α and β would agree; e.g. both of them would consider that the proposition representing the fact "*We are watching a football match on TV*" is true in s.

As far as beliefs are concerned, we argue that, in this situation, α should be capable of stating that $B_\alpha P$ (α has seen the fault and has noticed that it was made inside the penalty area; thus, it believes so). It would not seem very acceptable a situation in which α perceived the fault to have been made inside the penalty area and defended that it did not believe that a penalty kick should have been awarded (the only possible explanation being that α is a strong supporter of the offending team). It also seems reasonable to say that α cannot fail to notice that it believes that the fault was made inside the penalty area; thus, α may also assert in s that $B_\alpha B_\alpha P$. In a similar way, in this situation β cannot state that $B_\beta P$ (β cannot defend that it believes that the referee is right, in a situation in which it perceived the fault to have been made outside the penalty area). Thus, it seems clear that each agent's point of view on a situation strongly influences (or we could say even *determines*) its positive and negative beliefs in that situation.

In our framework we want to include the intuition that agents are smart enough to know that other agents may not perceive reality in the same way as they do. In the previous example, without further information (e.g. α shouting "*Penalty!*"), β should not be capable of supporting (or rejecting) that $B_\alpha P$; analogously, α could not affirm (or deny) that $B_\beta P$. That means that the communication between the agents is the main way in which an agent may attain beliefs about other agent's beliefs. We could have chosen other alternatives; for instance, we could have stated that an agent believes that the other agents perceive reality in the same way as they do, provided that they do not have information that denies that fact. If that were the case α would assume that β also believes that P is true, as far as it does not have any reason not to think so (e.g. β saying "*This referee is really blind*").

---

[2] It could be argued that we are somehow neglecting the need of a *justification* for the belief in order for it to become knowledge (as knowledge is usually defined in the philosophical literature as *true justified belief*). But, what could possibly count more as a justification that each agent's own direct perception of the situation?

A final reflection on the meaning of the accessibility relation between situations for $Agent_i$ (R_i) is necessary. It will be assumed that an agent cannot have any doubts about its own perceptions and beliefs in a given state. E.g. if, in situation s, α looks at the match and thinks P, then it surely must realise this fact and believe P in s (and even believe that it believes P, were it to think about that). Thus, if R_α links s with all those situations that α cannot tell apart from s, it must be the case that α also perceives P as true in all those states as well (otherwise, those states would be clearly distinguishable by α, because in some of them it would support P whereas in some of them P would be rejected). The only uncertainty that α may have is *about the perception of s by the other agents*. In the example, α does not know whether it is in a situation in which β supports P or in a situation in which β rejects P. Therefore, α's accessibility relation must reflect this uncertainty.

Summarising, the main points that have been illustrated with the previous discussion are the following:

- A situation may be considered not as an entity that may be objectively described, but as a piece of reality that may be perceived in different ways by different agents. Thus, it is necessary to think of a *subjective* way of representing each situation, in which each agent's point of view is taken into account. In the previous example, the description of s should include the fact that α is willing to support P, whereas β isn't.
- An agent's beliefs in each situation also depend on its point of view.
  In the situation of the example, $B_\alpha P$ would hold from α's perspective, whereas it would not be either supported or rejected by β. Thus, we argue that it does not make sense to ask whether $B_\alpha P$ holds in s or not; that question must be referred *to a particular agent's point of view*.
- The interpretation of the meaning of each agent's accessibility relation is slightly different from the usual one.
  Each accessibility relation R_i will keep its traditional meaning, *i.e.* it will represent the uncertainty of $Agent_i$ with respect to the situation in which it is located. However, our intuition is that an agent may only be uncertain about the other agents' perception of the present state, not about its own perception.

3   Formalization of Subjective Situations

These intuitive ideas are formalized in the *structures of subjective situations*:

Definition 1 (Structure of Subjective Situations)
An structure of subjective situations for n agents is a tuple

$$< S, R_1, ..., R_n, T_1, ...T_n, \mathcal{F}_1, ..., \mathcal{F}_n >, where$$

- *S is the set of possible situations.*
- *R_i is the accessibility relation between situations for $Agent_i$.*
- *T_i is a function that returns, for each situation s, the set of propositional formulae that are perceived as true by $Agent_i$ in s.*

- $\mathcal{F}_i$ *is a function that returns, for each situation s, the set of propositional formulae that are perceived as false by Agent$_i$ in s.*

$\mathcal{E}$ *is the set of all structures of subjective situations.*

The presence of T_i and $\mathcal{F}_i$ allows *Agent$_i$* to consider *partial* situations (those in which *Agent$_i$* does not have any reason to support or to reject a given formula) as well as *inconsistent* situations (those in which *Agent$_i$* may have reasons to support and to reject a given formula). This kind of situations was already considered by Levesque in his *logic of explicit and implicit beliefs* ([6]). A detailed comparison of our proposal and that of Levesque is offered in section 5.

The accessibility relation between situations for *Agent$_i$* has to reflect its uncertainty about the way in which the actual situation is perceived by the other agents. Thus, R_i has to link all those states that *Agent$_i$* perceives in the same way but that may be perceived in different ways by other agents. This intuition is formalized in the following condition:

Definition 2 (Condition on Accessibility Relations)

$$\forall s,t \epsilon S, \, (sR_i t) \text{ if and only if } (T_i(s) = T_i(t)) \text{ and } (\mathcal{F}_i(s) = \mathcal{F}_i(t))$$

This condition implies that the accessibility relations are equivalence relations. This result links this approach with the classical $S5$ modal system, in which this condition also holds. In $S5$ the presence of this condition makes true axiom *4* (positive introspection), axiom *5* (negative introspection) and axiom *T* (the axiom of knowledge); the modal operators of the system proposed in this article will have similar properties, as will be shown in section 4.

3.1 Satisfiability Relations

A simplified version of the doxastic propositional language for n agents is considered, as shown in the following definition:

Definition 3 (Doxastic Modal Language $\mathcal{L}$)
    Consider a set of modal belief operators for n agents (B_1, ..., B_n). $\mathcal{L}$ is the language formed by all propositional formulae (built in the standard way from a set $\mathcal{P}$ of basic propositions and the logical operators $\neg, \vee, \wedge, \rightarrow$), preceded by a (possibly empty) sequence of (possibly negated) modal operators. $\mathcal{L}_{PC}$ is the subset of $\mathcal{L}$ that contains those formulae that do not have any modal operator. The modal formulae of $\mathcal{L}$ are called linearly nested.

Thus, the language $\mathcal{L}$ contains formulae such as P, $B_3 Q$, $B_1 B_5 (R \vee T)$, $B_3 \neg B_2 S$ and $\neg B_1 B_1 \neg T$, but it is not expressive enough to represent formulae such as $(B_2 P \rightarrow B_3 Q)$ or $(P \vee B_5 Q)$. In most practical applications, an agent in a multi-agent system will only need to represent what it believes (or not) to be the case in the world and what it believes (or not) that the other agents believe (or not). This is just the level of complexity offered by *linearly nested* formulae.

In an structure of subjective situations each $Agent_i$ may have *positive* and *negative* information about some propositional formulae (given by T_i and F_i, respectively). This allows us to define two relations (of satisfiability, $\models_i$, and unsatisfiability, $=|_i$) between situations and formulae *for each* $Agent_i$. Given an structure of subjective situations E and a situation s, the expression $E, s \models_i \phi$ should hold whenever $Agent_i$ has some reason to think that ϕ is true in situation s. Similarly, $E, s =|_i \phi$ should hold whenever $Agent_i$ has some reason to reject ϕ in situation s.

Notice that $E, s \not\models_i \phi$ should not imply that $E, s =|_i \phi$ (i.e. $Agent_i$ not having any reason to support ϕ does not mean that it must have reasons to reject it). In the same spirit, $E, s \models_i \phi$ should not imply that $E, s \neq|_i \phi$ ($Agent_i$ could have reasons both to support and to reject a certain formula in a given situation). These facts will indeed be true, as will be seen in the next section, due to the presence of partial and inconsistent situations commented above.

The clauses that define the behaviour of these relations are shown in the following definition:

Definition 4 (Relations $\models_i$ and $=|_i$)

- $\forall E \epsilon \mathcal{E}, \forall s \epsilon S, \forall agent\ i, \forall \phi \epsilon \mathcal{L}_{PC}$

$$E, s \models_i \phi \Leftrightarrow \phi \epsilon T_i(s)$$
$$E, s =|_i \phi \Leftrightarrow \phi \epsilon F_i(s)$$

- $\forall E \epsilon \mathcal{E}, \forall s \epsilon S, \forall agents\ i, j, \forall \phi \epsilon \mathcal{L}$

$$E, s \models_i B_j \phi \Leftrightarrow \forall t \epsilon S\ ((sR_i t)\ implies\ E, t \models_j \phi)$$
$$E, s =|_i B_j \phi \Leftrightarrow \exists t \epsilon S\ ((sR_i t)\ and\ E, t =|_j \phi)$$

- $\forall E \epsilon \mathcal{E}, \forall s \epsilon S, \forall agents\ i, j, \forall \phi \epsilon \mathcal{L}$

$$E, s \models_i \neg B_j \phi \Leftrightarrow E, s =|_i B_j \phi$$
$$E, s =|_i \neg B_j \phi \Leftrightarrow E, s \models_i B_j \phi$$

A propositional formula ϕ is supported in a given situation s by an $Agent_i$ if and only if $Agent_i$ has reasons to think that ϕ is true in s. Analogously, ϕ will be rejected if and only if there are reasons that support its falsehood (recall that a formula may be both supported and rejected in a given situation). As far as beliefs are concerned, in a given situation s, $Agent_i$ supports that $Agent_j$ believes ϕ just in case $Agent_j$ supports ϕ in all the situations that are considered possible by $Agent_i$ in s ($Agent_i$'s doxastic alternatives). Similarly, $Agent_i$ may reject the fact that $Agent_j$ believes ϕ if it may think of a possible situation in which $Agent_j$ rejects ϕ. Finally, $Agent_i$ will support that $Agent_j$ does not believe ϕ if it may reject the fact that $Agent_j$ believes ϕ. We do not need more clauses to define the behaviour of the satisfiability and unsatisfiability relationships due to the restriction to linearly nested formulae imposed in definition 3.

4   Properties of the Belief Operators

The definition of an structure of subjective situations, the fact that the accessibility relations are equivalence relations and the clauses that describe the behaviour of the satisfiability (and unsatisfiability) relations compose a framework in which the modal belief operator of each $Agent_i$ has several interesting logical properties (that, in our opinion, make it an appropriate operator to model the notion of belief for a non-ideal agent). Some of these properties are described in this section.

4.1   General Results

Proposition 1 (Lack of Logical Omniscience)
   In the framework of subjective situations, none of the following forms of logical omniscience ([8]) holds:

- *Full logical omniscience.*
- *Belief of valid formulae.*
- *Closure under logical implication.*
- *Closure under logical equivalence.*
- *Closure under material implication.*
- *Closure under valid implication.*
- *Closure under conjunction.*
- *Weakening of beliefs.*
- *Triviality of inconsistent beliefs.*

   Proof: Let us take a state s in which $T_i(s) = \{P, (P \to Q), \neg P\}$ and $F_i(s) = \{P\}$. Consider an structure for subjective situations E that only contains the situation s.

- $E, s \models_i B_i P$ and $E, s \models_i B_i(P \to Q)$ hold, but $E, s \models_i B_i Q$ does not hold. Therefore, neither full logical omniscience nor closure under material implication hold.
- $E, s \models_i B_i(Q \vee \neg Q)$ does not hold. Therefore, there is no belief of valid formulae.
- $E, s \models_i B_i P$ holds, but $E, s \models_i B_i(P \vee Q)$ does not hold. Therefore, closure under logical implication and weakening of beliefs do not hold.
- $E, s \models_i B_i(P \to Q)$ holds, but $E, s \models_i B_i(\neg Q \to \neg P)$ does not. Therefore, beliefs are not closed under logical equivalence or under valid implication.
- $E, s \models_i B_i P$ and $E, s \models_i B_i(P \to Q)$ hold, but the expression $E, s \models_i B_i(P \wedge (P \to Q))$ does not hold. Therefore, there is no closure under conjunction.
- $E, s \models_i B_i P$ and $E, s \models_i B_i \neg P$ hold, but $E, s \models_i B_i Q$ does not hold. Therefore, there is no triviality of inconsistent beliefs.   □

   There are two basic reasons that account for the failure of all these properties:

- T_i and F_i are defined on sets of (arbitrary) formulae (not on basic propositions).
- T_i and F_i are unrelated. Thus, a given formula may belong to both sets, to only one of them or to none of them.

It is possible to impose any of the above properties on the belief operators by requiring these sets of formulae to satisfy some conditions (for instance, if $(\phi \wedge \psi)\epsilon T_i(s)$ implies that $\phi\epsilon T_i(s)$ and $\psi\epsilon T_i(s)$, then $Agent_i$'s belief set would be closed under conjunction).

Proposition 2 (Relation between $\models_i$ and $=|_i$)

> *For any linearly nested formula ϕ*
> $E, s \not\models_i \phi$ *does not imply* $E, s =|_i \phi$
> $E, s \models_i \phi$ *does not imply* $E, s \neq|_i \phi$

Proof: Take the structure of subjective situations E described in the proof of the previous proposition. It is easy to check these facts:

- $E, s \not\models_i B_i R$ and $E, s \neq|_i B_i R$. Therefore, $E, s \not\models_i \phi$ does not imply $E, s =|_i \phi$.
- $E, s \models_i B_i P$ and $E, s =|_i B_i P$. Therefore, $E, s \models_i \phi$ does not imply $E, s \neq|_i \phi$. $\qquad\square$

4.2   Results on Positive Introspection

Proposition 3 (Characterization of positive beliefs)
For any linearly nested formula ϕ,

$$E, s \models_i \phi \text{ if and only if } E, s \models_i B_i\phi$$

Proof: The *if* side of the formula coincides with proposition 4. The *only if* side may be proven as follows:
$E, s \models_i B_i\phi \implies \forall t(sR_it), (E, t \models_i \phi)$. As R_i is reflexive, (sR_is); therefore, $E, s \models_i \phi$. $\qquad\square$
This result states that $Agent_i$ believes ϕ in state s if and only if ϕ is one of the facts that is supported by $Agent_i$ in that state[3]. Thus, in our framework the difference between *belief* and *knowledge* vanishes: both concepts have to be understood as the propositional attitude that the agents adopt towards those formulae that they perceive to be true in the environment. Therefore, the (rather philosophical) difference between those beliefs that are true in the real world (that constitute knowledge) and those that are not (*plain* beliefs) is not taken into account.

Proposition 4 (Belief of supported formulae)

> *For any linearly nested formula ϕ,*
> $E, s \models_i \phi$ *implies* $E, s \models_i B_i\phi$

Proof: There are five cases to be considered:

- ϕ is a propositional formula.
  $E, s \models_i \phi$ and ϕ is propositional $\implies \phi\epsilon T_i(s) \implies \forall t(sR_it), \phi\epsilon T_i(t)$
  $\implies \forall t(sR_it), E, t \models_i \phi \implies E, s \models_i B_i\phi$

---

[3] The "only if" side of the proposition is the classical axiom of knowledge, axiom T.

- If ϕ is a modal formula that starts with an affirmed belief operator B_i (i.e. $\phi = B_i\psi$), this fact is exactly the next proposition.
- If ϕ is a modal formula that starts with an affirmed belief operator B_j (i.e. $\phi = B_j\psi$), this statement coincides with proposition 6, that will be proved later.
- If ϕ is a modal formula that starts with a negated belief operator B_i (i.e. $\phi = \neg B_i\psi$), this fact is the one proved as proposition 10.
- If ϕ is a modal formula that starts with a negated belief operator B_j (i.e. $\phi = \neg B_j\psi$), this fact is the one proved as proposition 11. $\square$

This proposition is telling us that an agent believes all formulae that it has reasons to support, as suggested in the motivating example. However, this proposition has an added value over our intuitions, because it refers to any kind of linearly nested formulae, and not only to propositional formulae.

Proposition 5 (Single-agent positive introspection)

$$\text{For any linearly nested formula } \phi,$$
$$E, s \models_i B_i\phi \text{ implies } E, s \models_i B_i B_i\phi$$

Proof: If $E, s \models_i B_i\phi$, that means that $E, s \models_i \phi$ holds in all the situations R_i-related to s. Being R_i an equivalence relation, these situations are exactly the ones included in the equivalence class of s induced by R_i. This class is also the set of situations that may be accessed from s in two steps (in fact, in any number of steps) via R_i, and ϕ is supported by $Agent_i$ in all of them. Thus, $\forall s'(sR_i s')\forall s''(s'R_i s'')E, s'' \models_i \phi$, and $E, s \models_i B_i B_i\phi$ also holds. $\square$

This proposition states that axiom 4 (the classical axiom of positive introspection) holds for each belief operator B_i (i.e. every agent has introspective capabilities on its own positive beliefs).

Proposition 6 (Generation of positive beliefs)

$$E, s \models_i B_j\phi \text{ implies } E, s \models_i B_i B_j\phi$$

Proof: $E, s \models_i B_j\phi \implies \forall t(sR_i t), E, t \models_j \phi$. Thus, $E, t \models_j \phi$ holds in all the worlds t that belong to the same equivalence class that s (considering the partition defined by R_i). Therefore, in all the worlds accessible from s via R_i in any number n of steps, $E, t \models_j \phi$. Taking the case $n = 2$, we obtain that $E, s \models_i B_i B_j\phi$. $\square$

If an agent has reasons to support a certain belief of another agent, then that belief will be included in its belief set.

Proposition 7 (Inter-agent positive introspection)

$$E, s \models_i B_j\phi \text{ implies } E, s \models_i B_j B_j\phi$$

Proof: $E, s \models_i B_j\phi \implies \forall t(sR_i t), E, t \models_j \phi$. Using the result given in proposition 4, that formula implies that $\forall t(sR_i t), E, t \models_j B_j\phi$; thus, $E, s \models_i B_j B_j\phi$. $\square$

This result is more general (proposition 5 reflected the case $i = j$). It states that each agent is aware of the fact that the other agents also have introspective capabilities.

Proposition 8 (Multi-agent positive introspection)
It does not hold (for three different agents $Agent_i$, $Agent_j$ and $Agent_k$ and a linearly nested formula ϕ) that

$$E, s \models_i B_j\phi \text{ implies } E, s \models_i B_k B_j\phi$$

Proof: We will show a counterexample. Take an structure for subjective situations E with two situations, s and t, such that $(sR_k t)$ holds, but $(sR_i t)$ and $(sR_j t)$ do not. Take a formula ϕ such that $\phi \epsilon T_j(s)$ and $\phi \not\epsilon T_j(t)$. In this state of affairs, $E, s \models_i B_j\phi$ holds but $E, s \models_i B_k B_j\phi$ does not hold.   □

This proposition states a negative result. It is telling that even if $Agent_i$ has reasons to support that $Agent_j$ believes something, that is not enough for $Agent_i$ to think that any other $Agent_k$ will have that belief. This proposition is essentially expressing the uncertainty of $Agent_i$ about the beliefs of a different $Agent_k$.

4.3   Results on Negative Introspection

Proposition 9 (Characterization of negative beliefs)
For any linearly nested formula ϕ,

$$E, s \models_i \phi \text{ if and only if } E, s \models_i \neg B_i\phi$$

Proof: The *if* side of the proposition may be proven as follows. As we know that $E, s \models_i \phi$ and $(sR_i s)$, it may be said that $\exists t(sR_i t), E, t \models_i \phi$. Therefore, $E, s \models_i B_i\phi$, which is equivalent to $E, s \models_i \neg B_i\phi$.

The *only if* side of the proposition (i.e. $E, s \models_i \neg B_i\phi$ implies $E, s \models_i \phi$) will be proved considering five different cases (as we did in the proof of proposition 4):

- ϕ is a propositional formula.
  $E, s \models_i \neg B_i\phi \Longrightarrow E, s \models_i B_i\phi \Longrightarrow \exists t(sR_i t), E, t \models_i \phi$. As ϕ is propositional, $E, t \models_i \phi$ implies that $\phi \epsilon F_i(t)$; as $(sR_i t)$, $\phi \epsilon F_i(s)$. Therefore, $E, s \models_i \phi$.
- ϕ is a modal formula that starts with an affirmed belief operator B_i (i.e. $\phi = B_i\psi$).

  $$E, s \models_i \neg B_i\phi \Longrightarrow E, s \models_i \neg B_i B_i\psi \Longrightarrow E, s \models_i B_i B_i\psi \Longrightarrow$$
  $$\exists t(sR_i t), E, t \models_i B_i\psi \Longrightarrow \exists t, u(sR_i t), (tR_i u), E, u \models_i \psi.$$

  As R_i is transitive, $(sR_i t)$ and $(tR_i u)$ imply that $(sR_i u)$. Thus, we may state that $\exists u(sR_i u), E, u \models_i \psi$. Therefore, $E, s \models_i B_i\psi$, which is equal to $E, s \models_i \phi$.
- ϕ is a modal formula that starts with an affirmed belief operator B_j (i.e. $\phi = B_j\psi$).

  $$E, s \models_i \neg B_i\phi \Longrightarrow E, s \models_i \neg B_i B_j\psi \Longrightarrow E, s \models_i B_i B_j\psi \Longrightarrow$$
  $$\exists t(sR_i t), E, t \models_i B_j\psi \Longrightarrow \exists t, u(sR_i t), (tR_i u), E, u \models_j \psi.$$

  As R_i is transitive, $(sR_i t)$ and $(tR_i u)$ imply that $(sR_i u)$. Thus, we may state that $\exists u(sR_i u), E, u \models_j \psi$. Therefore, $E, s \models_i B_j\psi$, which is equal to $E, s \models_i \phi$.
- ϕ is a modal formula that starts with a negated belief operator B_i (i.e. $\phi = \neg B_i\psi$).

$$E, s \models_i \neg B_i \phi \implies E, s \models_i \neg B_i \neg B_i \psi \implies E, s \models_i B_i \neg B_i \psi \implies$$
$$\exists t(sR_i t), E, t \models_i \neg B_i \psi \implies \exists t(sR_i t), E, t \models_i B_i \psi \implies$$
$$\exists t(sR_i t) \forall u(tR_i u), E, u \models_i \psi.$$

In this expression, t is a world that belongs to the same class of equivalence than s (according to the partition defined by R_i), and u represents all the worlds that belong to t's class of equivalence; thus, u ranges over all the worlds belonging to s's class of equivalence (all the worlds that are accessible from s via R_i in any number n of steps). If we take $n = 1$, we get that $\forall t(sR_i t), E, t \models_i \psi$. Thus, $E, s \models_i B_i \psi$, which is equivalent to $E, s \models_i \neg B_i \psi$. Therefore, $E, s \models_i \phi$.

– ϕ is a modal formula that starts with a negated belief operator B_j (i.e. $\phi = \neg B_j \psi$).

$$E, s \models_i \neg B_i \phi \implies E, s \models_i \neg B_i \neg B_j \psi \implies$$
$$E, s \models_i B_i \neg B_j \psi \implies \exists t(sR_i t), E, t \models_i \neg B_j \psi \implies$$
$$\exists t(sR_i t), E, t \models_i B_j \psi \implies \exists t(sR_i t) \forall u(tR_i u), E, u \models_j \psi.$$

In this expression, t is a world that belongs to the same class of equivalence than s (according to the partition defined by R_i), and u represents all the worlds that belong to t's class of equivalence; thus, u ranges over all the worlds belonging to s's class of equivalence (all the worlds that are accessible from s via R_i in any number n of steps). If we take $n = 1$, we get that $\forall t(sR_i t), E, t \models_j \psi$. Thus, $E, s \models_i B_j \psi$, which is equivalent to $E, s \models_i \neg B_j \psi$. Therefore, $E, s \models_i \phi$.   □

Agent$_i$ does not believe ϕ at s if and only if ϕ is one the facts that is rejected by i at s. Again, this proposition agrees with the intuitions that we had in the example that was used to motivate the need for the framework of subjective situations.

Proposition 10 (Single-agent negative introspection)

$$E, s \models_i \neg B_i \phi \text{ implies } E, s \models_i B_i \neg B_i \phi$$

Proof: $E, s \models_i \neg B_i \phi \implies E, s \models_i B_i \phi \implies \exists t(sR_i t), (E, t \models_i \phi)$. Thus, there exists at least one world (say w) such that $(sR_i w)$ and $E, w \models_i \phi$. In order to prove the proposition, we have to notice that R_i is Euclidean (i.e. whenever $(sR_i t)$ and $(sR_i u)$, $(tR_i u)$ also holds)[4]. Therefore, w is R_i accessible from all worlds that are R_i accessible from s, and we may state that $\forall t(sR_i t), (tR_i w)$ and $E, w \models_i \phi$. Thus, $\forall t(sR_i t) \exists u(tR_i u)E, u \models_i \phi$. Thus, $\forall t(sR_i t) E, t \models_i B_i \phi$, which is equivalent to $\forall t(sR_i t) E, t \models_i \neg B_i \phi$. Therefore, we have shown that $E, s \models_i B_i \neg B_i \phi$.   □

This proposition states that axiom 5 (the classical axiom of negative introspection) holds for each belief operator B_i (i.e. every agent has introspective capabilities on its own negative beliefs).

Proposition 11 (Generation of negative beliefs)

$$E, s \models_i \neg B_j \phi \text{ implies } E, s \models_i B_i \neg B_j \phi$$

---

[4] It is easy to prove that any relation that is symmetric and transitive is also Euclidean.

Proof: $E, s \models_i \neg B_j \phi \Longrightarrow E, s =\!|_i B_j \phi \Longrightarrow \exists t(sR_it), E, t =\!|_j \phi$. Let us call w to any of the worlds referred to by this existential quantifier. Being R_i Euclidean, we know that $\forall t(sR_it), (tR_iw)$; therefore, we may say that $\forall t(sR_it) \exists u(tR_iu), E, u =\!|_j \phi$. Thus, $\forall t(sR_it), E, t =\!|_i B_j \phi$, which is equivalent to $\forall t(sR_it), E, t \models_i \neg B_j \phi$. Therefore, $E, s \models_i B_i \neg B_j \phi$.   □

This proposition is expressing the fact that $Agent_i$ can make positive introspection on negated beliefs of other agents.

4.4   Summary of the Main Properties

Summarising the main results shown in this section:

- All forms of logical omniscience are avoided.
  None of the restricted forms of logical omniscience usually considered in the literature holds in the framework of subjective situations. This result is due to the presence of partial and inconsistent situations and to the fact that the description of a situation is formed with positive and negative information about propositional formulae (and not about basic propositions).
- Each agent is aware of its positive and negative beliefs, and is also aware of the fact that the other agents enjoy this introspective capability.
  However, an agent is uncertain about the way the present situation is perceived by other agents and, therefore, it is unable to know anything about the other agent's beliefs.
- The positive and negative beliefs of an agent in an state reflect, as our intuitions suggested, the facts that are taken as true or false by the agent in that state.
  Thus, an agent's perception determines its beliefs in a given situation, as it might be expected.

5   Comparison with Previous Proposals

The most outstanding difference of our proposal with previous works ([8]) is the idea of considering *subjective* situations, that may be perceived in different ways by different agents. Technically, this fact implies two differences of our approach with respect to others:

- A situation is described with two functions ($\mathcal{T}_i$ and $\mathcal{F}_i$) for each $Agent_i$.
  Thus, we take into account each agent's perception of the actual situation, considering a *subjective* description of each state.
- Two satisfiability and unsatisfiability relations between situations and formulae ($\models_i$ and $=\!|_i$) are also defined for each agent.
  Having a *subjective* description of each state, it makes sense to consider satisfiability relations that depend on each agent.

The rest of the section is devoted to the comparison of our proposal with the two approaches to the problem of logical omniscience with which it shares more similarities: Levesque's *logic of explicit and implicit beliefs* ([6]) and Thijsse's *hybrid sieve systems* ([7]).

5.1   Levesque's Logic of Implicit and Explicit Beliefs

Levesque uses a language with two modal operators: B for *explicit* beliefs and L for *implicit* beliefs. These operators are not allowed to be nested in the formulae of the language. A *structure for explicit and implicit beliefs* is defined as a tuple $M=(S, B, T, F)$, where S is the set of primitive situations, B is a subset of S that represents the situations that could be the actual one and T and F are functions from the set of primitive propositions into subsets of S. Intuitively, $T(P)$ contains all the situations that support the truth of P, whereas $F(P)$ contains the ones that support the falsehood of P. A situation s can be *partial* (if there is a primitive proposition which is neither true nor false in s) and/or *incoherent* (if there is a proposition which is both true and false in s). A situation is *complete* if it is neither partial nor incoherent. A complete situation s is *compatible* with a situation t if s and t agree in all the points in which t is defined. B^* is the set of all complete situations of S that are compatible with some situation in B.

The relations $\models_T$ and $\models_F$ between situations and formulae are defined as follows:

- $M,s \models_T P$, where P is a primitive proposition, if and only if $s \in T(P)$
- $M,s \models_F P$, where P is a primitive proposition, if and only if $s \in F(P)$
- $M,s \models_T \neg\varphi$ if and only if $M,s \models_F \varphi$
- $M,s \models_F \neg\varphi$ if and only if $M,s \models_T \varphi$
- $M,s \models_T (\varphi \wedge \psi)$ if and only if $M,s \models_T \varphi$ and $M,s \models_T \psi$
- $M,s \models_F (\varphi \wedge \psi)$ if and only if $M,s \models_F \varphi$ or $M,s \models_F \psi$
- $M,s \models_T B\varphi$ if and only if $M,t \models_T \varphi$ $\forall t \epsilon B$
- $M,s \models_F B\varphi$ if and only if $M,s \not\models_T B\varphi$
- $M,s \models_T L\varphi$ if and only if $M,t \models_T \varphi$ $\forall t \epsilon B^*$
- $M,s \models_F L\varphi$ if and only if $M,s \not\models_T L\varphi$

There are some similarities between our approach and Levesque's logic of implicit and explicit beliefs. However, they are more apparent than real, as shown in this listing:

- Levesque also considers a satisfiability and an unsatisfiability relation between situations and doxastic formulae.
  However, these relations are not considered for each agent.
- Levesque also describes each situation with two functions T and $\mathcal{F}$.
  These functions are not indexed by each agent, as our functions are (Levesque considers an objective description of what is true and what is false in each situation). Another important difference is that Levesque's functions deal with basic propositions, and not with formulae as our functions do.
- Both approaches allow the presence of *partial* or *inconsistent* situations.
  However note that, in our case, it is not the (objective) description of the situation that is partial or inconsistent, but the *subjective* perception that an agent may have of it. Thus, the notions of partiality and inconsistency have a much more natural interpretation in our framework.
- Both approaches avoid all the forms of logical omniscience.
  The reason is different in each case, though. In Levesque's logic of explicit and implicit beliefs, it is the presence of incoherent situations that prevents logical omniscience. In our proposal, there is no need to have inconsistent situations to avoid logical omniscience. In fact, we solve that problem by defining T_i and $\mathcal{F}_i$ over arbitrary sets of formulae, and not over basic propositions.

- There are accessibility relations between situations for each agent in both systems. Levesque's accessibility relation between situations is left implicit; our accessibility relations are explicit. Furthermore, the intuition underlying these relations is somewhat different, as explained in section 2.

Other differences with Levesque's approach are:

- Levesque only considers one agent, and does not allow nested beliefs. Thus, his agents do not have any introspective capabilities.
- Levesque defines *explicit* and *implicit* beliefs, whereas we do not make this distinction.
- Even though Levesque avoids logical omniscience, his agents must necessarily believe all those tautologies that are formed by *known* basic propositions (those propositions P for which the agent believes $(P \vee \neg P)$), regardless of their complexity. This is not the case in our approach, because we deal directly with formulae.
- There is a different treatment of the unsatisfiability relation when applied to beliefs, because he transforms $\models$ into $\not\models$, whereas we do not.

5.2 Thijsse's Hybrid Sieve Systems

Thijsse ([7]) proposes a way of using *partial logics* to deal with various forms of logical omniscience. He defines a *partial model* as a tuple $(W, \mathcal{B}_1, \ldots, \mathcal{B}_n, V)$, where W is a set of worlds, $\mathcal{B}_i$ is the accessibility relation between worlds for $Agent_i$ and V is a *partial* truth assignment to the basic propositions in each world. $\top$ is a primitive proposition that is always interpreted as *true*. Truth ($\models$) and falsity ($\dashv$) relations are defined in the following way:

- $M,w \models \top$
- $M,w \not\models \top$
- $M,w \models P$, where P is a primitive proposition, iff $V(P, w) = 1$
- $M,w \dashv P$, where P is a primitive proposition, iff $V(P, w) = 0$
- $M,w \models \neg\varphi$ iff $M,w \dashv \varphi$
- $M,w \dashv \neg\varphi$ iff $M,w \models \varphi$
- $M,w \models (\varphi \wedge \psi)$ iff $M,w \models \varphi$ and $M,w \models \psi$
- $M,w \dashv (\varphi \wedge \psi)$ iff $M,w \dashv \varphi$ or $M,w \dashv \psi$
- $M,w \models B_i\varphi$ iff $M,v \models \varphi$ $\forall v$ such that $(w, v) \in \mathcal{B}_i$
- $M,w \dashv B_i\varphi$ iff $\exists v$ s.t. $(w, v) \in \mathcal{B}_i$ and $M,v \dashv \varphi$

The most important similarities between our approach and Thijsse's are:

- n agents and n explicit accessibility relations are considered.
  However, as in Levesque's case, there are no restrictions on these relations, and the intuitive meaning of our accessibility relations is slightly different.
- Two relations (of satisfiability and unsatisfiability) are defined. Moreover, a similar clause is used to provide a meaning to the unsatisfiability relation with respect to the belief operator.
  As before, the main difference is that we provide two relations *for each agent*.

- There are no tautologies in Thijsse's system; therefore, he does not have to care about some forms of logical omniscience (closure under valid implication and belief of valid formulae).
- Closure under material implication and closure under conjunction do not hold in Thijsse's approach either.

The main difference with Thijsse's proposal is that he uses *partial* assignments of truth values *over basic propositions* for each state; thus, a proposition may be true, false or undefined in each state. We deal with formulae, not with basic propositions, and each formula may be supported *and/or* rejected by *each agent* in each state. Therefore, Thijsse's approach is three-valued, whereas ours is more of a four-valued kind, such as Levesque's.

6 Summary

In this paper it has been argued that each agent perceives its actual situation in a particular way, which may be different from that of other agents located in the same situation. The vision that an agent has of a situation determines its (positive and negative) beliefs in that situation. This intuitive idea has been formalized with the notion of *subjective situations*. These entities are the base of a doxastic logic, in which the meaning of the belief operators seems to fit with the general intuitions about how the doxastic attitude of a non-ideal agent should behave. In particular, logical omniscience is avoided while some interesting introspective properties are maintained. A detailed comparison of this approach with Levesque's *logic of implicit and explicit beliefs* ([6]) and Thijsse's *hybrid sieve systems* ([7]) has also been provided.

References

1. W. van der Hoek, J.-J. Meyer, *Epistemic logic for AI and Computer Science*, Cambridge Tracts in Theoretical Computer Science 41, Cambridge University Press, 1995.
2. J. Hintikka, *Knowledge and belief*, Cornell University Press, Ithaca, N.Y., 1962.
3. S. Kripke, A semantical analysis of modal logic I: normal modal propositional calculi, *Zeitschrift für Mathematische Logik und Grundlagen Mathematik* 9 (1963), 67-96.
4. J. Hintikka, Impossible possible worlds vindicated, *J. of Phil. Logic* 4 (1975), 475-484.
5. A.Moreno, U.Cortés, T.Sales, *Subjective situations*, in Multi-Agent System Engineering, F.J.Garijo and M.Boman (Eds.), Lecture Notes in Artificial Intelligence 1647, pp. 210-220, 1999.
6. H.J. Levesque, A logic of implicit and explicit belief, *Proceedings of the Conference of the American Association for Artificial Intelligence, AAAI-84* (1984), 198-202.
7. E. Thijsse, Combining partial and classical semantics. A hybrid approach to belief and awareness, in P. Doherty (Ed.), *Partiality, modality, and nonmonotonicity*, Studies in Logic, Language and Information, Center for the Study of Language and Information Publications (1996), 223-249.
8. A. Moreno, Avoiding logical omniscience and perfect reasoning: a survey, *AI Communications* 2 (1998), 101-122.

9. A. Moreno, T. Sales, Dynamic belief analysis, in J. Müller, M. Wooldridge, N. Jennings (Eds.), *Intelligent Agents III: Agent theories, architectures and languages*, Lecture Notes in Artificial Intelligence 1193, Springer Verlag (1997), 87-102.
10. A.Moreno, U.Cortés, T.Sales, Inquirers: a general model of non-ideal agents, *International Journal of Intelligent Systems* 15 (3), pp. 197-215, 2000.

Multi-agent $\mathcal{VSK}$ Logic

Michael Wooldridge[1] and Alessio Lomuscio[2]

[1] Department of Computer Science
University of Liverpool
Liverpool L69 7ZF, United Kingdom
M.J.Wooldridge@csc.liv.ac.uk
[2] Department of Computing
Imperial College of Science, Technology and Medicine
London SW7 2BZ, United Kingdom
A.Lomuscio@doc.ic.ac.uk

Abstract. We present a formalism for reasoning about the information properties of multi-agent systems. Multi-agent $\mathcal{VSK}$ logic allows us to represent what is *objectively true* of some environment, what is *visible*, or *accessible* of the environment to individual agents, what these agents actually *perceive*, and finally, what the agents actually *know* about the environment. The semantics of the logic are given in terms of a general model of multi-agent systems, closely related to the interpreted systems of epistemic logic. After introducing the logic and establishing its relationship to the formal model of multi-agent systems, we systematically investigate a number of possible interaction axioms, and characterise these axioms in terms of the properties of agents that they correspond to. Finally, we illustrate the use of the logic through a case study, and discuss issues for future work.

1 Introduction

Consider the following scenario:

A number of autonomous mobile robots are working in a factory, collecting and moving various goods around. All robots are equipped with sonars, which enable them to detect obstacles. To ensure that potentially costly collisions are avoided, a number of crash-avoidance techniques are used. First, all robots adhere to a convention that, if they detect a potential collision, they must take evasive action either when they detect that other agents have right of way or when they know that regardless of the convention of the right of way this is the only way to avoid a collision. Second, a "supervisor" agent C is installed in the factory, which monitors all data feeds from sonars. In the event of an impending collision, this agent is able to step in and override the control systems of individual agents. At some time, two robots, A and B, are moving towards each other in a narrow corridor; robot A has the right of way. Robot B's sonar is faulty, and as a result, B fails to notice the potential collision

M. Ojeda-Aciego et al. (Eds.): JELIA 2000, LNAI 1919, pp. 300–312, 2000.
© Springer-Verlag Berlin Heidelberg 2000

and does not give way to robot A. Robot A, using its sonar, detects the presence of robot B. Robot A recognises that B has not taken evasive action when it should have done, and reasons that B must be faulty; as a consequence, it takes additional evasive action. Meanwhile, the supervisor agent C, observing the scenario, also deduces that B must be faulty, and as a consequence shuts B down.

The aim of this scenario is not to suggest an architecture for multi-agent robotics, but to illustrate the utility of reasoning about the information that agents can and do perceive, their knowledge about their environment, and the actions that they perform. We argue that the ability to perform such reasoning will be of great value if autonomous agents are to be successfully deployed.

In this paper, we develop a formalism that will allow us to represent and reason about such aspects of multi-agent systems. We present *multi-agent $\mathcal{VSK}$ logic*, a multi-agent extension of $\mathcal{VSK}$ logic [9]. This logic allows us to represent what is *objectively true* of an environment, what is *visible*, or *knowable* about the environment to individual agents within it, what agents *perceive* of their environment, and finally, what agents actually *know* about their environment. Syntactically, $\mathcal{VSK}$ logic is a propositional multi-modal logic, containing three sets of indexed unary modal operators "$\mathcal{V}_i$", "$\mathcal{S}_i$", and "$\mathcal{K}_i$", one for each agent i. A formula $\mathcal{V}_i\varphi$ means that the information φ is accessible to agent i; $\mathcal{S}_i\varphi$ means that agent i perceives information φ; and $\mathcal{K}_i\varphi$ means that agent i knows φ.

An important feature of multi-agent $\mathcal{VSK}$ logic is that its semantics are given with respect to a general model of agents and their environments. We are able to characterise possible axioms of multi-agent $\mathcal{VSK}$ logic with respect to this semantic model. Consider, for example, the $\mathcal{VSK}$ formula $\mathcal{V}_i\varphi \Rightarrow \mathcal{S}_j\mathcal{V}_i\varphi$, which says that if information φ is accessible to agent i, then agent j sees (perceives) that φ is accessible to i. Intuitively, this formula says that agent j is able to see at least as much as agent i; we are able to show this formally by proving correspondence results with respect to a semantic description of agents and environments, as well as the Kripke frames they generate.

The remainder of this paper is structured as follows. We begin in section 2 by introducing the semantic framework that underpins multi-agent $\mathcal{VSK}$ logic. We then formally introduce the syntax and semantics of $\mathcal{VSK}$ logic in section 3, and in particular, we show how the semantics of the logic relate to the formal model of multi-agent systems introduced in section 2. In section 4, we discuss and formally characterise various *interaction axioms* of $\mathcal{VSK}$ logic. In section 5, we return to the case study presented above, and show how we can use multi-agent $\mathcal{VSK}$ logic to capture and reason about

Finally, in section 6, we present some conclusions.

2   A Semantic Framework

In this section, we present a semantic model of agents and the environments they occupy. This model plays the role in $\mathcal{VSK}$ logic that *interpreted systems* play in epistemic logic [2, pp103–107].

A *multi-agent $\mathcal{VSK}$ system* is assumed to be comprised of a collection Ag_1, ..., Ag_n of *agents*, together with an *environment*. We formally define environments below, but for the moment, it is assumed that an environment can be in any of a set E of *instantaneous states*. We adopt a quite general model of agents, which makes only a minimal commitment to an agent's internal architecture. One important assumption we do make is that agents have an internal state, although we make no assumptions with respect to the actual structure of this state. Agents are assumed to be composed of three functional components: some sensor apparatus, an action selection function, and a next-state function.

Formally, an agent Ag_i is a tuple $Ag_i = \langle L_i, Act_i, see_i, do_i, \tau_i, l_i \rangle$, where:

- $L_i = \{l_i^1, l_i^2, \ldots\}$ is a set of *instantaneous local states* for agent i.
- $Act_i = \{\alpha_i^1, \alpha_i^2, \ldots\}$ is a set of *actions* for agent i.
- $see_i : 2^E \rightarrow Perc_i$ is the *perception function* for agent i, mapping sets of environment states (*visibility sets*) to percepts for agent i.
  Elements of the set $Perc_i$ will be denoted by $\rho_i^1, \rho_i^2, \ldots$ and so on. If see_i is an injection into $Perc_i$ then we say that see_i is *perfect*, otherwise we say it is *lossy*.
- $do_i : L_i \rightarrow Act_i$ is the *action selection function* for agent i, mapping local states to actions available to agent i.
- $\tau_i : L_i \times Perc_i \rightarrow L_i$ is the *state transformer function* for agent i.
  We say τ_i is *complete* if for any

$$g = (e, \tau_1(l_1, \rho_1), \ldots, \tau_n(l_n, \rho_n))$$

and

$$g' = (e', \tau_1(l'_1, \rho'_1), \ldots, \tau_n(l'_n, \rho'_n))$$

we have that
$$\tau_i(l_i, \rho_i) = \tau_i(l'_i, \rho'_i) \quad \text{implies} \quad \rho_i = \rho'_i.$$

We say τ_i is *local* if for any

$$g = (e, \tau_1(l_1, \rho_1), \ldots, \tau_n(l_n, \rho_n))$$

and

$$g' = (e', \tau_1(l'_1, \rho'_1), \ldots, \tau_n(l'_n, \rho'_n))$$

we have that
$$\tau_i(l_i, \rho_i) = \tau_i(l'_i, \rho_i).$$

We say that an agent has *perfect recall* if the function τ_i is an injection.
- $l_i \in L$ is the *initial state* for agent i.

Perfect perception functions distinguish between all visibility sets; lossy perception functions are so called because they can map different visibility sets to the same percept, thereby losing information. We say that an agent has *perfect recall* of its history if it changes its local state at every tick of the clock (cf. [2, pp128–131]).

Following [2], we use the term "environment" to denote all the components of a system external to the agents that occupy it. Sometimes, environments can be represented as just another agent of the system; more often they serve a special purpose, as they can be used to model communication architectures, etc. We model an environment as a tuple containing a set of possible *instantaneous states*, a *visibility function* for each agent, which characterises the information available to an agent in every environment state, a *state transformer* function, which characterises the effects that an agent's actions have on the environment, and, finally, an *initial state*.

Formally, an environment *Env* is a tuple

$$Env = \langle E, vis_1, \ldots, vis_n, \tau_e, e_0 \rangle$$

where:

- $E = \{e_1, e_2, \ldots\}$ is a *set of instantaneous local states* for the environment.
- $vis_i : E \to 2^E$ is the *visibility function of agent* i. It is assumed that vis_i partitions E into mutually disjoint sets and that $e \in vis_i(e)$, for any $e \in E$. Elements of the codomain of the function *vis* are called *visibility sets*. We say that vis_i is *transparent* if for any $e \in E$ we have that $vis_i(e) = \{e\}$.
- $\tau_e : E \times Act_1 \times \cdots \times Act_n \to 2^E$ is a total *state transformer function* for the environment (cf. [2, p154]), which maps environment states and tuples of actions, one for each agent, to the set of environment states that could result from the performance of these actions in this state.
- $e_0 \in E$ is the *initial state* of *Env*.

Modelling an environment in terms of a set of states and a state transformer is quite conventional (see, e.g., [2]). The use of the visibility function, however, requires some explanation. Before we do this, let us define the concept of global state. The *global states* $G = \{g, g', \ldots\}$ of a $\mathcal{VSK}$ system are a subset of $E \times L_1 \times \cdots \times L_n$.

The visibility function defines what is in principle knowable about a $\mathcal{VSK}$ system; the idea is similar to the notion of "partial observability" in POMDPs [6]. Intuitively, not all the information in an environment state is in general accessible to an agent. So, in a global state $g = (e, l_1, \ldots, l_n)$, $vis_i(e) = \{e, e', e''\}$ represents the fact that the environment states e, e', e'' are indistinguishable to agent i from e. This is so regardless of agent i's efforts in performing the observation — it represents the maximum amount of information that is in principle available to i when observing state e. The concept of transparency, as defined above, captures "perfect" scenarios, in which all the information in a state is accessible to an agent. Note that visibility functions are *not* intended to capture the everyday notion of visibility as in "object x is visible to the agent".

A *multi-agent VSK system* is a structure $S = \langle Env, Ag_1, \ldots, Ag_n \rangle$, where *Env* is an environment, and $Ag_1, \ldots, Ag_n$ are agents. The class of VSK systems is denoted by $\mathcal{S}$.

Although the logics we discuss in this paper may be used to refer to *static* properties of knowledge, visibility, and perception, the semantic model naturally allows us to account for the temporal evolution of a VSK system. The behaviour of a VSK system can be summarised as follows. Each agent i starts in state l_i, the environment starts in state e_0. At this point every agent i "synchronises" with the environment by performing an initial observation through the visibility function vis_i, and generates a percept $\rho_i^0 = see_i(vis_i(e_0))$. The internal state of the agent is then updated, and becomes $\tau_i(l_i, \rho_i^0)$. The synchronisation phase is now over and the system starts its run from the initial state $g_0 = (e_0, \tau_1(l_1, \rho_1^0), \ldots, \tau_n(l_n, \rho_n^0))$. An action $\alpha_i^0 = do(\tau_i(l_i, \rho_i^0))$ is selected and performed by each agent i on the environment, whose state is updated into $e_1 = \tau_e(e_0, \alpha_1^0, \ldots, \alpha_n^0)$. Each agent enters another cycle, and so on.

A *run* of a system is thus a (possibly infinite) sequence of global states. A sequence $(g_0, g_1, g_2, \ldots)$ over G represents a run of a system $\langle Env, Ag_1, \ldots, Ag_n \rangle$ iff

 - $g_0 = (e_0, \tau_1(l_1, see_1(vis_1(e_0))), \ldots, \tau_n(l_n, see_n(vis_n(e_0))))$, and
 - for all u, if $g_u = (e, l_1, \ldots, l_n)$ and $g_{u+1} = (e', l_1', \ldots, l_n')$ then:

$$e' \in \tau_e(e_u, \alpha_1, \ldots, \alpha_n) \quad \text{and}$$
$$l_i' = \tau_i(l_i, see_i(vis_i(e')))$$

where $\alpha_i = do_i(l_i)$.

Given a multi-agent VSK system $S = \langle Env, Ag_1, \ldots, Ag_n \rangle$, we say $G_S \subseteq G$ is the set of global states *generated* by S if $g \in G_S$ occurs in a run of S.

3   Multi-agent VSK Logic

We now introduce a language $\mathcal{L}$, which will enable us to represent the information properties of multi-agent VSK systems. In particular, it will allow us to represent first what is true of the VSK system, then what is *visible*, or *knowable* of the system to the agents within it, then what these agents *perceive* of the system, and finally, what each agent *knows* of the system. $\mathcal{L}$ is a propositional multi-modal language, containing three sets of indexed unary modal operators, for visibility, perception, and knowledge respectively. Given a set P of propositional atoms, the language $\mathcal{L}$ of VSK logic is defined by the following BNF grammar:

$$\langle ag \rangle ::= 1 \mid \cdots \mid n$$
$$\langle wff \rangle ::= \textbf{true} \mid \text{any element of } P \mid \neg \langle wff \rangle \mid \langle wff \rangle \wedge \langle wff \rangle$$
$$\mid \mathcal{V}_{\langle ag \rangle} \langle wff \rangle \mid \mathcal{S}_{\langle ag \rangle} \langle wff \rangle \mid \mathcal{K}_{\langle ag \rangle} \langle wff \rangle$$

The modal operator "$\mathcal{V}_i$" will allow us to represent the information that is instantaneously visible or knowable about the state of the system to agent i. Thus suppose the formula $\mathcal{V}_i\varphi$ is true in some state $g \in G$. The intended interpretation of this formula is that the property φ is *accessible* to agent i when the system is in state g. This means that not only φ is true of the environment, but agent i, if it was equipped with suitable sensor apparatus, would be able to perceive φ. If $\neg\mathcal{V}_i\varphi$ were true in some state, then no matter how good agent i's sensor apparatus was, it would be unable to perceive φ.

The fact that something is visible to an agent does not mean that the agent actually sees it. What an agent *does* see is determined by its sensors. The modal operator "$\mathcal{S}_i$" will be used to represent the information that agent i "sees". The idea is as follows. Suppose agent i's sensory apparatus (represented by the see_i function in our semantic model) is a video camera, and so the percepts being received by agent i take the form of a video feed. Then $\mathcal{S}_i\varphi$ means that an impartial observer would say that the video feed currently being supplied by i's video camera carried the information φ — in other words, φ is true in all situations where i received the same video feed.

Finally, we can represent the *knowledge* possessed by agents within a system. We represent agent i's knowledge by means of a modal operator "$\mathcal{K}_i$". In line with the tradition that started with Hintikka [4], we write $\mathcal{K}_i\varphi$ to represent the fact that agent i has knowledge of the formula represented by φ. Our model of knowledge is that popularised by Halpern and colleagues [2]: agent i is said to know φ when in local state l if φ is guaranteed to be true whenever i is in state l. As with visibility and perception, knowledge is an *external* notion — an agent is said to know φ if an impartial, omniscient observer would say that the agent's state carried the information φ.

We now proceed to interpret our formal language. We do so with respect to the equivalence Kripke frames *generated* (see [2]) by $\mathcal{VSK}$ systems. Given a $\mathcal{VSK}$ system $S = \langle Env, Ag_1, \ldots, Ag_n \rangle$, the Kripke frame

$$F_S = \langle W, \sim_1^\nu, \sim_1^s, \sim_1^k, \ldots, \sim_n^\nu, \sim_n^s, \sim_n^k \rangle$$

generated by S is defined as follows:

- $W = G_S$ (recall that G_S is the set of global states reachable by system S),
- For every $i = 1, \ldots, n$, the relation $\sim_i^\nu \subseteq W \times W$ is defined by: $(e, l_1, \ldots, l_n) \sim_i^\nu (e', l_1', \ldots, l_n')$ if $e' \in vis_i(e)$,
- For every $i = 1, \ldots, n$, the relation $\sim_i^s \subseteq W \times W$ is defined by: $(e, l_1, \ldots, l_n) \sim_i^s (e', l_1', \ldots, l_n')$ if $see_i(vis_i(e)) = see_i(vis_i(e'))$,
- For every $i = 1, \ldots, n$, the relation $\sim_i^k \subseteq W \times W$ is defined by: $(e, l_1, \ldots, l_n) \sim_i^k (e', l_1', \ldots, l_n')$ if $l_i = l_i'$.

The class of frames generated by a $\mathcal{VSK}$ system S will be denoted by $\mathcal{F}_S$. As might be expected, all frames generated by systems in S are equivalence frames.

Lemma 1. *Every frame $F \in \mathcal{F}_S$ is an equivalence frame, i.e., all the relations in F are equivalence relations.*

We have now built a bridge between $\mathcal{VSK}$ systems and Kripke frames. In what follows, we assume the standard definitions of satisfaction and validity for Kripke

Axiom	$\mathcal{VSK}$ Class
$\mathcal{V}_i\varphi \Rightarrow \varphi$	none (valid in all systems)
$\varphi \Rightarrow \mathcal{V}_i\varphi$	vis_i is transparent
$\mathcal{S}_i\varphi \Rightarrow \mathcal{V}_i\varphi$	none (valid in all systems)
$\mathcal{V}_i\varphi \Rightarrow \mathcal{S}_i\varphi$	see_i is perfect
$\mathcal{K}_i\varphi \Rightarrow \mathcal{S}_i\varphi$	τ_i has perfect recall
$\mathcal{S}_i\varphi \Rightarrow \mathcal{K}_i\varphi$	τ_i is local

Table 1. Single-agent interaction axioms in $\mathcal{VSK}$ logic.

frames and Kripke models — we refer the reader to [5,3] for a detailed exposition of the subject. Following [2] and [7], we define the concepts of truth and validity on Kripke models that are *generated* by $\mathcal{VSK}$ systems.

Given an interpretation $\pi : W \to 2^P$, we say that a formula $\varphi \in \mathcal{L}$ is satisfied at a point $g \in G$ on a $\mathcal{VSK}$ system S if the model $M_S = \langle F_S, \pi \rangle$ built on the generated frame F_S by use of π is such that $M_S \models_g \varphi$. The propositional connectives are assumed to be interpreted as usual, and the modal operators $\mathcal{V}_i$, $\mathcal{S}_i$, and $\mathcal{K}_i$ are assumed to be interpreted in the standard way (see for example [5]) by means of the equivalence relations $\sim_i^v$, $\sim_i^s$, and $\sim_i^k$ respectively.

We are especially interested in the properties of a $\mathcal{VSK}$ system as a whole. The notion of validity is appropriate for this analysis. A formula $\varphi \in \mathcal{L}$ is valid on a class $\mathcal{S}$ of $\mathcal{VSK}$ systems if for any system $S \in \mathcal{S}$, we have that $F_S \models \varphi$.

4   Interaction Axioms in Multi-agent $\mathcal{VSK}$ Logic

In this section we will study some basic interaction axioms that can be specified within $\mathcal{VSK}$ logic. Interaction axioms are formulas in which different modalities are present; they specify a form of "binding" between the attitudes corresponding to the modal operators.

Note that, in previous work, we have studied and given semantic characterisations for *single-agent* interaction axioms (i.e., axioms in a $\mathcal{VSK}$ logic where there is only one $\mathcal{V}$ operator, only one $\mathcal{S}$ operator, and only one $\mathcal{K}$ operator) [9]. For example, we were able to show that the axiom schema $\mathcal{V}\varphi \Rightarrow \mathcal{S}\varphi$ characterised a particular property of an agent's perception function: namely, that it was *perfect*, in the sense that we defined in section 2. We summarise these results in table 1.

In this paper we analyse some *multi-agent* interaction axioms. The most obvious form that these interaction axioms may have is the following:

$$\Box_i^1\varphi \Rightarrow \Box_j^2\varphi \qquad \text{where } \Box_i^1 \in \{\mathcal{S}_i, \mathcal{V}_i, \mathcal{K}_i\}, \Box_j^2 \in \{\mathcal{S}_j, \mathcal{V}_j, \mathcal{K}_j\}. \tag{1}$$

If we assume $i \neq j$ (the case $i = j$ was dealt with in [9]), Axiom (1) generates nine possible interaction axioms in total, as summarised in table 2. The second column of table 1 gives the conditions on Kripke models that correspond (in the sense of [1]) to the axiom. The third column gives the first-order condition on

Axiom	Kripke Condition	$\mathcal{VSK}$ Condition
$\mathcal{V}_i\varphi \Rightarrow \mathcal{V}_j\varphi$	$\sim_j^\nu \subseteq \sim_i^\nu$	$vis_j(e) \subseteq vis_i(e)$
$\mathcal{V}_i\varphi \Rightarrow \mathcal{S}_j\varphi$	$\sim_j^s \subseteq \sim_i^\nu$	$sv_j(e) = sv_j(e') \to vis_i(e) = vis_i(e')$
$\mathcal{V}_i\varphi \Rightarrow \mathcal{K}_j\varphi$	$\sim_j^k \subseteq \sim_i^\nu$	$l_j = l'_j \to vis_i(e) = vis_i(e')$
$\mathcal{S}_i\varphi \Rightarrow \mathcal{V}_j\varphi$	$\sim_j^\nu \subseteq \sim_i^s$	$vis_j(e) = vis_j(e') \to sv_i(e) = sv_i(e')$
$\mathcal{S}_i\varphi \Rightarrow \mathcal{S}_j\varphi$	$\sim_j^s \subseteq \sim_i^s$	$sv_j(e) = sv_j(e') \to sv_i(e) = sv_i(e')$
$\mathcal{S}_i\varphi \Rightarrow \mathcal{K}_j\varphi$	$\sim_j^k \subseteq \sim_i^s$	$l_j = l'_j \to sv_i(e) = sv_i(e')$
$\mathcal{K}_i\varphi \Rightarrow \mathcal{V}_j\varphi$	$\sim_j^\nu \subseteq \sim_i^k$	$vis_j(e) = vis_j(e') \to l_i = l'_i$
$\mathcal{K}_i\varphi \Rightarrow \mathcal{S}_j\varphi$	$\sim_j^s \subseteq \sim_i^k$	$sv_j(e) = sv_i(e') \to l_i = l'_i$
$\mathcal{K}_i\varphi \Rightarrow \mathcal{K}_j\varphi$	$\sim_j^k \subseteq \sim_i^k$	$l_j = l'_j \to l_i = l'_i$

Table 2. Some multi-agent interaction axioms in multi-agent $\mathcal{VSK}$ logic. Note that in the table the function $sv_i : E \to Perc_i$ stands for $see_i \circ vis_i$.

$\mathcal{VSK}$ systems that corresponds to the interaction axioms. (Note that in these conditions each variable is assumed to be universally quantified: for example, the third axiom $\mathcal{V}_i\varphi \Rightarrow \mathcal{K}_j\varphi$ corresponds to systems S in which for all $g = (e, l_1, \ldots, l_n)$ and $g' = (e', l'_1, \ldots, l'_n)$, we have that $l_j = l'_j$ implies $vis_i(e) = vis_i(e')$.)

We begin our analysis with the schema which says that if φ is visible to i, then φ is visible to j.

$$\mathcal{V}_i\varphi \Rightarrow \mathcal{V}_j\varphi \tag{2}$$

This axiom says that everything visible to i is also visible to j. Note that the first-order condition corresponding to Axiom 2 implies that at least as much information is accessible to agent j as agent i.

$$\mathcal{V}_i\varphi \Rightarrow \mathcal{S}_j\varphi \tag{3}$$

Axiom (3) says that j sees everything visible to i. It is easy to see that in systems that validate this schema, since j sees everything i sees, it must be that everything visible to i is also visible to j. In other words, $\mathcal{VSK}$ systems that validate Axiom (3) will also validate (2).

$$\mathcal{V}_i\varphi \Rightarrow \mathcal{K}_j\varphi \tag{4}$$

Axiom (4) says that everything visible to i is known to j.

$$\mathcal{S}_i\varphi \Rightarrow \mathcal{V}_j\varphi \tag{5}$$

Axiom (5) says that everything i sees is visible to j. Intuitively, this means that the percepts i receives are part of the environment that is visible to j.

$$\mathcal{S}_i\varphi \Rightarrow \mathcal{S}_j\varphi \tag{6}$$

Axiom (6) says that j sees everything i sees. Since we know from [9] that any system S validates the axiom $S_j\varphi \Rightarrow V_j\varphi$, it follows that any VSK system validating Axiom (6) will also validate Axiom (5). Note that from table 2, it follows that

$$|see_j(vis_j(E))| \leq |see_i(vis_i(E))|$$

So, since agent i has more perception states at its disposal than agent j, it has a finer grain of perception.

$$S_i\varphi \Rightarrow K_j\varphi \tag{7}$$

Axiom (7) says that if i sees φ then j knows φ; in other words, j knows everything that i sees.

$$K_i\varphi \Rightarrow V_j\varphi \tag{8}$$

Axiom (8) says that if i knows φ, then φ is visible to j. Intuitively, this means that i's local state is visible to j. Axiom (8) thus says that entity j has "read access" to the state of another entity i.

$$K_i\varphi \Rightarrow S_j\varphi \tag{9}$$

Axiom (9) captures a more general case than that of (8), where entity j not only has read access to the state of i, but that it actually does read this state. Note that any system that validates (9) will also validate (8).

$$K_i\varphi \Rightarrow K_j\varphi \tag{10}$$

This final schema says that j knows everything that i knows. Note that from the corresponding condition on VSK systems in table 2, it follows that

$$|L_j| \leq |L_i|$$

So, since agent i has more local states, it has a finer grain of knowledge than agent j. If we also have the converse of (10), then we would have $K_i\varphi \Leftrightarrow K_j\varphi$ as valid; an obvious interpretation of this schema would be that i and j had the *same* state.

All these considerations lead us to the following:

Theorem 1. *For any axiom ψ of table 2 and any VSK system S we have that the following are equivalent:*

1. *The system S validates ψ, i.e., $S \models \psi$;*
2. *The generated frame F_S satisfies the corresponding Kripke condition R_ψ;*
3. *The system S satisfies the corresponding VSK condition S_ψ.*

Axiom	Kripke Condition	$\mathcal{VSK}$ Condition
$\mathcal{V}_i\varphi \Rightarrow \mathcal{V}_j\mathcal{V}_i\varphi$	$\sim_j^v \subseteq \sim_i^v$	$vis_j(e) \subseteq vis_i(e)$
$\mathcal{V}_i\varphi \Rightarrow \mathcal{S}_j\mathcal{V}_i\varphi$	$\sim_j^s \subseteq \sim_i^v$	$sv_j(e) = sv_j(e') \rightarrow vis_i(e) = vis_i(e')$
$\mathcal{V}_i\varphi \Rightarrow \mathcal{K}_j\mathcal{V}_i\varphi$	$\sim_j^k \subseteq \sim_i^v$	$l_j = l_j' \rightarrow vis_i(e) = vis_i(e')$
$\mathcal{S}_i\varphi \Rightarrow \mathcal{V}_j\mathcal{S}_i\varphi$	$\sim_j^v \subseteq \sim_i^s$	$vis_j(e) = vis_j(e') \rightarrow sv_i(e) = sv_i(e')$
$\mathcal{S}_i\varphi \Rightarrow \mathcal{S}_j\mathcal{S}_i\varphi$	$\sim_j^s \subseteq \sim_i^s$	$sv_j(e) = sv_j(e') \rightarrow sv_i(e) = sv_i(e')$
$\mathcal{S}_i\varphi \Rightarrow \mathcal{K}_j\mathcal{S}_i\varphi$	$\sim_j^k \subseteq \sim_i^s$	$l_j = l_j' \rightarrow sv_i(e) = sv_i(e')$
$\mathcal{K}_i\varphi \Rightarrow \mathcal{V}_j\mathcal{K}_i\varphi$	$\sim_j^v \subseteq \sim_i^k$	$vis_j(e) = vis_j(e') \rightarrow l_i = l_i'$
$\mathcal{K}_i\varphi \Rightarrow \mathcal{S}_j\mathcal{K}_i\varphi$	$\sim_j^s \subseteq \sim_i^k$	$sv_j(e) = sv_i(e') \rightarrow l_i = l_i'$
$\mathcal{K}_i\varphi \Rightarrow \mathcal{K}_j\mathcal{K}_i\varphi$	$\sim_j^k \subseteq \sim_i^k$	$l_j = l_j' \rightarrow l_i = l_i'$

Table 3. Other interaction axioms in multi-agent $\mathcal{VSK}$ logic.

Proof (Outline.). Given any axiom ψ in table 2, it is a known result that $F_S \models \psi$ if and only if F_S has the Kripke property R_φ shown in table 2 (see [7] for details). But since validity on a $\mathcal{VSK}$ system S is defined in terms of the generated frame F_S, the equivalence between items 1 and 2 follows.

For each line of the table, the equivalence between 2 and 3 can be established by re-writing the relational properties on Kripke frames in terms of the $\mathcal{VSK}$ conditions on $\mathcal{VSK}$ systems.

Other Interaction Axioms Before we leave our study of $\mathcal{VSK}$ interaction axioms, it is worth noting that there are many other possible interaction axioms of interest [7]. The most important of these have the following general form.

$$\Box_i^1\varphi \Rightarrow \Box_j^2\Box_i^1\varphi \quad \text{where } \Box_i^1 \in \{\mathcal{S}_i, \mathcal{V}_i, \mathcal{K}_i\}, \Box_j^2 \in \{\mathcal{S}_j, \mathcal{V}_j, \mathcal{K}_j\}, i \neq j. \quad (11)$$

It is easy to see that schema (11) generates nine possible interaction axioms. We can prove the following general result about such interaction axioms.

Lemma 2. *For any system S, we have that the generated frame F_S satisfies the following property.*

$$F_S \models \Box_i^1 p \Rightarrow \Box_j^2\Box_i^1 p \text{ if and only if } \sim_i^{\Box^1} \subseteq \sim_j^{\Box^2}$$

where $\Box_i^1 \in \{\mathcal{S}_i, \mathcal{V}_i, \mathcal{K}_i\}, \Box_j^2 \in \{\mathcal{S}_j, \mathcal{V}_j, \mathcal{K}_j\}$ and $\sim_i^{\Box^1}$ (respectively $\sim_j^{\Box^2}$) is the equivalence relation corresponding to the modal operator $\Box_i^1$ (respectively $\Box_j^2$).

Proof. Follows from the results presented in [7, Lemma A.11].

Thanks to the above result we can prove that the classes of $\mathcal{VSK}$ systems analysed above are also characterised by the axioms discussed in this section. Indeed we have the following.

Corollary 1. *For any axiom ψ of table 3 and any VSK system S we have that the following are equivalent:*

1. *The system S validates ψ, i.e., $S \models \psi$;*
2. *The generated frame F_S satisfies the corresponding Kripke condition R_ψ;*
3. *The system S satisfies the corresponding VSK condition S_ψ.*

Proof. Follows from Lemma 2 and Theorem 1.

5   A Case Study

In order to illustrate the use of multi-agent VSK logic, we consider again the scenario presented in section 1. While the scenario can be equally explored by means of VSK semantics, here we focus on the axiomatic side of the formalism.

As discussed in section 1, we have three robotic agents A, B, C involved in a coordination problem in a navigation scenario. We suppose the autonomous robots A, B to be equipped with sonars that can perfectly perceive the environment, up to a certain distance of, say, 1 metre; so their visibiliy function is not transparent (see Table 1). We further admit that within 1 metre of distance of the object the pairing sonar/environment is perfect; hence within this distance the environment is fully visible. For the ease with which we assume it is possible to process signals from sensors, we further assume that if the sensors are adequately working, then the agents have perfect perception, i.e. they are semantically described by a perfect *see* function as in Table 1. We also assume that agents know everything they see, i.e. that their τ function is local.

Further assume that the robots A, B follow the following rule: if they know that there is a moving object apparently about to collide with them, then they must take evasive action either when this is the only way to avoid a collision, or in case the object is another robot, when this has right of way. This rules are commonly known, or at least that they hold however nested in a number of $\mathcal{K}$ operators. The superuser has access to the sensors of all the agents (it therefore sees what the agents see and knows what is visible to the agents — see previous section) plus some fixed sensors in the environment they inhabit. Hence we model agent C by supposing that it has perfect perception of the environment, that the environment is completely visible to it and that all its perceptions are known by it.

We can now tailor the specification above to the scenario currently in analysis. We have that agents A, B are in a collision course with A having right of way, that this is visible both to agent B and to agent A, except that while agent A does see this, agent B does not. Formally:

$$\vdash coll \wedge V_A\, coll \wedge V_B\, coll \wedge \neg S_B\, coll \wedge r\text{-}o\text{-}w_A.$$

Given the assumptions on the agents presented above, it is possible to show that it follows that agent A will take evasive action and that agent B will be shut down by the controller agent C. A proof of this is as follows:

1. $\mathcal{V}_B\,coll \wedge \neg\mathcal{S}_B\,coll \wedge \mathcal{V}_A\,coll \wedge r\text{-}o\text{-}w_A$ [Given]
2. $\mathcal{V}_C(\mathcal{V}_B\,coll \wedge \neg\mathcal{S}_B\,coll) \Rightarrow \mathcal{S}_C(\mathcal{V}_B\,coll\wedge$
   $\neg\mathcal{S}_B\,coll) \Rightarrow \mathcal{K}_C(\mathcal{V}_B\,coll \wedge \neg\mathcal{S}_B\,coll)$ [Perfect Perception]
3. $\mathcal{K}_C(\mathcal{V}_B\,coll \wedge \neg\mathcal{S}_B\,coll) \Rightarrow shutdown_B$ [Given]
4. $(\mathcal{V}_B\,coll \wedge \neg\mathcal{S}_B\,coll) \Rightarrow \mathcal{K}_C(\mathcal{V}_B\,coll \wedge \neg\mathcal{S}_B\,coll)$ [Given]
5. $shutdown_B$ [1,3,4 + Taut]
6. $\mathcal{K}_A((\neg ev\text{-}act_B \wedge r\text{-}o\text{-}w_A) \Rightarrow \neg\mathcal{K}_B\,coll)$ [Given + Taut]
7. $\neg ev\text{-}act_B \Rightarrow \mathcal{S}_A\neg ev\text{-}act_B \Rightarrow \mathcal{K}_A\neg ev\text{-}act_B$ [Perfect Perception]
8. $\mathcal{K}_A\neg\mathcal{K}_B\,coll$ [6, 7, K]
9. $\mathcal{K}_A(coll \wedge \neg\mathcal{K}_B\,coll) \Rightarrow ev\text{-}act_A$ [Given]
10. $\mathcal{V}_A\,coll \Rightarrow \mathcal{S}_A\,coll \Rightarrow \mathcal{K}_A\,coll$ [Perfect Perception]
11. $\mathcal{K}_A r\text{-}o\text{-}w_A$ [1, Perfect Perception]
12. $ev\text{-}act_A$ [1, 8, 9, 10, 11, K]

6 Conclusions

In order to design or understand the behaviour of many multi-agent systems, it is necessary to reason about the *information properties* of the system — what information the agents within it have access to, what they actually perceive, and what they know. In this paper, we have presented a logic for reasoning about such properties, demonstrated the relationship of this logic to an abstract general model of multi-agent systems, and investigated various interaction axioms of the logic. Many issues suggest themselves as candidates for future work: chief among them is completeness. In [8], we proved completeness for a mono-modal fragment of $\mathcal{VSK}$ logic. In particular, we proved completeness not simply with respect to an abstract class of Kripke frames, but with respect to the class of Kripke frames corresponding to our model of agents and environments. It is reasonable to expect the proof to transfer to multi-agent settings. However, when interaction axioms of the form studied in section 4 are present, matters naturally become more complicated, and an analysis for each different system is required. This is future work, as are such issues as temporal extensions to the logic, and complexity results.

References

1. J. van Benthem. Correspondence theory. In D. Gabbay and F. Guenthner, editors, *Handbook of Philosophical Logic, Volume II: Extensions of Classical Logic*, volume 165 of *Synthese Library*, chapter II.4, pages 167–247. D. Reidel Publ. Co., Dordrecht, 1984.
2. R. Fagin, J. Y. Halpern, Y. Moses, and M. Y. Vardi. *Reasoning About Knowledge*. The MIT Press: Cambridge, MA, 1995.
3. R. Goldblatt. *Logics of Time and Computation, Second Edition, Revised and Expanded*, volume 7 of *CSLI Lecture Notes*. CSLI, Stanford, 1992. Distributed by University of Chicago Press.
4. J. Hintikka. *Knowledge and Belief*. Cornell University Press: Ithaca, NY, 1962.

5. G. E. Hughes and M. J. Cresswell. *A New Introduction to Modal Logic*. Routledge, New York, 1996.
6. L. P. Kaelbling, M. L. Littman, and A. R. Cassandra. Planning and acting in partially observable stochastic domains. *Artificial Intelligence*, 101:99–134, 1998.
7. A. Lomuscio. *Knowledge Sharing among Ideal Agents*. PhD thesis, School of Computer Science, University of Birmingham, Birmingham, UK, June 1999.
8. M. Wooldridge and A. Lomuscio. A logic of visibility, perception, and knowledge: Completeness and correspondence results. In *Formal and Applied Practical Reasoning – Proceedings of the Third International Conference (FAPR-2000)*, 2000.
9. Michael Wooldridge and Alessio Lomuscio. Reasoning about visibility, perception and knowledge. In N.R. Jennings and Y. Lespérance, editors, *Intelligent Agents VI — Proceedings of the Sixth International Workshop on Agent Theories, Architectures, and Languages (ATAL-99)*, Lecture Notes in Artificial Intelligence. Springer-Verlag, Berlin, 2000.

New Tractable Cases in Default Reasoning from Conditional Knowledge Bases

Thomas Eiter and Thomas Lukasiewicz

Institut und Ludwig Wittgenstein Labor für Informationssysteme, TU Wien
Favoritenstraße 9–11, A-1040 Wien, Austria
{eiter, lukasiewicz}@kr.tuwien.ac.at

Abstract. We present new tractable cases for default reasoning from conditional knowledge bases. In detail, we introduce q-Horn conditional knowledge bases, which allow for a limited use of disjunction. We show that previous tractability results for ε-entailment, proper ε-entailment, and z- and z^+-entailment in the Horn case can be extended to the q-Horn case. Moreover, we present feedback-free-Horn conditional knowledge bases, which constitute a new, meaningful class of conditional knowledge bases. We show that the maximum entropy approach and lexicographic entailment are tractable in the feedback-free-Horn case. Our results complement and extend previous results, and contribute in refining the tractability/intractability frontier of default reasoning from conditional knowledge bases.

1 Introduction

A conditional knowledge base consists of a collection of strict statements in classical logic and a collection of defeasible rules (also called defaults). The former are statements that must always hold, while the latter are rules $\phi \rightarrow \psi$ that read as "generally, if ϕ then ψ." For example, the knowledge "penguins are birds" and "penguins do not fly" can be represented by strict sentences, while the knowledge "birds fly" should be expressed by a defeasible rule (since penguins are birds that do not fly).

The semantics of a conditional knowledge base KB is given by the set of all defaults that are plausible consequences of KB. The literature contains several different proposals for plausible consequence relations and extensive work on their desired properties. The core of these properties are the rationality postulates proposed by Kraus, Lehmann, and Magidor [17], which constitute a sound and complete axiom system for several classical model-theoretic entailment relations under uncertainty measures on worlds. More precisely, they characterize classical model-theoretic entailment under preferential structures, infinitesimal probabilities, possibility measures, and world rankings. Moreover, they characterize an entailment relation based on conditional objects. A survey of all these relationships is given in [4]. We will use the notion of ε-entailment to refer to these equivalent entailment relations.

Mainly to solve problems with irrelevant information, the notion of rational closure as a more adventurous notion of entailment has been introduced by Lehmann [20]. It is equivalent to entailment in system Z by Pearl [22] (which is generalized to variable strength defaults in system Z^+ by Goldszmidt and Pearl [15,16]), to the least specific

M. Ojeda-Aciego et al. (Eds.): JELIA 2000, LNAI 1919, pp. 313–328, 2000.

possibility entailment by Benferhat et al. [4], and to a conditional (modal) logic-based entailment by Lamarre [18]. Finally, mainly to solve problems with property inheritance from classes to exceptional subclasses, the maximum entropy approach was proposed by Goldszmidt et al. [13] (and recently generalized to variable strength defaults by Bourne and Parsons [7]); lexicographic entailment was introduced by Lehmann [19] and Benferhat et al. [3]; and conditional entailment was proposed by Geffner [11,12].

However, while the semantic aspects of these formalisms are quite well understood, about their computational properties only partial results have been known so far. In previous work [9], we were filling some of these gaps by drawing a precise picture of the complexity of major formalisms for default reasoning from conditional knowledge bases. The main goal of this paper now is to complement this work by finding meaningful cases in which default reasoning from conditional knowledge bases is tractable. In particular, we aim at identifying nontrivial restrictions that can be checked efficiently and that guarantee sufficient expressiveness.

The main contributions of this paper can be summarized as follows:

- We introduce q-Horn conditional knowledge bases, which enrich in the spirit of [5] Horn conditional knowledge bases by allowing limited use of disjunction in both strict statements and defeasible rules. For example, a default $saturday \rightarrow hiking \lor shopping$, which informally expresses that on Saturday, someone is normally out for hiking or shopping, can be expressed in a q-Horn KB, but not in a Horn KB.

- We show that previous tractability results for ε-entailment [20,16], proper ε-entailment [14], and z- and z^+-entailment [16] in the Horn case can be extended to the q-Horn case. Thus, in all these approaches, tractability is retained under a limited use of disjunction.

- We present feedback-free-Horn conditional knowledge bases, which restrict the literal-Horn case (where default rules are Horn-like) by requesting that, roughly speaking, default consequents do not fire back into the classical knowledge of KB and that the defaults can be grouped into non-interfering clusters of bounded size. We give some examples from the literature that underline the importance of the feedback-free-Horn case. In particular, we show that taxonomic hierarchies that are augmented by default knowledge can be expressed in the feedback-free-Horn case.

- We show that in the feedback-free-Horn case, default reasoning under z^*-entailment [13], z_s^*-entailment [7], lex-entailment [3], and lex_p-entailment [19] is tractable. To our knowledge, no or only limited tractable cases [8] for these notions of entailment from conditional knowledge bases have been identified so far.

- Our tractability results for the feedback-free-Horn case are complemented by our proof that without a similar restriction on literal-Horn defaults, all the respective semantics remain intractable. In particular, this applies to the 1-literal-Horn case, in which each default is literal-Horn and has at most one atom in its antecedent, and the strict knowledge consists of Horn-clauses having at most two literals.

Note that detailed proofs of all results in this extended abstract are given in [10].

2 Preliminaries

2.1 Conditional Knowledge Bases

We assume a set of propositional *atoms* $At = \{p_1, \ldots, p_n\}$ with $n \geq 1$. We use $\bot$ and $\top$ to denote the propositional constants *false* and *true*, respectively. The set of *classical formulas* is the closure of $At \cup \{\bot, \top\}$ under the Boolean operations $\neg$ and $\wedge$. We use $(\phi \Rightarrow \psi)$ and $(\phi \vee \psi)$ to abbreviate $\neg(\phi \wedge \neg\psi)$ and $\neg(\neg\phi \wedge \neg\psi)$, respectively, and adopt the usual conventions to eliminate parentheses. A *literal* is an atom $p \in At$ or its negation $\neg p$. A *Horn clause* is a classical formula $\phi \Rightarrow \psi$, where ϕ is either $\top$ or a conjunction of atoms, and ψ is either $\bot$ or an atom. A *definite Horn clause* is a Horn clause $\phi \Rightarrow \psi$, where ψ is an atom.

A *conditional rule* (or *default*) is an expression $\phi \rightarrow \psi$, where ϕ and ψ are classical formulas. A *conditional knowledge base* is a pair $KB = (L, D)$, where L is a finite set of classical formulas and D is a finite set of defaults. Informally, L contains facts and rules that are certain, while D contains defeasible rules. In case $L = \emptyset$, we call KB a *default knowledge base*. A default $\phi \rightarrow \psi$ is *Horn* (resp., *literal-Horn*), if ϕ is either $\top$ or a conjunction of atoms, and ψ is a conjunction of Horn clauses (resp., ψ is a literal). A *definite literal-Horn default* is a literal-Horn default $\phi \rightarrow \psi$, where ψ is an atom.

Given a conditional knowledge base $KB = (L, D)$, a *strength assignment* σ on KB is a mapping that assigns each $d \in D$ an integer $\sigma(d) \geq 0$. A *priority assignment* on KB is a strength assignment π on KB with $\{\pi(d) \mid d \in D\} = \{0, 1, \ldots, k\}$ for some $k \geq 0$.

An *interpretation* (or *world*) is a truth assignment $I: At \rightarrow \{\textbf{true}, \textbf{false}\}$, which is extended to classical formulas as usual. We use $\mathcal{I}_{At}$ to denote the set of all worlds for At. The world I *satisfies* a classical formula ϕ, or I is a *model* of ϕ, denoted $I \models \phi$, iff $I(\phi) = \textbf{true}$. I satisfies a default $\phi \rightarrow \psi$, or I is a *model* of $\phi \rightarrow \psi$, denoted $I \models \phi \rightarrow \psi$, iff $I \models \phi \Rightarrow \psi$. I satisfies a set K of classical formulas and defaults, or I is a *model* of K, denoted $I \models K$, iff I satisfies every member of K. The world I *verifies* a default $\phi \rightarrow \psi$ iff $I \models \phi \wedge \psi$. I *falsifies* a default $\phi \rightarrow \psi$, iff $I \models \phi \wedge \neg\psi$ (that is, $I \not\models \phi \rightarrow \psi$). A set of defaults D *tolerates* a default d *under* a set of classical formulas L iff $D \cup L$ has a model that verifies d. A set of defaults D is *under L in conflict* with a default $\phi \rightarrow \psi$ iff all models of $D \cup L \cup \{\phi\}$ satisfy $\neg\psi$.

A *world ranking* κ is a mapping $\kappa: \mathcal{I}_{At} \rightarrow \{0, 1, \ldots\} \cup \{\infty\}$ such that $\kappa(I) = 0$ for at least one world I. It is extended to all classical formulas ϕ as follows. If ϕ is satisfiable, then $\kappa(\phi) = \min\{\kappa(I) \mid I \in \mathcal{I}_{At}, I \models \phi\}$; otherwise, $\kappa(\phi) = \infty$. A world ranking κ is *admissible* with a conditional knowledge base (L, D) iff $\kappa(\neg\phi) = \infty$ for all $\phi \in L$, and $\kappa(\phi) < \infty$ and $\kappa(\phi \wedge \psi) < \kappa(\phi \wedge \neg\psi)$ for all defaults $\phi \rightarrow \psi \in D$. A *default ranking* σ on D maps each $d \in D$ to a nonnegative integer.

2.2 Semantics for Conditional Knowledge Bases

ε-Semantics (Adams [1] and Pearl [21]). We describe the notions of ε-consistency, ε-entailment, and proper ε-entailment in terms of world rankings.

A conditional knowledge base KB is *ε-consistent* iff there exists a world ranking that is admissible with KB. It is *ε-inconsistent* iff no such a world ranking exists.

A conditional knowledge base KB ε-entails a default $\phi \rightarrow \psi$ iff either $\kappa(\phi) = \infty$ or $\kappa(\phi \wedge \psi) < \kappa(\phi \wedge \neg\psi)$ for all world rankings κ admissible with KB. Moreover, KB properly ε-entails $\phi \rightarrow \psi$ iff KB ε-entails $\phi \rightarrow \psi$ and KB does not ε-entail $\phi \rightarrow \bot$.

The next theorem is a simple generalization of a result by Adams [1].

Theorem 2.1 (essentially [1]). *A conditional knowledge base* (L, D) ε-*entails a default* $\phi \rightarrow \psi$ *iff the conditional knowledge base* $(L, D \cup \{\phi \rightarrow \neg\psi\})$ *is* ε-*inconsistent.*

Systems Z and Z^+ (Pearl [22] and Goldszmidt and Pearl [15,16]). Entailment in system Z^+ applies to ε-consistent conditional knowledge bases $KB = (L, D)$ with strength assignment σ on KB. It is linked to a default ranking z^+ and a world ranking κ^+, which are the unique solution of the following system of equations:

$$z^+(d) = \sigma(d) + \kappa^+(\phi \wedge \psi) \qquad \text{(for all } d = \phi \rightarrow \psi \in D) \quad (1)$$

$$\kappa^+(I) = \begin{cases} \infty & \text{if } I \not\models L \\ 0 & \text{if } I \models L \cup D \\ 1 + \max_{d \in D: \, I \not\models d} z^+(d) & \text{otherwise} \end{cases} \qquad \text{(for all } I \in \mathcal{I}_{At}) \quad (2)$$

A default $\phi \rightarrow \psi$ is z^+-entailed by (KB, σ) at strength τ iff either $\kappa^+(\phi) = \infty$ or $\kappa^+(\phi \wedge \psi) + \tau < \kappa^+(\phi \wedge \neg\psi)$.

Entailment in system Z is a special case of entailment in system Z^+. It applies to ε-consistent conditional knowledge bases KB. A default $\phi \rightarrow \psi$ is z-entailed by KB iff $\phi \rightarrow \psi$ is z^+-entailed by (KB, σ) at strength 0, where $\sigma(d) = 0$ for all $d \in D$.

Maximum Entropy (Goldszmidt et al. [13] and Bourne and Parsons [7]). The notion of z_s^*-entailment applies to ε-consistent conditional knowledge bases $KB = (L, D)$ with positive strength assignment σ. It is defined whenever the following system of equations (3) and (4) has a unique solution z_s^*, κ_s^* with positive z_s^*:

$$\kappa_s^*(\phi \wedge \neg\psi) = \sigma(\phi \rightarrow \psi) + \kappa_s^*(\phi \wedge \psi) \qquad \text{(for all } d = \phi \rightarrow \psi \in D) \quad (3)$$

$$\kappa_s^*(I) = \begin{cases} \infty & \text{if } I \not\models L \\ 0 & \text{if } I \models L \cup D \\ \sum_{d \in D: \, I \not\models d} z_s^*(d) & \text{otherwise} \end{cases} \qquad \text{(for all } I \in \mathcal{I}_{At}) \quad (4)$$

The uniqueness of z_s^* and κ_s^* is guaranteed by assuming that κ_s^* is *robust* [7], which is the following property: for all distinct defaults $d_1, d_2 \in D$, it holds that all models I_1 and I_2 of L having smallest ranks in κ_s^* such that $I_1 \not\models d_1$ and $I_2 \not\models d_2$, respectively, are different. That is, d_1 and d_2 do not have a common minimal falsifying model under L. We say KB is *robust* iff the system of equations given by (3) and (4) has a unique solution z_s^*, κ_s^* such that z_s^* is positive and κ_s^* is robust. A default $\phi \rightarrow \psi$ is z_s^*-*entailed* by (KB, σ) at strength τ iff either $\kappa_s^*(\phi) = \infty$ or $\kappa_s^*(\phi \wedge \psi) + \tau \leq \kappa_s^*(\phi \wedge \neg\psi)$.

The notion of z^*-entailment is a special case of z_s^*-entailment. It applies to ε-consistent minimal-core conditional knowledge bases $KB = (L, D)$ without strength assignment, where KB is *minimal-core* iff for each default $d \in D$ there exists a model I of $L \cup (D - \{d\})$ that falsifies d. A default $\phi \rightarrow \psi$ is z^*-*entailed* by KB iff $\phi \rightarrow \psi$ is z_s^*-entailed by (KB, σ) at strength 1, where $\sigma(d) = 1$ for all $d \in D$.

Lexicographic Entailment (Lehmann [19] and Benferhat et al. [3]). Lexicographic entailment in [3] applies to conditional knowledge bases $KB = (L, D)$ with a priority assignment π on KB, which defines an ordered partition $(D_0, \ldots, D_k)$ of D by $D_i = \{d \in D \mid \pi(d) = i\}$, for all $i \leq k$. It is used to define a *preference ordering* on worlds as follows. A world I is π-*preferable* to a world I' iff there exists some $i \in \{0, \ldots, k\}$ such that $|\{d \in D_i \mid I \models d\}| > |\{d \in D_i \mid I' \models d\}|$ and $|\{d \in D_j \mid I \models d\}| = |\{d \in D_j \mid I' \models d\}|$ for all $i < j \leq k$. A model I of a set of classical formulas $\mathcal{F}$ is a π-*preferred* model of $\mathcal{F}$ iff no model of $\mathcal{F}$ is π-preferable to I. A default $\phi \to \psi$ is *lex$_p$-entailed* by (KB, π) iff ψ is satisfied in every π-preferred model of $L \cup \{\phi\}$.

The notion of lexicographic entailment in [19] is a special case of lexicographic entailment as above. It applies to ε-consistent conditional knowledge bases KB, and uses the default ranking z of KB in system Z as priority assignment. That is, a default $\phi \to \psi$ is *lex-entailed* by KB iff $\phi \to \psi$ is *lex$_p$-entailed* by (KB, z).

2.3   Example

Consider the following conditional knowledge base $KB = (L, D)$, adapted from [16], which represents the strict knowledge "all penguins are birds", and the defeasible rules "generally, birds fly", "generally, penguins do not fly", "generally, birds have wings", "generally, penguins live in the arctic", and "generally, flying animals are mobile".

$$L = \{penguin \Rightarrow bird\},$$
$$D = \{bird \to fly,\ penguin \to \neg fly,\ bird \to wings,\ penguin \to arctic,\ fly \to mobile\}.$$

We would like KB to entail "generally, birds are mobile" (as birds generally fly, and flying animals are generally mobile) and "generally, red birds fly" (as the property "red" is not mentioned at all in KB and should thus be considered irrelevant to the flying ability of birds). Moreover, KB should entail "generally, penguins have wings" (as the set of all penguins is a subclass of the set of all birds, and thus penguins should inherit all properties of birds), and "generally, penguins do not fly" (as properties of more specific classes should override inherited properties of less specific classes).

The corresponding behavior of ε-, z-, z^*-, and lex-entailment is shown in Table 1. In detail, $bird \to mobile$ is a plausible consequence of KB under all notions of entailment except for ε-entailment. Moreover, in this example, every notion of entailment except for ε-entailment ignores irrelevant information, while every notion of entailment except for ε- and z-entailment shows property inheritance from the class of all birds to the exceptional subclass of all penguins. Finally, the default $penguin \to \neg fly$ is entailed by KB under all notions of entailment.

For instance, let us verify that $penguin \to \neg fly$ is ε-entailed by KB. By Theorem 2.1, we have to check that $(L, D \cup \{penguin \to fly\})$ is ε-inconsistent. But this is indeed the case, since there is no world ranking κ that satisfies $\kappa(penguin) < \infty$ as well as $\kappa(penguin \wedge \neg fly) < \kappa(penguin \wedge fly)$ and $\kappa(penguin \wedge fly) < \kappa(penguin \wedge \neg fly)$.

Table 1. Plausible consequences of KB under different semantics

	$bird \rightarrow mobile$	$red \wedge bird \rightarrow fly$	$penguin \rightarrow wings$	$penguin \rightarrow \neg fly$
ε-entailment	−	−	−	+
z-entailment	+	+	−	+
$z^\star$-entailment	+	+	+	+
lex-entailment	+	+	+	+

3 Overview of Tractability Results

3.1 Problem Statements

A *default reasoning problem* is a pair (KB, d), where $KB = (L, D)$ is a conditional knowledge base and d is a default. It is *Horn* (resp., *literal-Horn*) iff L is a finite set of Horn clauses, D is a finite set of Horn (resp., literal-Horn) defaults, and d is a Horn (resp., literal-Horn) default. In case of z^+- and $z_s^\star$-entailment, we assume that KB and d have additionally a strength assignment $\sigma(KB)$ and a strength $\tau(d)$, respectively. In case of lex_p-entailment, KB has in addition a priority assignment $\pi(KB)$.

Informally, a default reasoning problem represents the input for the entailment problem under a fixed semantics S. We tacitly assume that KB satisfies any preconditions that the definition of S-entailment in the previous section may request.

We consider the following problems:

- ENTAILMENT: Given a default reasoning problem (KB, d), decide whether KB entails d under some fixed semantics S. In case of z^+- and $z_s^\star$-entailment, decide whether d is z^+- and $z_s^\star$-entailed, respectively, by $(KB, \sigma(KB))$ at strength $\tau(d)$. In case of lex_p-entailment, we are asked whether d is lex_p-entailed by $(KB, \pi(KB))$.
- RANKING: Given a conditional knowledge base KB, compute the default ranking R of KB according to some fixed semantics S (that is, the rank of each $d \in D$).
- RANK-ENTAILMENT: Same as entailment, but the (unique) default ranking R of KB according to some fixed semantics S is part of the problem input.

3.2 Previous Tractability Results

Previous results on the tractability/intractability frontier can be described as follows.

Deciding ε-entailment is intractable in the general case [20] and tractable in the Horn case [20,16]. Similarly, deciding proper ε-entailment is intractable in the general case [9] and tractable in the Horn case [14]. Moreover, the problems ENTAILMENT, RANKING, and RANK-ENTAILMENT for systems Z and Z^+ are intractable in the general case [9] and tractable in the Horn case [16].

The problems ENTAILMENT, RANKING, and RANK-ENTAILMENT for the semantics $z^\star$ and $z_s^\star$ are intractable even in the literal-Horn case [9]. Moreover, also deciding lex- and lex_p-entailment is intractable in the literal-Horn case [9]. To our knowledge, no or only limited tractable cases for these notions of entailment have been identified so far (a limited tractable case for lex_p-entailment has been presented in [8]).

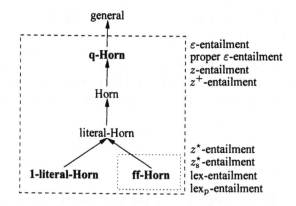

Fig. 1. Tractability of ENTAILMENT

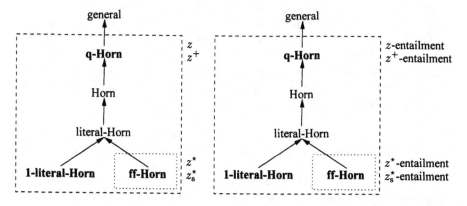

Fig. 2. Tractability of RANKING and RANK-ENTAILMENT

3.3   New Tractability Results

It would be interesting to know whether the tractability results for ε-entailment, proper ε-entailment, and z- and z^+ entailment in the Horn case can be extended to more expressive classes of problems. Moreover, it would be interesting to know whether there are meaningful tractable classes of problems for z^*-, z_s^*-, lex-, and $\mathrm{lex_p}$-entailment.

Concerning the first issue, we introduce the class of q-Horn conditional knowledge bases. This class generalizes Horn conditional knowledge bases syntactically by allowing a restricted use of disjunction, and contains instances that cannot be represented in Horn conditional knowledge bases. As we show, the tractability results for Horn conditional knowledge bases extend to q-Horn conditional knowledge bases.

Finding meaningful tractable cases for the more sophisticated semantics for conditional knowledge bases is more challenging. A natural attempt is to show that a further restriction of the literal-Horn case leads to tractability. An obvious candidate restriction is bounding the size of the bodies in the strict and classical rules to at most one atom. Unfortunately, this does not buy tractability. An analysis of our proof reveals that the interaction of the defaults among each other and with the classical background knowledge must be controlled such that interferences have a local effect. This leads us to the class of feedback-free-Horn (ff-Horn) default reasoning problems. As we show, tractability is gained on this class for all the intractable semantics discussed here.

The hierarchy of all classes of conditional knowledge bases considered in this paper, along with the corresponding tractability results, is given in Figures 1–2.

4   Q-Horn

4.1   Motivating Example

Q-Horn conditional knowledge bases generalize Horn conditional knowledge bases by allowing a limited form of disjunction, which is illustrated by the following example.

Example 4.1. Assume that John is looking for Mary. Unfortunately, he did not find her at home. So, he is wondering where she might be. He knows that Mary might have tea with her friends, that she might be in the library, or that she might play tennis. He also knows that these scenarios are pairwise exclusive and not exhaustive. Moreover, John knows that "generally, in the afternoon, Mary is having tea with her friends or she is in the library" and that "generally, on Friday afternoon, Mary plays tennis". This knowledge can be expressed by the following $KB = (L, D)$:

$$L = \{\neg tea \lor \neg library, \ \neg tea \lor \neg tennis, \ \neg library \lor \neg tennis\},$$
$$D = \{afternoon \rightarrow tea \lor library, \ Friday \land afternoon \rightarrow tennis\}.$$

Assume that it is Friday afternoon and that John is wondering whether he should go to the library to look for Mary. That is, does KB entail $Friday \land afternoon \rightarrow library$?

4.2   Definitions

A *clause* is a disjunction of literals. A default $\phi \rightarrow \psi$ is *clausal* iff ϕ is either $\top$ or a conjunction of literals, and ψ is a conjunction of clauses. A conditional knowledge base $KB = (L, D)$ is *clausal* iff L is a finite set of clauses and D is a finite set of clausal defaults. A default reasoning problem (KB, d) is *clausal* iff both KB and d are clausal.

A classical formula ϕ is in *conjunctive normal form* (or *CNF*) iff ϕ is either $\top$ or a conjunction of clauses. We use the operator $\sim$ to map each atom a to its negation $\neg a$, and each negated atom $\neg a$ to a. We define a mapping $\mathcal{N}$ that associates each clausal default d with a classical formula in CNF as follows. If d is of the form $\top \rightarrow c_1 \land \cdots \land c_n$ with clauses $c_1, \ldots, c_n$, then $\mathcal{N}(d) = c_1 \land \cdots \land c_n$. If d is of the form $l_1 \land \cdots \land l_m \rightarrow c_1 \land \cdots \land c_n$ with literals $l_1, \ldots, l_m$ and clauses $c_1, \ldots, c_n$, then $\mathcal{N}(d)$ is the conjunction of all $\sim l_1 \lor \cdots \lor \sim l_m \lor c_i$ with $i \in \{1, \ldots, n\}$. We extend $\mathcal{N}$ to classical formulas in CNF ϕ by $\mathcal{N}(\phi) = \phi$. We extend $\mathcal{N}$ to finite sets K of classical formulas in CNF and clausal defaults as follows. Let K' denote the set of all $k \in K$ with $\mathcal{N}(k) \neq \top$. If $K' \neq \emptyset$, then $\mathcal{N}(K)$ is the conjunction of all $\mathcal{N}(k)$ with $k \in K'$. Otherwise, $\mathcal{N}(K) = \top$.

A *partial assignment* S is a set of literals such that for every atom $a \in At$ at most one of the literals a and $\neg a$ is in S. A classical formula in CNF ϕ is *q-Horn* [5] iff there exists a partial assignment S such that (i) each clause in ϕ contains at most two literals outside of S, and (ii) if a clause in ϕ contains exactly two literals $u, v \notin S$, then neither $\sim u$ nor $\sim v$ belongs to S. Note that every conjunction of Horn clauses is q-Horn.

A finite set K of classical formulas in CNF and clausal defaults is *q-Horn* iff $\mathcal{N}(K)$ is q-Horn. A conditional knowledge base $KB = (L, D)$ is *q-Horn* iff KB is clausal and $L \cup D$ is q-Horn. Clearly, every Horn KB is q-Horn, but not vice versa. A default reasoning problem (KB, d) is *q-Horn*, if KB is q-Horn and d is a clausal default.

Example 4.2. The conditional knowledge base $KB = (L, D)$ of Example 4.1 is q-Horn. In detail, the classical formula $\mathcal{N}(L \cup D)$ associated with KB is given by:

$$\mathcal{N}(L \cup D) = (\neg tea \vee \neg library) \wedge (\neg tea \vee \neg tennis) \wedge (\neg library \vee \neg tennis) \wedge$$
$$(\neg afternoon \vee tea \vee library) \wedge (\neg Friday \vee \neg afternoon \vee tennis).$$

A partial assignment that satisfies (i) and (ii) is given by $\{\neg Friday, \neg afternoon\}$. That is, $\mathcal{N}(L \cup D)$ is q-Horn. Since KB is also clausal, it thus follows that KB is q-Horn.

Note that KB can be made Horn by "renaming" atoms, in particular, by replacing the atom *library* by a negated new atom $\overline{library}$, where $\overline{library}$ stands for $\neg library$. However, if the scenarios were exhaustive and thus the clause *library* $\vee$ *tea* $\vee$ *tennis* is in KB, then no Horn renaming of KB is possible. But, the resulting KB is still q-Horn.

The *size* of a classical formula in CNF ϕ, denoted $\|\phi\|$, is defined as the number of occurrences of literals in ϕ. We use $|\phi|$ to denote the number of clauses in ϕ. The *size* of a clausal default $d = \phi \rightarrow \psi$, denoted $\|d\|$, is defined as $\|\phi\| + \|\psi\|$. The *size* of a finite set of clauses L, denoted $\|L\|$, is defined as the size of $\mathcal{N}(L)$. The *size* of a clausal $KB = (L, D)$, denoted $\|KB\|$, is defined as the size of $\mathcal{N}(L \cup D)$. We use $|D|$ to denote the cardinality of D.

4.3 Q-Horn Formulas

The problems of deciding whether a q-Horn formula is satisfiable and of recognizing q-Horn formulas are both tractable and can in fact be solved in linear time.

Proposition 4.3 (see [5,6]). *a) Given a q-Horn formula ϕ, deciding whether ϕ is satisfiable can be done in time $O(\|\phi\|)$. b) Given a classical formula in CNF ϕ, deciding whether ϕ is q-Horn can be done in time $O(\|\phi\|)$.*

By this result, it follows easily that also q-Horn conditional knowledge bases can be recognized in linear time.

Theorem 4.4. *Given a clausal conditional knowledge base $KB = (L, D)$, deciding whether KB is q-Horn can be done in time $O(\|KB\|)$.*

4.4 ε-Semantics

The following theorem shows that deciding whether a q-Horn KB is ε-consistent is tractable. The proof of this result is based on the fact that checking the ε-consistency of KB is reducible to a polynomial number of classical satisfiability tests. By closure properties of q-Horn formulas, it then follows that for q-Horn KB, each satisfiability test is done on a q-Horn formula and thus possible in polynomial time.

Theorem 4.5. *Given a q-Horn conditional knowledge base* $KB = (L, D)$, *deciding whether KB is ε-consistent is possible in time* $O(|D|^2 \|KB\|)$.

The next result shows that deciding ε-entailment is tractable in the q-Horn case.

Theorem 4.6. *Given a q-Horn default reasoning problem* $(KB, d) = ((L, D), \phi \rightarrow \psi)$, *deciding whether KB ε-entails d is possible in time* $O((\|\phi\| + |\psi|) |D|^2 (\|KB\| + \|\psi\|))$.

Finally, deciding proper ε-entailment is also tractable in the q-Horn case.

Theorem 4.7. *Given a q-Horn default reasoning problem* $(KB, d) = ((L, D), \phi \rightarrow \psi)$, *deciding whether KB properly ε-entails d is possible in time* $O((\|\phi\| + |\psi|) |D|^2 (\|KB\| + \|\psi\|))$.

4.5 Systems Z and Z^+

We next focus on entailment in systems Z and Z^+. The following result, which can be proved in a similar way as Theorems 4.5–4.7, shows that computing the default ranking z^+ is tractable in the q-Horn case. Since system Z^+ properly generalizes system Z, this result shows also that computing the default ranking z is tractable in the q-Horn case.

Theorem 4.8. *Given an ε-consistent q-Horn conditional knowledge base* $KB = (L, D)$ *with strength assignment σ, the default ranking z^+ can be computed in polynomial time.*

Finally, the following theorem shows that deciding z^+-entailment is tractable in the q-Horn case. Again, since system Z^+ properly generalizes system Z, this result shows also that deciding z-entailment is tractable in the q-Horn case. Trivially, these tractability results remain true when z^+ and z, respectively, are part of the input, that is, for RANK-ENTAILMENT.

Theorem 4.9. *Given a q-Horn default reasoning problem* $(KB, d) = ((L, D), \phi \rightarrow \psi)$, *where KB is ε-consistent and has a strength assignment σ, deciding whether (KB, σ) z^+-entails d at a given strength $\tau \geq 0$ can be done in polynomial time.*

5 Feedback-Free-Horn

5.1 Intractability Results for 1-Literal-Horn Case

How do we obtain tractability of deciding s-entailment, where $s \in \{z^*, z_s^*, \text{lex}, \text{lex}_p\}$? In particular, are there any syntactic restrictions on default reasoning problems that give tractability? We could, for example, further restrict literal-Horn defaults by limiting the number of atoms in the antecedent of each default as follows. A default $\phi \rightarrow \psi$ is *1-literal-Horn* iff ϕ is either $\top$ or an atom, and ψ is a literal. A *1-Horn clause* is a classical formula $\phi \Rightarrow \psi$, where ϕ is either $\top$ or an atom, and ψ is a literal. A conditional knowledge base $KB = (L, D)$ is *1-literal-Horn* iff L is a finite set of 1-Horn clauses and D is a finite set of 1-literal-Horn defaults. A default reasoning problem (KB, d) is 1-literal-Horn iff both KB and d are 1-literal-Horn.

Unfortunately, the following theorem shows that deciding z^*-entailment is still (presumably) intractable even for this very restricted kind of default reasoning problems.

Theorem 5.1. *Given a 1-literal-Horn KB, which is ε-consistent and minimal-core, and a 1-literal-Horn default d, deciding whether KB z^*-entails d is co-NP-hard.*

Informally, this intractability is due to the fact that the default knowledge generally does not fix a unique instantiation of the atoms to truth values, in particular, when defaults "fire back" into the bodies of other defaults, and when defaults are logically related through their heads.

Since z_s^*-entailment is a proper generalization of z^*-entailment, it immediately follows that deciding z_s^*-entailment is (presumably) intractable in the 1-literal-Horn case.

Corollary 5.2. *Given a 1-literal-Horn conditional knowledge base KB, which is ε-consistent and robust, a strength assignment σ on KB, a 1-literal-Horn default d, and a strength τ, deciding whether (KB, σ) z_s^*-entails d at strength τ is co-NP-hard.*

The following theorem shows that also deciding lex- and lex_p-entailment is (presumably) intractable in the 1-literal-Horn case.

Theorem 5.3. *a) Given an ε-consistent 1-literal-Horn conditional knowledge base KB and a 1-literal-Horn default d, deciding whether KB lex-entails d is co-NP-hard.*

b) Given a 1-literal-Horn conditional knowledge base KB with priority assignment π and a 1-literal-Horn default d, deciding whether (KB, π) lex_p-entails d is co-NP-hard.

5.2 Motivating Examples

We will see that deciding s-entailment, where $s \in \{z^*, z_s^*, \text{lex}, \text{lex}_p\}$, becomes tractable, if we assume that the default reasoning problems can be sensibly decomposed into smaller problems of size bounded by a constant. We now give some examples to illustrate the main ideas behind this kind of decomposability. In the following examples, we assume that conditional knowledge bases are implicitly associated with a strength assignment σ (resp., priority assignment π), when $s = z_s^*$ (resp., $s = \text{lex}_p$).

Example 5.4. Take again $KB = (L, D)$ of Section 2.3. Assume that we are wondering whether KB s-entails *penguin* $\rightarrow$ *fly*, *red* $\wedge$ *bird* $\rightarrow$ *fly*, *bird* $\rightarrow$ *mobile*, *penguin* $\rightarrow$ *arctic*, or *penguin* $\rightarrow$ *wings*, where $s \in \{z^*, z_s^*, \text{lex}, \text{lex}_p\}$. As it turns out, each of these problems can be reduced to one classical reasoning problem and one default reasoning problem. More precisely, the former is done w.r.t. the set of atoms {*penguin, bird, red*}, which refers to the atoms in L and the antecedent of the query default, while the latter is done w.r.t. the sets of atoms {*fly, mobile*}, {*arctic*}, and {*wings*}, respectively, by sensibly eliminating irrelevant defaults and simplifying the remaining defaults by instantiating atoms to truth values. For instance, deciding whether KB s-entails *red* $\wedge$ *bird* $\rightarrow$ *fly* is reduced to the classical reasoning problem of computing the least model of $L \cup \{red \wedge bird\}$ and the default reasoning problem of deciding whether ({*red, bird, ¬penguin*}, {*bird* $\rightarrow$ *fly, fly* $\rightarrow$ *mobile*}) s-entails *red* $\wedge$ *bird* $\rightarrow$ *fly*.

We next consider a taxonomic hierarchy adorned with some default knowledge [2].

Example 5.5. The strict knowledge "all birds and fish are animals", "all penguins and sparrows are birds", "no bird is a fish", "no penguin is a sparrow", and the defeasible knowledge "generally, animals do not swim", "generally, fish swims", and "generally, penguins swim" can be represented by the following $KB = (L, D)$:

$$L = \{bird \Rightarrow animal,\ fish \Rightarrow animal,\ penguin \Rightarrow bird,$$
$$sparrow \Rightarrow bird,\ bird \Rightarrow \neg fish,\ penguin \Rightarrow \neg sparrow\},$$
$$D = \{animal \rightarrow \neg swims,\ fish \rightarrow swims,\ penguin \rightarrow swims\}.$$

Do sparrows generally swim? That is, does KB s-entail $sparrow \rightarrow swims$, where $s \in \{z^\star, z_s^\star, \text{lex}, \text{lex}_p\}$? This default reasoning problem can be reduced to one classical reasoning problem w.r.t. the set of atoms $\{animal, bird, fish, sparrow, penguin\}$ and one default reasoning problem w.r.t. the set of atoms $\{swims\}$. In detail, we first compute the least model of $L \cup \{sparrow\}$ and then decide whether $(\{sparrow, bird, animal, \neg fish, \neg penguin\}, \{animal \rightarrow \neg swims\})$ s-entails $sparrow \rightarrow swims$.

5.3   Definitions

Suppose that for a literal-Horn conditional knowledge base $KB = (L, D)$, there exists a set of atoms $At_a \subseteq At$ such that L is defined over At_a and that all consequents of definite literal-Horn defaults in D are defined over $At - At_a$. The greatest such At_a, which clearly exists, is called the *activation set* of KB. Intuitively, in any "context" given by L and ϕ, where ϕ is either $\top$ or a conjunction of atoms from At, all those atoms in At_a that are not logically entailed by $L \cup \{\phi\}$ can be safely set to false in the preferred models of $L \cup \{\phi\}$.

For At_a, there is a greatest partition $\{At_1, \ldots, At_n\}$ of $At - At_a$ such that every $d \in D$ is defined over some $At_a \cup At_i$ with $i \in \{1, \ldots, n\}$, which we call the *default partition* of KB. We say $KB = (L, D)$ is *k-feedback-free-Horn* (or *k-ff-Horn*) iff it is literal-Horn, it has an activation set At_a, and it has a default partition $\{At_1, \ldots, At_n\}$ such that every At_i with $i \in \{1, \ldots, n\}$ has a cardinality of at most k.

Example 5.6. The conditional knowledge base KB of Example 5.5 is 1-ff-Horn. More precisely, its activation set (resp., default partition) is given by $\{animal, bird, fish, sparrow, penguin\}$ (resp., $\{\{swims\}\}$).

Moreover, KB of Example 5.4 is 2-ff-Horn. Its activation set (resp., default partition) is given by $\{penguin, bird, red\}$ (resp., $\{\{fly, mobile\}, \{arctic\}, \{wings\}\}$).

For sets of Horn clauses L, we use L^+ to denote the set of all definite Horn clauses in L. For sets of literal-Horn defaults D, we use D^+ to denote the set of all definite literal-Horn defaults in D. Assume additionally that $d = \phi \rightarrow \psi$ is a literal-Horn default. Then, a literal-Horn default $\alpha \rightarrow a$ (resp., $\alpha \rightarrow \neg a$) with $a \in At$ is *active* w.r.t. (L, D) and d iff $L^+ \cup D^+ \cup \{\phi\} \models \alpha$ (resp., $L^+ \cup D^+ \cup \{\phi\} \models \alpha \wedge a$).

A default reasoning problem $(KB, d) = ((L, D), \phi \rightarrow \psi)$ is *k-ff-Horn*, where $k \geq 1$, iff (i) it is literal-Horn, and (ii) $(L, D_a \cup \{d\})$ has an activation set At_a and a default partition $\{At_1, \ldots, At_n\}$ such that d is defined over some $At_a \cup At_j$ with $|At_j| \leq k$, where D_a is the set of all active defaults in D w.r.t. KB and d. The class k-ff-Horn consists of all k-ff-Horn default reasoning problems; we define the class *feedback-free-Horn* (or *ff-Horn*) by ff-Horn $= \bigcup_{k \geq 1} k$-ff-Horn.

Example 5.7. Consider the literal-Horn default reasoning problem (KB, d) with $KB = (L, D)$ as in Example 5.4 and $d = red \wedge bird \rightarrow fly$. The set D_a of active defaults in D w.r.t. KB and d is given by $D_a = \{bird \rightarrow fly, bird \rightarrow wings, fly \rightarrow mobile\}$.

Now, $(L, D_a \cup \{d\})$ has the activation set $At_a = \{penguin, bird, red, arctic\}$ and the default partition $\{At_1, At_2\}$, where $At_1 = \{fly, mobile\}$ and $At_2 = \{wings\}$. Moreover, d is defined over $At_a \cup At_1$ with $|At_1| = 2$. That is, (KB, d) is 2-ff-Horn.

For Horn conditional knowledge bases $KB = (L, D)$ with activation set At_a, and classical formulas α that are either $\top$ or conjunctions of atoms from At, we define the classical formula α^* as follows. If $L \cup \{\alpha\}$ is satisfiable, then α^* is the conjunction of all $b \in At$ with $L \cup \{\alpha\} \models b$ and all $\neg b$ with $b \in At_a$ and $L \cup \{\alpha\} \not\models b$. Otherwise, we define $\alpha^* = \bot$. Moreover, for satisfiable $L \cup \{\alpha\}$, we define the world I_α^* over the activation set At_a by $I_\alpha^*(b) = \mathbf{true}$ iff $L \cup \{\alpha\} \models b$, for all $b \in At_a$.

5.4 Recognizing Feedback-Free-Horn

Both recognizing k-ff-Horn conditional knowledge bases, and computing their activation set and default partition are efficiently possible using standard methods.

Theorem 5.8. *a) Given a literal-Horn conditional knowledge base KB and an integer $k \geq 1$, deciding whether KB is k-ff-Horn can be done in linear time.*

b) Given a k-ff-Horn conditional knowledge base KB, computing the activation set At_a and the default partition $\{At_1, \ldots, At_n\}$ can be done in linear time.

Moreover, recognizing k-ff-Horn default reasoning problems is also efficiently possible.

Theorem 5.9. *a) Given a literal-Horn default reasoning problem (KB, d), and an integer $k \geq 1$, deciding whether (KB, d) is k-ff-Horn can be done in linear time.*

b) Given a k-ff-Horn default reasoning problem (KB, d) with $KB = (L, D)$, computing the set D_a of active defaults in D w.r.t. KB and d can be done in linear time.

5.5 Maximum Entropy Semantics

In the sequel, let $KB = (L, D)$ be an ε-consistent k-ff-Horn conditional knowledge base with positive strength assignment σ. Let At_a denote the activation set of KB, and let $(At_1, \ldots, At_n)$ be the default partition of KB. Let z_s^* be a ranking that maps each $d \in D$ to a positive integer, and let κ_s^* be defined by (4).

For each $i \in \{1, \ldots, n\}$, let D_i denote the set of all defaults in D that are defined over $At_a \cup At_i$. Let the function $\kappa_{s,i}^*$ on worlds I over At be defined as follows:

$$\kappa_{s,i}^*(I) = \begin{cases} \infty & \text{if } I \not\models L \\ 0 & \text{if } I \models L \cup D_i \\ \sum_{d \in D_i \,:\, I \not\models d} z_s^*(d) & \text{otherwise.} \end{cases} \tag{5}$$

In order to compute the default ranking z_s^*, we have to compute ranks of the form $\kappa_s^*(\alpha \wedge \beta_1 \wedge \cdots \wedge \beta_n)$, where α is either $\top$ or a conjunction of atoms from At_a, and each β_i is either $\top$ or a conjunction of literals over At_i. It can now be shown that such $\kappa_s^*(\alpha \wedge \beta_1 \wedge \cdots \wedge \beta_n)$ coincide with $\sum_{i=1}^{n} \kappa_{s,i}^*(\alpha^* \wedge \beta_i)$.

Using this result, it can be shown that computing the default ranking z_s^* is tractable in the k-ff-Horn case. Since z_s^* is a proper generalization of z^*, this result shows also that computing the default ranking z^* is tractable in the k-ff-Horn case.

Theorem 5.10. *Let $k > 0$ a fixed integer. Given an ε-consistent k-ff-Horn $KB = (L, D)$ with positive strength assignment σ, computing the default ranking z_s^* for KB, if KB is robust, and returning nil otherwise, can be done in polynomial time.*

In the sequel, let $(KB, d) = ((L, D), \phi \rightarrow \psi)$ be a k-ff-Horn default reasoning problem with ε-consistent and robust KB. Let σ be a positive strength assignment on KB. Let D_a be the set of active defaults in D w.r.t. KB and d, let At_a be the activation set of $(L, D_a \cup \{d\})$, and let $(At_1, \ldots, At_n)$ be the default partition of $(L, D_a \cup \{d\})$. Let z_s^*, κ_s^* be the unique solution of (3) and (4).

For every $i \in \{1, \ldots, n\}$, let $\overline{D}_i$ denote the set of all defaults in D_a that are defined over $At_a \cup At_i$, and let σ_i be the restriction of σ to $\overline{D}_i$. Let $\overline{z}_{s,i}^*$ map each default in $\overline{D}_i$ to a positive integer, and let the function $\overline{\kappa}_{s,i}^*$ on worlds I over At be defined by:

$$\overline{\kappa}_{s,i}^*(I) = \begin{cases} \infty & \text{if } I \not\models L \\ 0 & \text{if } I \models L \cup \overline{D}_i \\ \displaystyle\sum_{d \in \overline{D}_i \,:\, I \not\models d} \overline{z}_{s,i}^*(d) & \text{otherwise.} \end{cases} \quad (6)$$

It can be shown that in order to decide whether KB z_s^*-entails d at given strength $\tau > 0$, it is sufficient to know all $z_s^*(d)$ with $d \in \overline{D}_j$, where $j \in \{1, \ldots, n\}$ such that d is defined over $At_a \cup At_j$. Moreover, it can be shown that the restriction of z_s^* to $\overline{D}_j$ coincides with the default ranking for $(L, \overline{D}_j)$ under the strength assignment σ_j.

Using these results, it can be shown that deciding z_s^*-entailment is tractable in the k-ff-Horn case. Again, since z_s^* properly generalizes z^*, this result shows also that deciding z^*-entailment is tractable in the k-ff-Horn case. Trivially, these tractability results remain true when z_s^* and z^*, respectively, are part of the input.

Theorem 5.11. *Let $k > 0$ be fixed. Given a k-ff-Horn default reasoning problem $(KB, d) = ((L, D), \phi \rightarrow \psi)$, where KB is ε-consistent and robust, and a positive strength assignment σ on KB, deciding whether (KB, σ) z_s^*-entails d at given strength $\tau > 0$ can be done in polynomial time.*

Example 5.12. Let the 2-ff-Horn default reasoning problem (KB, d) be given by $KB = (L, D)$ of Example 5.4 and $d = red \wedge bird \rightarrow fly$. Let $\sigma(\delta) = 1$ for all $\delta \in D$.

Now, d is z_s^*-entailed by (KB, σ) at strength τ iff either (i) $L \cup \{red \wedge bird, \neg fly\}$ is unsatisfiable, or (ii) both $L \cup \{red \wedge bird, fly\}$ and $L \cup \{red \wedge bird, \neg fly\}$ are satisfiable, and $\kappa_s^*(red \wedge bird \wedge fly) + \tau \leq \kappa_s^*(red \wedge bird \wedge \neg fly)$. It can be shown that the latter is equivalent to $\overline{\kappa}_{s,1}^*((red \wedge bird)^* \wedge fly) + \tau \leq \overline{\kappa}_{s,1}^*((red \wedge bird)^* \wedge \neg fly)$, that is, $\overline{\kappa}_{s,1}^*(red \wedge bird \wedge fly) + \tau \leq \overline{\kappa}_{s,1}^*(red \wedge bird \wedge \neg fly)$, where $\overline{\kappa}_{s,1}^*$ is defined through the default ranking $\overline{z}_{s,1}^*$ for $(L, D_1) = (L, \{bird \rightarrow fly, fly \rightarrow mobile\})$ under $\sigma_1 = \sigma|_{D_1}$.

It is now easy to verify that $\overline{z}_{s,1}^*(d_1) = 1$ for all $d_1 \in D_1$, that both $L \cup \{red \wedge bird, fly\}$ and $L \cup \{red \wedge bird, \neg fly\}$ are satisfiable, that $\overline{\kappa}_{s,1}^*(red \wedge bird \wedge fly) = 0$, and that $\overline{\kappa}_{s,1}^*(red \wedge bird \wedge \neg fly) = 1$. Thus, (KB, σ) z_s^*-entails $red \wedge bird \rightarrow fly$ at strength 1.

5.6 Lexicographic Entailment

We now focus on lexicographic entailment. In the sequel, let $(KB, d) = ((L, D), \phi \to \psi)$ be a k-ff-Horn default reasoning problem. Let π be a priority assignment on KB. Let D_a denote the set of all active defaults w.r.t. KB and d, let At_a be the activation set of $(L, D_a \cup \{d\})$, and let $(At_1, \ldots, At_n)$ be the default partition of $(L, D_a \cup \{d\})$.

For every $i \in \{1, \ldots, n\}$, let D_i denote the set of all defaults in D_a that are defined over $At_a \cup At_i$, and let $KB_i = (L, D_i)$. Let π_i be the unique priority assignment on KB_i that is consistent with π on KB (that is, $\pi_i(d) < \pi_i(d)$ iff $\pi(d) < \pi(d)$, for all $d \in D_i$). Let $j \in \{1, \ldots, n\}$ such that d is defined over $At_a \cup At_j$.

In order to decide whether (KB, π) lex$_p$-entails d, we must check whether every π-preferred model of $L \cup \{\phi\}$ satisfies ψ. It can now be shown that we can equivalently check whether every π_j-preferred model of $L \cup \{\phi^*\}$ satisfies ψ.

Using this result, it can be shown that deciding lex$_p$-entailment is tractable in the k-ff-Horn case. Moreover, as computing the z-partition for ε-consistent conditional knowledge bases KB is tractable in the Horn case [16], this result shows also that deciding lex-entailment is tractable in the k-ff-Horn case.

Theorem 5.13. *Let $k > 0$ be fixed. Given a k-ff-Horn default reasoning problem $(KB, d) = ((L, D), \phi \to \psi)$ and a priority assignment π on KB, deciding whether (KB, π) lex$_p$-entails d can be done in linear time.*

Example 5.14. Let the 2-ff-Horn default reasoning problem (KB, d) be given by $KB = (L, D)$ of Example 5.4 and $d = red \land bird \to fly$. Let $\pi(\delta) = 0$, if $\delta \in \{bird \to fly, bird \to wings, fly \to mobile\}$, and $\pi(\delta) = 1$, if $\delta \in \{penguin \to \neg fly, penguin \to arctic\}$.

It can be shown that (KB, π) lex$_p$-entails $red \land bird \to fly$ iff either $L \cup \{red \land bird\}$ is unsatisfiable, or all π_1-preferred models of $L \cup \{(red \land bird)^*\} = L \cup \{red \land bird\}$ satisfy fly, where π_1 is the priority assignment on $KB_1 = (L, D_1) = (L, \{bird \to fly, fly \to mobile\})$ that maps each element of D_1 to 0. It is now easy to verify that this is indeed the case. That is, (KB, π) lex$_p$-entails $red \land bird \to fly$.

Acknowledgments. This work has been partially supported by the Austrian Science Fund Project N Z29-INF and a DFG grant.

References

1. E. W. Adams. *The Logic of Conditionals*, volume 86 of *Synthese Library*. D. Reidel, Dordrecht, Netherlands, 1975.
2. F. Bacchus, A. Grove, J. Y. Halpern, and D. Koller. From statistical knowledge bases to degrees of beliefs. *Artif. Intell.*, 87:75–143, 1996.
3. S. Benferhat, C. Cayrol, D. Dubois, J. Lang, and H. Prade. Inconsistency management and prioritized syntax-based entailment. In *Proceedings IJCAI-93*, pages 640–645, 1993.
4. S. Benferhat, D. Dubois, and H. Prade. Nonmonotonic reasoning, conditional objects and possibility theory. *Artif. Intell.*, 92(1–2):259–276, 1997.
5. E. Boros, Y. Crama, and P. L. Hammer. Polynomial-time inference of all valid implications for Horn and related formulae. *Ann. Math. Artif. Intell.*, 1:21–32, 1990.

6. E. Boros, P. L. Hammer, and X. Sun. Recognition of q-Horn formulae in linear time. *Discrete Appl. Math.*, 55:1–13, 1994.
7. R. A. Bourne and S. Parsons. Maximum entropy and variable strength defaults. In *Proceedings IJCAI-99*, pages 50–55, 1999.
8. C. Cayrol, M.-C. Lagasquie-Schiex, and T. Schiex. Nonmonotonic reasoning: From complexity to algorithms. *Ann. Math. Artif. Intell.*, 22(3–4):207–236, 1998.
9. T. Eiter and T. Lukasiewicz. Complexity results for default reasoning from conditional knowledge bases. In *Proceedings KR-2000*, pages 62–73, 2000.
10. T. Eiter and T. Lukasiewicz. Default reasoning from conditional knowledge bases: Complexity and tractable cases, 2000. Manuscript.
11. H. Geffner. *Default Reasoning: Causal and Conditional Theories*. MIT Press, 1992.
12. H. Geffner and J. Pearl. Conditional entailment: Bridging two approaches to default reasoning. *Artif. Intell.*, 53(2–3):209–244, 1992.
13. M. Goldszmidt, P. Morris, and J. Pearl. A maximum entropy approach to nonmonotonic reasoning. *IEEE Trans. Pattern Anal. Mach. Intell.*, 15(3):220–232, 1993.
14. M. Goldszmidt and J. Pearl. On the consistency of defeasible databases. *Artif. Intell.*, 52(2):121–149, 1991.
15. M. Goldszmidt and J. Pearl. System Z^+: A formalism for reasoning with variable strength defaults. In *Proceedings AAAI-91*, pages 399–404, 1991.
16. M. Goldszmidt and J. Pearl. Qualitative probabilities for default reasoning, belief revision, and causal modeling. *Artif. Intell.*, 84(1–2):57–112, 1996.
17. S. Kraus, D. Lehmann, and M. Magidor. Nonmonotonic reasoning, preferential models and cumulative logics. *Artif. Intell.*, 14(1):167–207, 1990.
18. P. Lamarre. A promenade from monotonicity to non-monotonicity following a theorem prover. In *Proceeding KR-92*, pages 572–580, 1992.
19. D. Lehmann. Another perspective on default reasoning. *Ann. Math. Artif. Intell.*, 15(1):61–82, 1995.
20. D. Lehmann and M. Magidor. What does a conditional knowledge base entail? *Artif. Intell.*, 55(1):1–60, 1992.
21. J. Pearl. Probabilistic semantics for nonmonotonic reasoning: A survey. In *Proceedings KR-89*, pages 505–516, 1989.
22. J. Pearl. System Z: A natural ordering of defaults with tractable applications to default reasoning. In *Proceedings TARK-90*, pages 121–135, 1990.

Monodic Epistemic Predicate Logic

Holger Sturm[1], Frank Wolter[1], and Michael Zakharyaschev[2]

[1] Institut für Informatik, Universität Leipzig,
Augustus-Platz 10-11, 04109 Leipzig, Germany
[2] Division of Artificial Intelligence, School of Computer Studies
University of Leeds, Leeds LS2 9JT, UK.

Abstract. We consider the monodic formulas of common knowledge predicate logic, which allow applications of epistemic operators to formulas with at most one free variable. We provide finite axiomatizations of the monodic fragment of the most important common knowledge predicate logics (the full logics are known to be not recursively enumerable) and single out a number of their decidable fragments. On the other hand, it is proved that the addition of the equality symbol to the monodic fragment makes it not recursively enumerable.

1 Introduction

Ever since it became common knowledge that intelligent behaviour of an agent is based not only on her knowledge about the world but also on knowledge about both her own and other agents' knowledge, logical formalisms designed for reasoning about knowledge have attracted attention in artificial intelligence, computer science, economic theory, and philosophy (cf. e.g. the books [5,16,13] and the seminal works [8,1]). In all these areas, one of the most successful approaches is to supply classical—propositional or first-order—logic with an explicit epistemic operator K_i for each agent i under consideration. $K_i\varphi$ means that *agent i knows (or believes)* φ, $K_1 K_2 \varphi$ says then that *agent 1 knows that agent 2 knows* φ, and the schema of positive introspection $K_i\psi \to K_i K_i \psi$ states that *agent i knows what she knows*. In the first-order case this language is capable of formalizing the distinction between 'knowing that' and 'knowing what' (i.e., modalities *de dicto* and *de re*): the formula $K_i \exists x$ name(x, y) stands for '*i knows that y has a name*,' while $\exists x K_i$name(x, y) means '*i knows a name of y*.'

There can be different interpretations of the knowledge operators (e.g. with or without positive or negative introspection), and for many of them transparent axiomatic representations have been found (cf. e.g. [7,5]). On the other hand, the possible worlds semantics [8] provided a framework to interpret this language: in a world w agent i knows φ if and only if φ holds in all worlds that i regards possible in w (the difference between various understandings of K_i is reflected by different accessibility relations among the worlds).

The situation becomes much more complicated when—in order to describe the behavior of *multi-agent* systems—we extend the language with one more modal operator, C, to capture the *common knowledge* of a group of agents. Such

M. Ojeda-Aciego et al. (Eds.): JELIA 2000, LNAI 1919, pp. 329–344, 2000.

an operator was required for analyzing conventions [14], coordinizations in multi-agent systems [5], common sense reasoning [15], agreement [1,2], etc.[1] Although the intended meaning of the *common knowledge operator* involves infinity: $C\varphi$ stands for the infinite conjunction of the form

$$K_1\varphi \wedge K_1 K_2\varphi \wedge K_2 K_1 K_2\varphi \wedge \dots,$$

both natural possible worlds semantics and clear inductive axiomatizations have been found for *propositional* common knowledge logics [7]. (The new operator, however, considerably increases the computational complexity of these logics—from PSPACE to EXPTIME; consult [5].)

But real problems arise when we try to combine the common knowledge operator with the first-order quantifiers. First, no common knowledge predicate logic with both a finitary (or at least recursive) axiomatization *and* a reasonable semantics has ever been constructed! And second, the common knowledge predicate logics determined by the standard possible worlds semantics *are known* to be not recursively axiomatizable (and so not recursively enumerable) [17]. Thus, similar to second-order logic or first-order temporal logic, it is impossible to characterize common knowledge predicate logics syntactically. In some sense this means that neither we nor the Turing machine have the capacity of understanding the interaction between common knowledge and quantifiers. Moreover, this is true of even very small fragments of the logics, say, the monadic or two-variable fragments (see [17]).

Does it mean that we should completely abandon the idea of using common knowledge predicate logic? Still there exist manageable fragments with non-trivial interaction between the common knowledge operator and quantifiers.

A promising approach to singling out non-trivial decidable fragments of first-order modal and temporal logics has been proposed in [9,20]. The idea is to restrict attention to the class of *monodic*[2] formulas which allow applications of modal or temporal operators only to formulas with at most one free variable. In the epistemic context, monodicity means, in particular, that

— we have the full expressive power of first-order logic as far as we do not apply epistemic operators to open formulas;
— we can reason about agents' knowledge of properties, for instance,

$$\forall x \ (C \, \mathsf{loves}(John, x) \vee C \, \neg \mathsf{loves}(John, x))$$

('for every object x, it is a common knowledge whether John loves x'); however, we are not permitted to reason about agents' knowledge of relations, say

$$\forall x, y \ (C \, \mathsf{loves}(x, y) \vee C \, \neg \mathsf{loves}(x, y))$$

('for all pairs x, y, it is a common knowledge whether x loves y').

---

[1] An alternative approach adds infinitary operators to the language, see [10,11].
[2] *Monody* is a composition with only one melodic line.

The main aim of this paper is to show that the monodic fragment of common knowledge predicate logic turns out to be quite manageable. First, we show that for almost all interesting interpretations of the operators K_i the *monodic fragment* of the valid formulas (*without equality*) can be finitely axiomatized. Moreover, we observe that a number of natural subclasses of the monodic fragment, say, with only monadic predicates or two variables, are decidable. On the other hand, it is proved that the addition of the equality symbol to the monodic fragment makes it not recursively enumerable.

2 First-Order Logics of Common Knowledge

The logics we deal with in this paper are all based on the language we call $\mathcal{CL}$, which extends the standard first-order language (without equality) with a number of epistemic operators, including the operator expressing common knowledge. The alphabet of $\mathcal{CL}$ consists of:

- *predicate symbols* $P_0, P_1, \ldots,$
- *individual variables* $x_0, x_1, \ldots,$
- *individual constants* $c_0, c_1, \ldots,$
- the *booleans* $\wedge, \neg,$
- the *universal quantifier* $\forall x$ for each individual variable x,
- a finite number of *knowledge operators* $K_1, \ldots, K_n$, $n \geq 1$, and
- the *common knowledge operator* C.

We assume that the set of predicate symbols is non-empty and that each of them is equipped with some fixed arity; 0-ary predicates are called *propositional variables* and denoted by $p_0, p_1, \ldots$. The individual variables together with the individual constants form the set of $\mathcal{CL}$-terms. The set of $\mathcal{CL}$-formulas is defined as follows:

- if P is an n-ary predicate symbol and $\tau_1, \ldots, \tau_n$ are terms, then $P(\tau_1, \ldots, \tau_n)$ is a formula;
- if φ and ψ are formulas, then so are $\varphi \wedge \psi$ and $\neg\varphi$;
- if φ is a formula and x a variable, then $\forall x \varphi$ is a formula;
- if φ is a formula and $i \leq n$, then $K_i \varphi$ and $C\varphi$ are formulas.

Throughout the paper we make use of the following abbreviations: $\top, \bot, \varphi \vee \psi,$ $\varphi \rightarrow \psi, \varphi \leftrightarrow \psi,$ and $\exists x \varphi,$ which are defined as usual, as well as $E\varphi$ ('*everyone knows φ*') which stands for $K_1 \varphi \wedge \cdots \wedge K_n \varphi$.

The language $\mathcal{CL}$ is interpreted in *first-order Kripke models* which are structures of the form $\mathfrak{M} = \langle \mathfrak{F}, D, I \rangle$, where $\mathfrak{F} = \langle W, R_1, \ldots, R_n \rangle$ is the underlying *Kripke frame* ($W \neq \emptyset$ is a set of worlds and the R_i are binary relations on W), D is a nonempty set, the *domain* of $\mathfrak{M}$, and I a function associating with every world $w \in W$ a first-order structure

$$I(w) = \left\langle D, P_0^{I(w)}, \ldots, c_0^{I(w)}, \ldots \right\rangle,$$

in which $P_i^{I(w)}$ is a predicate on D of the same arity as P_i (for a propositional variable p_i, the predicate $p_i^{I(w)}$ is either $\top$ or $\bot$), and $c_i^{I(w)}$ is an element in D such that $c_i^{I(u)} = c_i^{I(v)}$ for any $u, v \in W$. The latter means that constants are treated as rigid designators in the sense that they designate the same object in every world. To simplify notation we will omit the superscript I and write P_i^w, p_i^w, c_i^w, etc., if I is clear from the context.

Remark 1. Note that we assume domains to be constant. Axiomatizations for the case of expanding or varying domains can easily be obtained from our results.

An *assignment* in D is a function $\mathfrak{a}$ from the set *var* of variables to D. The value $\tau^{\mathfrak{M},\mathfrak{a}}$ (or simply $\tau^{\mathfrak{a}}$ if understood) of a term τ under $\mathfrak{a}$ in $\mathfrak{M}$ is $\mathfrak{a}(\tau)$, if τ is a variable, and $\tau^{I(w)}$ otherwise, where w is some (any) world in W. The *truth-relation* $(\mathfrak{M}, w) \models^{\mathfrak{a}} \varphi$ (or simply $w \models^{\mathfrak{a}} \varphi$) in the model $\mathfrak{M}$ in the world w under the assignment $\mathfrak{a}$ is defined inductively as follows:

- $w \models^{\mathfrak{a}} P_i(\tau_1, \ldots, \tau_n)$ iff $(\tau_1^{\mathfrak{a}}, \ldots, \tau_n^{\mathfrak{a}}) \in P_i^w$; this fact will also be written as $I(w) \models^{\mathfrak{a}} P_i(\tau_1, \ldots, \tau_n)$;
- $w \models^{\mathfrak{a}} \psi \wedge \chi$ iff $w \models^{\mathfrak{a}} \psi$ and $w \models^{\mathfrak{a}} \chi$;
- $w \models^{\mathfrak{a}} \neg\psi$ iff $w \not\models^{\mathfrak{a}} \psi$;
- $w \models^{\mathfrak{a}} \forall x \psi(x, y_1, \ldots, y_n)$ iff $w \models^{\mathfrak{b}} \psi(x, y_1, \ldots, y_n)$ for every assignment $\mathfrak{b}$ in D that may differ from $\mathfrak{a}$ only on x;
- $w \models^{\mathfrak{a}} K_i\psi$ iff $v \models^{\mathfrak{a}} \psi$ for all $v \in W$ such that wR_iv;
- $w \models^{\mathfrak{a}} C\psi$ iff $v \models^{\mathfrak{a}} \psi$ for all v such that $w(\bigcup_{i \leq n} R_i)^+v$, where the superscript $^+$ means taking the transitive closure of $\bigcup_{i \leq n} R_i$.

For a set of formulas Γ, a model $\mathfrak{M}$, a world w and an assignment $\mathfrak{a}$, we write $w \models^{\mathfrak{a}} \Gamma$ to say that $w \models^{\mathfrak{a}} \varphi$ for every $\varphi \in \Gamma$. In this case Γ is said to be *satisfied* in $\mathfrak{M}$. By $\mathfrak{F} \models \Gamma$ we mean that Γ is *valid* in $\mathfrak{F}$, i.e., $(\mathfrak{M}, w) \models^{\mathfrak{a}} \Gamma$ holds for every model $\mathfrak{M}$ based on $\mathfrak{F}$, every assignment $\mathfrak{a}$ in it, and every world w in $\mathfrak{F}$.

Different epistemic logics correspond to different classes of frames. Usually these classes are determined by combinations of the following properties: reflexivity (denoted by r), transitivity (t), seriality (s), and euclideanness (e). We denote by F^r the class of all reflexive frames, by F^{re} the class of all reflexive and euclidean frames (i.e., the class of frames with equivalence relations), etc. $\mathsf{F}^{\mathfrak{a}}$ is the class of all frames.

For a class F of frames, we define $L(\mathsf{F})$, the *logic of* F, to be the set of all $\mathcal{CL}$-formulas that are valid in all $\mathfrak{F} \in \mathsf{F}$. Here is a list of standard logics of common knowledge: $\mathbf{K}_n^C = L(\mathsf{F}^{\mathfrak{a}})$, $\mathbf{T}_n^C = L(\mathsf{F}^r)$, $\mathbf{KD}_n^C = L(\mathsf{F}^s)$, $\mathbf{K4}_n^C = L(\mathsf{F}^t)$, $\mathbf{S4}_n^C = L(\mathsf{F}^{rt})$, $\mathbf{KD45}_n^C = L(\mathsf{F}^{ste})$, $\mathbf{S5}_n^C = L(\mathsf{F}^{re})$.

3 Axiomatizing the Monodic Fragment

As was shown in [17], none of the logics listed above is recursively axiomatizable. Moreover, the restriction of these logics to such 'orthodox' fragments as the monadic or two-variable formulas does not bring a relief: they are still not recursively enumerable. By analyzing proofs of these 'negative' results, one can observe that all of them make use of formulas asserting that some agents know relations between two objects. On the other hand, the results of [18] establishing decidability of epistemic description logics (in which epistemic operators are applicable only to unary predicates) give some hope that the fragment without such formulas can be more manageable.

Definition 1 (monodic formulas). *Denote by $\mathcal{CL}_1$ the set of all $\mathcal{CL}$-formulas φ such that any subformula of φ of the form $K_i\psi$ or $C\psi$ has at most one free variable. Such formulas will be called* monodic. *For a class* F *of frames, let $L_1(\mathsf{F}) = L(\mathsf{F}) \cap \mathcal{CL}_1$. In other words, $L_1(\mathsf{F})$ is the monodic fragment of the logic $L(\mathsf{F})$.*

From now on all formulas are assumed to be monodic.

In this section we give axiomatizations of the monodic fragments of the epistemic logics defined above. (These axiomatizations are first-order extensions of those in [7].) To begin with, we axiomatize the monodic fragment of $\mathbf{K}_n^C$, i.e., $L_1(\mathsf{F}^\mathrm{a})$. This axiomatic system, denoted by K_n^C, has the following axiom schemata and inference rules:

Axiom schemata (over formulas in $\mathcal{CL}_1$):

- the set of axiom schemata from some axiomatization of classical first-order logic,
- $K_i(\varphi \to \psi) \to (K_i\varphi \to K_i\psi)$, for $i \leq n$,
- $C\varphi \to E(\varphi \wedge C\varphi)$,
- $K_i\forall x\psi \leftrightarrow \forall x K_i\psi$.

Inference rules (over formulas in $\mathcal{CL}_1$):

- the rules of classical first-order logic,
- $\dfrac{\varphi}{K_i\varphi}$, for $i \leq n$,
- $\dfrac{\varphi \to E(\psi \wedge \varphi)}{\varphi \to C\psi}$.

The monodic fragments of the remaining logics are axiomatized by adding to K_n^C the corresponding standard axiom schemata:

A_D: $K_i\varphi \to \neg K_i\neg\varphi$, $i \leq n$,
A_T: $K_i\varphi \to \varphi$, $i \leq n$,
A_4: $K_i\varphi \to K_iK_i\varphi$, $i \leq n$,
A_5: $\neg K_i\varphi \to K_i\neg K_i\varphi$, $i \leq n$.

Namely, T_n^C, KD_n^C and $K4_n^C$ as the axiomatic systems obtained by adding to K_n^C the schemata A_T, A_D, and A_4, respectively. $S4_n^C$ is $K4_n^C$ plus A_T. $KD45_n^C$ is $K4_n^C$ extended by A_D and A_5, and $S5_n^C$ is $KD45_n^C$ plus A_T.

Given an axiomatic system S, we denote by $\vdash_S$ its consequence relation. Our aim now is to prove that the defined systems indeed axiomatize the monodic fragments of our common knowledge logics. That is, we are going to show that for every monodic formula φ, we have $\vdash_{K_n^C} \varphi$ iff $\varphi \in \mathbf{K}_n^C$ iff φ is valid in all frames from F^a, and similar claims for the other logics.

The easy 'only if' part of these claims, i.e., correctness, follows from well-known results (consult e.g. [5,3]) and the almost obvious fact that the rule $\varphi \to E(\psi \wedge \varphi)/\varphi \to C\psi$ preserves validity. The 'if' part, i.e., completeness, is much more complicated. It will be proved in the next section.

4   Completeness

Given a set Γ of $\mathcal{CL}_1$-formulas, we denote by $con(\Gamma)$ and $sub(\Gamma)$ the sets of all constants and subformulas of formulas in Γ, respectively; $sub_C(\Gamma)$ is defined as:

$$sub_C(\Gamma) = sub(\Gamma) \cup \{E(\psi \wedge C\psi), \psi \wedge C\psi, K_i(\psi \wedge C\psi) : C\psi \in sub(\Gamma), i \le n\}.$$

Let $sub_C^{\neg}(\Gamma) = \{\neg\psi : \psi \in sub_C(\Gamma)\} \cup sub_C(\Gamma)$ and let $sub_n(\Gamma)$ be the subset of $sub_C^{\neg}(\Gamma)$ containing only formulas with $\le n$ free variables. For instance, $sub_0(\Gamma)$ denotes the set of sentences in $sub_C^{\neg}(\Gamma)$. (Note that $sub_n(\Gamma)$ is not necessarily closed under subformulas and that modulo equivalence we may assume that $sub_n(\Gamma)$ is closed under $\neg$.) In what follows we will not be distinguishing between a finite set Γ of formulas and the conjunction $\bigwedge \Gamma$ of formulas in it.

Let x be a variable not occurring in Γ. Put

$$sub_x(\Gamma) = \{\psi\{x/y\} : \psi(y) \in sub_1(\Gamma)\} \cup \{\neg K_i \neg\bot, K_i \neg\bot, \neg\bot, \bot : i \le n\}$$

For the rest of this section we fix an arbitrary $\mathcal{CL}_1$-sentence φ.

Definition 2 (type). *By a* type *for φ we mean a boolean-saturated subset t of $sub_x(\varphi)$, i.e.,*

- $\psi \wedge \chi \in t$ *iff* $\psi \in t$ *and* $\chi \in t$, *for every* $\psi \wedge \chi \in sub_x(\varphi)$;
- $\neg\psi \in t$ *iff* $\psi \notin t$, *for every* $\neg\psi \in sub_x(\varphi)$.

We say that two types t and t' agree on $sub_0(\varphi)$ if $t \cap sub_0(\varphi) = t' \cap sub_0(\varphi)$. Given a type t for φ and a constant $c \in con(\varphi)$, the pair $\langle t, c \rangle$ will be called an indexed type *for φ (indexed by c) and denoted by $t_c(x)$ or simply t_c.*

Definition 3 (state candidate). *Suppose T is a set of types for φ that agree on $sub_0(\varphi)$, and $T^{con} = \{\langle t, c \rangle : c \in con(\varphi)\}$ a set of indexed types such that $\{t : \langle t, c \rangle \in T^{con}\} \subseteq T$ and for each $c \in con(\varphi)$, T^{con} contains exactly one pair of the form $\langle t, c \rangle$. The pair $\mathfrak{C} = \langle T, T^{con} \rangle$ is called then a* state candidate *for φ. A* pointed state candidate *for φ is the pair $\mathfrak{P} = \langle \mathfrak{C}, t \rangle$, where t is a type in T, called the* point *of $\mathfrak{P}$. With $\mathfrak{C}$ and $\mathfrak{P}$ we associate the formulas*

$$\alpha_{\mathfrak{C}} = \bigwedge_{t \in T} \exists x\, t(x) \wedge \forall x \bigvee_{t \in T} t(x) \wedge \bigwedge_{\langle t,c \rangle \in T^{con}} t(c), \qquad \beta_{\mathfrak{P}} = \alpha_{\mathfrak{C}} \wedge t.$$

In what follows S ranges over the axiomatic systems introduced in Section 3. We remind the reader that a formula χ is said to be S-consistent if $\nvdash_S \neg\chi$.

Definition 4 (suitable pairs). *(1) A pair (t_1, t_2) of types for φ is called i-suitable for S, $i \leq n$, if the formula $t_1 \wedge \neg K_i \neg t_2$ is S-consistent.*

(2) A pair $(\mathfrak{C}_1, \mathfrak{C}_2)$ of state candidates is i-suitable for S, $i \leq n$, if $\alpha_{\mathfrak{C}_1} \wedge \neg K_i \neg \alpha_{\mathfrak{C}_2}$ is S-consistent.

(3) A pair $(\mathfrak{P}_1, \mathfrak{P}_2)$ of pointed state candidates is i-suitable for S, $i \leq n$, if $\beta_{\mathfrak{P}_1} \wedge \neg K_i \neg \beta_{\mathfrak{P}_2}$ is S-consistent. In this case we write $\mathfrak{P}_1 \prec_i \mathfrak{P}_2$.

Lemma 1. (i) *For every finite S-consistent set Ψ of CL_1-formulas, there is a pointed state candidate $\mathfrak{P} = \langle \mathfrak{C}, t \rangle$ for φ such that $\bigwedge \Psi \wedge \beta_{\mathfrak{P}}$ is S-consistent. Moreover, if $\psi \in sub_x(\varphi)$ and $\psi \in \Psi$, then $\psi \in t$.*

(ii) *Suppose Ψ is a finite set of CL_1-formulas and $\neg K_i \theta$ is a formula in $sub_x(\varphi)$ such that $\bigwedge \Psi \wedge \neg K_i \theta$ is S-consistent. Then there exists a pointed state candidate $\mathfrak{P} = \langle \mathfrak{C}, t \rangle$ for φ such that $\neg \theta \in t$ and $\bigwedge \Psi \wedge \neg K_i \neg \beta_{\mathfrak{P}}$ is S-consistent.*

Proof. (i) Denote by β_φ the disjunction of all formulas $\beta_{\mathfrak{P}}$, $\mathfrak{P}$ a pointed state candidate for φ. As β_φ is classically valid, it is provable in S, hence $\bigwedge \Psi \wedge \beta_\varphi$ is S-consistent. It follows that there is a disjunct $\beta_{\mathfrak{P}}$ of β_φ such that $\bigwedge \Psi \wedge \beta_{\mathfrak{P}}$ is S-consistent. Now, if $\psi \in \Psi \cap sub_x(\varphi)$ and t is the point of $\mathfrak{P}$, then $\psi \in t$, for otherwise $\neg \psi \in t$, which is a contradiction.

(ii) If $\bigwedge \Psi \wedge \neg K_i \theta$ is S-consistent, then so is $\bigwedge \Psi \wedge \neg K_i \neg (\neg \theta \wedge \beta_\varphi)$. It follows that there is a pointed state candidate $\mathfrak{P}$ with point t such that $\bigwedge \Psi \wedge \neg K_i \neg (\neg \theta \wedge \beta_{\mathfrak{P}})$ is S-consistent. Clearly, $\neg \theta \in t$, and we are done.

Note that Lemma 1 will hold true if we replace x by some constant c.

Lemma 2. (i) *If a pair $(\mathfrak{C}_1, \mathfrak{C}_2)$ of state candidates for φ is i-suitable for S, $i \leq n$, then:*

1. *for every $t \in T_1$ there exists a $t' \in T_2$ such that (t, t') is i-suitable for S;*
2. *for every $t' \in T_2$ there exists a $t \in T_1$ such that (t, t') is i-suitable for S.*

(ii) *Suppose that a pair of types (t, t') is i-suitable for S. Then:*

1. *$\psi \in t'$ whenever $K_i \psi \in t$;*
2. *if $A_4 \in S$, then $K_i \psi \in t'$ whenever $K_i \psi \in t$;*
3. *if $\{D, A_5\} \subseteq S$, then $K_i \psi \in t$ whenever $K_i \psi \in t'$;*
4. *if $\{D, A_4, A_5\} \subseteq S$ or $\{T, A_5\} \subseteq S$, then $K_i \psi \in t$ iff $K_i \psi \in t'$.*

(iii) *Suppose (t, t') is i-suitable for S. Then $C\psi \in t$ implies $C\psi \in t'$. If $\{A_4, A_5\} \subseteq S$, then $C\psi \in t$ iff $C\psi \in t'$.*

Proof. (i) Suppose that $t \in T_1$ but there is no $t' \in T_2$ for which (t, t') is i-suitable for S. This means that $\vdash_S t \rightarrow K_i \neg t'$, for each $t' \in T_2$, and so $\vdash_S t \rightarrow K_i \neg \bigvee_{t' \in T_2} t'$. Then we have $\vdash_S \exists x t \rightarrow K_i \exists x \neg \bigvee_{t' \in T_2} t'$. Since $\vdash_S \exists x \neg \bigvee_{t' \in T_2} t' \rightarrow \neg \alpha_{\mathfrak{C}_2}$ and $\vdash_S \alpha_{\mathfrak{C}_1} \rightarrow \exists x t$, we finally obtain $\vdash_S \alpha_{\mathfrak{C}_1} \rightarrow K_i \neg \alpha_{\mathfrak{C}_2}$, contrary to S-consistency of $\alpha_{\mathfrak{C}_1} \wedge \neg K_i \neg \alpha_{\mathfrak{C}_2}$. Claim (i.2) is proved in a similar way.

(ii) Suppose that $K_i\psi \in t$ but $\psi \notin t'$. Then $\neg\psi \in t'$, $\vdash_S K_i\psi \to K_i\neg t'$, and so $t \wedge \neg K_i\neg t'$ is S-inconsistent, which is a contradiction.

Now suppose that S contains A_4, $K_i\psi \in t$, but $K_iK_i\psi \notin t'$. Then $\neg K_i\psi \in t'$. Hence $\vdash_S K_iK_i\psi \to K_i\neg t'$. It follows from A_4 that $\vdash_S K_i\psi \to K_i\neg t'$, and so $K_i\psi \wedge \neg K_i\neg t'$ is S-inconsistent, contrary to S-consistency of (t, t'). Claims (i.3) and (i.4) are proved analogously.

(iii) Suppose $C\psi \in t$. Then $E(\psi \wedge C\psi) \in t$ and so $K_i(\psi \wedge C\psi) \in t$, for $i \leq n$. By (ii.1), $\psi \wedge C\psi \in t'$, from which $C\psi \in t'$.

If $C\psi \in t'$ then, as we know, $K_i(\psi \wedge C\psi) \in t'$, for $i \leq n$. So if $\{A_4, A_5\} \subseteq S$, then we have by (ii.4), $\psi \wedge C\psi \in t$, and so $C\psi \in t$.

Definition 5 (basic tree). *Let $\mathfrak{T} = \langle W, <_1, \ldots, <_n \rangle$ be a structure with pairwise disjoint binary relations $<_i$ on W such that $\langle W, \bigcup_{i \leq n} <_i \rangle$ is an intransitive tree.[3] By a basic tree for φ we mean the pair $\langle \mathfrak{T}, \sigma \rangle$, where σ is a map associating with every $w \in W$ a state candidate $\sigma(w) = \langle T_w, T_c^w \rangle$ for φ. A basic tree is called a basic S-tree if $\alpha_{\sigma(w)}$ is S-consistent, for every $w \in W$, and the pair $(\sigma(w_1), \sigma(w_2))$ is i-suitable for S whenever $w_1 <_i w_2$.*

Definition 6 (run). *A run r in a basic S-tree $\langle \mathfrak{T}, \sigma \rangle$ is a map associating with every $w \in W$ a type $r(w) \in T_w$ such that*

- *the pair $(r(w_1), r(w_2))$ is i-suitable for S whenever $w_1 <_i w_2$;*
- *if $\neg K_i\psi \in r(w)$ then $\psi \notin r(w')$ for some $w' >_i w$;*
- *if $\neg C\psi \in r(w)$ then $\psi \notin r(w')$ for some w' such that $w(\bigcup_{i \leq n} <_i)^+ w'$.*

Definition 7 (quasimodel). *A basic S-tree $\langle \mathfrak{T}, \sigma \rangle$ is called an S-quasimodel for φ if*

- *for all $w \in W$ and $t \in T_w$ $(\sigma(w) = \langle T_w, T_c^w \rangle)$, there exists a run r in $\langle \mathfrak{T}, \sigma \rangle$ such that $r(w) = t$;*
- *for every constant $c \in con(\varphi)$, the function r_c defined by $r_c(w) = t$, for $\langle t, c \rangle \in T_w^{con}$, $w \in W$, is a run in $\langle \mathfrak{T}, \sigma \rangle$.*

We say φ is satisfied in $\langle \mathfrak{T}, \sigma \rangle$ if there exists $w \in W$ such that $\alpha_{\sigma(w)} \wedge \varphi$ is S-consistent.

Theorem 1. *If φ is satisfiable in an S-quasimodel for φ, then φ is satisfiable in a model based on a frame for S.*

---

[3] We remind the reader that $\mathfrak{G} = \langle W, \lhd \rangle$ is an intransitive *tree* if (i) $\mathfrak{G}$ is *rooted*, i.e., there is $w_0 \in W$ (a root of $\mathfrak{G}$) such that $w_0 \lhd^* w$ for every $w \in W$, where $\lhd^*$ is the transitive and reflexive closure of $\lhd$, (ii) for every $w \in W$, the set $\{v \in W : v \lhd^* w\}$ is finite and linearly ordered by $\lhd^*$, (iii) every world v in $\mathfrak{G}$, save its root, has precisely one predecessor, i.e., $|\{u \in W : u \lhd v\}| = 1$, and (iv) the root w_0 is *irreflexive*, i.e., $\neg w_0 \lhd w_0$.

Proof. For every monodic formula $\psi(y)$ of the form $K_i\chi(y)$ or $C\chi(y)$ with one free variable y, we reserve a unary predicate $P_\psi(y)$. Likewise, for every sentence $\psi = K_i\chi$ or $\psi = C\chi$ we fix a propositional variable p_ψ. $P_\psi(y)$ and p_ψ will be called the *surrogates* for $\psi(y)$ and ψ.

Given a monodic formula ψ, we denote by $\overline{\psi}$ the formula that results from ψ by replacing all subformulas of the form $K_i\chi(y)$, $K_i\chi$, $C\chi(y)$, and $C\chi$, which are not within the scope of another epistemic operator, with their surrogates. Thus, $\overline{\psi}$ contains no occurrences of epistemic operators, i.e., it is a purely first-order formula; we will call $\overline{\psi}$ the *$\mathcal{L}$-reduct* of ψ. For a set of $C\mathcal{L}_1$-formulas Γ, let $\overline{\Gamma} = \{\overline{\psi} : \psi \in \Gamma\}$.

Now suppose φ is satisfied in an S-quasimodel $\langle \mathfrak{T}, \sigma \rangle$, $\mathfrak{T} = \langle W, <_1, \ldots, <_n \rangle$. So there is $w^* \in W$ such that $\varphi \wedge \alpha_{\sigma(w^*)}$ is S-consistent. It follows that the $\mathcal{L}$-reduct $\overline{\varphi} \wedge \overline{\alpha}_{\sigma(w^*)}$ is consistent with respect to classical first-order logic. Moreover, by Definition 5, $\alpha_{\sigma(w)}$ is S-consistent and $\overline{\alpha}_{\sigma(w)}$ is first-order consistent, for every $w \in W$. So, for each $w \in W$, we can find a structure $I(w) \models \overline{\alpha}_{\sigma(w)}$. We may also assume that $I(w^*) \models^{\mathfrak{a}^*} \overline{\varphi}$, for some assignment $\mathfrak{a}^*$.

Take a cardinal $\kappa \geq \aleph_0$ exceeding the cardinality of the set Ω of all runs in $\langle \mathfrak{T}, \sigma \rangle$ and put

$$D = \{\langle r, \xi \rangle : r \in \Omega, \xi < \kappa\}.$$

Without loss of generality we can assume that D is the domain of the first-order structures $I(w)$ satisfying the $\overline{\alpha}_{\sigma(w)}$, that $c^w = \langle r_c, 0 \rangle$, and that

$$r(w) = \{\psi \in sub_x(\varphi) : I(w) \models \overline{\psi}[\langle r, \xi \rangle]\}, \tag{1}$$

for all runs r and $\xi < \kappa$. (Note that the underlying first-order language does not contain equality; for details see [9], Lemma 9.)

Let us now define the underlying frame $\mathfrak{F}$ of the model we are constructing. Its set of worlds is W. The accessibility relations R_i depend on S. Namely, we define R_i to be

- $<_i$ if $S = K_n^C$ or $S = KD_n^C$;
- $<_i \cup \{\langle w, w \rangle : w \in W\}$ if $S = T_n^C$;
- the transitive closure of $<_i$ if $S = K4_n^C$;
- the reflexive and transitive closure of $<_i$ if $S = S4_n^C$;
- $<_i^+ \cup \{\langle w, w' \rangle : \exists v \in W(v <_i^+ w \ \& \ v <_i^+ w' \ \& \ \neg \exists u \ u <_i v)\}$ if $S = KD45_n^C$;
- the reflexive, symmetric and transitive closure of $<_i$ if $S = S5$.

Note that for $S = KD_n^C$ the R_i are serial because in this case every S-consistent type for φ contains at least one formula of the form $\neg K_i\psi$. For $S = KD45_n^C$ the R_i are clearly serial. Suppose $w_1 R_i w$ and $w R_i w_2$. If w_1 has no $<_i$-predecessor, then $w_1 <_i^+ w_2$ and so $w_1 R_i w_2$. Otherwise, there are v_j, for $j = 1, 2$, such that $v_j <_i^+ w_j$, $v_j <_i^+ w$ and the v_j have no $<_i$-predecessors. Since $\mathfrak{T}$ is an irreflexive tree, we get $v_1 = v_2$. Thus $v_1 <_i^+ w_1$ and $v_1 <_i^+ w_2$. By the definition of R_i, it follows that $w_1 R_i w_2$. Hence R_i is transitive. Similarly one can show that R_i is euclidean.

Thus we have the model $\mathfrak{M} = \langle \mathfrak{F}, R_1, \ldots, R_n, I \rangle$. By induction on the construction of $\psi \in sub(\varphi)$ we will show now that for every assignment $\mathfrak{a}$

$$I(w) \models^{\mathfrak{a}} \overline{\psi} \text{ iff } (\mathfrak{M}, w) \models^{\mathfrak{a}} \psi.$$

The basis of induction, i.e., the case where $\psi = P_i(\tau_1, \ldots, \tau_m)$ is clear; for then $\psi = \overline{\psi}$. The induction step for $\psi = \psi_1 \wedge \psi_2$, $\psi = \neg \psi_1$, and $\psi = \forall y \psi_1$ follows by the induction hypothesis from the equations $\overline{\psi_1 \wedge \psi_2} = \overline{\psi_1} \wedge \overline{\psi_2}$, $\overline{\neg \psi_1} = \neg \overline{\psi_1}$, $\overline{\forall y \psi_1} = \forall y \overline{\psi_1}$. Let $\psi = K_i \chi(y)$ and assume that $\mathfrak{a}(y) = \langle r, \xi \rangle$. (If ψ is a sentence, y is any variable.) We then have:

$$
\begin{aligned}
I(w) \models^{\mathfrak{a}} \overline{K_i\chi(y)} &\Leftrightarrow_1 I(w) \models^{\mathfrak{a}} P_{K_i\chi}(y) \\
&\Leftrightarrow_2 K_i\chi(x) \in r(w) \\
&\Leftrightarrow_3 \forall v \, (wR_i v \rightarrow \chi(x) \in r(v)) \\
&\Leftrightarrow_4 \forall v \, (wR_i v \rightarrow I(v) \models^{\mathfrak{a}} \overline{\chi}(y)) \\
&\Leftrightarrow_5 \forall v \, (wR_i v \rightarrow (\mathfrak{M}, v) \models^{\mathfrak{a}} \chi(y)) \\
&\Leftrightarrow_6 (\mathfrak{M}, w) \models^{\mathfrak{a}} K_i\chi(y).
\end{aligned}
$$

Equivalence $\Leftrightarrow_1$ holds by the definition of $\overline{\psi}$; $\Leftrightarrow_2$ and $\Leftrightarrow_4$ are consequences of (1). The induction hypothesis yields $\Leftrightarrow_5$, and $\Leftrightarrow_6$ holds by definition. The only non-trivial case is $\Leftrightarrow_3$.

($\Rightarrow_3$) Suppose $K_i\chi \in r(w)$ and wR_iw'. So if $w <_i w'$ the claim follows by Definition 6. If $S = K$ or $S = D$ then we are done, because $R_i =<_i$.

Let $S = KD45$. By the definition of R_i, we have either $w <_i^+ w'$ or there exists v such that $v <_i^+ w$, $v <_i^+ w'$ and $\neg \exists u \, u <_i v$. The former case is easy; we leave it to the reader and consider the latter one here. We have some $m \in \omega$ and worlds $v_0, \ldots, v_{m+1}$ such that $v_0 = v$, $v_{m+1} = w$ and $v_j <_i v_{j+1}$ for every $j \leq m$. By Definition 6, $(r(v_j), r(v_{j+1}))$ is i-suitable for S whenever $j \leq m$. By Lemma 2, $K_i\chi(x) \in r(v_0) = r(v)$. Similarly, using $v <_i^+ w'$ we obtain worlds $u_0, \ldots, u_{l+1}$ such that $u_0 = v$, $u_{l+1} = w'$ and $u_j <_i u_{j+1}$ for every $j \leq l$. Again, $(r(u_j), r(u_{j+1}))$ is i-suitable for S whenever $j \leq l$, and by Lemma 2, $K_i\chi(x) \in r(u_l)$. Using the same lemma once again, we obtain $\chi(x) \in r(u_{l+1}) = r(w')$. ($\Leftarrow_3$) is an immediate consequence of Definition 6. Other cases for S are treated analogously.

Finally, let $\psi = C\chi(y)$ and $\mathfrak{a}(y) = \langle r, \xi \rangle$. Since the proof is similar to the foregoing one, we leave it to the reader.

Thus, to prove completeness of our axiom system S, it suffices to construct an S-quasimodel satisfying φ whenever φ is S-consistent.

Lemma 3. *Let $\mathfrak{P} = \langle \mathfrak{C}, t \rangle$ be a pointed state candidate for φ such that $\beta_{\mathfrak{P}}$ is S-consistent.*

(i) If $\neg K_i\psi \in t$, then there exists $\mathfrak{P}' = \langle \mathfrak{C}', t' \rangle$ such that $\mathfrak{P} \prec_i \mathfrak{P}'$ and $\psi \notin t'$. Moreover, if $\langle t, c \rangle \in \mathfrak{C}$ for some constant c, then we can choose $\mathfrak{P}' = \langle \mathfrak{C}', t' \rangle$ with $\mathfrak{P} \prec_i \mathfrak{P}'$ and $\psi \notin t'$ so that $\langle t', c \rangle \in \mathfrak{C}'$.

(ii) *If* $\neg C\psi \in t$, *then there are pointed state candidates* $\mathfrak{P}_j = \langle \mathfrak{C}_j, t_j \rangle$, $j \leq k$, *with* $\mathfrak{P}_0 = \mathfrak{P}$ *and*

$$\mathfrak{P}_0 \prec_{i_1} \mathfrak{P}_1 \prec_{i_2} \cdots \prec_{i_k} \mathfrak{P}_k,$$

for some $i_1, \ldots, i_k \leq n$, *such that* $\neg\psi \in t_k$. *Moreover, if* $\langle t, c \rangle \in \mathfrak{C}$, *then we can choose such a sequence with* $\langle t_j, c \rangle \in \mathfrak{C}_j$ *for all* $j \leq k$.

Proof. (i) follows from Lemma 1. So let us prove (ii). Suppose that such a sequence does not exist. Let $\mathcal{T}$ be the minimal set of pointed state candidates such that

- $\mathfrak{P} \in \mathcal{T}$,
- if $\mathfrak{D}_1 \in \mathcal{T}$ and $\mathfrak{D}_1 \prec_i \mathfrak{D}_2$ for some i, then $\mathfrak{D}_2 \in \mathcal{T}$.

Let $\vartheta = \bigvee_{\mathfrak{D} \in \mathcal{T}} \beta_{\mathfrak{D}}$. Then $\vdash_S \vartheta \to K_i\vartheta$, for all $i \leq n$. Indeed, suppose otherwise. Then $\vartheta \wedge \neg K_i\vartheta$ is S-consistent for some $i \leq n$. But then, by Lemma 1, $\vartheta \wedge \neg K_i \neg \beta_{\mathfrak{P}'}$ is S-consistent for some pointed state candidate $\mathfrak{P}' \notin \mathcal{T}$. This, however, contradicts the definition of $\mathcal{T}$, since we would have $\mathfrak{D} \prec_i \mathfrak{P}'$ for some disjunct $\beta_{\mathfrak{D}}$ of ϑ. Hence $\vdash_S \vartheta \to E\vartheta$. Clearly, $\psi \in s$ for every $\langle \mathfrak{C}^*, s \rangle \in \mathcal{T}$, for otherwise we could construct a sequence satisfying condition (ii). Thus, $\vdash_S \vartheta \to \psi$ and so $\vdash_S \vartheta \to E(\psi \wedge \vartheta)$. By the inference rule for C we obtain $\vdash_S \vartheta \to C\psi$, and so $\vdash_S \beta_{\mathfrak{P}}(x) \to C\psi$, since $\mathfrak{P} \in \mathcal{T}$. But then $\beta_{\mathfrak{P}}$ is S-inconsistent, which is a contradiction.

We are in a position now to prove the main result of this section.

Theorem 2. *If S is one of the axiomatic systems defined above and φ an S-consistent monodic formula, then φ is satisfiable in a model based on a frame for S.*

Proof. In view of Theorem 1, it suffices to construct an S-quasimodel satisfying φ. By Lemma 1, we can find a state candidate $\mathfrak{C}^*$ such that $\varphi \wedge \alpha_{\mathfrak{C}^*}$ is S-consistent. We are going to construct the required quasimodel as the limit of a sequence

$$\langle \mathfrak{I}_m, \sigma_m \rangle = \langle \langle W_m, <_1^m, \ldots, <_n^m \rangle, \sigma_m \rangle,$$

of basic S-trees, $m \in \omega$.

Let $W_0 = \{w^*\}$ for some point w^* and let $\sigma_0(w^*) = \mathfrak{C}^*$. Suppose now that $\langle \mathfrak{I}_m, \sigma_m \rangle$ has been already defined. For every $w \in W_m - W_{m-1}$ we shall construct a number of new points 'saturating' $\sigma_m(w)$ ($W_{-1} = \emptyset$). Let $\mathfrak{C} = \sigma_m(w)$, $\mathfrak{C} = \langle T, T^{con} \rangle$. Pick some $t \in T$ and do the following:

(a) For every $\chi = \neg K_i\psi \in t$ we take *two* points a_χ and b_χ, add them to W_m, put $w <_i^{m+1} a_\chi$, $w <_i^{m+1} b_\chi$, and $\sigma_{m+1}(a_\chi) = \sigma_{m+1}(b_\chi) = \mathfrak{C}'$, for some $\mathfrak{C}'$ underlying a pointed state candidate $\mathfrak{P}' = \langle \mathfrak{C}', t' \rangle$ with $\langle \mathfrak{C}, t \rangle \prec_i \mathfrak{P}'$ and $\psi \notin t'(x)$. That such a $\mathfrak{P}'$ exists is guaranteed by Lemma 3. If $\langle t, c \rangle \in T^{con}$ for some constant c, then we take for t' the type s with $\langle s, c \rangle$ in $\mathfrak{C}'$.

(b) For every $\chi = \neg C\psi \in t$ we take *two* sequences $a_\chi^1, \ldots, a_\chi^k$ and $b_\chi^1, \ldots, b_\chi^k$ and put

$$w <_{i_1}^{m+1} a_\chi^1 <_{i_2}^{m+1} \cdots <_{i_k}^{m+1} a_\chi^k, \qquad w <_{i_1}^{m+1} b_\chi^1 <_{i_2}^{m+1} \cdots <_{i_k}^{m+1} b_\chi^k,$$

and
$$\sigma_{m+1}(a_\chi^j) = \sigma_{m+1}(b_\chi^j) = \mathfrak{C}^j, \text{ for all } j \leq k,$$

where the $\langle \mathfrak{C}^j, t^j \rangle$ form a sequence of pointed state candidates with

$$\langle \mathfrak{C}, t \rangle \prec_{i_1} \langle \mathfrak{C}^1, t^1 \rangle \prec_{i_2} \cdots \prec_{i_k} \langle \mathfrak{C}^k, t^k \rangle$$

and $\neg \psi \in t^k$. Again Lemma 3 ensures the existence of such a sequence. If $\langle t, c \rangle \in T^{con}$ for some constant c, then we take a sequence with $\langle t^j, c \rangle$ from $\mathfrak{C}^j$ for all $j \leq k$.

In the same manner we consider the remaining types in T and then the remaining worlds $v \in W_m - W_{m-1}$. W_{m+1} is defined as the (disjoint) union of W_m and the constructed new points. The relations $<_i^{m+1}$ coincide with $<_i^m$ on W_m. For the new points their extension is defined above. The function σ_{m+1} coincides with σ_m on W_m and is defined above for the new worlds. Thus we have constructed $\langle \mathfrak{T}_{m+1}, \sigma_{m+1} \rangle$.

Finally, put $\langle \mathfrak{T}, \sigma \rangle = \langle \langle W, <_1^m, \ldots, <_n^m \rangle, \sigma \rangle$, where

$$W = \bigcup_{m < \omega} W_m, \quad <_i = \bigcup_{m < \omega} <_i^m, \quad \sigma = \bigcup_{m < \omega} \sigma_m.$$

It remains to show that $\langle \mathfrak{T}, \sigma \rangle$ is an S-quasimodel. It should be clear that the functions r_c are runs. So it suffices to show that, for all $w \in W$ and t from $\sigma(w)$, there exists a run r with $r(w) = t$.

First, using Lemma 2 we find a sequence

$$w^* = w_0 <_{i_1} w_1 <_{i_2} \cdots <_{i_k} w_k = w$$

and types t_j from $\sigma(w_j)$, $0 \leq j \leq k$, such that $t_k = t$ and $t_j(x) \wedge \neg K_{i_{j+1}} \neg t_{j+1}(x)$ is S-consistent for all $j < k$.

Let $r(w_j) = t_j$ and $V_0 = \{w_0, \ldots, w_{k+1}\}$. Define by induction an increasing chain of sets $V_i \supseteq W_i$ with $V_i - W_i \subseteq V_0$, on which we define r. Suppose V_n is defined. For every $w \in W_n - W_{n-1}$ with $r(w) = t$ we do the following:

- If $\neg K_i \psi \in t$, then take $v \in W_{n+1} - V_n$ with $w <_i v$ and t' from $\sigma(v)$ such that $t \wedge \neg K_i \neg t'$ is S-consistent and $\psi \notin t'$. This can be done because we always took *two* saturating worlds in the construction above. Put $r(v) = t'$.
- If $\neg C \psi \in t$, then take a sequence $v_1, \ldots, v_k$ from $W_{n+1} - V_n$ such that $w <_{i_0} v_1, v_1 <_{i_1} \cdots <_{i_{k-1}} v_{i_k}$, and types t_j from $\sigma(v_j)$, $1 \leq j \leq k$, such that
  - (t, t_1) is i_0-suitable for S,
  - (t_j, t_{j+1}) is i_j-suitable for S, $1 \leq j < k$,
  - $\psi \notin t_k$.

Again, this can be done since we always took two saturating sequences. Put $r(v_j) = t_j$ for all $1 \leq j \leq k$.

Finally, we have to define r for all $v \in W_{n+1}$ where r was not defined above. This can be done recursively as follows. Suppose $r(v)$ is not defined yet for some $v \in W_{n+1}$. If r is defined already for the (unique) v' such that $v' <_i v$, then take a t from $\sigma(v)$ such that $(r(v'), t)$ is i-suitable for S and put $r(v) = t$ (t exists by Lemma 2). Otherwise consider first v' itself.

It is now straightforward to see that r is a run.

As a consequence we obtain:

Theorem 3. *Let* $S \in \{K_n^C, T_n^C, KD_n^C, K4_n^C, S4_n^C, KD45_n^C, S5_n^C\}$ *and let* F *be the class of frames for* S*. Then for every monodic formula* φ *it holds that* $\vdash_S \varphi$ *iff* $\varphi \in L_1(\mathsf{F})$.

5   Decidability

Another important algorithmic feature of the monodic formulas is that if, roughly speaking, we restrict the underlying purely first-order formulas to a decidable class, then the resulting monodic fragments of the epistemic logics under consideration will also be decidable. In particular, we have the following:

Theorem 4. *Let* F *be any of the frame classes mentioned at the end of Section 2. Then the following fragments are decidable:*

- *the monadic fragment of* $L_1(\mathsf{F})$,
- *the two-variable fragment of* $L_1(\mathsf{F})$,
- *the guarded fragment of* $L_1(\mathsf{F})$.

(Note, however, that the guarded fragment of $L(\mathsf{F}^a)$ is undecidable.) For more details and an idea of the proof the reader is referred to [9,20]. Actually, no non-trivial decidable fragments of epistemic predicate logics have been constructed before.

It maybe also of interest to note that these decidability results make it possible to construct various decidable description logics with common knowledge and other epistemic operators applicable to concepts and formulas (but not to roles; see [19]). Weaker epistemic description logics were proposed in [6,12].

6   Adding Equality

In this section we show that the addition of equality to the language of monodic formulas restores the 'status quo,' namely, that all the fragments considered above become non-enumerable. Let $\mathcal{CL}_1^=$ be the language $\mathcal{CL}_1$ extended with the equality symbol interpreted in first-order structures as identity.

Theorem 5. *Let* F *be any of the frame classes defined at the end of Section 2. Then the logic* $L_1(\mathsf{F})$ *in the language* $\mathcal{CL}_1^=$ *is not recursively enumerable.*

Proof. Define ψ to be the conjunction of the following $\mathcal{CL}_1^=$-formulas:

$$\psi_1 = \exists x P(x) \wedge \forall x \forall y\, (P(x) \wedge P(y) \rightarrow x = y),$$
$$\psi_2 = C\forall x \forall y\, (\neg K_i \neg P(x) \wedge \neg K_i \neg P(y) \wedge \neg P(x) \wedge \neg P(y) \rightarrow x = y), \text{ for } i \leq 2,$$
$$\psi_3 = \neg C \neg \forall x\, (P(x) \leftrightarrow \neg C \neg P(x)),$$
$$\psi_4 = \forall x\, (\neg C \neg P(x) \rightarrow C \neg C \neg P(x)),$$
$$\psi_5 = \forall x\, (Q(x) \leftrightarrow \neg C \neg P(x)).$$

First we notice that for all models $\mathfrak{M} = \langle \mathfrak{F}, D, I \rangle$ with $\mathfrak{F} = \langle W, R_1, \ldots, R_n \rangle$ and all $w \in W$, if $(\mathfrak{M}, w) \models \psi$ then $Q^{I(w)}$ is finite. Indeed, by ψ_1, the set $P^{I(w)}$ is a singleton. From ψ_3 we get some w' for which $w(\bigcup_{i \leq 1} R_i)^+ w'$ and $w' \models \forall x \, (P(x) \leftrightarrow \neg C \neg P(x))$. Hence there are $w_0, \ldots, w_{m+1} \in W$ such that $w_0 = w$, $w_{m+1} = w'$ and for every $j \leq m$ there is some $i_j \leq 2$ for which $w_j R_{i_j} w_{j+1}$ holds. In view of ψ_2, $|P^{I(w_{j+1})} - P^{I(w_j)}| \leq 1$ for every $j \leq m$, which yields $|P^{I(w')}| \leq m + 1$. Thus it remains to show that $Q^{I(w)} \subseteq P^{I(w')}$. Suppose $a \in Q^{I(w)}$. Then, by ψ_5, $w \models \neg C \neg P[a]$ and by ψ_4 we obtain $w \models C \neg C \neg P[a]$, from which $w' \models P[a]$.

Second, we show that for every first-order sentence θ containing neither P nor Q the following are equivalent:

(a) θ is true in all *finite* first-order structures;
(b) $\psi \to \theta^Q$ is valid in all frames in F.

(Here θ^Q is the *relativisation of* θ to Q, i.e., $\theta^Q = \theta$ if θ is atomic, ^{Q} commutes with the booleans, and $(\forall x \theta_1)^Q = \forall x \, (Q(x) \to \theta_1^Q)$.)

(a) $\Rightarrow$ (b). Suppose there is a model $\mathfrak{M}$ and a world w in it such that $w \models \psi$ but $w \not\models \theta^Q$. Define a finite first-order structure J with domain $E = Q^{I(w)}$ and predicates $P_k^J = P_k^{I(w)} \cap E$. It can be easily shown by induction that for every formula χ and every assignment $\mathfrak{a}$ in E, we have $J \models^{\mathfrak{a}} \chi$ iff $w \models^{\mathfrak{a}} \chi^Q$. In particular, $J \not\models \theta$.

(a) $\Leftarrow$ (b). Let us show first that for every natural number $m > 0$, there are $\mathfrak{M}_m = \langle \mathfrak{F}_m, D_m, I_m \rangle$ based on a frame $\mathfrak{F}_m \in \mathsf{F}^{\mathrm{re}}$ and w in $\mathfrak{F}_m$ such that $|W| = m$ and $w \models \psi$. Put $W_m = \{w_1, \ldots, w_m\}$, $w_i \neq w_j$ whenever $i \neq j$, $D_m = \mathbb{N}$, and

$$R_1 = \{\langle w_k, w_{k+1} \rangle, \langle w_{k+1}, w_k \rangle : k < m \ \& \ \exists l \ k = 2l + 1\} \cup \{\langle w_k, w_k \rangle : k \leq m\},$$
$$R_2 = \{\langle w_k, w_{k+1} \rangle, \langle w_{k+1}, w_k \rangle : k < m \ \& \ \exists l \ k = 2l\} \cup \{\langle w_k, w_k \rangle : k \leq m\}.$$

Finally, for each $k \leq m$, put $P^{I(w_k)} = \{0, \ldots, k-1\}$ and $Q^{I(w_k)} = \{0, \ldots, m-1\}$ (see Fig. 1). It is easy to see that the model is reflexive and euclidean, and $w_1 \models \psi$.

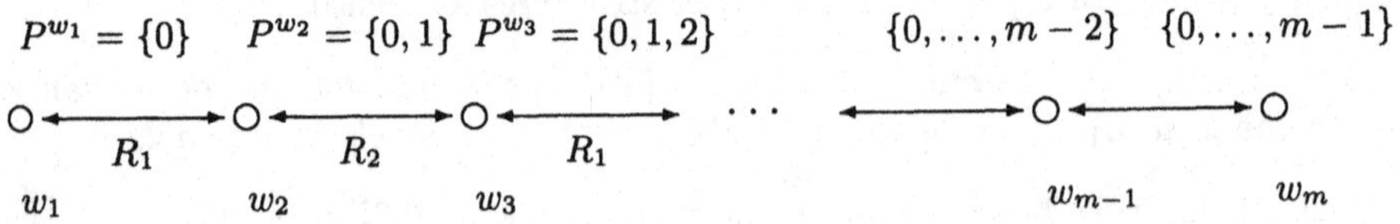

$P^{w_1} = \{0\} \quad P^{w_2} = \{0, 1\} \quad P^{w_3} = \{0, 1, 2\} \qquad \{0, \ldots, m-2\} \quad \{0, \ldots, m-1\}$

Fig. 1.

Now, to complete the proof, suppose that there is a finite first-order structure J with domain D such that $|D| = m$ and $J \not\models \theta$. Take the model $\mathfrak{M}_m$ and the world $w \not\models \psi$ constructed above. Without loss of generality we can assume that

$Q^{I_m(w)} = D$. Now expand $\mathfrak{M}_m$ to a model $\mathfrak{M}'_m$ by interpreting each predicate symbol P_i as follows: $P_i^{I'_m(w')} = P_i^J$, for each $w' \in W_m$. Now, for every first-order formula χ (without P, Q) and every assignment a in D, we have $J \models^a \chi$ iff $(\mathfrak{M}'_m, w) \models^a \chi^Q$. Therefore, $(\mathfrak{M}'_m, w) \not\models \theta^Q$, and so $\psi \to \theta^Q$ is not valid in F^{re} (which is contained in all our frame classes).

It remains to recall that, by Trakhtenbrot's theorem (see [4]), the set of first-order sentences that are valid in finite structures is not recursively enumerable.

Acknowledgments. The authors would like to thank Nobu-Yuki Suzuki for his helpful comments. While carrying out this research the first author was supported by the Deutsche Forschungsgemeinschaft.

References

1. R.J. Aumann. Agreeing to disagree. *The Annals of Statistics*, 4:1236–1239, 1976.
2. M. Bacharach. The epistemic structure of a theory of game. *Theory and Decision*, 37:7–48, 1994.
3. A.V. Chagrov and M.V. Zakharyaschev. *Modal Logic*. Clarendon Press, Oxford, 1997.
4. H. Ebbinghaus and J. Flum. *Finite Model Theory*. Springer, 1995.
5. R. Fagin, J. Halpern, Y. Moses, and M. Vardi. *Reasoning about Knowledge*. MIT Press, 1995.
6. A. Gräber, H. Bürckert, and A. Laux. Terminological reasoning with knowledge an belief. In A. Laux and H. Wansing, editors, *Knowledge and Belief in Philosophy and Artificial Intelligence*, pages 29–61. Akademie Verlag, 1995.
7. J. Halpern and Yo. Moses. A guide to completeness and complexity for modal logics of knowledge and belief. *Artificial Intelligence*, 54:319–379, 1992.
8. J. Hintikka. *Knowledge and Belief: An Introduction to the Logic of Two Notions*. Cornell University Press, 1962.
9. I. Hodkinson, F. Wolter, and M. Zakharyaschev. Decidable fragments of first-order temporal logics. *Annals of Pure and Applied Logic*, 2000.
10. M. Kaneko and T. Nagashima. Game logic and its applications 1. *Studia Logica*, 57:325–354, 1996.
11. M. Kaneko and T. Nagashima. Game logic and its applications 2. *Studia Logica*, 58:273–303, 1997.
12. A. Laux. Beliefs in multi-agent worlds: a terminological approach. In *Proceedings of the 11th European Conference on Artificial Intelligence*, pages 299–303, Amsterdam, 1994.
13. W. Lenzen. Recent work in epistemic logic. *Acta Philosophica Fennica*, 30:1–219, 1978.
14. D. Lewis. *Convention. A Philosophical Study*. Harvard University Press, Cambridge, Massachusets, 1969.
15. J. McCarthy, M. Sato, T. Hayashi, and S. Igarishi. On the model theory of knowledge. Technical Report STAN-CS-78-657, Stanford University, 1979.
16. J.J. Meyer and W. van der Hoek. *Epistemic Logic for AI and Computer Science*. Cambrigde University Press, 1995.
17. F. Wolter. Fragments of first-order common knowledge logics. *Studia Logica*, 2000.

18. F. Wolter and M. Zakharyaschev. Satisfiability problem in description logics with modal operators. In *Proceedings of the sixth Conference on Principles of Knowledge Representation and Reasoning*, Montreal, Canada, 1998. Morgan Kaufman.

19. F. Wolter and M. Zakharyaschev. Modal description logics: modalizing roles. *Fundamenta Informaticae*, 39:411–438, 1999.

20. F. Wolter and M. Zakharyaschev. Decidable fragments of first-order modal logics. 2000.

Updates plus Preferences*

José Júlio Alferes and Luís Moniz Pereira

Centro de Inteligência Artificial, Fac. Ciências e Tecnologia, Univ. Nova de Lisboa,
P-2825-114 Caparica, Portugal,
Voice: +351 21 294 8533, Fax: +351 21 294 8541
jja,lmp@di.fct.unl.pt

Abstract. The aim of this paper is to combine, into a single logic programming framework, the hitherto separate forms of reasoning of preferences and updating. More precisely, we define a language capable of considering sequences of logic programs that result from the consecutive updates of an initial program, where it is possible to define a priority relation among the rules of all successive programs. Moreover, within the framework, the priority relation can itself be updated.
In order to define a declarative semantics for the language, we start by reviewing the declarative semantics of updates of [1], and by presenting a definition of a semantics for preferences, shown equivalent to the one in [5], in a form suitable for its integration with the updates one.
Before the conclusions and mention of future work, we present two illustrative examples of application of the framework.

1 Introduction

In recent times, there has been a spate of work on reasoning with preferences and also, but separately, another spate of work on knowledge updating, both of which in the logic programming context. This interest has followed in the wake of a more general examination of flexible and dynamic forms of non-monotonic reasoning within artificial intelligence (AI). The present writing aims at combining these two heretofore separate forms of reasoning, preferring and updating, again in the purview of logic programming. We shall show how they complement each other, in that preferences select among pre-existing models, and updates actually create new models. Moreover, preferences may be enacted on the results of updates, and updates may be pressed into service for the purpose of changing preferences.

Forms of preference which have been intensely studied include specificity in taxonomic defaults, authority as well as temporal overriding in legal reasoning, priority of effect rules over inertia rules in causal reasoning, more likely faults in model-based diagnosis, preferred configurations in system synthesis, and scenario considerations in decision making. Many prioritized versions of existing non-monotonic formalisms have, already for some time, been developed, namely for circumscription, for hierarchical auto-epistemic logic, for default logic, for belief revision, and for abduction. In the case of logic programming (LP), research on

---

* Work partly supported by PRAXIS XXI project #2/2.1/TIT/1593/95 MENTAL.

M. Ojeda-Aciego et al. (Eds.): JELIA 2000, LNAI 1919, pp. 345–360, 2000.

the topic of preferences is much more recent. Cf. [4,5] for additional motivation, comparisons, applications, and references. Here, we expressly adopt the stable models based semantic framework of Brewka and Eiter [5], though replacing it with an equivalent formulation to bring it in line with our own stable models based update framework, with which we enmesh it. Another paramount reason for this choice of preference semantics are the two desirable principles (cf. Section 3) and the properties that semantics obeys, as spelled out by their authors.

In what concerns updates, its significance for AI has long been the object of much study [14,9,7]. In the LP setting, the accomplishments in this topic have likewise been garnered at a much later date [1,6,11,13,12,15]. Herein we adopt the stable models based update framework of [1] for the purpose of expanding it with the aforesaid preferences one. Sample prototypical applications of LP updates have included legal knowledge evolution [2], modelling of actions [3], taxonomic inheritance [6], and software development.

Preferences and updates are different forms of reasoning and serve different goals and applications. Preferences are used along with incomplete knowledge, when this is modeled with default rules. In such a setting, due to the incompleteness of the knowledge, several models may be possible. Preferences act by choosing among those possible models. A classical example is the birds-fly problem, where the incomplete knowledge contains the rules that birds normally fly and penguins normally don't. Given an individual which is both a penguin and a bird, two models are possible: one, using the one rule, where the individual flies; another, using the other more specific rule, where it doesn't. Preferences among rules can then be used to choose which one.

Updates are used to model dynamically evolving worlds. The problem arising here being, given a piece of knowledge describing the world, and given a change in the world (be it a rule or fact), how to modify the knowledge to cope with that change. The knowledge may itself be complete or incomplete: that's not the key issue in updates; rather, the key issue is about the process of accomodating, in the represented knowledge, any changes in the world. In this setting it may well happen that change in the world contradicts previous knowledge, i.e. the union of the previous knowledge with the representation of the new knowledge has no model. It is up to updates to remove from the prior knowledge representation a piece that changed, and to replace it by the new one. In this respect, mark well the distinction between update and revision of the knowledge, well broughtout e.g. in [14]. Whereas in the former knowledge changes due to changes in the world, in the latter incomplete knowledge is changed due to additional information (further completing the knowledge) about a static world view. These processes are different, and lead to different results. For example, suppose that your knowledge consists of a single rule stating that you have a flight booked for London, that is either for Heathrow or for Gatwick. If new information, stating that it is not for Heathrow, arrives thereby completing this knowledge (e.g. a call from your travel agency, clarifying this issue), then you should conclude that the flight is booked for Gatwick. If, the same information ($\neg Heathrow$) arrives due to a change in the world (e.g. you heard on the radio that all flights for

Heathrow have been cancelled), then you should not conclude now that your flight is booked for Gatwick.

One way to look at revision is to consider any prior rules as defeasible, add the new knowledge to the previous one, and assign preference to this new knowledge over the old one (revision as chronological preference). This stance is justifiable in revision: our knowledge is incomplete; to make it less incomplete, when some new information arrives it should be given preference over the previous. But a similar rationale makes little sense in updates. Suppose that at some point we know that normally quakers are pacifists, and that the republican Nixon is a quaker. Forthwith we can conclude that Nixon is a pacifist. Now something happens in the world so that republicans tend to be belicists. A new rule, stating that normally republicans are belicists, is added as an update. What should we conclude about Nixon? In our opinion, nothing different from a situation where both rules are given at the same time: for Nixon, there is a conflict, and two models exist - one where he is considered pacifist, and the other where he isn't. It may well happen that, given the conflict among such defeasible rules in our incomplete knowledge, one may want to give preference to the quakers-pacisfist rule over the other rule.

In many real applications one is bound to have just incomplete knowledge about the world, default rules, and may want to be able to deal with a dynamically evolving world, where these rules may change in time. In such a situation preferences may be needed to choose among various possible models of the world, whereas updates are needed to deal with the knowledge on the evolution of the world. In this evolution, preferences themselves may change in time. Thus, a combination of both reasoning forms into a single framework is needed.

Consider the following example, where default rules as well as preferences change over time, which requires a combination of preferences and updates, including the updating of preferences themselves.

Example 1 (A sad story). (1) In the initial situation I am living and working everyday in the city. (2) Next, as I have received some monies, I conjure up other, alternative but more costly, living scenarios, namely travelling, settling up on a mountain, or living by the beach. And, to go with them, also the attending preferences, but still in keeping with the work context, namely that the city is better for that purpose than any of the new scenarios, which are otherwise incomparable amongst themselves. (3) Consequently, I decide to quit working and go on vacation, supported by my increased wealth, and hence to define my vacation priorities. To wit, the mountain and the beach are each preferable to travel, which in turn gainsays the city. (4) Next, I realize my preferences keep me all the while undecided between the mountain and the beach, and opt for the former. (5) Forthwith, I venture up the mountain, only to become ill on account of the height, and a physician advises me against too much sun exposure, be it at the mountain or the beach level. (6) So, I update my knowledge regarding health, and my concomitant priorities, and thus travel becomes *the* choice par excellence. (7) I finally run out of money for travel and return, still ill, to the city, cannot work, and continue my sad vacation there.

Despite their differences, the preference and the update LP approaches we adopt are also similar, in that both can be envisaged as wiping out rules. In the preference setting, one wipes out less preferred rules in order to select only some among the available stable models. In the update setting, one wipes out rules that are overruled by new rules, thereby engendering new models, including cases when there were none before the update took place. Looking at both in a similar way facilitates their coming together under one same framework. For preferences it makes all the sense to employ some (strict) but partial order on rules, for there are cases where one wishes to allow incomparable rules to defeat but not wipe out one another. For updates, a linear temporal order is employed, and alternative results may be obtained via distinct but nevertheless linear updating sequences, to produce a tree. A root node always exists, if need be the initial empty program.

The sequel is organized as follows. First, we recap the fixpoint semantics of updates, which relies on erasing rules rejected by an update. Second, we define a fixpoint semantics for preferences which resorts to erasing unpreferred rules. Third, on the basis of these, we proffer a joint fixpoint semantics for both updates and preferences. Finally, conclusions and future work are brought out.

2   Dynamic Logic Programs

In this section we recall the framework of Dynamic Logic Programming (DLP) [1] that, as motivated above, can be used to model the evolution of logic program through sequences of updates.

To represent negative information in logic programs and their updates, DLP allows for the presence of default negation in rule heads[1].

Definition 1 (Generalized logic program). *A generalized logic program in the language $\mathcal{L}$ is a finite or infinite set of ground rules r of the form:*

$$L_0 \leftarrow L_1, \ldots, L_n. \quad n \geq 0$$

where each L_i is a literal in $\mathcal{L}$ (i.e. an atom or a default literal not A where A is an atom). By head(r) we mean L_0, by body(r) the set of literals $\{L_1, \ldots, L_n\}$, by body$_{pos}(r)$ the set of all atoms in body(r), and by body$_{neg}(r)$ the set of all default literals in body(r). We refer to body$_{pos}(r)$ as the prerequisites of r. Whenever L is of the form not A, not L stands for the atom A.

The semantics of generalized logic programs is then defined as a generalization of the stable models semantics [8]. First note that, instead of using the fixpoint operator $\Gamma(M)$, one may take default literals in rule bodies as new propositional variables, add a fact *not A* for every $A \notin M$, and then compute

---

[1] See [1] for an explanation on why default negation is needed in rule heads, rather than explicit negation. Note that a default negated atom in a rule's head means that the atom should no longer be assumed true, whilst an explicit negated atom would mean that the atom should become false. In an update context this difference is similar to the difference between deleting a fact and asserting its complement.

the least model of the resulting definite program. It is easy to check that the resulting set of atoms, not of the form $not\,A$, will be exactly the same as in $\Gamma(M)$. Moreover, for every fixpoint of $\Gamma(M)$, $A \notin M$ iff all rules of the program with head A have a false body in M. Thus, if one is only interested in fixpoints, instead one may add $not\,A$ for just every A having no rule with a true body in M. This approach views stable models as deriving $not\,A$ for every atom A which is not "supported" in the program by the model.

Now, since one can have default literals in rule heads, there are more ways of deriving them. But the previous one remains. This is the basic intuition behind the definition of stable models for generalized programs: given a model M, first add facts $not\,A$ for every A with no rule with true body in M; M is a stable model if the least model obtained after such additions coincides with M, where M has been enlarged with new propositional variables $not\,A$ for every $A \notin M$.

Definition 2 (Default assumptions). *Let M be a model of P. Then:*

$$Default(P, M) = \{not\,A \mid \nexists r \in P : head(r) = A \wedge M \models body(r)\}$$

Definition 3 (Stable Models of Generalized Programs). *A model M is a stable model of the generalized program P iff $M = least(P \cup Default(P, M))$*

For normal programs, this definition is equivalent to the original definition of stable models [8]. As shown in [1], it also coincides with the semantics presented in [10] when the latter is restricted to the language of generalized programs.

In DLP, sequences of generalized programs $P_1 \oplus \ldots \oplus P_n$ are given. Intuitively a sequence may be viewed as the result of, starting with program P_1, updating it with program P_2, ..., and updating it with program P_n. In such a view, dynamic logic programs are to be used in knowledge bases that evolve. New rules (coming from new, or newly acquired, knowledge) can be added at the end of the sequence, bothering not whether they conflict with previous knowledge. The rôle of dynamic programming is to ensure that these newly added rules are in force, and that previous rules are still valid (by inertia) as far as possible, i.e. they are kept for as long as they do not conflict with newly added ones.

The semantics of dynamic logic programs is defined according to the rationale above. Given a model M of the last program P_n, start by removing all the rules from previous programs whose head is the complement of some later rule with true body in M (i.e. by removing all rules which conflict with more recent ones). All other persist through by inertia. Then, as for the stable models of a single generalized program, add facts $not\,A$ for all atoms A which have no rule at all with true body in M, and compute the least model. If M is a fixpoint of this construction, M is a stable model of the sequence up to P_n.

Other possible views on and usage of DLP, justify slight generalizations of the above informally described language and semantics. In general, the distinguished programs represent knowledge true at some state s, where different states may stand for different stages of knowledge in the linear evolution of the knowledge base (as above), but also for different time points in possible future evolutions of the knowledge, or even for knowledge of ever more specific objects organized in a hierarchy. In the latter case, each program contains the rules that are specific to the object under consideration, and rules from programs above in the hierarchy are inherited just as long as they do not conflict with the more specific information (for more on this stance see [6]). These other views justify a tree-like structure of programs (rather than a sequence), and also that dynamic programs can be queried at any state, rather than only at the last one.

Definition 4 (Dynamic Logic Program). *Let S be an ordered set with a smallest element s_0 and with the property that every $s \in S$ other than s_0 has an immediate predecessor $s - 1$ and that $s_0 = s - n$ for some finite n. Then $\bigoplus\{P_i : i \in S\}$ is a Dynamic Logic Program, where each of the P_is is a generalized logic program.*

Definition 5 (Rejected rules). *Let $\bigoplus\{P_i : i \in S\}$ be a Dynamic Logic Program, let $s \in S$, and let M be a model of P_s. Then:*

$$Reject(s, M) = \{r \in P_i \mid \exists r' \in P_j, \ head(r) = not\,head(r') \wedge \ i < j \leq s \wedge$$
$$M \models body(r')\}$$

To allow for querying a dynamic program at any state s, the definition of stable model is parameterized by the state:

Definition 6 (Stable Models of a DLP at state s). *Let $\bigoplus\{P_i : i \in S\}$ be a Dynamic Logic Program, let $s \in S$, and let $\mathcal{P} = \bigcup_{i \leq s} P_i$. A model M of P_s is a stable model of $\bigoplus\{P_i : i \in S\}$ at state s iff:*

$$M = least([\mathcal{P} - Reject(s, M)] \cup Default(\mathcal{P}, M))$$

It is clear from the definitions that stable models of dynamic programs are a generalization of stable models of generalized and normal programs, i.e. if the dynamic program consists of a single generalized (resp. normal) program then its semantics is the same as that of the stable models of generalized (resp. normal) programs. It is also shown in [1] that dynamic logic programs generalize the interpretation updates of [11].

In [1] a transformational semantics for dynamic programs is also presented. According to this equivalent definition, a sequence of programs is translated into a single generalized program (with one new argument added to all predicates) whose stable models are in one-to-one correspondence with the stable models of the dynamic program. This transformational semantics is the basis of an existing implementation of dynamic logic programming[2].

3   Preferred Stable Models

In this section we recall the preferences approach of [5], and set forth a definition of preferred stable models for generalized logic programs (rather than

---

[2] Publicly available from: `http://centria.di.fct.unl.pt/~jja/updates/`

for extended logic programs as in [5]) in a form suitable for integration with the above described updates. In [5], logic programs are supplied with priority information, given in the form of a strict partial ordering on program rules[3].

Definition 7 (Prioritized generalized logic program). *Let P be a generalized program and let $<$ be a strict partial order over the rules of P, where $r_1 < r_2$ means r_1 is preferred to r_2. Then $(P, <)$ is a prioritized generalized program[4].*

Intuitively, the priority information is used to prefer among the various stable models of the program. The question here is what stable models to prefer in the face of a given priority relation among rules. To respond to this question, the authors in [5] start by formulating two principles all preference system should satisfy. The first (*Principle I*), is envisaged as a minimal requirement for preference handling, and states that if a stable model M_1 is generated by a set of rules[5] $R \cup \{r_1\}$, and another stable model M_2 is generated by $R \cup \{r_2\}$, where $r_1, r_2 \notin R$, then, if $r_1 < r_2$, M_2 cannot be preferred. The second (*Principle II*), captures a notion of relevance. It affirms that adding a rule which is not applicable in a preferred model can never render this model unpreferred.

With these two principles in mind, [5] defines a criterion for preferring among stable models, given a priority relation on rules. Their basic idea is that a stable model M can only be preferred if, for each rule in the program, whenever its (positive) prerequisites are true in M and its head is false in M, then there must be some *not A* in its body which is false in M, and there is a more preferred rule generating A. I.e. for a rule with true prerequisites not to be applied, there must be a more prioritary rule preventing its application.

Before presenting our equivalent definition of preferred stable models, let us first briefly review the formal definition of preferred answer sets of [5] specialized for the case where the program is ground. A preferred answer-set is a model of the program simultaneously satisfying two conditions: it must be a stable model (i.e. be a fixpoint of the Γ Gelfond-Lifschitz operator); it must satisfy a fixpoint equation which, intuitively, guarantees that the rules are being applied in observance of the partial order, i.e. that the criterion described above is met.

Adopting the view that rules are applied one at a time, a partial ordering on rules should be viewed as a representative of all its possible refinements into total orderings. These, defined in [5], are dubbed *full prioritizations* of prioritized programs. A program is said *fully prioritized* if it coincides with its single full prioritization. The fixpoint construction guaranteeing that rules of a fully prioritized program are applied in the correct order is carried out in two steps. First, all (positive) atoms in the body are preprocessed away on the basis of their truth

---

[3] For a comparison with approaches ordering atoms rather than rules see [5].

[4] Note that, in contradistinction to [5], our priority relation is defined for ground programs. To define the relation directly on non-ground programs, the methodology given in [5], using well-orderings, could just as well be applied to our case. However, for simplicity, we will not consider it in this paper.

[5] The set of rules that generate a stable model is made up of all the rules in the program whose body is true in the stable model.

value in the model. More precisely, this so called dual Gelfond-Lifschitz reduction $^M\mathcal{R}$ is obtained from $\mathcal{R}$ by first deleting every rule having a prerequisite A such that $A \notin M$, and then removing from the remaining rules all prerequisites. All bodies of rules now exhibit only default literals.

The correct order of applying rules is then checked in the thus obtained prerequisite-free program. Informally, this is achieved by, following rule order, adding the heads of those rules that are not defeated by a rule having higher priority (whose head has been added). Formally:

Definition 8 (Defeating of rules). *A rule r is defeated by a set of literals S iff $\exists\ not\ A \in body(r) : A \in S$.*

Definition 9 ($C_\mathcal{R}$ operator [5]). *Let $\mathcal{R} = (P, <)$ be a prerequisite-free fully prioritized logic program, and let M be a set of ground literals. $C_\mathcal{R}(M)$ is the least fixpoint of the sequence S_α (where α ranges over the rules of the fully prioritized P, according to their (total) ordering):*

$$S_\alpha = \begin{cases} \bigcup_{\beta < \alpha} S_\beta & \text{if } r_\alpha \text{ is defeated by } \bigcup_{\beta < \alpha} S_\beta \text{ or} \\ & \quad r_\alpha \text{ is defeated by } M \text{ and } head(r_\alpha) \in M; \\ \bigcup_{\beta < \alpha} S_\beta \cup \{head(r_\alpha)\} & \text{otherwise} \end{cases}$$

Definition 10 (Preferred Answer Set). *Let $\mathcal{R} = (P, <)$ be a prioritized logic program and let $\mathcal{R}_f = (P, <_f)$ be a full prioritization of $\mathcal{R}$. A model M of P is a preferred answer set of $\mathcal{R}$ iff $M = \Gamma_P(M)$ and $M = C_{M\mathcal{R}_f}(M)$.*

As motivated in the Introduction, and in order to facilitate the capture of both preferences and updates in one single framework, it is our goal in this section to devise a declarative semantics for prioritized generalized programs based on the removal of (less preferred) rules, inasmuch our update framework hinges likewise on the removal of rules; this maneuver is crucial for fusing the two. Moreover, we require this semantics to coincide with the one in [5] on normal programs. The main issue in so doing rests in determining criteria for which rules to remove, in order to obtain exactly the same semantics. Before presenting its definition, we begin by reporting, with small but illustrative examples, on the problems involved in finding them[6]. Like in [5], we start with the case of prerequisite-free programs.

Example 2. Consider the program: (1) $a \leftarrow not\,b$ (2) $b \leftarrow not\,a$,   where rule (1) is preferred over rule (2). Its stable models are $M_1 = \{a\}$ and $M_2 = \{b\}$, the preferred one being M_1. Intuitively, since (1) < (2) and the head of rule (1) defeats (2), in order to obtain the preferred stable model, one should remove rule (2). Indeed, M_1 is the single stable model of the program after the excision.

---

[6] This account is important here because for lack of space, the proof of equivalence with [5] does not fit. The problems depicted below form the core issues dealt with by the proof.

Mark that the reasoning brought out in this example concurs with the definition of the $C_{\mathcal{R}}$ operator. According to it, the head of a rule is not added if the rule is defeated by the previously constructed set. But this set is formed precisely by the heads of the more preferred rules. Instead of not adding the head to the set, the same effect can be achieved by removing the rule, i.e. by removing all rules defeated by the head of a more preferred rule, which has not itself in turn been removed.

Example 3. Consider now: (1) $b \leftarrow not\,c$ (2) $c \leftarrow not\,d$ (3) $a \leftarrow not\,b$ (4) $b \leftarrow not\,a$, where a rule (i) is preferred over rule (j) iff $i < j$. Its stable models are $M_1 = \{a, c\}$ and $M_2 = \{b, c\}$. According to Principle I above, since M_1 is generated by rules (2) and (3), and M_2 by rules (2) and (4), M_2 should not be preferred. But, resorting to the reasoning explained above, rule (3) is removed (as it is defeated by the head of rule (1)), and the only stable model of the resulting program becomes M_2. Why shouldn't rule (1) remove (3)? Because rule (1) is defeated in whichever model. This is in line with Definition 9 (2nd line of S_α) where heads of rules true in the model, whose body is defeated by the model, are not added to the set. Accordingly, given some model, all such rules are removed. Hereafter, we refer to them as "unsupported rules".

Consequently, in model M_2 rule (1) is removed, as well as rule (4) (the latter is defeated by the head of the more preferred and non-removed rule (3)). And M_2 is not a stable model of the program after those rules are withdrawn.

The two above criteria for deleting rules (viz. deleting less preferred rules defeated by the head of some more preferred rule, and deleting "unsupported rules") concur with the definition of the $C_{\mathcal{R}}$ operator. However, as evidenced by the example below, they are not enough.

Example 4. Consider now: (1) $a \leftarrow not\,b$ (2) $b \leftarrow not\,c$ (1) < (2), whose only stable model is $M = \{b\}$, which according to [5] is not preferred. This is so because rule (1) is neither unsupported (a is not true in M) nor defeated by a more preferred rule, so a is added in the construction of $C_{\mathcal{R}}(M)$, and M cannot thereafter be a fixpoint of the operator. However, using only the two above criteria none of these two rules is eliminated, and M would be preferred.

To obtain the effect achieved by [5], one must guarantee that, in spite of rule removal, a is enforced in the preferred models of the reduced program. This is accomplished by removing any rules less preferred than the one for a, which, if otherwise were not removed, would cause a not to be in the preferred models. In other words, one is required to remove all rules having true body in the model, whose heads defeat a more preferred rule. Mark well that if the body of the less preferred rule is not actually true in the model, then the defeating is only a potential but not effective one, and the rule must not be eliminated. Indeed, its preservation will permit it to defeat, and cause to remove, rules less preferred than itself even if they attack it. When considering programs with prerequisites, one must further insist that the more preferred rule is not deleted by the dual reduct transformation. This is ensured by verifying that the positive part of the body of the more preferred rule is actually true in the model. A similar reasoning applies to the other two criteria explained above. These three criteria suffice for formalizing, in Definition 12, the set of unpreferred rules.

Definition 11 (Unsupported rules). *Let M be a model of P. Then:*

$$Unsup(P, M) = \{r \in P : M \models \{head(r)\} \cup body_{pos}(r) \wedge M \not\models body_{neg}(r)\}$$

Definition 12 (Unpreferred rules). *Let M be a model of P. The set of unpreferred rules, $Unpref(P, M)$, is the least set of rules that includes $Unsup(P, M)$, and every r in P such that:*

$$\exists r' \in P - Unpref(P, M) : r' < r \wedge M \models body_{pos}(r') \wedge$$
$$[not\, head(r') \in body_{neg}(r) \quad \vee \quad (not\, head(r) \in body_{neg}(r') \wedge M \models body(r))]$$

Lack of space prevents us from showing that such a least set always exists. Indeed, it can be contructed by iterating the definition of unpreferred rule according to rules' ordering, starting from the set of unsupported rules.

For programs with positive atoms in rule bodies, the effect of the dual reduction operation of [5] is obtained by adding to the program facts *not A* for every A with no rule in the original program with true body in the model, and thereafter computing the least model.

Definition 13 (Preferred Stable Models). *A model M of program P is a preferred stable model of the prioritized generalized program $(P, <)$ iff:*

$$M = least([P - Unpref(P, M)] \cup Default(P, M))$$

This guarantees that the preferred models obtained after removing all unpreferred rules are also stable models of P, and so only one fixpoint equation is needed in this definition, as desired. Verily:

Proposition 1. *Let M be a preferred stable model of $(P, <)$. Then M is also a stable model of P, i.e. $M = least(P \cup Default(P, M))$.*

Now, as expected, as this was one of our primary goals for the definition of preferred stable models, in programs where both the preferred answer sets of [5] and our preferred stable models can be applied (i.e. in normal programs), their results coincide. For an extensive study of the properties of preferred answer-sets, its intuitions, examples, and comparisons with related approaches see [5].

Theorem 1. *Let P be a ground normal logic program, and let $<$ be a strict partial order over the rules of P. M is a preferred stable model of $(P, <)$ iff M is a preferred answer-set of $(P, <)$ in the sense of Brewka and Eiter [5].*

4   Updating Logic Programs with Preferences

Having separately defined both updates and preferences in an analogous way, in this section we combine both concepts into an unified framework. Moreover, as motivated in the Introduction, the combined framework must also allow for the updating of the priority relation itself.

Leaving, for now, the issue of updating the priority relation, we must consider sequences of generalized programs $P_1 \oplus \ldots \oplus P_n$, viewed as sequences of updates of an original program, plus some priority relation among rules. One first basic question is in order: where to define the priority relation? Among the rules for the same program? Or among rules in the union of all programs in the sequence? More formally, should there be a strict partial order $<_i$ for each of the P_i in the sequence, or should there be a single strict partial order $<$ defined over the rules of $\bigcup_{i \in S} P_i$? Clearly, the latter approach is more general than the former: it does not prevent limiting the priority relation to rules in the same P_i, while the former does prevent priority relations between rules from different P_is. Furthermore, the extra generality is useful. For instance, in the situation of Example 1, one may want to say at a given state that I go to the beach unless I go to the mountain, and later say that I go to the mountain unless I go to the beach, and establish a priority over these rules. Note that the rules were introduced at different update stages, and so the priority relation is to be established between rules of different P_is. Accordingly, in our framework we consider a single priority relation defined on the rules of $\bigcup_{i \in S} P_i$, which can evolve as new rules are introduced.

To cope with the possibility of updating the priority relation, it cannot be fixed. Rather it must be described in some language that allows for the possibility of its evolution, via updates. One such language is precisely DLP and, for uniformity, that is what is used in our framework. Thus, instead of a sequence of programs representing knowledge, we have a sequence of pairs: of programs representing knowledge, and of programs describing the priority relation among rules of the knowledge representation. In general, an update of the priority relation may depend on some other predicate (e.g. in Example 1, I may want to say that, if I have to work, then I prefer the rule advising me to stay in the city). To permit this generality, we allow rules in programs describing the priority relation to refer to predicates defined in the programs that represent knowledge.

Definition 14 (Dynamic Prioritized Programs). *Let* $\mathcal{P} = \{P_s : s \in S\}$ *be a dynamic logic program whose alphabet does not contain the strict partial order arity 2 predicate symbol* $<$*, and let* $\mathcal{R} = \{R_s : s \in S\}$ *be another dynamic logic program whose alphabet contains at least the predicate symbol* $<$*, and whose sets of constants includes all the rules in the union of all Ps in* $\mathcal{P}$*. Then* $\bigoplus\{(P_s, R_s) : s \in S\}$ *is a Dynamic Prioritized Program.*

Given the very deliberate definition forms of the semantics of preferences and of updates, it is not difficult to combine both in a single one, as per the above delineated framework. Given a model M of the last program in the sequence (or, in the general setting, of the program state we want to query), for testing stability we have first to remove all the rejected rules according to updates, and then all the unpreferred rules according to preferences. Note, however, that if both sets of rules (rejected and unpreferred) were removed simultaneosly, then a rule which is rejected by an update, might serve for unprefering some other rule. This would lead to counterintuitive results. In fact, updates have precedence over preferences. If a previous rule is invalidated by a subsequently introduced rule, then the former should no longer be available in the preferences setting. Accordingly, the set of unpreferred rules must be determined on the basis of the program obtained after removing those rules rejected by any updates. In general, the union of all programs in the sequence may be inconsistent, and it would make no sense to apply preferences to this inconsistent set of rules; updates are applied first (by rejecting rules) and allow you to come up with a consistent set of rules; preferences then intervene to choose among the various models of that consistent set of rules.

Since the priority relation is itself defined by the dynamic prioritized program, models must also take into account the $<$ predicate, i.e. one has to entertain models of the union of P_n with R_n. Moreover, in the definition of unpreferred rules, the priority relation must be checked in regard to the model under consideration:

Definition 15 (Unpreferred rules). *$Unpref(P, M)$ is the least set of rules including $Unsup(P, M)$ and rules r in P such that:*

$$\exists r' \in P - Unpref(P, M) : M \models r' < r \wedge M \models body_{pos}(r') \wedge$$
$$[\,not\,head(r') \in body_{neg}(r) \quad \vee \quad (not\,head(r) \in body_{neg}(r') \wedge M \models body(r))\,]$$

In the definition of preferred stable model, it is crucial that the priority relation be a strict partial order (i.e. irreflexive and transitive). In our framework, since the user can write any rules for describing predicate $<$, it may well happen that its extention be a relation not complying with those properties. The definition of the semantics must prevent this being the case, i.e. must only consider models where the extension of predicate $<$ is indeed a strict partial order. Thus:

Definition 16 (Preferred Stable Models at state s). *Let $\bigoplus\{(P_i, R_i) : i \in S\}$ be a Dynamic Prioritized Logic Program, let $s \in S$, and let $\mathcal{PR} = \bigcup_{i \leq s}(P_i \cup R_i)$. A model M of $P_s \cup R_s$ is a preferred stable model at state s iff:*

$$- \; \forall r : (r < r) \notin M \; and \; \forall r_1, r_2, r_3 : \{r_1 < r_2, r_2 < r_3\} \subseteq M \Rightarrow (r_1 < r_3) \in M$$
$$- \; M = least(\, [\mathcal{PR} - Reject(s, M) - Unpref(\mathcal{PR} - Reject(s, M), M)]$$
$$\cup \quad Default(\mathcal{PR}, M) \qquad\qquad\qquad)$$

This definition makes it clear that dynamic prioritized programs generalize both dynamic logic programs and prioritized logic programs. In fact, if all the R_is are empty, then Definition 16 is clearly equivalent to Definition 6. And if there is a single pair (P, R) in the sequence, then Definition 16 is equivalent to Definition 13, the priority relation of the prioritized program being the least model of R. We now illustrate the overall framework with two examples:

Example 5. The first 4 stages in the "sad story" of Example 1 can be modelled by the dynamic prioritized program $(P_1, R_1) \oplus \ldots \oplus (P_4, R_4)$ (where, for simplicity, we adopt unique numbers for rules, instead of the rules themselves in the priority relation, and where c stands for "living in the city", mt for "settling on a mountain", b for "living by the beach", t for "travelling", wk for "work", vac for "vacations", and mo for "possessing money"):

$P_1 :$ (1) $c \leftarrow not\, mt, not\, b, not\, t$ $R_1 : X < Y \leftarrow X < Z, Z < Y$
    (2) $wk \leftarrow$ $R_2 :$ (1) $< (4) \leftarrow wk$
    (3) $vac \leftarrow not\, wk$ (1) $< (5) \leftarrow wk$
$P_2 :$ (4) $mt \leftarrow not\, c, not\, b, not\, t, mo$ (1) $< (6) \leftarrow wk$
    (5) $b \leftarrow not\, mt, not\, c, not\, t, mo$ $R_3 :$ (4) $< (6) \leftarrow vac$
    (6) $t \leftarrow not\, mt, not\, b, not\, c, mo$ (5) $< (6) \leftarrow vac$
    (7) $mo \leftarrow$ (6) $< (1) \leftarrow vac$
$P_3 :$ (8) $not\, wk \leftarrow$
$P_4 :$ $\{\}$ $R_4 :$ (4) $< (5)$

For example, the only preferred stable model at state 4 is:

$$\{mt, vac, mo, (4) < (5), (4) < (6), (4) < (1), (5) < (6), (5) < (1), (6) < (1)\}$$

and the preferred stable models at state 3 are two:

$$\{mt, vac, mo, (4) < (6), (4) < (1), (5) < (6), (5) < (1), (6) < (1)\}$$
$$\{b, vac, mo, (4) < (6), (4) < (1), (5) < (6), (5) < (1), (6) < (1)\}$$

Note in this example how the inertia of the transitivity rule (added in R_1) enforces transitivity on the priority relation in all the subsequent states.

Example 6. Consider the following situation (adapted from an example of qualitative decision making in [5]). You want to buy a car and, for that purpose, you have collected the following information about different types of cars: $safe(volvo)$, $fast(chevrolet)$, $expensive(chevrolet)$, $safe(chevrolet)$, and $fast(porsche)$. Let's assume you like fast cars, and your budget does not allow you to purchase an expensive one. Moreover, you cannot afford more than one car.

This situation can be modelled by P_1 which, besides the facts above, has[7]:

(1) $not\, buy(X) \leftarrow avoid(X)$
(2) $avoid(X) \leftarrow not\, buy(X), expensive(X)$
(3) $buy(X) \leftarrow not\, avoid(X), fast(X)$
(4) $avoid(Y) \leftarrow fast(X), buy(X), Y \neq X$

See [5] for an explanation on how to come up with this program given the described situation, in particular the need for rule (4) in modelling the fact that you may not buy two cars[8]. Since there is not much you can do with your

---

[7] Rules with variables simply stand for their (finite) ground instances.
[8] In fact, the coding of this piece of knowledge by itself is not related to updates, and the rules above are just those present in [5] where $\neg buy(X)$ is here replaced by $avoid(X)$, and (1) encodes the relation between these two predicates.

restricted budget, rule (2) has priority over rules (3) and (4). So $R_1 = \{(2) < (3), (2) < (4)\}$.

The reader can check that the only preferred stable model of (P_1, R_1) includes $\{buy(porsche), avoid(volvo), avoid(chevrolet)\}$, besides the facts and the priority relation, and you should buy the Porsche.

Now your "significant other" insists that you should consider buying a safe car. Moreover, as a gentleperson, you ascribe priority to your partner's suggestion. To assimilate this new information you update your knowledge with:

$$P_2 : (5) \quad buy(X) \leftarrow not\, avoid(X), safe(X)$$
$$(6)\; avoid(Y) \leftarrow safe(X), buy(X), Y \neq X$$

and $R_2 = \{(5) < (3), (5) < (4), (6) < (3), (6) < (4), (2) < (5), (2) < (6)\}$. Now the only preferred stable model (at state 2) includes $buy(volvo)$, $avoid(porsche)$ and $avoid(chevrolet)$, and you should buy the Volvo instead.

Now suppose you discover Volvos are out of stock, and so you cannot buy one so soon. For that you add $P_3 = \{not\, buy(volvo)\}$, plus an empty R_3. With this new update, rule (5) is now rejected, and the only stable model at state 3 this time includes $\{buy(porsche), avoid(volvo), avoid(chevrolet)\}$.

5   Conclusions and Future Work

We have motivated the need for coupling preferences with updates, and shown how to accomplish it within the logic programming paradigm. We did so by devising a unified framework that combines the hitherto separate approaches to each aspect, and allows for preferences themselves to be updated. The framework coincides with [5] when a single program is given in the sequence, and with [1] when the preference relation is empty. Thus, for comparisons of this framework with others with preferences alone see [5], and for that with others with updates alone see [1].

To the best of our knowledge, [15] is the only work considering some combination of preferences and updates. However, the generality of the combination of both reasoning mechanisms in [15] is far from that of the present paper. In fact, [15]'s concern is with updates alone, and mainly considers the process of updating one program by another program, with mechanisms similar to those of [1] (i.e. removing rules from the initial program which "somehow" contradict rules from the update program, and retaining all others by inertia). Additionally, at the end, all rules from the update program are given preference over all retained rules of the initial program. No other preference ordering is considered there. And, as argued in the Introduction, updates alone do not necessarily force such preferences. In our framework, the user can state that more recent rules are preferred over older ones, but is also free to state differently. Moreover, in our framework the preference relation itself can be updated. The greater generality of our approach stems as well from our usage of [1] as the basis for updates. In fact, note that [1] considers arbitrary sequences of updates whereas [15] simply considers the update of one program by another. In [15] some semantical properties of their system are investigated. However, all such properties address only updating, and not some combination of it with preferences. It remains to be studied what generic principles any system combining preferences and updates (not necessarily in logic programming) should comply with. Those principles would help the comparison with [15] and possible other systems. Such generic study, and the verification of the principles in our framework, is work we are now developing.

Several other topics cry out for subsequent development. First, we are working on a transformational semantics of preferences into logic programs, to be coupled with the extant aforementioned one for updates. This will readily propitiate an implementation of the overall framework, as well as serve as a basis for the study of its computational properties.

An outstanding issue, on which some effort needs deploying, concerns how to automatically ensure irreflexivity and transitivity of the partial order, as it is being updated. For the moment this responsibility is wholly relegated to the updater. As it stands, in case of infringement there will simply be no model, as per Definition 16. It is in our plans to study the adequacy of the update mechanism on rules for predicate < so as to automatically guarantee irreflexivity and transitivity. In this respect, note in Example 5, how transitivity is always guaranteed by adding one rule to the initial program.

Finally, we also intend to explore application areas such as e-commerce, legal reasoning, and rational agents. They will certainly provide valuable opportunities and hints for the evolution of the topics broached in this paper.

References

1. J. J. Alferes, J. A. Leite, L. M. Pereira, H. Przymusinska, and T. Przymusinski. Dynamic updates of non-monotonic knowledge bases. *Journal of Logic Programming*, 2000. To appear. A short version titled *Dynamic Logic Programming* appeared in A. Cohn and L. Schubert (eds.), *KR'98*, Morgan Kaufmann.

2. J. J. Alferes, L. M. Pereira, H. Przymusinska, and T. Przymusinski. LUPS – a language for updating logic programs. In M. Gelfond, N. Leone, and G. Pfeifer, editors, *LPNMR'99*. Springer, 1999.

3. J. J. Alferes, L. M. Pereira, T. Przymusinski, H. Przymusinska, and P. Quaresma. Preliminary exploration on actions as updates. In *AGP'99*, 1999.

4. G. Brewka. Well-founded semantics for extended logic programs with dynamic preferences. *Journal of Artificial Intelligence Research*, 4, 1996.

5. G. Brewka and T. Eiter. Preferred answer sets for extended logic programs. *Artificial Intelligence*, 109, 1999. A short version appeared in A. Cohn and L. Schubert (eds.), *KR'98*, Morgan Kaufmann.

6. F. Buccafurri, W. Faber, and N. Leone. Disjunctive logic programs with inheritance. In D. De Schreye, editor, *ICLP'99*. MIT Press, 1999.

7. M. Dekhtyar, A. Dikovsky, S. Dudakov, and N. Spyratos. Monotone expansion of updates in logical databases. In M. Gelfond, N. Leone, and G. Pfeifer, editors, *LPNMR'99*. Springer, 1999.

8. M. Gelfond and V. Lifschitz. The stable model semantics for logic programming. In R. Kowalski and K. A. Bowen, editors, *ICLP'88*. MIT Press, 1988.

9. H. Katsuno and A. Mendelzon. On the difference between updating a knowledge base and revising it. In J. Allen, R. Fikes, and E. Sandewall, editors, *KR'91*. Morgan Kaufmann, 1991.

10. V. Lifschitz and T. Woo. Answer sets in general non-monotonic reasoning (preliminary report). In B. Nebel, C. Rich, and W. Swartout, editors, *KR'92*. Morgan-Kaufmann, 1992.

11. V. Marek and M. Truszczynski. Revision specifications by means of programs. In C. MacNish, D. Pearce, and L. M. Pereira, editors, *JELIA'94*. Springer, 1994.

12. T. Przymusinski and H. Turner. Update by means of inference rules. In V. Marek, A. Nerode, and M. Truszczynski, editors, *LPNMR'95*. Springer, 1995.

13. C. Sakama and K. Inoue. Updating extended logic programs through abduction. In M. Gelfond, N. Leone, and G. Pfeifer, editors, *LPNMR'99*. Springer, 1999.

14. M. Winslett. Reasoning about action using a possible models approach. In *AAAI'88*, 1988.

15. Y. Zhang and N. Foo. Updating logic programs. In H. Prade, editor, *ECAI'98*. Morgan Kaufmann, 1998.

A Framework for Belief Update

Paolo Liberatore

Dipartimento di Informatica e Sistemistica
Università di Roma "La Sapienza"
Via Salaria 113, I-00198 Roma, Italy
liberato@dis.uniroma1.it

Abstract. In this paper we show how several different semantics for belief update can be expressed in a framework for reasoning about actions. This framework can therefore be considered as a common core of all these update formalisms, thus making it clear what they have in common. This framework also allows expressing scenarios that are problematic for the classical formalization of belief update.

1 Introduction

Belief update and reasoning about actions are two well studied areas of research about the evolution of knowledge over time. The similarities between these two fields have already been pointed out by some researchers: for example del Val and Shoham [4] use a theory of action to derive a semantics for belief update; Li and Pereira [8] use a Ginsberg-like semantics for updating a theory of actions.

In this paper we present a very simple action description language [6] with narratives that allows expressing several different update semantics. The basic principles of this language has already been investigated in the literature. Indeed, the basic semantics of this language can be seen as a proper restriction of the language $\mathcal{L}$ by Baral et al. [1]. What is new in this paper is not the language itself, but rather the way it is able to express update semantics.

To introduce the language, we consider an example similar to the evergreen Yale Shooting Problem.

initially *Loaded*

initially *Alive*

Alive holds at 3

Shoot happens at 2

Unload causes ¬*Loaded*

Shoot causes ¬*Alive* if *Loaded*

Short explanation of the syntax: at time 0 Fred is alive, and the gun is loaded. Fred is still alive at time 3. This is the meaning of the initially and holds at propositions. The last two propositions specify the effect of actions: the

M. Ojeda-Aciego et al. (Eds.): JELIA 2000, LNAI 1919, pp. 361–375, 2000.

action *Unload* causes the gun to be loaded no longer, while the action *Shoot* causes Fred to die, if the gun is loaded.

According to the original semantics of the basic action description language $\mathcal{A}$ [6], this domain description is inconsistent. This can be intuitively explained as follows: at time 0 the gun is loaded. Since nothing happens between time 0 and 2, the gun remains loaded. As a result, the effect of shooting at Fred at time 2 causes him to die, since the gun is still loaded.

Such inconsistent scenarios are very common in the field of belief revision and update. Suppose for example we have loaded the gun at time 0. Then, we have done nothing modifying the domain of interest (e.g. we go out for a walk, we have a nap, we just do nothing at all, etc.) When we shoot the gun, Fred does not die. This is surprising, since we expected the gun to be still loaded. However, it is very easy to find an explanation: someone unloaded the gun while we was not looking at it. Such conclusion can be drawn assuming that some actions may take place at some time points, and this is initially not known.

In languages with narratives, such that the language $\mathcal{AU}$ introduced in this paper, such a deduction is possible. Note that it is not only a matter of finding an explanation of already known facts. For example, we can conclude that $\neg Loaded$ holds at 2 from the domain description above. Such an inference is clearly impossible in the basic action description language $\mathcal{A}$.

The example describes a prototypical scenario of belief update: we have a set of facts which are known to holds at a certain time point (e.g. the gun is loaded and Fred is alive at time 0). In a subsequent time point something is observed (e.g. Fred is alive at time 3). The possible inconsistency between the facts and the observation is explained as due to changes happened in the world. In this paper, the assumption is that all changes are caused by actions.

The formalization of change given in belief update is very simple. If T is a set of known facts, and P is an observation, $T * P$ denotes the result of updating T with P, that is, our knowledge after the observation of P. The use of this notation seemed the natural choice to the first researchers in the field, since what we want to formalize is indeed the update of T with P.

This notation is very simple, but sometimes it does not allow to express enough information. The example of the gun contains information that cannot be formalized using the star notation. For example, there is no way to express the fact that it is impossible that Fred becomes alive, once it is dead. Such information cannot be represented using the notation $T * P$, since the only information expressed in this way is the old set of facts T and the observation P. Another problem is the impossibility of deciding what is true in time points before the update. In the example, *Loaded* is false at time 2. However, $T * P$ only expresses the result of the update, that is, what is known at the time of the observation (in this case, at time 3). As a result, there is no way to even ask what is true at time 1, or 2, etc. Finally, there are problems in formalizing the process of iterated update. For example, $(T * P_1) * P_2$ is different from the intuitive result of incorporating two observations P_1 and P_2 (for an explanation of why, we refer the reader to the borrowed car example [5]).

All these issues have already been pointed out by researchers in the field, for example Boutilier [2, 3] and Li and Pereira [8]. However, most of these formalisms employ an ad-hoc syntax and semantics. The framework introduced in this paper allows for formalizing all those forms of update. The semantics of the language generalizes many semantics given for belief update.

The benefits of $\mathcal{AU}$ are twofold: it is at the same time a useful extension of action theories with narratives, and it allows an easy and intuitive formalization (in a standard way) of theories of belief update.

The paper is organized as follows: in the next section we describe the syntax and the semantics of the language $\mathcal{AU}$. The syntax of $\mathcal{AU}$ is similar to that of action description languages with narratives. As a result, we can define a "classical" semantics for it, as well as a semantics that formalizes actions that are not known to be happened. We prove that many belief update semantics can be captured this way. Finally, we compare our approach with other ones dealing with updates and action theories, and discuss possible extensions of this work.

2   The Language $\mathcal{AU}$

2.1   Syntax

The alphabet of the language is composed by three mutually disjoint sets: the set of actions, the set of fluents, and the set of time points. In this paper we assume that the set of time points is the set of non-negative integers.

A fluent literal is a fluent possibly preceded by the negation symbol $\neg$. A fluent expression is a propositional formula over the alphabet of fluents. Thus, all the fluent literals are also fluent expressions, and if E_1 and E_2 are fluent expressions, so are $E_1 \wedge E_2$, $E_1 \vee E_2$, and $\neg E_1$.

A domain description is composed of three parts: behavioral, historical, and actual. If D is a domain description then D_B, D_H, and D_A are its behavioral, historical, and actual parts, respectively.

Behavioral Part. Is the set of effect propositions, and is the part of the domain that specifies how the domain behaves in response to actions. An effect proposition is as follows:

$$A \text{ causes } F \text{ if } P_1, \ldots, P_m$$

where F is a fluent *literal*, $P_1, \ldots, P_m$ are fluent expressions, and A is an action. The meaning is that the action A causes the fluent literal F to become true, if the fluent expressions $P_1, \ldots, P_m$ are currently true. For this reason, the fluent expressions $P_1, \ldots, P_m$ are called the preconditions of the proposition, and F is called the effect.

Historical Part. Is the specification of the actions that are known to have been executed. A happens proposition is a statement of the form

$$A \text{ happens at } t$$

where A is an action and t is a time point. The meaning is clear: the action A is executed at time point t.

Actual Part. Is the set of propositions that specify the status of a fluent at a certain time point.

$$E \text{ after } A_1; \ldots; A_m \text{ from } t$$

where E is a fluent expression, $A_1; \ldots; A_m$ are actions and t is a time point. The meaning is that the fluent expression E is true after executing the actions $A_1; \ldots; A_m$ in sequence starting from the time point t. This propositions allows to specify both the status of a "real" time point, and the status of hypothetical situations. When $m = 0$ (i.e. no actions) the proposition is written E holds at t, its meaning being that the fluent expression E is true in the time point t. On the other hand, when $t = 0$, we write E after $A_1; \ldots; A_m$. What is the difference between propositions like E holds at t and E after $A_1; \ldots; A_m$? The first one refers to a specific time point t. The second one refers to a sequence of actions. It is possible that the actions executed from 0 are *not* the sequence $A_1; \ldots; A_m$. If this is the case, E after $A_1; \ldots; A_m$ is a form of conditional knowledge: if the actions $A_1; \ldots; A_m$ *were* executed then E would be true. On the other hand, E holds at t refers to the real status of the world at a certain time point.

2.2   Classical Semantics

In this section we present the semantics of the language, according to the hypothesis that all the actions that are executed are known.

A state is a set of fluent names. A fluent literal without negation F is true in the state σ if $F \in \sigma$, false otherwise. A fluent expression $\neg E$ is true in σ if and only if E is false in σ. A fluent expression $E_1 \wedge E_2$ is true in σ if both E_1 and E_2 are true in σ. A fluent expression $E_1 \vee E_2$ is true in σ if either E_1 is true in σ or E_2 is true in σ.

A transition function Φ is a function from the set of pairs (A, σ), where A is an action and σ a state, to the set of states. With $\Phi(A, \sigma)$ we want to represent the state obtained performing the action A in the state σ. We abbreviate $\Phi(A_m, \Phi(A_{m-1}, \ldots, \Phi(A_1, \sigma) \ldots))$ as $\Phi(A_1; \ldots; A_m, \sigma)$. This is the state obtained after executing the sequence of actions $A_1; \ldots; A_m$ in σ.

Let $V_{D_B}^+(A, \sigma)$ be the set of the fluent names F (i.e. positive fluent literals) such that there exists an effect proposition A causes F if $P_1, \ldots, P_m$ in the behavioral part of the domain description D and $P_1, \ldots, P_m$ are true in σ. Intuitively, $V_{D_B}^+(A, \sigma)$ represents the set of fluents whose value must became true when the action A is performed in the state σ.

In a similar manner, $V_{D_B}^-(A, \sigma)$ is the set of fluents whose value must became false, and thus is defined as the set of fluent names F such that there exists an effect proposition A causes $\neg F$ if $P_1, \ldots, P_m$ in D_B and $P_1, \ldots, P_m$ are true in σ.

The transition function associated to a behavioral part D_B is the (partial) function Ψ_{D_B} defined as

$$\Psi_{D_B}(A,\sigma) = \begin{cases} (\sigma \cup V_{D_B}^+(A,\sigma)) \backslash V_{D_B}^-(A,\sigma) & \text{if } V_{D_B}^+(A,\sigma) \cap V_{D_B}^-(A,\sigma) = \emptyset \\ \text{undefined} & \text{otherwise} \end{cases}$$

We used the subscript D_B here to stress the fact that the transition function of a domain description is determined by its behavioral part only. We assume that the transition function associated to a domain description D is always total. This can be verified in polynomial time.

The sequence of actions associated to a time point t is defined as the sequence of actions $B_1; \ldots; B_k$ that have been happened before t. Formally, given a set of happens propositions H, we define

$$S(H,t) = B_1; \ldots; B_k \text{ such that}$$

1. $\{B_1 \text{ happens at } t_1, \ldots, B_k \text{ happens at } t_k\} \subseteq H$
2. $0 \le t_1 < t_2 < \cdots < t_k < t$
3. there is no other proposition C happens at t'
   in H such that $0 \le t' < t$

$S(H,t)$ is the sequence of actions that have took place in the time interval between the time points 0 and t.

We define interpreted structures and models as follows.

Definition 1. *An interpreted structure is a 3-tuple* $M = (\sigma_0, \Phi, H)$*, where* σ_0 *is a state,* Φ *is a transition function, and* H *is a set of* happens at *propositions.*

Definition 2. *An interpreted structure* $M = (\sigma_0, \Phi, H)$ *is a model of a domain description* $D = D_B \cup D_H \cup D_A$ *(written* $M \models D$*) if and only if*

1. $\Phi = \Psi_{D_B}$
2. $H = D_H$
3. *for each pair of actions A_1 and A_2, and each time point t, it does not hold A_1 happens at $t \in H$ and A_2 happens at $t \in H$ (non-concurrency).*
4. *for each proposition E after $A_1; \ldots; A_m$ from t in D_A, the fluent expression E is true in the state $\Phi(S(H,t); A_1; \ldots; A_m, \sigma_0)$.*

A domain description is consistent if it has models. A domain description entails a proposition E after $A_1; \ldots; A_m$ from t if and only if, for each $M = (\sigma_0, \Phi, H)$ such that $M \models D$, the fluent expression E is true in the state $\Phi(S(H,t); A_1; \ldots; A_m, \sigma_0)$. If this is the case, we write $D \models E$ after $A_1; \ldots; A_m$ from t.

2.3   Update Semantics

The semantics of the previous section does not take into account actions that happened, but of which we have no knowledge. For example, the domain description

$$D = \{A \text{ causes } F, \ \neg F \text{ holds at } 0, \ F \text{ holds at } 1\}$$

is not consistent. This is because the value of a fluent remains unchanged if there is no action modifying it. Since there is no happens proposition specifying that an action happened in the time point 0, the value of the fluent F at 1 should be the same of that at 0. Instead, the truth value of the fluent is changed.

Intuitively, it is clear that the action A happens at 0, and this causes the fluent F to become true. However, such an inference is not allowed in the semantics of the previous section, which assumes that the only actions that have been happened are those specified in the domain description.

In this section we present a semantics that allows the inference of statements about actions which are not known to be happened. First of all, we define a model with abduced actions as follows.

Definition 3. *An interpreted structure $M = (\sigma_0, \Phi, H)$ is a model with abduced actions for the domain description $D = D_B \cup D_H \cup D_A$ (written $M \models_A D$) if and only if:*

1. *$\Phi = \Psi_{D_B}$*
2. *$D_H \subseteq H$*
3. *for each pair of different actions A_1 and A_2, and each time point t, it does not hold A_1 happens at $t \in H$ and A_2 happens at $t \in H$ (non-concurrency).*
4. *for each proposition E after $A_1; \ldots; A_m$ from t in D_A, the fluent expression E is true in the state $\Phi(S(H, t); A_1; \ldots; A_m, \sigma_0)$.*

The only difference between this definition and the one given in the previous section is the fact that H can be a superset of D_H, rather than D_H itself. Of course, this way arbitrarily large sets of happens propositions are allowed to be part of H. To this extent, a definition of minimality is needed. We assume that there is an ordering $\preceq$ between interpreted structures.

Definition 4. *A minimal model M of a domain description D is a minimal (w.r.t. $\preceq$) model with abduced actions of D.*

Thus, "minimal model" is indeed a shorthand. We define a domain description D to be consistent if it has at least one minimal model. A domain description D entails a proposition E after $A_1; \ldots; A_m$ from t if and only if, for each minimal model M of D, the fluent expression E is true in the state $\Phi(S(H, t); A_1; \ldots; A_m, \sigma_0)$. If this is the case, we write:

$$D \models_A E \text{ after } A_1; \ldots; A_m \text{ from } t$$

The last point to be defined is the ordering $\preceq$. The choice of $\preceq$ depends on the knowledge about the domain. A general principle is that a model with less happens statements should be preferred (i.e. should be lower than, according to $\preceq$) over models with more happens statements. This leads to the following definition.

Definition 5. *The standard ordering $\preceq_S$ is defined as:*

$$(\sigma_0, \Psi_0, H_0) \preceq_S (\sigma_1, \Psi_1, H_1) \quad \text{iff} \quad \begin{cases} \sigma_0 = \sigma_1 \\ \Psi_0 = \Psi_1 \\ H_0 \subseteq H_1 \end{cases}$$

Using this specific ordering, $\mathcal{AU}$ can be seen as a fragment of the logic $\mathcal{L}$ by Baral et al. [1]. What makes $\mathcal{AU}$ interesting is the fact that, using different orderings, it allows for expressing different update semantics, thus characterizing a number of natural processes of abducting execution of actions.

The entailment relation $\models_A$ obtained from the standard ordering $\preceq_S$ can be used to express the scenario of the example described in the introduction. Indeed, one can prove that the domain description entails for example $\neg Loaded$ holds at 2, which is intuitively the only possible reason of why Fred is still alive. Note that it is also possible to formalize the similar scenario in which we *know* that nothing happens between time 0 and 2: just add an action *Nop*, without effects, and two happens propositions *Nop* **happens at** 0 and *Nop* **happens at** 1 to the domain description. This new domain description is inconsistent: in this case, this is the intuitive outcome.

3   Belief Update Using $\mathcal{AU}$

In this section we show how several definitions of belief update can be formalized in a domain of actions using the language $\mathcal{AU}$. The motivation for doing so is twofold. The first is that this formalization allows for a new interpretation of the definitions of update. For example, Winslett's update can be expressed by introducing an action that change the value of a variable, and minimizing the set of actions happened.

Moreover, by giving definitions of the ordering $\preceq$, we solve the problem of not complete specification of the entailment relation $\models_A$. Indeed, the ordering defined could be used for domain descriptions different from those given from the formalization of update.

We consider the following update definitions: Winslett's update [14], Katsuno and Mendelzon's updates [7], and Boutilier's abduction-based update [2]. We do not consider Boutilier's event based update [3] due to the lack of space, but this update can be expressed in the formalism.

We use the following notations: if P is a propositional formula, then $Mod(P)$ is the set of its models. Conversely, if A is a set of models, then $Form(A)$ is a propositional formula whose set of models is A. Thus, $Form$ is a multi-valued function, since there are many formulas sharing the same set of models. This is not a problem in this work.

3.1   Winslett's Update

Consider a propositional formula T representing the state of the world. This information is assumed to be correct, but not (necessarily) complete. When a change in the world occurs, this description of the world must be modified. The assumption behind belief update is that what we know about the change is a propositional formula P that is true in the new situation. Winslett's approach is model-based, that is, the result of the update $T *_W P$ is defined in terms of the sets of models of T and P.

The underlying assumption in belief revision and update is that of minimal change: the knowledge base T should be changed as little as possible, in the process of incorporation of the update P.

Winslett's update [14] operates on a model by model base. Let I be an interpretation, and let $\leq_I$ be the ordering on interpretations defined as

$$J \leq_I Z \text{ iff } Diff(I, J) \subseteq Diff(I, Z)$$

where $Diff(I, J)$ is the set of variable on which I and J disagree. Intuitively, $J \leq_I Z$ means that, since J and I have more literals assigned to the same truth value than Z and I, the interpretation J must be considered to be closer to I that Z.

The update of the k.b. T when a new formula P becomes true after a change is defined considering each model of T separately.

$$Mod(T *_W P) = \bigcup_{I \in Mod(T)} \min(Mod(P), \leq_I))$$

We show that Winslett's update can be easily expressed in our framework. Let X be the alphabet of T and P. We define a domain description as follows. The set of fluents is the set of variables X. The intuitive explanation is: the set of fluent is the set of facts that may change over time, and this is also the meaning of the fluents in reasoning about actions. For each variable x_i there is an actions A_i. This action formalizes the change of value of the variable x_i between time points.

The domain description is built as follows. For each variable x_i there are two effect propositions:

$$D_B = \bigcup_{x_i \in X} \{A_i \text{ causes } x_i \text{ if } \neg x_i, \ A_i \text{ causes } \neg x_i \text{ if } x_i\}$$

The historical part of the domain is empty: $D_H = \emptyset$. Let $n = |X|$, that is, the number of variables. The actual part of the domain description is composed of two propositions:

$$D_A = \{T \text{ holds at } 0, \ P \text{ holds at } n\}$$

Thus, $D = D_B \cup D_A$. This formalization is a very intuitive one: the fluents are facts, and each action changes the value of a fact. This definition captures Winslett's semantics of update.

Theorem 1. *Let D be the domain description corresponding to T and P. Then, for each propositional formula Q over the alphabet X, it holds $T *_W P \models Q$ if and only if $D \models_A Q$ holds at n (using the standard ordering $\preceq_S$).*

P is assumed to hold at time n because we do not allow concurrent action.

3.2 Katsuno and Mendelzon's Update

Katsuno and Mendelzon [7] defined a family of updates, rather than a specific operator. They also proved that Winslett's operator is a sub-case of their definition.

Let $O = \{\leq_I \mid I$ is an interpretation$\}$ be a family of partial orderings over the set of the interpretations, one for each interpretation I. In other words, for each interpretation I there is a partial ordering $\leq_I$ over the set of the interpretations. An interpretation I represents a complete description of the world. $J \leq_I Z$ means that the situation represented by the interpretation J is considered more plausible than the situation of Z. As a result, assuming that there has been a transition from I to J requires less change than the change from I to Z. Thus, assuming that I represents the current state, the result of the update should be:

$$Mod(Form(I) *_{KM} P) = \min(Mod(P), \leq_I)$$

If the current k.b. is not composed of a single interpretation, this must be done for each $I \in Mod(T)$:

$$Mod(T *_{KM} P) = \bigcup_{I \in Mod(T)} \min(Mod(P), \leq_I)$$

Note that Katsuno and Mendelzon define a set of update operators rather than a single one: indeed, each family of orderings define a specific KM operator. As a result, in order to specify an actual update, a family of orderings must be defined.

There is a simple way to capture and Katsuno and Mendelzon's update in our framework. Given a family of orderings (one for each interpretation) we define the domain description as the one given in the previous section. The ordering used is defined as follows.

Definition 6. *Given a family of partial ordering $O = \{\leq_I\}$, one for each interpretation I, we define an ordering over interpreted structures $\preceq_{KM}$ as $(\sigma_0, \Phi_0, H_0) \preceq_{KM} (\sigma_1, \Phi_1, H_1)$ if and only if*

1. $\sigma_0 = \sigma_1$.
2. $\Phi_0 = \Phi_1$.
3. $\Phi_0(S(H_0, n), \sigma_0) \leq_{\sigma_0} \Phi_1(S(H_1, n), \sigma_1)$.

Note that there is an ordering $\preceq_{KM}$ for each family of orderings over the interpretations. Thus, the formally correct notation should be $\preceq_O$, but we use $\preceq_{KM}$ for simplicity. The following theorem shows that we are indeed formalizing the Katsuno and Mendelzon updates.

Theorem 2. *For each Katsuno and Mendelzon update, and for each 3-tuple of propositional formulas T, P, and Q, it holds $T * P \models Q$ if and only if $D \models_A Q$ holds at n, using the ordering $\preceq_{KM}$ as in Definition 6.*

3.3   Abduction-Based Update

The rationale of the abduction-based update [3] is that the events that change the world can be modeled by an abductive semantics. Some of these events may be more plausible than others. In order to explain the change, we choose only the ones we consider to be more plausible.

Since this update requires the specification of the outcome of events, and their plausibility, the current knowledge base T and the update P do not suffice to evaluate the updated k.b.. This kind of updates, in which some extra information is required is called *update schema*. It can be viewed as a family of updates, one of each set of events and their plausibility. Giving the events and their plausibility is equivalent to selecting a specific update of the family.

We now give the formal definition of the update. A more detailed explanation can be found in the paper where this update is introduced [3]. In order to explain the changes, we have a set of *events* E. Each event e is a function from interpretations to sets of interpretations. Thus, for each interpretation I, $e(I)$ is a set of interpretations. The meaning of $J \in e(I)$ is that the possible world represented by the interpretation J is one of the possible outcomes of the event e, if this event occur in the world represented by the interpretation I. An event e is said to be deterministic if $e(I)$ is always composed of a single interpretation.

As seen in the informal explanation above, not all the events are considered equally plausible. To represent the relative plausibility of events we have a family of preorders $O = \{\leq_I \mid I \in \mathcal{M}\}$, one for each interpretation I. When $e \leq_I s$ the event e is considered more likely to happen that s, in the world represented by the interpretation I. We denote by $e <_I s$ the fact that e is *strictly* more likely than s; formally, that $e \leq_I s$ but not $s \leq_I e$.

Let T be the current k.b. and P the update. The set of explanations of P is the set of events whose occurrence can explain the fact that P is now true. There are two possible definitions.

Definition 7. *The set of weak explanations of P is*

$$\mathrm{Expl}(I, P) = \min(\{e \mid e(I) \cap Mod(P) \neq \emptyset\}, \leq_I)$$

The set of predictive explanations of P is

$$\mathrm{Expl}_p(I, P) = \min(\{e \mid e(I) \subseteq Mod(P)\}, \leq_I)$$

The outcome of the update is defined in terms of the *progression* of a possible world I.

Definition 8. *The progression of an interpretation I is the set*

$$\mathrm{Prog}(I, P) = \bigcup \{e(I) \cap Mod(P) \mid e \in \mathrm{Expl}(I, P)\}$$

The progression of an interpretation can be defined also for predictive explanations. The updated k.b. is defined as the union of all the progressions.

Definition 9. *The result of updating T with P is*[1]

$$Mod(T *_{ABD} P) = \bigcup \{\operatorname{Prog}(I, P) \mid I \in Mod(T)\}$$

In this definition we assume the use of the weak explanations. A similar definition can be given using predictive explanations instead.

We show how the abduction based update can be expressed in $\mathcal{AU}$. We assume that all the events are deterministic. This is a natural assumption, since the actions of the language $\mathcal{AU}$ are always deterministic. Under the assumption of determinism, weak and predictive explanations are the same.

Let $E = \{e_1, \ldots, e_m\}$ be the set of events. The corresponding action theory has m actions $A_1, \ldots, A_m$. The behavioral part of the domain is determined by the events in the following manner. For each event e_j and interpretation I, if x_i is true in $e_j(I)$ we have the effect proposition

$$A_j \text{ causes } x_i \text{ if } \left(\bigwedge_{x_k \in I} x_k \wedge \bigwedge_{x_k \notin I} \neg x_k \right)$$

otherwise the effect proposition to add is

$$A_j \text{ causes } \neg x_i \text{ if } \left(\bigwedge_{x_k \in I} x_k \wedge \bigwedge_{x_k \notin I} \neg x_k \right)$$

The behavioral part D_B of the domain description is the union of all these effect propositions, for each event e, interpretation I and atom x_i.

The actual part is composed by two propositions only:

$$D_A = \{T \text{ holds at } 0, \ P \text{ holds at } 1\}$$

The historical part of the domain description is empty: $D_H = \emptyset$. The ordering $\preceq_A$ is defined as follows.

Definition 10. *The ordering $\preceq_A$ is defined as: $(\sigma_0, \Phi_0, H_0) \preceq (\sigma_1, \Phi_1, H_1)$ if and only if*

1. $\sigma_0 = \sigma_1$
2. $\Phi_0 = \Phi_1$
3. *it holds e_0* happens at $0 \in H_0$, e_1 happens at $0 \in H_1$, *and $e_0 \leq_{\sigma_0} e_1$.*

About the correctness of this definition, the following theorem relates the entailment in $\mathcal{AU}$ and the inference of $*_{ABD}$.

Theorem 3. *For each 3-tuple of propositional formulas T, P, and Q, it holds $T *_{ABD} P \models Q$ if and only if $D \models_A Q$ holds at 1 (using the ordering $\preceq_A$), where D is the domain description defined above.*

---

[1] In the original Boutilier's definition, the update is inconsistent if there is an $I \in Mod(T)$ such that $\operatorname{Prog}(I, P)$ is empty. For simplicity, we do not consider this case.

4 Related Work

In this section we compare our approach with others that use the similarities between reasoning about actions and update. The approach that is most similar to ours is the Possible Causes Approach (PCA) proposed by Li and Pereira [8]. Although it is based on similar principles, is different from our proposal in two aspects. First of all, update is not embedded into the temporal logic. Rather, given a domain description and an update, the aim of PCA is to consistently incorporate the update in the domain. In our semantics, the updating formula is expressed as a proposition of the domain description.

Another difference regards the KM postulates. Li and Pereira's approach does not obey the KM principle that models of the initial knowledge base must be updated separately. This implies, for example, that Winslett's update cannot be easily expressed into Li and Pereira's formalism.

The KM postulates, as our framework, provide a generalization of Winslett's approach to update. Due to the lack of space, we cannot make a detailed comparison between these two frameworks. Let us only say that, while KM postulates only generalizes Winslett's semantics, our approach is more general, as other update methods can be encoded in it.

Another approach which is somewhat related to ours is due to Peppas [10], which shows how epistemic entrenchment (a well-known notion in belief revision) can be used in the update framework as well.

The relationship between belief update and reasoning about actions have been also analyzed by del Val and Shoham. The key idea of their work can be summarized by the following quotation [4].

> The initial database is taken to describe a particular situation, and the update formula is taken to describe the effect of a particular action. A formal theory of action is then used to infer facts about the result of taking the particular action in the particular situation [...]. Finally, anything inferred about the resulting situation can be translated back to the timeless framework of belief update.

Their framework is used to derive a semantics for belief update. In order to do this, they translate a specific initial base and an update into a specific theory of actions. A single update is translated into a single action. From this point of view, our framework is exactly the opposite: we derive a semantics of a possibly inconsistent theory of actions by employing the idea of update. An update is indeed a fact that holds in some time point, and changes are caused by actions.

Winslett's update, as it was initially defined [13], was used in a similar way: the initial knowledge base is the state of the world at a certain time point, and the update is the effect of a complex action. The result of Winslett's update is used to determine the state of the world after that the action is performed. This way the frame problem is solved, if the effect of the action is a conjunction of literals.

In this context, an action is formalized by an update: as have shown, updates can be in turns formalized as the result of a number of simpler actions. Following this approach, and using Winslett's update, a complex action is considered to be equivalent to a set of elementary actions, each changing the truth value of a variable. In the case of non-disjunctive actions, having yet another solution from the frame problem is not really intersting at this point, as many other solutions already exist [11, 12]. The point of view offered by this approach may be of interest in the case of disjunctive actions.

5 Discussion

In this paper we have introduced the language $\mathcal{AU}$, that formalizes scenarios in which actions may take place, and of which the agent has no knowledge. The language $\mathcal{AU}$ is essentially a dialect of the language $\mathcal{A}$ for reasoning about actions [6] with narratives.

This formalism is also useful for the field of belief update. Indeed, the definitions given by Boutilier, Katsuno and Mendelzon, and Winslett can be easily encoded in $\mathcal{AU}$. This provides a way for comparing the semantics of these formalisms. For example, Winslett's update can be expressed in $\mathcal{AU}$ by assuming that the change that caused the updating formula to hold in a successive state is due to the effect of a sequence of simple actions, each causing the truth value of a variable to change. The actions we used to formalize Boutilier's abduction-based update are more complicated (i.e. involving more that one variable).

Regarding Boutilier's update, we also note that the translation given here is exponential-size. This can be explained by observing that Boutilier's events may be arbitrarily involved. In real scenarios, there should be a simple rule to determine the effect of events.

The language $\mathcal{AU}$ allows the integration of many features that are recognized by many researchers as fundamental in expressive theories of belief update.

1. It is possible to express which changes may take place (for example, the fact that Fred cannot become alive, once he is dead is formalized by the absence of actions that makes Fred alive, if he is dead).
2. In some situation, the observation at time 1 leads to modify our knowledge about time 0. This can be expressed in $\mathcal{AU}$.
3. It is possible to express multiple observations at different time points (iterated belief update).

An interesting feature of $\mathcal{AU}$ is that it allows inference of happens statements: a domain description D implies an happens proposition A happens at t if and only if the A happens at t is contained in all the models of D. This issue is of course trivial in classical action description languages, in which an happens proposition is implied by a domain description if and only if it is in the domain. In $\mathcal{AU}$ (with the update semantics) it is possible to infer that an action took place at time t if A happens at t is in all the models of the domain description.

So far for the benefits of this beautiful language $\mathcal{AU}$. Let now turn our attention to the possible extensions. A first open problem of this paper is a translation from domain description into abductive logic programs (or circumscription).

From a semantical point of view, $\mathcal{AU}$ itself can be extended in many ways: non-deterministic actions, concurrent actions, and the integration of revision and update.

Consider an extension of $\mathcal{AU}$ with non-deterministic actions. A first application is the incorporation of abduction-based update with non deterministic events in our formalism.

Another benefit regards the treatment of disjunctive information. This is a well-known benchmark problem: the initial knowledge base is $T = x \wedge \neg y$, and the update is $P = x \oplus y$. In such cases, the result of Winslett's update (as well as any other KM update) is $T * P = x \wedge \neg y$. This is sometimes correct, but there are scenarios in which this result is intuitively wrong. Let for example x be "the coin is on the head", and y be "the coin is on the tail". According to T, the head is currently on the head. When we toss the coin, the knowledge base is updated with $P = x \oplus y$, that is, what we know is that either the tail is on the head or it is on the tail. The result of updating T with P should be $T * P = x \oplus y$.

The addition of non-deterministic actions in our framework allows for solving such problems. Indeed, what is needed is a non-deterministic action A causes $x \oplus y$. Note that this is very different from the standard update $x \oplus y$ happens at n (this second scenario gives $x \wedge \neg y$ as the result of the update). In this formalism it is possible to provide enough information to decide whether we are in a situation when Winslett's treatment of disjunctive information is correct, and when it is not. This second case is essentially due to the existence of actions whose effect is the considered disjunction.

This use of non-deterministic actions is similar to that of del Val and Shoham [4]. However, in their formalism there is no way to distinguish scenarios in which the result must be equal to that of Winslett's update, and when it must be different. Indeed, there are scenario in which the result of Winslett's update is correct (i.e. the result of updating $T = x \wedge \neg y$ with $P = x \oplus y$ must be $T * P = x \wedge \neg y$) and others in which it is not. Del Val and Shoham's semantics does not give any hint on how to make a choice, which is left to the user. On the converse, in $\mathcal{AU}$ with non-deterministic actions the choice is simply determined by the actions that may happen and their effects. Del Val and Shoham's semantics maps both actions and updates into actions, and this leads to a loss of information.

A second possible extension is the addition of concurrent actions. Consider the formalization of Winslett's update in our framework. There is an action for each variable of the alphabet. This is reasonable, since the assumption is that the variables can change their value arbitrarily. What is not so intuitive is the fact that the observation P is formalized as the value proposition P holds at n. Since there is only a knowledge base about the initial time point T, and the observation P, there is no intuitive reason of the fact that P holds at 1 does not work as well. The technical reason is that the assumption of Winslett's update is that all the changes may happen simultaneously or, still better, between two time points it is always possible to perform an arbitrary number of changes. This can be expressed in our formalism by introducing concurrent actions.

Finally, the principles of $\mathcal{AU}$ can be used to extend the system BReLS [9] to deal with complex actions. BReLS has been introduced to deal with domains in which both revision and update are necessary. The semantics of BReLS are based on the principle of combining a measure of reliability of sources of information with the likeliness of events. The way in which events are formalized is so far quite simple: the only possible actions are those setting the value of a variable to a given value (true or false). The user can decide the likeliness of such actions, but cannot define more complex actions. Syntactically, this is done with a statement like change(i) : l, which means that the penalty (degree of unlikeliness) of the literal l becoming true is i. Extending the syntax is quite straightforward: change(i) : A means that the penalty of the action A to take place is i. The extension of the semantics is also quite easy: a model is composed by a set of static models (propositional interpretations), one for each time point, and a set of actions for any pair of consecutive time points. This model is consistent with the domain description if and only if the static model at time $t + 1$ is the result of applying the actions relative to the pair $\langle t, t + 1 \rangle$ to the static model of time t. The ordering between models can also be obtained by combining the degree of reliability of sources with the penalty associated to changes, as usual. Extending the implemented algorithms, on the other hand, seems to be not as simple.

References

[1] C. Baral, M. Gelfond, and A. Provetti. Representing actions: laws, observations, and hypothesis. *J. of Logic Programming*, 31, 1997.

[2] C. Boutilier. Generalized update: belief change in dynamic settings. In *Proc. of IJCAI'95*, pages 1550–1556, 1995.

[3] C. Boutilier. Abduction to plausible causes: an event-based model of belief update. *Artificial Intelligence*, 83:143–166, 1996.

[4] A. del Val and Y. Shoham. Deriving properties of belief update from theories of action. *J. of Logic, Language and Information*, 3:81–119, 1994.

[5] N. Friedman and J. Y. Halpern. A knowledge-based framework for belief change, part II: Revision and update. In *Proc. of KR'94*, pages 190–200, 1994.

[6] M. Gelfond and V. Lifschitz. Representing action and change by logic programs. *J. of Logic Programming*, 17:301–322, 1993.

[7] H. Katsuno and A. O. Mendelzon. Propositional knowledge base revision and minimal change. *Artificial Intelligence*, 52:263–294, 1991.

[8] R. Li and L. Pereira. What is believed is what is explained. In *Proc. of AAAI'96*, pages 550–555, 1996.

[9] P. Liberatore and M. Schaerf. BReLS: A system for the integration of knowledge bases. In *Proc. of KR 2000*, pages 145–152, 2000.

[10] P. Peppas. PMA epistemic entrenchment: The general case. In *Proc. of ECAI'96*, pages 85–89, 1996.

[11] E. Sandewall. *Features and Fluents*. Oxford University Press, 1994.

[12] M. Thielscher. Introduction to the fluent calculus. *ETAI*, 3, 1998.

[13] M. Winslett. Reasoning about actions using a possible models approach. In *Proc. of AAAI'88*, pages 89–93, 1988.

[14] M. Winslett. *Updating Logical Databases*. Cambridge University Press, 1990.

A Compilation of Brewka and Eiter's Approach to Prioritization

James P. Delgrande[1], Torsten Schaub[2*], and Hans Tompits[3]

[1] School of Computing Science, Simon Fraser University,
Burnaby, B.C., Canada V5A 1S6,
jim@cs.sfu.ca
[2] Institut für Informatik, Universität Potsdam,
Postfach 601553, D–14415 Potsdam, Germany,
torsten@cs.uni-potsdam.de
[3] Institut für Informationssysteme 184/3, Technische Universität Wien,
Favoritenstraße 9–11, A–1040 Wien, Austria,
tompits@kr.tuwien.ac.at

Abstract. In previous work, we developed a framework for expressing general preference information in default logic and logic programming. Here we show that the approach of Brewka and Eiter can be captured within this framework. Hence, the present results demonstrate that our framework is general enough to capture other independently-developed methodologies. As well, since the extended logic program framework has been implemented, we provide an implementation of the Brewka and Eiter approach via an encoding of their approach.

1 Introduction

In previous work [6], we presented a general framework based on default logic for expressing general preference information. There, we addressed the problem of representing preferences among individual and aggregated properties in default logic. In this approach, one begins with an ordered default theory, in which preferences are specified on default rules. This is transformed into a second, standard, default theory in which the preferences are respected, in the sense that the obtained default extensions contain just those conclusions that accord with the order expressed by the original preference information. The approach is fully general: One may specify preferences that hold by default, or give preferences among preferences, or give preferences among sets of defaults.

We adapted this approach in [8] for logic programming under the answer set semantics [11]. While the original approach is usable for full-fledged theorem provers for default logic, like DeReS [5], this subsequent approach applies to logic programming systems, such as dlv [10] or smodels [14]. In fact, we have provided an implementation of the approach in extended logic programs, serving as a front-end for dlv and smodels (see [9] for details).

In the context of default logic, our methodology involves the appropriate "decomposition" of default rules, so that one can detect the applicability conditions of default

---

* Affiliated with the School of Computing Science at Simon Fraser University, Canada.

M. Ojeda-Aciego et al. (Eds.): JELIA 2000, LNAI 1919, pp. 376–390, 2000.

rules and control their actual application. In our framework, this is carried out *within* a default theory. This is accomplished, first, by associating a unique name with each default rule, so that it can be referred to within a theory. Second, special-purpose predicates are introduced for detecting conditions in a default rule, and for controlling rule invocation. This in turn allows a fine-grained control over what default rules are applied and in what cases. By means of these named rules and special-purpose predicates, one can formalise various phenomena of interest.

Given an ordered default theory $(D, W, <)$, where $<$ is a strict partial order on D, the intuition is that one applies the $<$-maximal default(s), if possible, then the next $<$-greatest, and so on. Thus we adopt a *prescriptive* interpretation of the ordering, in that $<$ prescribes the order in which rules are applied. This can be contrasted with a *descriptive* interpretation, in which the preference order represents a ranking on desired outcomes: the desirable (or: preferred) situation is one where the most preferred default(s) are applied.

The approach of Brewka and Eiter [3], first developed with respect to extended logic programs and subsequently generalized for default logic in [4], arguably fits the "descriptive" interpretation. In common with previous work, Brewka and Eiter begin with a partial order on a rule base, but define preference with respect to total orders that conform to the original partial order. As well, answer sets or extensions, respectively, are first generated and the "prioritized" answer sets (extensions) are selected subsequently. In contrast, in our approach, we deal only with the original partial order, which is translated into the object theory. As well, only "preferred" extensions are produced in our approach; there is no need for meta-level filtering of extensions.

However, we show here that the approach of Brewka and Eiter is expressible in our framework. Consequently, this serves to show the scope and generality of our framework. As well, this result enables a straightforward implementation of the Brewka and Eiter approach.

In the next subsection we briefly introduce default logic, while Sections 3 and 4 introduce our approach and Brewka and Eiter's, respectively. Section 5 describes the translation of their approach expressed in default logic, while Section 6 does the same for the case of extended logic programs. Section 7 gives brief concluding remarks.

2   Background

Default logic [16] augments classical logic by *default rules* of the form

$$\frac{\alpha \; : \; \beta_1, \ldots, \beta_n}{\gamma}$$

where $\alpha, \beta_1, \ldots, \beta_n, \gamma$ are sentences of first-order or propositional logic. Here we mainly deal with *singular* defaults for which $n = 1$. A singular rule is *normal* if β is equivalent to γ; it is *semi-normal* if β implies γ. [12] shows that any default rule can be transformed into a set of semi-normal defaults. We sometimes denote the *prerequisite* α of a default δ by $Prereq(\delta)$, its *justification* β by $Justif(\delta)$, and its *consequent* γ by $Conseq(\delta)$. Accordingly, $Prereq(D)$ is the set of prerequisites of all default rules in D; $Justif(D)$ and $Conseq(D)$ are defined analogously. Empty components, such as no prerequisite or even no justifications, are assumed to be tautological (we speak in such cases of *prerequisite-free* and *justification-free* defaults, respectively). *Open defaults* with unbound variables are taken to stand for all corresponding instances. A set of default rules D and a set of sentences W form a *default theory* (D, W) that may induce a single, multiple, or even zero *extensions* in the following way:

Definition 1. *Let (D, W) be a default theory and let E be a set of sentences. Define $E_0 = W$ and for $i \geq 0$:*

$$GD_i = \left\{ \frac{\alpha : \beta_1, \ldots, \beta_n}{\gamma} \in D \;\middle|\; \alpha \in E_i, \neg\beta_1 \notin E, \ldots, \neg\beta_n \notin E \right\};$$
$$E_{i+1} = Th(E_i) \cup \{Conseq(\delta) \mid \delta \in GD_i\}.$$

Then, E is an extension for (D, W) iff $E = \bigcup_{i=0}^{\infty} E_i$.

($Th(E)$ refers to the logical closure of set E of sentences.) Any such extension represents a possible set of beliefs about the world at hand. The above procedure is not constructive since E appears in the specification of GD_i. We define $GD(D, E) = \bigcup_{i=0}^{\infty} GD_i$ as the set of *generating defaults* of extension E. An enumeration $\langle \delta_i \rangle_{i \in I}$ of default rules is *grounded* in a set of sentences W, if we have for every $i \in I$ that $W \cup Conseq(\{\delta_0, \ldots, \delta_{i-1}\}) \vdash Prereq(\delta_i)$.

For simplicity, we restrict our attention in what follows to finite, singular default theories, consisting of finite sets of default rules and sentences.

3   Preference-Handling in Standard Default Logic

For adding preferences among default rules, a default theory is usually extended with an ordering on the set of default rules. In accord with [4], we define:

Definition 2. *A prioritized default theory is a triple $(D, W, <)$ where (D, W) is a default theory and $<$ is a strict partial order on D.*

In contrast to [4], however, we use the ordering $<$ in the sense of "higher priority", i.e., $\delta < \delta'$ expresses that δ' has "higher priority" than δ.

The methodology of [6] provides a translation, $\mathcal{T}$, that takes such a prioritized theory $(D, W, <)$ and translates it into a regular default theory $\mathcal{T}((D, W, <)) = (D', W')$ such that the explicit preferences in $<$ are "compiled" into D' and W' and such that the extensions of (D', W') correspond to the "preferred" extensions of $(D, W, <)$. Moreover, the approach admits not only "static" preferences as discussed here—where the ordering of the defaults is specified at the meta-level—but also "dynamic" preferences *within the object language*.

In [6], to begin with, a unique name is associated with each default rule. This is done by extending the original language by a set of constants[1] N such that there is a bijective mapping $n : D \rightarrow N$. We write n_δ instead of $n(\delta)$ (and abbreviate n_{δ_i} by n_i to ease notation). Also, for default rule δ with name n, we sometimes write $n : \delta$ to render

---

[1] McCarthy effectively first suggested the naming of defaults using a set of *aspect* functions [13]; Theorist [15] uses atomic propositions to name defaults.

naming explicit. To encode the fact that we deal with a finite set of distinct default rules, we adopt a unique names assumption (UNA_N) and domain closure assumption (DCA_N) with respect to N. That is, for a name set $N = \{n_1, \ldots, n_m\}$, we add axioms

$$UNA_N : \quad (n_i \neq n_j) \text{ for all } n_i, n_j \in N \text{ with } i \neq j;$$
$$DCA_N : \quad \forall x.\ name(x) \equiv (x = n_1 \vee \cdots \vee x = n_m).$$

For convenience, we write $\forall x \in N.\ P(x)$ instead of $\forall x.\ name(x) \supset P(x)$.

Given $\delta_i < \delta_j$, we want to ensure that, before δ_i is applied, δ_j can be applied or found to be inapplicable.

More formally, we wish to exclude the case where $\delta_i \in GD_n$ but $\delta_j \notin GD_n$ although $\delta_j \in GD_m$ for some $m > n$ in Definition 1. For this purpose, we need to be able to (i) detect when a rule has been applied or when a rule is blocked, and (ii) control the application of a rule based on other antecedent conditions. For a default rule $\frac{\alpha : \beta}{\gamma}$ there are two cases for it to not be applied: it may be that the antecedent is not known to be true (and so its negation is consistent), or it may be that the justification is not consistent (and so its negation is known to be true). For detecting this case, we introduce a new, special-purpose predicate $bl(\cdot)$. Similarly we introduce a predicate $ap(\cdot)$ to detect when a rule has been applied. To control application of a rule we introduce predicate $ok(\cdot)$. Then, a default rule $\delta = \frac{\alpha : \beta}{\gamma}$ is mapped to

$$\frac{\alpha \wedge ok(n_\delta) : \beta}{\gamma \wedge ap(n_\delta)}, \quad \frac{ok(n_\delta) : \neg\alpha}{bl(n_\delta)}, \quad \frac{\neg\beta \wedge ok(n_\delta) :}{bl(n_\delta)}. \tag{1}$$

These rules are sometimes abbreviated by $\delta_a, \delta_{b_1}, \delta_{b_2}$, respectively. While δ_a is more or less the image of the original rule δ, rules δ_{b_1} and δ_{b_2} capture the non-applicability of the rule.

None of the three rules in the translation can be applied unless $ok(n_\delta)$ is true. Since $ok(\cdot)$ is a new predicate symbol, it can be expressly made true in order to potentially enable the application of the three rules in the image of the translation. If $ok(n_\delta)$ is true, the first rule of the translation may potentially be applied. If a rule has been applied, then this is indicated by asserting $ap(n_\delta)$. The last two rules give conditions under which the original rule is inapplicable: either the negation of the original antecedent α is consistent (with the extension) or the justification β is known to be false; in either such case $bl(n_\delta)$ is concluded.

We can assert that default $n_j : \frac{\alpha_j : \beta_j}{\gamma_j}$ is preferred to $n_i : \frac{\alpha_i : \beta_i}{\gamma_i}$ in the object language by introducing a new predicate, $\prec$, and then asserting that $n_i \prec n_j$. However, this translation so far does nothing to control the order of rule application. Nonetheless, for $\delta_i < \delta_j$ we can now control the order of rule application: we can assert that if δ_j has been applied (and so $ap(n_j)$ is true), or known to be inapplicable (and so $bl(n_j)$ is true), then it is ok to apply δ_i. The idea is thus to *delay* the consideration of less preferred rules until the applicability question has been settled for the higher ranked rules. Formally, this is realized by adding the axiom

$$\forall x \in N.\ \big[\forall y \in N.\ (x \prec y) \supset (bl(y) \vee ap(y))\big] \supset ok(x) \tag{2}$$

to the translation.

To summarize, let $T((D, W, <)) = (\tilde{D}, \tilde{W})$ be the translation obtained in this way, for a given prioritized default theory $(D, W, <)$. Then, the prioritized extensions of $(D, W, <)$ are determined by the (regular) extensions of $(\tilde{D}, \tilde{W})$, modulo the original language.

It is important to note that this translation schema is just one possible preference strategy. Changes to the conditions when a default is considered to be applicable (realized by the specific form of the decomposed defaults $\delta_a, \delta_{b_1}, \delta_{b_2}$ and axiom (2)) result in different preference strategies. Also, further rules and special-purpose predicates can be added, if needed. For instance, in Sections 5 and 6 we rely on an additional predicate $ko(\cdot)$ that aims at eliminating rules from the reasoning process.

4   Brewka and Eiter's Approach to Preference

We now describe the approach to dealing with a prioritized default theory introduced in [4]. First, partially ordered default theories are reduced to totally ordered ones.[2]

Definition 3. *A fully prioritized default theory is a prioritized default theory $(D, W, <)$ where $<$ is a total ordering.*

The general case of arbitrary prioritized default theories is reduced to this restricted case as follows.

Definition 4. *Let $(D, W, <)$ be a prioritized default theory. Then, E is a prioritized extension of $(D, W, <)$ iff E is a prioritized extension of some fully prioritized default theory $(D, W, <')$ such that $< \subseteq <'$.*

Conclusions of prioritized default theories are defined in terms of prioritized extensions, which are a subset of the regular extensions of a default theory, i.e., the extensions of (D, W) according to [16].

The construction of prioritized extensions relies on the notion of *activeness* [1, 2]. A default δ is *active* in a set of formulas S, if (i) $Prereq(\delta) \in S$, (ii) $\neg Justif(\delta) \notin S$, and (iii) $Conseq(\delta) \notin S$ hold. Intuitively, a default is active in S if it is applicable with respect to S but has not yet been applied.

Definition 5. *Let $\Delta = (D, W, <)$ be a fully prioritized prerequisite-free default theory. The operator C is defined as follows: $C(\Delta) = \bigcup_{i \geq 0} E_i$, where $E_0 = Th(W)$, and for every $i > 0$,*

$$E_i = \begin{cases} \bigcup_{j<i} E_j & \text{if no default from } D \text{ is active in } \bigcup_{j<i} E_j; \\ Th(\bigcup_{j<i} E_j \cup \{Conseq(\delta)\}) & \text{otherwise, where } \delta \in D \text{ is the maximal} \\ & \text{default (w.r.t. } <) \text{ active in } \bigcup_{j<i} E_j. \end{cases}$$

In the case of prerequisite-free, normal default theories, the operator C always produces an extension in the sense of [16] and thus can directly be used to define prioritized extensions:

---

[2] In fact, [4] deal with so-called *well-orderings*, which are generalised total orderings, needed for treating infinite domains.

Definition 6. *Let* $\Delta = (D, W, <)$ *be a fully prioritized prerequisite-free, normal default theory. Then,* E *is the prioritized extension of* Δ *iff* $E = C(\Delta)$.

The next definition addresses the more general class of prerequisite-free theories:

Definition 7. *Let* $\Delta = (D, W, <)$ *be a fully prioritized prerequisite-free default theory. Then, a set* E *of formulas is a prioritized extension of* Δ *iff* $E = C(\Delta^E)$, *where* $\Delta^E = (D^E, W, <)$ *and* $D^E = D \setminus \{\delta \in D \mid Conseq(\delta) \in E \text{ and } \neg Justif(\delta) \in E\}$.

That is, Δ^E is obtained from Δ by deleting all defaults whose consequents are in E and which are defeated in E. Clearly, this leaves normal rules unaffected. The purpose of this filter is illustrated in [4] by the following default theory:

$$\Delta_3 = (\{\, n_1 : \tfrac{:\neg B}{A}, n_2 : \tfrac{:\neg A}{\neg A}, n_3 : \tfrac{:A}{A}, n_4 : \tfrac{:B}{B} \,\}, \emptyset, \{\delta_j < \delta_i \mid i < j\}). \quad (3)$$

This theory has two regular extensions, $Th(\{A, B\})$ and $Th(\{\neg A, B\})$. Applying operator C to Δ_3 yields the first extension. However, it is argued in [4] that this extension does not preserve priorities because default δ_2 is defeated in E by applying a default which is less preferred than δ_2, namely default δ_3. This extension is ruled out by the filter in Definition 7 because $Th(\{A, B\}) \neq Th(\{\neg A, B\}) = C(\Delta_3^{Th(\{A,B\})})$. Theory Δ_3 has therefore no prioritized extension.

The next definition accounts for the general case by reducing it to the prerequisite-free one. For checking whether a given regular extension E is prioritized, Brewka and Eiter evaluate the prerequisites of the default rules according to the extension E. To this end, for a default δ, define $\delta^\top$ as the prerequisite-free version of δ, i.e., $\delta^\top$ results from δ by replacing $Prereq(\delta)$ by $\top$.

Definition 8. *Let* $\Delta = (D, W, <)$ *be a fully prioritized default theory and* E *a set of formulas. The default theory* $\Delta_E = (D_E, W, <_E)$ *is obtained from* Δ *as follows:*

1. $D_E = \{\delta^\top \mid \delta \in D \text{ and } Prereq(\delta) \in E\}$;
2. *for any* $\zeta_1, \zeta_2 \in D_E$, $\zeta_1 <_E \zeta_2$ *iff* $\delta_1 < \delta_2$ *where* $\delta_i = \max_< \{\delta \in D \mid \delta^\top = \zeta_i\}$.

In other words, D_E is obtained from D by (i) eliminating every default $\delta \in D$ such that $Prereq(\delta) \notin E$, and (ii) replacing $Prereq(\delta)$ by $\top$ in all remaining defaults δ.

Definition 9. *Let* $\Delta = (D, W, <)$ *be a fully prioritized default theory. Then,* E *is a prioritized extension of* Δ, *if* (i) E *is a classical extension of* Δ, *and* (ii) E *is a prioritized extension of* Δ_E.

That is, (ii) is equivalent to $E = C((\Delta_E)^E)$.

For illustration, consider [4, Example 4]:

$$\tfrac{:A}{A} < \tfrac{:\neg B}{\neg B} < \tfrac{A:B}{B}, \quad (4)$$

and where $W = \emptyset$. This theory, Δ, has two regular extensions: $E_1 = Th(\{A, B\})$ and $E_2 = Th(\{A, \neg B\})$. Δ_{E_1} amounts to $\tfrac{:A}{A} < \tfrac{:\neg B}{\neg B} < \tfrac{:B}{B}$. Clearly, $(\Delta_{E_1})^{E_1} = \Delta_{E_1}$. Also, we obtain that $C(\Delta_{E_1}) = E_1$, that is, E_1 is a prioritized extension. In contrast to this, E_2 is not prioritized. While $\Delta_{E_2} = \Delta_{E_1}$ and $(\Delta_{E_2})^{E_2} = \Delta_{E_1}$, we get $C((\Delta_{E_2})^{E_2}) = E_1 \neq E_2$. That is, $C((\Delta_{E_2})^{E_2})$ reproduces E_1 rather than E_2.

This example reveals the difference between the prescriptive methodology of [6] discussed in the previous section, and Brewka and Eiter's descriptive approach discussed here, insofar as the former method actually selects *no prioritized extension*. Intuitively, this can be explained by the observation that for the highest-ranked default $\frac{A:B}{B}$, neither applicability nor blockage can be asserted: Either of these properties relies on the applicability of lesser-ranked defaults, effectively resulting in a circular situation destroying any possible extension. Nonetheless, as we show next, the methodology of [6] is general enough to admit a suitable preference strategy enforcing the simulation of prioritized extensions in the sense of Definition 9.

5   Prioritized Extensions via Standard Default Logic

Given an alphabet $\mathcal{P}$ of some language $\mathcal{L}_\mathcal{P}$, we define a disjoint alphabet $\mathcal{P}'$ as $\mathcal{P}' = \{p' \mid p \in \mathcal{P}\}$ (so implicitly there is an isomorphism between $\mathcal{P}$ and $\mathcal{P}'$). Then, for $\alpha \in \mathcal{L}_\mathcal{P}$, we define $\alpha' \in \mathcal{L}_{\mathcal{P}'}$ as the result of replacing in α each proposition p from $\mathcal{P}$ by the corresponding proposition p' in $\mathcal{P}'$. This is defined analogously for sets of formulas, default rules and sets of default rules. We abbreviate $\mathcal{L}_\mathcal{P}$ and $\mathcal{L}_{\mathcal{P}'}$ by $\mathcal{L}$ and $\mathcal{L}'$, respectively.

We obtain the following translation mapping prioritized default theories in some language $\mathcal{L}$ onto standard default theories in the language $\mathcal{L}^\circ$ obtained by extending $\mathcal{L} \cup \mathcal{L}'$ by new predicates symbols $(\cdot \prec \cdot)$, $\mathrm{ok}(\cdot)$, $\mathrm{ko}(\cdot)$, $\mathrm{bl}(\cdot)$, and $\mathrm{ap}(\cdot)$, and a set of associated default names:

Definition 10. *Given a prioritized default theory $\Delta = (D, W, <)$ over $\mathcal{L}$ and its set of default names $N = \{n_\delta \mid \delta \in D\}$, define $T_{BE}(\Delta) = (D^\circ, W^\circ)$ over $\mathcal{L}^\circ$ by:*

$$D^\circ = D \cup \left\{ \frac{\mathrm{ok}(n_\delta) \wedge \alpha : \beta, \beta'}{\gamma' \wedge \mathrm{ap}(n_\delta)}, \ \frac{\mathrm{ok}(n_\delta) : \neg\alpha, \neg\alpha'}{\mathrm{bl}(n_\delta)}, \ \frac{\mathrm{ok}(n_\delta) \wedge \neg\beta \wedge \neg\beta' :}{\mathrm{bl}(n_\delta)} \ \middle| \ \delta = \frac{\alpha:\beta}{\gamma} \in D \right\} \quad (5)$$

$$\cup \left\{ \frac{: \neg(x \prec y)}{\neg(x \prec y)} \right\} \cup \left\{ \frac{\gamma \wedge \neg\beta :}{\mathrm{ko}(n_\delta)} \ \middle| \ \delta = \frac{\alpha:\beta}{\gamma} \in D \right\} \cup \left\{ \frac{: \exists x \in N. \ \neg\mathrm{ok}(x)}{\bot} \right\} \quad (6)$$

$$W^\circ = W \cup W' \quad (7)$$

$$\cup \ \{n_1 \prec n_2 \mid (\delta_1, \delta_2) \in <\} \cup \{\mathrm{DCA}_N, \mathrm{UNA}_N\} \quad (8)$$

$$\cup \ \{\forall x \in N. \ [\forall y \in N. \ \mathrm{ko}(y) \vee [(x \prec y) \supset (\mathrm{bl}(y) \vee \mathrm{ap}(y))]] \supset \mathrm{ok}(x)\} \quad (9)$$

We denote the second group of rules in (5) by δ_a°, $\delta_{b_1}^\circ$, and $\delta_{b_2}^\circ$; those in (6) are abbreviated by $\delta_\prec^\circ$, δ_{ko}°, and $\delta_\bot^\circ$, respectively.

It is important to note that the inclusions $D \subseteq D^\circ$ and $W \subseteq W^\circ$ hold. As we show in Theorem 2, this allows us to construct regular extensions of (D, W) within extensions of (D°, W°). Such an extension can be seen as the *guess* in a guess-and-check approach; it corresponds to Condition (*i*) in Definition 9.

The salient part of the corresponding *check*, viz. Condition (*ii*) in Definition 9, is accomplished by the second group of rules in (5) and the remaining facts in W°. Together with $W' \subseteq W^\circ$, the rules of form δ_a° aim at rebuilding the guessed extension in $\mathcal{L}'$. They form the prerequisite-free counterpart of the original default theory in $\mathcal{L}'$. In fact, the prerequisite of δ_a° refers via α to the guessed extension in $\mathcal{L}$; no formula in $\mathcal{L}'$ must be derived for applying δ_a°. This accounts for the elimination of prerequisites in

Condition (1) of Definition 8. Moreover, the elimination of rules whose prerequisites are not derivable is accomplished by rules of form $\delta_{b_1}^\circ$. Rules of form $\delta_{b_2}^\circ$ guarantee that defaults are only defeatable by rules with higher priority. In fact, it is $\neg\beta'$ that must be derivable in such a way only.

The application of rules according to the given preference information is enforced by axiom (9): For every n_i, we derive $\mathsf{ok}(n_i)$ whenever, for every n_j, either $\mathsf{ko}(n_j)$ is true, or, if $n_i \prec n_j$ holds, either $\mathsf{ap}(n_j)$ or $\mathsf{bl}(n_j)$ is true. This axiom allows us to derive $\mathsf{ok}(n_i)$, indicating that δ_i may potentially be applied, whenever we have for all δ_j with $\delta_i < \delta_j$ that δ_j has been applied or cannot be applied, or δ_j has already been eliminated from the preference handling process. This elimination of rules is in accord with Definition 7 and realized by δ_{ko}°. The preference information in (8) is rendered complete through rules of form $\delta_{\preceq}^\circ$. This completion is necessary for the formula in (9) to work properly: whenever $(\delta_i, \delta_j) \not\in <$, rule $\delta_{\preceq}^\circ$ allows us to conclude (in the extension) that $\neg(n_i \prec n_j)$ holds.

Lastly, $\delta_\perp^\circ$ rules out unsuccessful attempts in rebuilding the regular extension from $\mathcal{L}$ within $\mathcal{L}'$ according to the given preference information. In this way, we eliminate all regular extensions that do not respect preference.

For illustration, reconsider theory (4), viz.

$$n_3 : \frac{:A}{A} < n_2 : \frac{:\neg B}{\neg B} < n_1 : \frac{A:B}{B}$$

and $W = \emptyset$. Recall that this theory has two regular extensions: one containing $\{A, \neg B\}$ and another containing $\{A, B\}$; but that only the latter is a prioritized extension according to [3]. We get:

$$\frac{:A}{A} \qquad \frac{:\neg B}{\neg B} \qquad \frac{A:B}{B}$$

$$\frac{\mathsf{ok}(n_3):A,A'}{A'\wedge\mathsf{ap}(n_3)} \qquad \frac{\mathsf{ok}(n_2):\neg B,\neg B'}{\neg B'\wedge\mathsf{ap}(n_2)} \qquad \frac{\mathsf{ok}(n_1)\wedge A:B,B'}{B'\wedge\mathsf{ap}(n_1)} \qquad \frac{:\neg\mathsf{ok}(n_1)\vee\neg\mathsf{ok}(n_2)\vee\neg\mathsf{ok}(n_3)}{\perp}$$

$$\frac{\mathsf{ok}(n_1):\neg A,\neg A'}{\mathsf{bl}(n_1)}$$

$$\frac{\mathsf{ok}(n_3)\wedge\neg A\wedge\neg A':}{\mathsf{bl}(n_3)} \quad \frac{\mathsf{ok}(n_2)\wedge B\wedge B':}{\mathsf{bl}(n_2)} \quad \frac{\mathsf{ok}(n_1)\wedge\neg B\wedge\neg B':}{\mathsf{bl}(n_1)}$$

For brevity, we omit all defaults of form $\frac{\perp:}{\mathsf{ko}(n)}$.

First, suppose there is an extension with A and $\neg B$. Clearly, $\frac{:A}{A}$ and $\frac{:\neg B}{\neg B}$ contribute to such an extension. Having $\neg B$ denies the derivation of $\mathsf{ap}(n_1)$. Also, we do not get $\mathsf{bl}(n_1)$ since we can neither derive $\neg B'$ nor is $\neg A$ consistent. Therefore, we do not obtain $\mathsf{ok}(n_2)$; thus, $\neg\mathsf{ok}(n_2)$ is consistent and we obtain $\perp$ which destroys the putative extension at hand.

Next, consider a candidate extension with A and B. In this case, $\frac{:A}{A}$ and $\frac{A:B}{B}$ apply. Given $\mathsf{ok}(n_1)$ and A, we may derive $B' \wedge \mathsf{ap}(n_1)$. This gives $\mathsf{ok}(n_2)$ and then $\mathsf{ok}(n_2) \wedge B \wedge B'$, from which we get $\mathsf{bl}(n_2)$. Finally, we derive $\mathsf{ok}(n_3)$ and $A' \wedge \mathsf{ap}(n_3)$. Unlike the above, we cannot derive $\perp$ and we obtain an extension containing A and B.

For another example, consider the theory obtained from example (3):

$$\frac{:\neg B}{A} \qquad \frac{:\neg A}{\neg A} \qquad \frac{:A}{A} \qquad \frac{:B}{B}$$

$$\frac{ok(n_1):\neg B,\neg B'}{A'\wedge ap(n_1)} \quad \frac{ok(n_2):\neg A,\neg A'}{\neg A'\wedge ap(n_2)} \quad \frac{ok(n_3):A,A'}{A'\wedge ap(n_3)} \quad \frac{ok(n_4):B,B'}{B'\wedge ap(n_3)} \quad \frac{:\exists x \in N.\,\neg ok(x)}{\bot}$$

$$\frac{ok(n_1)\wedge B\wedge B':}{bl(n_1)} \quad \frac{ok(n_2)\wedge A\wedge A':}{bl(n_2)} \quad \frac{ok(n_3)\wedge \neg A\wedge \neg A':}{bl(n_3)} \quad \frac{ok(n_4)\wedge \neg B\wedge \neg B':}{bl(n_3)}$$

$$\frac{A\wedge B:}{ko(n_1)}$$

While this theory has two regular extensions, it has no prioritized extension under the ordering imposed in (3). Suppose there is a prioritized extension containing A and B. This yields $ko(n_1)$ and then (9) gives $ok(n_2)$. Having A excludes $(\delta_2)^\circ_a$. Moreover, we cannot apply $(\delta_2)^\circ_{b_2}$ since A' is not derivable (by higher-ranked rules). We thus cannot derive $ok(n_3)$, which leads to a destruction of the current extension through $\delta^\circ_\bot$.

The next theorem gives the major result of our paper.

Theorem 1. *Let $\Delta = (D, W, <)$ be a prioritized default theory over $\mathcal{L}$ and E a set of formulas over $\mathcal{L}$.*

E is a prioritized extension of Δ iff $E = F \cap \mathcal{L}$ and F is a (regular) extension of $\mathcal{T}_{BE}(\Delta)$.

In what follows, we elaborate upon the structure of the encoded default theories:

Theorem 2. *Let $\Delta = (D, W, <)$ be a prioritized default theory over $\mathcal{L}$ and let E° be a regular extension of $\mathcal{T}_{BE}(\Delta) = (D^\circ, W^\circ)$. Then, we have the following results:*

1. *$E^\circ \cap \mathcal{L}$ is a (regular) extension of (D, W);*
2. *$(E^\circ \cap \mathcal{L})' = E^\circ \cap \mathcal{L}'$ (or $\varphi \in E^\circ$ iff $\varphi' \in E^\circ$ for $\varphi \in \mathcal{L}$);*
3. *$\delta \in D \cap GD(D^\circ, E^\circ)$ iff $\delta^\circ_a \in GD(D^\circ, E^\circ)$;*
4. *$\delta \in D \setminus GD(D^\circ, E^\circ)$ iff $\delta^\circ_{b_1} \in GD(D^\circ, E^\circ)$ or $\delta^\circ_{b_2} \in GD(D^\circ, E^\circ)$;*
5. *if $\delta^\circ_{ko} \in GD(D^\circ, E^\circ)$, then $\delta^\circ_{b_2} \in GD(D^\circ, E^\circ)$.*

The last property shows that eliminated rules are eventually found to be inapplicable. This illustrates another choice of our translation: instead of using the second group of rules in (5), we could have used

$$\left\{ \frac{ok(n)\wedge \alpha : \beta,\beta',\neg ko(n)}{\gamma'\wedge ap(n)}, \ \frac{ok(n):\neg\alpha,\neg\alpha',\neg ko(n)}{bl(n)}, \ \frac{ok(n)\wedge \neg\beta\wedge \neg\beta':\neg ko(n)}{bl(n)} \ \middle| \ n:\frac{\alpha:\beta}{\gamma} \in D \right\}.$$

Although this renders the derivation of $ap(n)$, $bl(n)$, and $ko(n)$ mutually exclusive, the additional justification $\neg ko(n)$ is not needed. That is, it is sufficient to remove $\frac{\alpha:\beta}{\gamma}$ from the preference handling process; the rule is found to be blocked anyway.

The following theorem summarizes some technical properties of our translation:

Theorem 3. *Let E be a consistent extension of $\mathcal{T}_{BE}(\Delta)$ for prioritized default theory $\Delta = (D, W, <)$. We have for all $\delta, \delta' \in D$ that*

1. *$n_\delta \prec n_{\delta'} \in E$ iff $\neg(n_\delta \prec n_{\delta'}) \notin E$;*
2. *$ok(n_\delta) \in E$;*
3. *$ap(n_\delta) \in E$ iff $bl(n_\delta) \notin E$.*

The two last results reveal an alternative choice for $\delta^\circ_\bot$, namely $\dfrac{:\exists x \in N.\,\neg ap(x)\wedge \neg bl(x)}{\bot}$.

One may wonder how our translation avoids the explicit use of total extensions of the given partial order. The next theorem shows that these total extensions are reflected by the grounded enumerations of the second group of rules in (5):

Theorem 4. *Given the same prerequisites as in Theorem 2, let $\langle \delta_i^\circ \rangle_{i \in I}$ be some grounded enumeration of $GD(D^\circ, E^\circ)$. For all $\delta_1, \delta_2 \in D^{E^\circ \cap \mathcal{L}}$, define $\delta_1 \ll \delta_2$ iff $k_2 < k_1$ where $k_j = \min\{i \in I \mid \delta_i^\circ = (\delta_j)_x^\circ$ for $x \in \{a, b_1, b_2\}\}$ for $k = 1, 2$. Then, $\ll$ is a total ordering on $D^{E^\circ \cap \mathcal{L}}$ such that $\ll \subseteq (< \cap (D^{E^\circ \cap \mathcal{L}} \times D^{E^\circ \cap \mathcal{L}}))$.*

That is, whenever $\Delta = \Delta^E$ according to Definition 7, we have that $\ll$ is a total ordering on D such that $\ll \subseteq <$.

Finally, one may ask why we do not need to account for the "inherited" ordering in Condition 2 of Definition 8. In fact, this is taken care of through the "tags" $\mathrm{ap}(n_\delta)$ in the consequents of rules δ_a° that guarantee an isomorphism between D and D_E in Definition 8. More generally, such a "tagging of consequents" provides an effective correspondence between the applicability of default rules and the presence of their consequents in an extension at hand. As a side effect, this facilitates the notion of activeness in Section 4 by rendering Condition (iii) unnecessary.

6 Compiling Prioritized Answer Sets

In this section, we describe how Brewka and Eiter's preference approach [3] for extended logic programs can be encoded within standard answer set semantics, following the methodology developed in [8]. We commence with a recapitulation of the necessary concepts.

As usual, a *literal*, L, is an expression of the form p or $\neg p$, where p is an atom. The set of all literals is denoted by *Lit*. A *rule*, r, is an expression of the form

$$L_0 \leftarrow L_1, \ldots, L_m, not\ L_{m+1}, \ldots, not\ L_n, \tag{10}$$

where $n \geq m \geq 0$, and each L_i ($0 \leq i \leq n$) is a literal. The symbol "*not*" denotes *negation as failure*, or *weak negation*. Accordingly, the classical negation sign "$\neg$" is in this context also said to represent *strong negation*. The literal L_0 is called the *head* of r, and the set $\{L_1, \ldots, L_m, not\ L_{m+1}, \ldots, not\ L_n\}$ is the *body* of r. We use $head(r)$ to denote the head of rule r, and $body(r)$ to denote the body of r. Furthermore, let $body^+(r) = \{L_1, \ldots, L_m\}$ and $body^-(r) = \{L_{m+1}, \ldots, L_n\}$. The elements of $body^+(r)$ are referred to as the *prerequisites* of r. If $body^+(r) = \emptyset$, then r is a *prerequisite-free rule*; if $body(r) = \emptyset$, then r is a *fact*; if r contains no variables, then r is *ground*. We say that a rule r is *defeated* by a set of literals X iff $body^-(r) \cap X \neq \emptyset$. As well, each literal in $body^-(r) \cap X$ is said to *defeat* r. We define *not* X as the set $\{not\ L \mid L \in X\}$.

A set of literals X is *consistent* iff it does not contain a complementary pair $p, \neg p$ of literals. We say that X is *logically closed* iff it is either consistent or equals *Lit*.

A *rule base* is any collection of rules; an (*extended*) *logic program*, or simply a *program*, is a finite rule base. A rule base (program) is *prerequisite-free* (*ground*) if all rules in it are prerequisite-free (ground).

For a rule base R, we denote by R^* the ground instantiation of R over the Herbrand universe of the language $\mathcal{L}$ of R.

The answer set semantics interprets ground rules of the form (10) as defaults

$$\frac{L_1 \wedge \ldots \wedge L_m \ : \ \neg L_{m+1}, \ldots, \neg L_n}{L_0}. \tag{11}$$

A set X of ground literals is called an *answer set* of the ground program P iff X is of the form $E \cap Lit$, where E is an extension of the default theory obtained by identifying each rule $r \in P$ as a default of the form (11). Answer sets of programs not necessarily ground are obtained by taking the answer sets of the ground instantiation P^* of P.

A *prioritized logic program* is a pair $\Pi = (P, <)$, where P is a logic program and $<$ is a strict partial order. Following [3], the ground instantiation of a prioritized logic program $(P, <)$ is obtained as follows: Let P^* be the ground instantiation of P and define $r^* <^* s^*$ for $r^*, s^* \in P^*$ providing r^*, s^* are instances of $r, s \in P$, respectively, such that $r < s$. If $<^*$ is a strict partial order, then the pair $(P^*, <^*)$ defines the ground instantiation of $(P, <)$; otherwise, the ground instantiation of $(P, <)$ is undefined. In the sequel, we will be concerned with ground prioritized programs only.

A *fully prioritized logic program* is a prioritized logic program $(P, <)$ where $<$ is a total ordering. Prioritized answer sets of prioritized logic programs are defined similarly to prioritized extensions of prioritized default theories. That is to say, first the prerequisite-free case is treated, and afterwards the general case is addressed in terms of the prerequisite-free case.

For fully prioritized ground programs, Definitions 5 and 7 boil down to the following operator: Let $\Pi = (P, <)$ be a fully prioritized ground prerequisite-free logic program, $\langle r_i \rangle_{i \in I}$ be an enumeration of the ordering $<$, and X be a set of literals. Then, $C_\Pi(X)$ is the smallest logically closed set of literals containing $\bigcup_{i \in I} X_i$, where

$$X_i = \begin{cases} \bigcup_{j<i} X_j & \text{if } r_i \text{ is defeated by } \bigcup_{j<i} E_j, \text{ or} \\ & head(r_i) \in X \text{ and } r_i \text{ is defeated by } X; \\ \bigcup_{j<i} X_j \cup \{head(r_i)\} & \text{otherwise.} \end{cases}$$

As in the default logic case, this construction is unique in the sense that for a fully prioritized prerequisite-free ground program Π, there is at most one answer set X of P such that $C_\Pi(X) = X$ (cf. [3, Lemma 4.1]). Accordingly, this set is referred to as the *prioritized answer set* of Π, if it exists. Prioritized answer sets of an arbitrary (i.e., not necessarily prerequisite-free) ground fully prioritized program $\Pi = (P, <)$ are given by sets X of ground literals which are prioritized answer sets of the prioritized program $\Pi_X = (P_X, <_X)$, where $<_X$ is constructed just as the ordering $<_E$ of Definition 8, and P_X results from P by (i) deleting any rule $r \in P$ such that $body^+(r) \nsubseteq X$, and (ii) removing any prerequisites in the body of the remaining rules. Lastly, X is a prioritized answer set of a ground prioritized logic program $(P, <)$ iff (i) X is a (regular) answer set of P and (ii) X is a prioritized answer set of some fully prioritized program $(P, <')$ such that $< \subseteq <'$.

This concludes the review of prioritized answer sets according to [3]; we continue with a compilation of this approach in standard answer set semantics.

As in Section 5, given a ground prioritized program Π over language $\mathcal{L}$, we assume a disjoint language $\mathcal{L}'$ containing literals L' for each L in $\mathcal{L}$. Likewise, rule r' results from r by replacing each literal L in r by L'. We maintain for rules the same naming convention as for defaults, i.e., the term n_r serves as name for rule r, similarly writing $n : r$ as before. As well, the language $\mathcal{L}^\circ$ extends $\mathcal{L} \cup \mathcal{L}'$ by new ground atoms ($n_r \prec n_s$), $\mathsf{ok}(n_r)$, $\mathsf{ko}(n_r)$, $\mathsf{ry}(n_r, n_s)$, $\mathsf{bl}(n_r)$, and $\mathsf{ap}(n_r)$, for each r, s in Π.

Definition 11. *Let $\Pi = (P, <)$ be a prioritized ground logic program over $\mathcal{L}$ such that $P = \{r_1, \ldots, r_k\}$. Then, the logic program $T_{BE}^{lp}(\Pi)$ over $\mathcal{L}^\circ$ is given by*

$$P \cup \bigcup_{r \in P} \tau(r) \cup \{(n_1 \prec n_2) \leftarrow \mid (r_1, r_2) \in <\},$$

where $\tau(r)$ consists of the following collection of rules, for $L \in body^+(r)$, $K \in body^-(r)$, and $s \in P$:

$$
\begin{aligned}
a_1(r): \quad & head(r') \leftarrow \mathsf{ap}(n_r) \\
a_2(r): \quad & \mathsf{ap}(n_r) \leftarrow \mathsf{ok}(n_r), body(r), not\ body^-(r') \\
b_1(r, L): \quad & \mathsf{bl}(n_r) \leftarrow \mathsf{ok}(n_r), not\ L, not\ L' \\
b_2(r, K): \quad & \mathsf{bl}(n_r) \leftarrow \mathsf{ok}(n_r), K, K' \\[6pt]
c_1(r): \quad & \mathsf{ok}(n_r) \leftarrow \mathsf{ry}(n_r, n_{r_1}), \ldots, \mathsf{ry}(n_r, n_{r_k}) \\
c_2(r, s): \quad & \mathsf{ry}(n_r, n_s) \leftarrow not\ (n_r \prec n_s) \\
c_3(r, s): \quad & \mathsf{ry}(n_r, n_s) \leftarrow (n_r \prec n_s), \mathsf{ap}(n_s) \\
c_4(r, s): \quad & \mathsf{ry}(n_r, n_s) \leftarrow (n_r \prec n_s), \mathsf{bl}(n_s) \\
c_5(r, s): \quad & \mathsf{ry}(n_r, n_s) \leftarrow \mathsf{ko}(n_s) \\[6pt]
d(r): \quad & \bot \leftarrow not\ \mathsf{ok}(n_r) \\[6pt]
e(r, K): \quad & \mathsf{ko}(n_r) \leftarrow head(r), K
\end{aligned}
$$

The first group of rules in $\tau(r)$ expresses applicability and blocking conditions of r and contains the counterparts of the defaults δ_a°, $\delta_{b_1}^\circ$, and $\delta_{b_2}^\circ$ in Definition 10, respectively. To wit, applicability of r is captured by the two rules $a_1(r)$ and $a_2(r)$, while k rules of the form $b_1(r, L)$ and $b_2(r, K)$ detect blockage of r, where k is the number of literals in $body(r)$. The second group of rules unfolds axiom (9) and relies on auxiliary atoms $\mathsf{ry}(\cdot, \cdot)$ ("ready"), taking care of instantiating the quantification over names expressed in (9). Finally, rules $d(r)$ and $e(r, K)$ correspond to δ_{ko}°, and $\delta_\bot^\circ$, respectively.

We obtain the following result corresponding to Theorem 1:

Theorem 5. *Let $\Pi = (P, <)$ be a prioritized ground logic program over $\mathcal{L}$ and X a set of literals over $\mathcal{L}$.*

X is a prioritized answer set of Π iff $X = Y \cap \mathcal{L}$ and Y is a (regular) answer set of $T_{BE}^{lp}(\Pi)$.

Additionally, given suitable concepts for the present case, analogous results to Theorems 2, 3, and 4 can be shown. We just note the counterpart of Theorem 3:

Theorem 6. *Let X be a consistent answer set of $T_{BE}^{lp}(\Pi)$ for prioritized logic program $\Pi = (P, <)$. We have for all $r \in P$ that*

1. *$\mathsf{ok}(n_\delta) \in X$;*
2. *$\mathsf{ap}(n_\delta) \in X$ iff $\mathsf{bl}(n_\delta) \notin X$.*

388 James P. Delgrande, Torsten Schaub, and Hans Tompits

The approach is implemented in Prolog and serves as a front-end to the logic programming systems dlv [10] and smodels [14]. Our current prototype, called plp, is available at http://www.cs.uni-potsdam.de/~torsten/plp/. This URL contains also diverse examples taken from the literature. The implementation differs from the approach described here, in that the translation applies to named rules only; it thus leaves unnamed rules unaffected.

For illustration, consider the logic programming counterpart of Example (4) in the syntax of plp :

```
 b :- name(1), not -b, a.
-b :- name(2), not b.             2<1.
 a :- name(3), not -a.            3<2.
```

We use '-' (or 'neg') for classical negation and 'not' (or '~') for negation as failure. Furthermore, name(·) is used to identify rule names; and natural numbers serve as names. Note that our implementation handles transitivity implicitly, so that there is no need to specify 3<1.

This is then translated into the following (intermediate) standard program:

```
(1)    b :- not neg b, a.
(2)    b1 :- ap(1).
(3)    ap(1) :- name(1), ok(1), not neg b, not neg b1, a.
(4)    bl(1) :- ok(1), neg b, neg b1.
(5)    bl(1) :- ok(1), not a, not a1.
(6)    ko(1) :- b, neg b.
(7)    neg b :- not b.
(8)    neg b1 :- ap(2).
(9)    ap(2) :- name(2), ok(2), not b, not b1.
(10)   bl(2) :- ok(2), b, b1.
(11)   ko(2) :- neg b, b.
(12)   a :- not neg a.
(13)   a1 :- ap(3).
(14)   ap(3) :- name(3), ok(3), not neg a, not neg a1.
(15)   bl(3) :- ok(3), neg a, neg a1.
(16)   ko(3) :- a, neg a.
(17)   2 < 1.
(18)   3 < 2.
(19)   neg M < N :- name(N), name(M), N < M.
(20)   N < M :- name(N), name(M), name(O), N < O, O < M.
(21)   ok(N) :- name(N), ry(N, 1), ry(N, 2), ry(N, 3).
(22)   ry(N, M) :- name(N), name(M), not N < M.
(23)   ry(N, M) :- name(N), name(M), N < M, ap(M).
(24)   ry(N, M) :- name(N), name(M), N < M, bl(M).
(25)   ry(N, M) :- name(N), name(M), ko(M).
(26)   false :- name(N), not ok(N).
```

The original rules, viz. r_1, r_2, and r_3, are given by (1), (7), and (12). The additional encoding of, e.g., rule (1) is given by (2) to (6). We append the symbol '1' for priming here, e.g., b1 is the primed version of b. In detail, (2) and (3) correspond to $a_1(r_1)$ and $a_2(r_1)$, (4) and (5) correspond to $b_2(r_1, B)$ and $b_1(r_1, A)$, and finally

(6) corresponds to $e(r_1, B)$. Rules (19) and (20) are additional rules enforcing a strict partial order. Rules (21) to (25) account for $c_1(r)$ to $c_5(r, s)$. Lastly, (26) implements $d(r)$.

The above program is then refined once more in order to account for some special features of dlv and smodels, like implementation of classical negation 'neg' and 'false'. Also, an extensional database for rule names is provided.

Calling one of these provers with the respective input corresponding to the above program, we obtain the desired prioritized answer set containing the literals A and B (i.e., represented by a and b).

7 Conclusion

We have shown how the approach of Brewka and Eiter, both with respect to extended logic programs [3] and to default logic [4], can be expressed in our general framework for preferences [6, 8]. On the one hand, this illustrates the generality of our framework; on the other hand, it sheds light on Brewka and Eiter's approaches, since it provides a translation and encoding of their approaches into extended logic programs and default logic, respectively. As well, our encoding allows a straightforward implementation of [3] via a translation into extended logic programs.

Lastly, we note that our approach described in [8] used *dynamic* preference information, in that preferences were expressed within a logic program. As well, in the case of default logic, [6] also describes the incorporation of dynamic preferences. Thus in these approaches, preferences can be encoded as holding only in specific contexts, holding by default, and so on. Such a dynamic setting was also sketched in [4]. It is a straightforward matter to extend Definitions 10 and 11 to handle this dynamic case as well.

Acknowledgements The first author was partially supported by a Research Grant from the Natural Sciences and Engineering Research Council of Canada. The second author was partially supported by the German Science Foundation (DFG) under grant FOR 375/1-1, TP C. The third author was partially supported by the Austrian Science Fund (FWF) under grants N Z29-INF and P13871-INF.

References

[1] F. Baader and B. Hollunder. How to prefer more specific defaults in terminological default logic. In *Proceedings of the International Joint Conference on Artificial Intelligence*, pages 669–674, 1993.

[2] G. Brewka. Reasoning about priorities in default logic. In *Proceedings of the AAAI National Conference on Artificial Intelligence*, volume 2, pages 940–945. AAAI Press/The MIT Press, 1994.

[3] G. Brewka and T. Eiter. Preferred answer sets for extended logic programs. *Artificial Intelligence*, 109(1-2):297–356, 1999.

[4] G. Brewka and T. Eiter. Prioritizing default logic. In St. Hölldobler, editor, *Intellectics and Computational Logic — Papers in Honour of Wolfgang Bibel*. Kluwer Academic Publishers, 2000.

[5] P. Chołewinski, V. W. Marek, and M. Truszczyński. Default reasoning system DeReS. In *Proceedings KR '96*, pages 518-528, 1996.

[6] J. Delgrande and T. Schaub. Compiling reasoning with and about preferences into default logic. In M. Pollack, editor, *Proceedings of the International Joint Conference on Artificial Intelligence*, pages 168–174. Morgan Kaufmann Publishers, 1997.

[7] J. Delgrande and T. Schaub. The role of default logic in knowledge representation. In J. Minker, editor, *Logic-Based Artificial Intelligence*. Kluwer Academic Publishers, 2000.

[8] J. Delgrande, T. Schaub, and H. Tompits. Logic programs with compiled preferences. In C. Baral and M. Truszczyński, editors, *Proceedings of the Eighth International Workshop on Non-Monotonic Reasoning*, 2000.

[9] J. Delgrande, T. Schaub, and H. Tompits. A compiler for ordered logic programs. In C. Baral and M. Truszczyński, editors, *Proceedings of the Eighth International Workshop on Non-Monotonic Reasoning*, 2000.

[10] T. Eiter, N. Leone, C. Mateis, G. Pfeifer, and F. Scarcello. A deductive system for nonmonotonic reasoning. In J. Dix, U. Furbach, and A. Nerode, editors, *Proceedings LPNMR '97*, pages 363–374. Springer Verlag, 1997.

[11] M. Gelfond and V. Lifschitz. Classical negation in logic programs and deductive databases. *New Generation Computing*, 9:365–385, 1991.

[12] T. Janhunen. Classifying semi-normal default logic on the basis of its expressive power. In M. Gelfond, N. Leone, and G. Pfeifer, editors, *Proceedings LPNMR '99*, pages 19–33. Springer Verlag, 1999.

[13] J. McCarthy. Applications of circumscription to formalizing common-sense knowledge. *Artificial Intelligence*, 28:89–116, 1986.

[14] I. Niemelä and P. Simons. Smodels: An implementation of the stable model and well-founded semantics for normal logic programs. In J. Dix, U. Furbach, and A. Nerode, editors, *Proceedings LPNMR '97*, pages 420–429. Springer Verlag, 1997.

[15] D. Poole. A logical framework for default reasoning. *Artificial Intelligence*, 36:27–47, 1988.

[16] R. Reiter. A logic for default reasoning. *Artificial Intelligence*, 13(1-2):81–132, 1980.

A Logic for Modeling Decision Making with Dynamic Preferences

Marina De Vos* and Dirk Vermeir

Dept. of Computer Science. Free University of Brussels, VUB
Pleinlaan 2, Brussels 1050, Belgium.
Tel: +32 2 6293308; Fax: +32 2 6293525;
{marinadv,dvermeir}@vub.ac.be

Abstract. We present a framework for decision making with the possibility to express circumstance-dependent preferences among different alternatives for a decision. This new formalism, Ordered Choice Logic Programs (OCLP), builds upon choice logic programs to define a preference/specialization relation on sets of choice rules. We show that our paradigm is an intuitive extension of both ordered logic and choice logic programming such that decisions can comprise more than two alternatives which become only available when a choice is actually forced. The semantics for OCL programs is based on stable models for which we supply a characterization in terms of assumption sets and a fixpoint algorithm. Furthermore we demonstrate that OCLPs allow an elegant translation of finite extensive games with perfect information such that the stable models of the program correspond, depending on the transformation, to either the Nash equilibria or the subgame perfect equilibria of the game.

1   Introduction

Preferences among defaults or alternatives play an important role in nonmonotonic reasoning, especially when modeling the complex way people reason in every day live. In case of conflict, humans prefer the default or alternative which provides more reliable, more specific or more important information.

For the last two decades, a lot of research in the nonmonotonic reasoning community has concentrated on bringing preference into the different paradigms: for example logic programming ([6,9,12]), extended logic programming ([3]), extended disjunctive logic programming ([1]) and prioritized circumscription ([7]). We will discuss some of these systems in more detail later on in this paper when we compare them to our approach.

These systems have demonstrated their usage in a wide variety of applications like law, object orientation, model based diagnosis or configuration tasks. They are especially suitable for working with exceptions to defaults.

In this paper we present a formalism that enables us to reason about decisions with more than two alternatives where the preference between alternatives depends on the situation. The systems mentioned above do not support such dynamic preferences: they either use the preferences when the model is already being computed, which means that

---

* The author wishes to thank the FWO-Vlaanderen for its support.

the decisions are already made, or they only support preferences between rules with opposite consequences, leaving out the possibility to have decisions with more than two alternatives. Another problem of the latter type of systems is that the alternatives (i.e. complementary literals) are fixed even before writing the program. We feel that alternatives should emerge only when a choice between them is required. Let us illustrate this with the following example.

Example 1 (Tommy's Birthday). Today it is Tommy's birthday. Six years old, time goes fast. To celebrate this, his mother agreed to invite some of his friends over for a party. Sitting in his room he is dreaming about his own private party: "A huge birthday cake with lots of candles, of course not forgetting the icing. Lots of candy and biscuits. We just have to make sure that there is plenty, you can never have enough treats. But no matter what, there definitely has to be that big cake. Hopefully my mum will let me decide, that way I can have everything my heart desires. I know that if she starts interfering, she will force me to choose. That is what mums always do."
Intuitively, one would expect two possible outcomes for this party:

- Tommy's Birthday, Tommy is planning, Tommy and his friends having cake, biscuits and candy.
- Tommy's Birthday, Tommy's mother does the planning, Tommy and his friends only having cake.

Thus, in the first solution $cake, biscuit$ and $candy$ are not considered alternatives of which only one has to be selected, while in second they are because Tommy's mother forces him to make this difficult choice.
To allow this kind of reasoning, two things need to be added to logic programming. First of all we need a mechanism to represent the possible decisions. As argued in [4,5], choice logic programs are an intuitive tool to represent conditional decisions, as the semantics make sure that only one alternative is chosen. Thus, choice logic programs will be the fundaments on which we build our new formalism. Now only a mechanism for denoting preference/order amongst different alternatives is missing. To this end, we will use a generalization to multiple alternatives of the ideas behind Ordered Logic [6]. Our formalism, called Ordered Choice Logic Programs, defines a partial order amongst choice logic programs, called components. Each component inherits, like in object orientation, the rules of the less specific components. Normal model semantics is used until alternatives for the same decision are in conflict. Then, the most specific alternative is decided upon.
These extensions offer a new view point to the above mentioned application domains. For example it is possible to reason about which method overrides the others in a subclassing chain, where with the previous systems one could only detect whether a method was overridden or not. Also applications in AI & law can be envisaged: e.g. lawyer can work out a whole strategy by taking into account the possible actions of the other parties.
We are also able to add a new application domain to this list: Game Theory[1][8]. We will show that ordered choice logic programs are capable of naturally representing finite extensive games with perfect information such that the stable models of the former

---

[1] Game Theory has proven its usefulness in domains such as economics and computer science.

correspond with, depending on the transformation, either the Nash equilibria or the subgame perfect equilibria of the latter.

The outline of the rest of the paper is as follows: In Sect. 2 we introduce ordered choice logic programs. The stable model semantics for such programs is presented in Sect. 3. Sect. 4 is used for discussing an application in game theory while Sect. 5 compares ordered choice logic programs with some alternative approaches.

2   Ordered Choice Logic Programs

The basis of Ordered Choice Logic Programs are, as the name already might have indicated, choice logic programs[4,5].
We identify these choice logic program with their grounded version, i.e. the set of all ground instances of its clauses. This keeps the program finite as we do not allow function symbols (i.e. we stick to datalog).

Definition 1 ([4,5]). *A **Choice Logic Program**, CLP for short, is a finite set of rules of the form $A \leftarrow B$ where A and B are finite sets of atoms*

Intuitively, atoms in A are assumed to be xor'ed together while B is read as a conjunction (note that A may be empty, i.e. constraints are allowed). In examples, we often use "$\oplus$" to denote exclusive or, while "," is used to denote conjunction.

The Herbrand Base and interpretations for a choice logic programs are defined in the usual way, except that we will only consider total interpretations in this paper.

Definition 2 ([4,5]). *Let P be a CLP. The **Herbrand Base** of P, denoted $\mathcal{B}_P$, is defined as the set of all atoms appearing in the program. An **interpretation** I is any subset of the Herbrand Base of P, i.e. $I \subseteq \mathcal{B}_P$. An atom in I is assumed to be true while an atom in $\mathcal{B}_P \setminus I$ is considered false. We denote the set of all false atoms wrt I as $\overline{I}$.*

Definition 3. *An **Ordered Choice Logic Program**, or OCLP, is a pair $\langle C, \preccurlyeq \rangle$ where C is a finite set of choice logic programs, called **components**, and "$\preccurlyeq$" is a partial order on C. In this paper we assume that C contains a minimal element $C_\perp$ such that $C_\perp \preccurlyeq X$ for all $X \in C$. Furthermore, we assume that a rule appears in at most one component of C[2].*

For two components $C_1, C_2 \in C$, $C_1 \prec C_2$ implies that C_2 contains more general information than C_1[3]. Also $[A, B]$ is used to denote the set $\{X \mid A \preccurlyeq X \preccurlyeq B\}$. Similarly, $[A, B[$ denotes the set $\{X \mid A \preccurlyeq X \prec B\}$.
Throughout the examples, we will often represent an OCLP P by means of a directed acyclic graph (dag) in which the nodes represent the components and the arcs the relation "$\prec$".

---

[2] This is only a technical restriction that considerably simplifies the notation.

[3] As usual, "$\prec$" denotes the restriction of "$\preccurlyeq$" to all the pairs of distinct components.

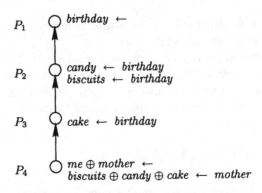

P_1 *birthday* ←

P_2 *candy* ← *birthday*
         biscuits ← *birthday*

P_3 *cake* ← *birthday*

P_4 *me* ⊕ *mother* ←
         biscuits ⊕ *candy* ⊕ *cake* ← *mother*

Fig. 1. Tommy's Birthday Dream.

Example 2. Tommy's Birthday dream can easily be translated into the OCLP depicted in Fig. 1 where the choice rules in P_4 correspond with Tommy specifically knowing that either he or his mother will do the organization and that in case his mother will be in charge, he will be forced to choose between all the goodies. The order, together with the rules of P_2 and P_3, expresses that Tommy is more in favor of cake than any of the other treats. Finally P_1 introduces the general fact that it is Tommy's birthday.

As more specific components "inherit" the rules from more general components, we also need, when defining an interpretation, to consider the atoms mentioned in those less specific parts.

Definition 4. *Given an OCLP P and a component $A \in C$ of P. An **interpretation for** P **in** A is any interpretation of A^*, where A^* denotes the CLP $\{r \mid r \in B \in C \text{ and } A \preccurlyeq B\}$. An interpretation for P is called **global** if it is an interpretation in $C_\perp$.*
*We say that a rule r is **applicable** in I if $B_r \subseteq I$ [4] and that r is **applied** in I if r is applicable and $|H_r \cap I| = 1$ [5].*

We argued in the introduction that choice rules represent a choice between the head elements once the precondition, the body, is satisfied (e.g. the rule is applicable). From that moment on, we can consider those elements as alternatives. With this we can define the alternatives for an atom a from a viewpoint B known in a specific component A, called horizon, as those atoms that appear together with a in the head of an applicable choice rule in a component C at least as specific as B but not more so than A (e.g. $C \in [A, B]$).

Definition 5. *Let P be an OCLP, let $A, B \in C$ be components of P and let I be an interpretation in A. For any rule $r \in A^*$, we use $c(r)$ to denote its component. The set of all **alternatives for** an atom $a \in \mathcal{B}_{A^*}$ **in** $[A, B]$, wrt I, denoted $\Omega^I_{[A,B]}(a)$, is defined as:*

$$\Omega^I_{[A,B]}(a) = \{b \mid \exists r \in A^* \cdot c(r) \in [A, B] \wedge B_r \subseteq I \wedge a, b \in H_r \text{ with } a \neq b\} .$$

---

[4] For a rule $r \equiv Q \leftarrow R$, we use H_r to denote its **head** Q while B_r denotes its **body** R.
[5] $|A|$ denotes the number of elements in the set A.

Now we are in a position to demonstrate that OCLPs are really dynamic when considering the alternatives for a decision.

Example 3. Reconsider Tommy's Dream OCLP of example 2. Let I and J be the following global interpretations: $I = \{birthday, me\}$ and $J = \{birthday, mother\}$ The set of alternatives for $biscuit$ in $[C_\perp, P_2]$ wrt I equals: $\Omega^I_{[C_\perp, P_2]}(biscuit) = \emptyset$, while the one wrt J is $\Omega^J_{[C_\perp, P_2]}(biscuit) = \{cake, candy\}$. In words, this means that $biscuits$ is not part of any decision when considering I, while it is if you are using J instead.

Deciding upon different alternatives can vary depending on the one who is making the decision or on the kind of decision. In all cases, when one alternative is preferred over all others, the choice is easily made: you simply take that alternative and leave out the others. But what happens if some alternatives are equally preferred (or incomparable)? One possible way of dealing with this dilemma is just making an objective choice between those alternatives. In this case, one is at least sure that there is a solution to the problem. This is the credulous[6] way of looking at the world.

In this context we say, intuitively, that a rule is defeated if there exist(s) some applied rule(s) containing head alternatives that are not less preferred than the ones defeated in the head of the defeated rule.

Definition 6. *Let P be an OCLP, let $A \in C$ be a component of P and let I be an interpretation in A. A rule $r \in A^*$ is **defeated in A wrt I** iff*

$$\forall a \in H_r \cdot \exists r' \in A^* \cdot c(r) \not\prec c(r') \wedge r' \text{ is applied} \wedge H_{r'} \subseteq \Omega^I_{[A, c(r)]}(a) .$$

*The rules r' are called **defeaters**.*

The following two examples illustrate the two possible ways that a rule can be defeated: a rule can either be defeated by a single rule containing only alternatives for each head element, or by a number of rules containing only alternatives for some of the head elements, but together they offer alternatives for the whole lot.

Example 4. Consider the following OCLP $\langle C, \preccurlyeq \rangle$ with:

$$P_1 : r_1 : a \leftarrow \qquad P_2 : r_2 : a \oplus b \leftarrow \qquad P_3 : r_3 : b \leftarrow$$

such that $C = \{P_1, P_2, P_3\}$ and $P_3 \prec P_2 \prec P_1$. Let $I = \{b\}$ be an interpretation in P_3. For this interpretation, the rule r_1 is defeated by the more specific rule r_3 as a has a more specific alternative b, due to the more specific rule r_2.

---

[6] There exists also a more skeptical way of facing alternatives that are equally preferred or incomparable. Whereas in the credulous approach a choice between the alternatives is acceptable, one remains undecided in the skeptical one. Although most results in this paper also hold for the skeptical semantics, we will only use the credulous approach in this paper.

Example 5. Consider the following OCLP $\langle C, \preccurlyeq \rangle$ with:

$$P_1 : r_1 : a \oplus b \leftarrow \qquad P_2 : r_2 : a \leftarrow \qquad P_3 : r_3 : b \leftarrow$$

such that $C = \{P_1, P_2, P_3\}$, $P_3 \prec P_2$ and $P_3 \prec P_1$. Assume the global interpretation $I = \{a, b\}$. The atoms a and b are alternatives of each other in $[P_1, P_1]$ wrt I and r_2 and r_3 together defeat r_1 in P_3 wrt I Notice also that r_3 does not defeat r_2, as a and b are no longer alternatives in $[P_3, P_2]$.

A model for a program P in a component A is an interpretation that satisfies every rule in one way or another. We extend the usual satisfaction criteria for choice logic programs with the possibility that rules may also be defeated in order to be satisfied.

Definition 7. *Let P be an OCLP and let $A \in C$ be a component of P. An interpretation I in A is a **model** in A iff every rule in A^* is either not applicable, applied or defeated in A wrt I. A model is **global** iff it is a model in $C_\perp$.*

Example 6. The program of example 2 has two global models, which correspond to the intuition given in example 1, namely: $M_1 = \{birthday, candy, biscuits, cake, me\}$ and $M_2 = \{birthday, cake, mother\}$.

Facing a decision, one expects that, for obtaining a solution (model), a choice has to be made among the available alternatives.

Proposition 1. *Let P be an OCLP and let M be a model for P in a component $A \in C$. For every applicable rule $r \in A^*$:*

$$\forall a \in H_r \cdot a \in M \lor (\exists b \in \Omega^M_{[A,c(r)]}(a) \cdot b \in M)$$

3   The Stable Model Semantics

The simple semantics presented in the previous section is not always intuitive, as is illustrated by the following example.

Example 7. Consider the following OCLP P:

$$P_1 : a \oplus b \leftarrow \qquad\qquad P_2 : a \leftarrow b$$
$$b \leftarrow a$$

with $P_2 \prec P_1$.
This program has a single global minimal model $M = \{a, b\}$. Note that the presence of either a or b in M depends on the application of the defeated rule $a \oplus b \leftarrow$.

In this section, we will present the so-called stable model semantics which, while preserving minimality, will prevent unnatural models such as the one in example 7
Just as stable models for "normal" logic programs and disjunctive logic programs, our stable models are based on the notion of a Gelfond-Lifschitz transformation.

Definition 8. *Let M be an interpretation for an OCLP P in a component A. We define the **Gelfond-Lifschitz transformation** for P in A wrt M, denoted P_A^M, as the positive logic program with constraints obtained from A^* in the following way:*

1. *remove all defeated rules from A^*,*
2. *remove all false atoms from the head of the remaining rules with more than one atom in the head,*
3. *replace all rules r with more than one head atom with constraint rules: for each such rule r where $a, b \in H_r$ and $a \neq b$, we add a constraint*

$$\leftarrow B_r, a, b \ .$$

The introduction of constraints is necessary to assure that a non-defeated applicable choice rule with more than one head atom will be properly satisfied (i.e. only one head atom must be considered true).

Stable models for a program are then minimal models of the program obtained from applying the Gelfond-Lifschitz transformation.

Definition 9. *Let M be an interpretation for an OCLP P in a component A. M is called a **stable model** for P in A iff M is a minimal model for the positive logic program P_A^M.*

In example 6, both M_1 and M_2 are stable.

The next theorem confirms our earlier claim that the stable model semantics restricts the minimal model semantics.

Theorem 1. *Let M be a stable model for an OCLP P in a component A. Then, M is minimal model for P in A.*

The reverse is not true, as illustrated by the following example.

Example 8. Consider the program P from example 7 which has a unique minimal model $M = \{a, b\}$ in P_2. Applying the Gelfond-Lifschitz transformation on P in P_2 yields

$$P_{P_2}^M : \begin{array}{l} a \leftarrow b \\ b \leftarrow a \end{array}$$

This program has as a minimal model $\emptyset \neq M$, so M is not stable.

Looking back on example 7, we note that, for the minimal model $M = \{a, b\}$, at least one atom must have been produced only by a defeated rule. Intuitively, such atoms can be considered assumptions, because they lack a proper motivating rule to introduce them. The following definition makes this intuition more precise.

Definition 10. *Let I be an interpretation for an OCLP P in a component A. A set $X \subseteq \mathcal{B}_{A^*}$ is called an **assumption set** wrt I iff for each $a \in X$ one of the following conditions is satisfied:*

1. *$\exists r \equiv (a \oplus A \leftarrow B) \in A^* \cdot B \subseteq I \wedge A \cap I \neq \emptyset \wedge r$ is not defeated in A wrt I; or*
2. *$\exists r \equiv (\leftarrow B, a) \cdot B \subseteq I$; or*

3. $\forall r \in A^*$ where $a \in H_r$, one of the following conditions holds:
   (a) $B_r \not\subseteq I$; or
   (b) $B_r \cap X \neq \emptyset$; or
   (c) r is defeated in A wrt I; or
   (d) $H_r \cap B_r \neq \emptyset$.

The set of all assumption sets for P in A wrt I is denoted $\mathcal{A}_{P|A}(I)$. The **greatest assumption set** for P in A wrt I, denoted $\mathcal{GAS}_{P|A}(I)$, is the union of all assumption sets for P in A wrt I.

The first condition in Definition 10 expresses that, if there exists a non-defeated applicable rule with already a true atom in the head, then the interpretation does not need a to become/maintain a model. The second condition says that, if a constraint contains, besides the element one is considering, only true atoms, one should not assume that element to be true as well. The last condition states that if every rule with a in the head is either not applicable, defeated, containing assumptions in the body or sharing atoms both in the head and the body, then we know that the atom a is not involved in making the interpretation into a model.

The greatest assumption set is an assumption set.

Proposition 2. *Let I be an interpretation for an OCLP P in a component A. Then,* $\mathcal{GAS}_{P|A}(I) \in \mathcal{A}_{P|A}(I)$.

Assumption sets can be used to eliminate candidate models.

Proposition 3. *Let M be a model for an OCLP P in a component A. Then $\overline{M}$ is an assumption set, i.e. $\overline{M} \in \mathcal{A}_{P|A}(M)$.*

Checking the assumption-free property can be quite time consuming when one needs to verify every subset of $\mathcal{B}_{A^*}$. The following proposition implies that there is an easier way.

Proposition 4. *Let I be an interpretation for an OCLP P in a component A. I is assumption-free, i.e. $I \cap \mathcal{GAS}_{P|A}(I) = \emptyset$, iff no non-empty subset of I is an assumption set for P in A wrt I.*

Assumption sets characterize stable models.

Theorem 2. *Let M be a model for an OCLP P in a component A. Then, M is stable iff M is assumption-free for P in A wrt M, i.e. $M \cap \mathcal{GAS}_{P|A}(M) = \emptyset$.*

For choice logic programs we have that minimal models are unfounded-free, which equals assumption-free when the interpretation is total. For OCLP, this can no longer be maintained. A counter example was presented in example 7: the minimal model $\{a, b\}$ is not assumption-free (i.e., $\{a, b\} \in \mathcal{A}_P(\{a, b\})$).

Assumption sets are also useful to compute stable models: Fig. 2 contains a sketch of a backtracking fixpoint procedure BF such that $BF(\emptyset)$ generates all stable models (in the component A).

```
procedure BF(I :set<atom>) {
set<atom> X = GASP|A(I)
if (X ∩ I) ≠ ∅
        fail
if (X = Ī)
        I is a stable model
else {
        set<rule> R = {r | r ∈ A* applicable and not defeated and Hr ∩ I = ∅}
        set<atom> J = {a | a ∈ Hr ∧ r ∈ R ∧ a ∉ X}
        for each a ∈ J
                BF(I ∪ {a})
        }
}
```

Fig. 2. Computing stable models

4 An Application to Finite Extensive Games with Perfect Information

In this section we give a brief and informal overview of extensive games with perfect information ([8]) and demonstrate in more detail how OCLP's can be used to retrieve the games' equilibria from the transformed programs.

An extensive game is a detailed description of a sequential structure representing the decision problems encountered by agents (called *players*) in strategic decision making (agents are capable to reason about their actions in a rational manner). The agents in the game are perfectly informed of all events that previously occurred. Thus, they can decide upon their action(s) using information about the actions which have already taken place. This is done by means of passing *histories* of previous actions to the deciding agents. *Terminal histories* are obtained when all the agents/players have made their decision(s). Players have a preference for certain outcomes over others. Often, preferences are indirectly modeled using the concept of *payoff* where players are assumed to prefer outcomes where they receive a higher payoff.

Summarizing, a game is 4-tupple, denoted $\langle N, H, P, (\geq_i)_{i \in N} \rangle$, containing the players N of the game, the histories H, a player function P telling who's turn it is after a certain history and a preference relation $\geq_i$ for each player i over the set of terminal histories. For examples, we use a more convenient representation: a tree. The small circle at the top represents the initial history. Each path starting at the top represents a history. The terminal histories are the paths ending in the leafs. The numbers next to nodes represent the players while the labels of the arcs represent an action. The number below the terminal histories are payoffs representing the players' preferences (The first number is the payoff of the first player, the second number is the payoff of the second player, ...).

Example 9. Two people use the following procedure to share two desirable identical objects. One of them proposes an allocation, which the other either accepts or rejects. In the event of rejection, neither person receives either of the objects.

An extensive game with perfect information , $\langle N, H, P, (\geq_i)_{i \in N} \rangle$, that models the individuals' predicament is shown in its alternative representation in Fig. 3.

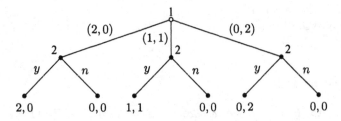

Fig. 3. The Sharing-an-Object game of example 9.

A *strategy* of a player in an extensive game is a plan that specifies the actions chosen by the player for every history after which it is her turn to move. A *strategy profile* contains a strategy for each player. E.g. $((2, 0), yyy)$ is a strategy profile where the first player intends to take both objects and the second player plans to accept (indicated by "y") any of the three possible proposals from the first player.

The first solution concept for an extensive game with perfect information ignores the sequential structure of the game; it treats the strategies as choices that are made once and for all before the actual game starts. A strategy profile is a *Nash equilibrium* if no player can unilaterally improve upon his choices. Put in another way, given the other players' strategies, the strategy stated for the player is the best this player can do[7].

Example 10. The extensive game with perfect information of example 9 has nine Nash equilibria: $((2, 0), yyy), ((2, 0), yyn), ((2, 0), yny), ((2, 0), ynn), ((1, 1), nyy),$ $((1, 1), nyn), ((0, 2), nny), ((2, 0), nny), ((2, 0), nnn)$.

The following transformation will be used to retrieve the Nash equilibria from the game as the stable models of the corresponding OCLP.

Definition 11. *Let* $\langle N, H, P, (\geq_i)_{i \in N} \rangle$ *be a extensive game with perfect information. The corresponding OCLP* P_n *can be constructed in the following way:*

- $C = \{C^t\} \cup \{C_u \mid \exists i \in N, h \in Z \cdot u = U_i(h)\}$;
- $C^t \prec C_u$ *for all* $C_u \in C$;
- $\forall C_u, C_w \in C \cdot C_u \prec C_w$ *iff* $u > w$;
- $\forall h \in (H \setminus Z) \cdot (\{a \mid ha \in H\} \leftarrow) \in C^t$;
- $\forall h = h_1 a h_2 \in Z \cdot a \leftarrow B \in C_u$ *with* $B = \{b \in [h]^8 \mid h = h_3 b h_4, P(h_3) \neq i\}$ *and* $u = U_{P(h_1)}(h)$.

The set of components consists of a component containing all the decisions that need to be considered and a component for each payoff. The order amongst the components is established according to their represented payoff (higher payoffs correspond to more specific components) with the decision component at the bottom of the hierarchy

---

[7] Note that the strategies of the other players are not actually known to i, as the choice of strategy has been made before the play starts. As stated before, no advantage is drawn from the sequential structure.

[8] We use $[h]$ to denote the set of actions appearing in a sequence h.

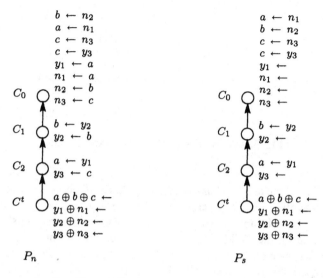

$$
\begin{array}{ll}
& b \leftarrow n_2 \\
& a \leftarrow n_1 \\
& c \leftarrow n_3 \\
& c \leftarrow y_3 \\
& y_1 \leftarrow a \\
& n_1 \leftarrow a \\
C_0 \quad & n_2 \leftarrow b \\
& n_3 \leftarrow c \\[4pt]
C_1 \quad & b \leftarrow y_2 \\
& y_2 \leftarrow b \\[4pt]
C_2 \quad & a \leftarrow y_1 \\
& y_3 \leftarrow c \\[4pt]
C^t \quad & a \oplus b \oplus c \leftarrow \\
& y_1 \oplus n_1 \leftarrow \\
& y_2 \oplus n_2 \leftarrow \\
& y_3 \oplus n_3 \leftarrow
\end{array}
$$

Fig. 4. The corresponding P_n and P_s OCLPs of the extensive game with perfect information of example 9.

(the most specific component). Since Nash equilibria do not take into account the sequential structure of the game, players have to decide upon their strategy before starting the game, leaving them to reason about both past and future. This is reflected in the rules: each rule in a payoff component is made out of a terminal history (path from top to bottom in the tree) where the head represents the action taken when considering the past and future created by the other players according to this history. The component of the rule corresponds with the payoff the deciding player would receive in case the history was carried out.

Example 11. Reconsider the Object-sharing game of example 9. The corresponding OCLP P_n is depicted on the left side of Fig. 4[9]. This program P_n has nine stable models which exactly correspond with the nine Nash equilibria of the game.

In the next theorem we show that there is indeed a correspondence between Nash equilibria and stable models.

Theorem 3. *Let $G = \langle N, H, P, (\geq_i)_{i \in N} \rangle$ be a finite extensive game with perfect information and let P_n be its corresponding OCLP. Then, s^* is a Nash equilibrium for G iff s^* is a global stable model for P_n.*

Although the Nash equilibria for an extensive game with perfect information are intuitive, they have, in some situations, undesirable properties due to not exploiting the sequential structure of the game. These undesirable properties are illustrated by the next example.

---

[9] To make the graph more readable we renamed the actions $(2,0)$, $(1,1)$ and $(0,2)$ as respectively a, b and c. We also labeled the responses of the second player to make the choices disjoint.

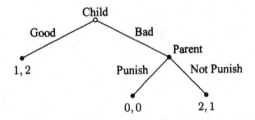

Fig. 5. The Child-Parent game of example 12.

Example 12. The game in Fig. 5 has two Nash equilibria: (Good, Punish) and (Bad, Not Punish), with payoff profiles (1,2) and (2,1). The strategy profile (Good, Punish) is an unintuitive Nash equilibrium because given that the Parent chooses Punish after history Bad, it is optimal for the Child to choose Good at the start of the game. So the Nash equilibrium is sustained by the "threat" of the Parent to choose Punish if the Child is Bad. However, this threat is not credible since the Parent has no way to commit herself to this choice. Thus the Child can be confident that the Parent will Not Punish him in case he is Bad; since the Child prefers the outcome (Bad, Not Punish) to the Nash equilibrium (Good, Punish), he has thus the incentive to deviate from the equilibrium and choose Bad. We will see that the notion of a subgame perfect equilibrium captures these considerations.

Because players are informe֑ about the previous actions they only need to reason about actions taken in the future. This philosophy is represented by subgames. A *subgame* is created by pruning the tree in the upwards direction. So, intuitively, a subgame represent a stage in the decision making process where irrelevant and already known information is removed.

Example 13. The two subgames of the game presented in example 12 are depicted in Fig. 6.

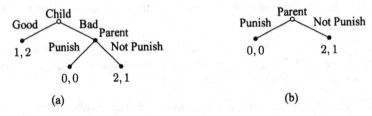

Fig. 6. The subgames of the Child-Parent game of example 13.

Instead of just demanding that the strategy profile is optimal at the beginning of the game, we require that for a *subgame perfect equilibrium* the strategy is optimal after every history. In other words, for every subgame, the strategy profile, restricted to this subgame, needs to be a Nash equilibrium. This can be interpreted as if the players revise their strategy after every choice made by them or an other player.

Example 14. The Child-Parent game of example 12 has one subgame perfect equilibrium, (Bad, Not Punish), corresponding to the non-credible threat of the Parent. The Object-sharing game of example 9 has two subgame perfect equilibrium :
$((2,0), yyy)$ and $((1,1), nyy)$.

The following transformation makes sure that subgame perfect equilibria correspond with the stable models of an OCLP.

Definition 12. *Let* $\langle N, H, P, (\geq_i)_{i \in N} \rangle$ *be an extensive game with perfect information. The corresponding OCLP* P_s *can be constructed as follows:*

- $C = \{C^t\} \cup \{C_u \mid \exists i \in N, h \in Z \cdot u = U_i(h)\}$;
- $C^t \prec C_u$ *for all* $C_u \in C$;
- $\forall C_u, C_w \in C \cdot C_u \prec C_w$ *iff* $u > w$;
- $\forall h \in (H \setminus Z) \cdot (\{a \mid ha \in H\} \leftarrow) \in C^t$;
- $\forall h = h_1 a h_2 \in Z : P(h_1) = i \cdot (a \leftarrow B) \in C_u$ *with* $B = \{b \in [h2] \mid h = h_3 b h_4, P(h_3) \neq i\}$ *and* $u = U_{P(h_1)}(h)$.

This transformation is quite similar to the one for obtaining the Nash equilibria. The only difference between the two is the creation of history-dependent rules: since subgame perfect equilibria take the sequential structure into account, players no longer need to reason about what happened before their decision. They can solely focus on the future.

Example 15. Consider once more the object-sharing game of example 9. The corresponding OCLP P_s is show on the right side of Fig. 4. This P_s has the subgame perfect equilibria $(a, y_1 y_2 y_3)$ and $(b, n_1 y_2 y_3)$ as its stable models.

Theorem 4. *Let* $G = \langle N, H, P, (\geq_i)_{i \in N} \rangle$ *be a extensive game with perfect information and let* P_s *be its corresponding OCLP. Then,* s^* *is a subgame perfect equilibrium of* G *iff* s^* *is a global stable model for* P.

Note that [10] proposes an alternative formalism to model strategic games using an extension of logic programming. However, in [10], the specification of choices is external to the program while, in our approach, we rely on nondeterminism (and priority) to represent alternatives and on the properties of the stable model semantics to obtain equilibria.

5   Relationships to Other Approaches

5.1   Ordered Logic ([6])

Ordered logic programs are a special, also semantically, case of OCLP's: all choices are restricted to 2 alternatives a and $\neg a$. This is confirmed by the following.

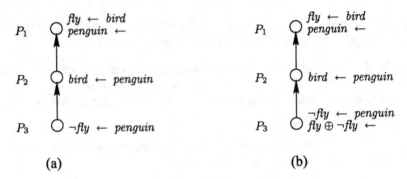

Fig. 7. a) The Ordered logic version of the Penguin problem. b) The corresponding Penguin OCLP P_{P_3} wrt component P_3.

Proposition 5. *Let $P = \langle C, \preccurlyeq \rangle$ be an ordered logic program in the sense of [6] and let $A \in C$ be a component for it. The corresponding OCLP P_A with respect to A equals $\langle C', \preccurlyeq \rangle$ where:*

$$C' = \{B \in C \mid B \neq A\} \cup \{A \cup \{a \oplus \neg a \leftarrow \mid a, \neg a \in \mathcal{B}_{A^*}\}\} \ .$$

An interpretation I in A is a model for P in A iff I is a model for P_A in A.

We illustrate this construction with the following well-known example:

Example 16 (Tweety, the penguin). The left side of Fig. 7 depicts the ordered logic program for the problem. The right hand side gives the corresponding OCLP wrt to component P_3. Both programs have only one model in component P_3, namely $M = \{bird, \neg fly, penguin\}$.

5.2   Other Approaches to Preference

Dynamic preference in extended logic programs is introduced in [3] in order to obtain a better suited well-founded semantics. Although preferences are called dynamic they are not dynamic in our sense. Instead of defining a preference relation on subsets of rules, preferences are incorporated as rules in the program. Moreover, a stability criterion may come into play to overrule preference information. Another difference with our approach is that the alternatives are static.

A totally different approach is proposed in [12]. Here the preferences are defined amongst atoms. Given these preferences, one can combine them to obtain preferences for sets of atoms. Defining models in the usual way, the preferences are then used to filter out the less preferred models. That way, this system is not convenient for decision making as the preferences cannot easily be made to depend on the situation.

In [1], preference in extensive disjunctive logic programming is considered. As far as overriding is concerned the technique corresponds rather well with our skeptical defeating, but alternatives are fixed as an atom and its (real) negation.

Outside the context of logic programming, [2] proposes to add priorities to the object language of default logic. Extensions are then required to be compatible with this information. OCLP and [2] support different intuitions on the notion of priority, as shown by the following example[10]:

Example 17.

$$P_1 : a \leftarrow \qquad P_2 : \quad \neg c \quad \leftarrow$$
$$P_3 : c \leftarrow a \qquad P_4 : c \oplus \neg c \leftarrow$$

with $P_4 \prec P_3 \prec P_2 \prec P_1$. With our approach, we obtain $\{a, c\}$ as the (stable) model of this program while [2] returns $\{a, \neg c\}$ as the extension for the default theory. [2] considers the knowledge of a coming from a more general rule insufficient (the rule from P_1) to favor the rule from P_4 over the one from P_3. We , and also [11], prefer to say that there is no counter evidence for a so we should exploit this knowledge as much as possible.

References

1. Francesco Buccafurri, Wolfgang Faber, and Nicola Leone. Disjunctive Logic Programs with Inheritance. In Danny De Schreye, editor, *International Conference on Logic Programming (ICLP)*, pages 79–93, Las Cruces, New Mexico, USA, 1999. The MIT Press.
2. Gerhardt Brewka. Reasoning about priorities in default logic. In Barbara Hayes-Roth and Richard Korf, editors, *Proceedings of the Twelfth National Conference on Artificial Intelligence*, pages 940–945, Menlo Park, California, 1994. American Association for Artificial Intelligence, AAAI Press.
3. Gerhard Brewka. Well-Founded Semantics for Extended Logic Programs with Dynamic Preferences. *Journal of Articficial Intelligence Research*, 4 (1996) 19–36.
4. Marina De Vos and Dirk Vermeir. Choice Logic Programs and Nash Equilibria in Strategic Games. In Jörg Flum and Mario Rodríguez-Artalejo, editors, *Computer Science Logic (CSL'99)*, volume 1683 of *Lecture Notes in Computer Science*, pages 266–276, Madrid, Spain, 1999. Springer Verslag.
5. Marina De Vos and Dirk Vermeir. On the Role of Negation in Choice Logic Programs. In Michael Gelfond, Nicola Leone, and Gerald Pfeifer, editors, *Logic Programming and Non-Monotonic Reasoning Conference (LPNMR'99)*, volume 1730 of *Lecture Notes in Artificial Intelligence*, pages 236–246, El Paso, Texas, USA, 1999. Springer Verslag.
6. D. Gabbay, E. Laenens, and D. Vermeir. Credulous vs. Sceptical Semantics for Ordered Logic Programs. In J. Allen, R. Fikes, and E. Sandewall, editors, *Proceedings of the 2nd International Conference on Principles of Knowledge Representation and Reasoning*, pages 208–217, Cambridge, Mass, 1991. Morgan Kaufmann.
7. Vladimir Lifschitz. Computing Circumscription. In *9th International Joint Conference on Artificial Intelligence (IJCAI-85)*, Los Angeles, 1985.
8. Martin J. Osborne and Ariel Rubinstein. *A Course in Game Theory*. The MIT Press, Cambridge, Massachusets, London, Engeland, third edition, 1996.
9. David Poole. On the Comparison of Theories: Preferring the Most Specific Explanation. In *9 th International Joint Conference on Artificial Intelligence (IJCAI-85)*, Los Angeles, 1985.
10. David Poole. The independent choice logic for modelling multiple agents under uncertainty. *Artificial Intelligence*, 94(1–2):7–56, 1997.

---

[10] For the ease of notation we simply denote the program in our formalism.

11. Henry Prakken and Giovanni Sartor. A system for defeasible argumentation, with defeasible priorities. In Dov M. Gabbay and Hans Jürgen Ohlbach, editors, *Proceedings of the International Conference on Formal and Applied Practical Reasoning (FAPR-96)*, volume 1085 of *LNAI*, pages 510–524, Berlin, June 3–7 1996. Springer.
12. Chiaki Sakama and Katsumi Inoue. Representing Priorities in Logic Programs. In Michael Maher, editor, *Proceedings of the 1996 Joint International Conference and Symposium on Logic Programming*, pages 82–96, Cambridge, September 2–6 1996. MIT Press.

Author Index

Lecture Notes in Computer Science

Lecture Notes in Artificial Intelligence (LNAI)